GEOGRAPHY

An Integrated Approach

David Waugh

Nelson

Thomas Nelson and Sons Ltd
Nelson House   Mayfield Road
Walton-on-Thames   Surrey
KT12 5PL   UK

51 York Place
Edinburgh
EH1 3JD   UK

Thomas Nelson (Hong Kong) Ltd
Toppan Building 10/F
22A Westlands Road
Quarry Bay   Hong Kong

Thomas Nelson Australia
102 Dodds Street
South Melbourne
Victoria 3205
Australia

Nelson Canada
1120 Birchmount Road
Scarborough   Ontario
M1K 5G4   Canada

© David Waugh 1990

First published by Thomas Nelson and Sons Ltd 1990

ISBN 0-17-444065-0
NPN 15 14 13 12 11 10

Acknowledgements

The author and publishers are grateful to the following for permission to reproduce copyright material.

Examining bodies (questions from 'A' level examination papers):
Joint Matriculation Board p. 107, (1981 Paper I), Q1, 2, 3, 4, 5, 6; p. 144, (Figure 6.52); p. 232, (Figure 10.26), Q3; p. 393, (Figure 16.17), Q1, 2; p. 437, Q1, 2;
London Board p. 129, (Figures 6.32, 6.33), Q1, 2, 3, 4, 5; p. 144, (Figure 6.52); p. 195, (Figure 9.42) Q1; p. 437, Q1, 2;

Page 9 BBC Publications, extract from *The Making of a Continent* by Ron Redfern, 1983; p. 10 BBC Publications, extract from *The Restless Earth* by Nigel Calder, 1972; p. 101 David and Charles, extract from *The Unquiet Landscape* by Denys Brunsden, John Doornkamp, 1977; p. 109 Cambridge University Press, extract from *The Coastline of England and Wales* by J.A. Steers, 1960; p. 158 Unwin Hyman Ltd, extract from *The Physical Geography of Landscape* by Roy Collard, 1988; p. 167 Collins Fontana, extract from *The Climatic Threat* by John Gribbin, 1978; p. 213 Unwin Hyman Ltd, extract from *Soil Processes* by Brian Knapp, 1979; p. 240 Pan Books, extract from *The Gaia Atlas of Planet Management* by James Lovelock, 1985; p. 261 Oxford University Press, extract from *The Ages of Gaia* by James Lovelock, 1989; p. 285 Pan Books, extract from *North–South: A Programme for Survival* by Willy Brandt, 1980; p. 350 BBC Enterprises Ltd, extract from *Only One Earth* by Lloyd Timberlake, 1987; p. 383 World Wildlife Fund in association with North-South Productions, extract from *Only One Earth* by Lloyd Timberlake, 1987; 418 (Figure 16.48), George Philip and Sons, extract from *New Geographical Digest* 1986; pp. 430, 455, Abacus Books, extract from *Small is Beautiful* by E.F. Schumacher, 1974.

Every effort has been made to contact holders of copyright to obtain permission to reproduce copyright material. However, if any have been inadvertently overlooked, the publishers will be pleased to make the necessary arrangements at the first opportunity.

The publishers are grateful to the following for their permission to reproduce photographs (listed alphabetically by source and by Figure number):
Hunting Aerofilms: 6.22, 6.43, 14.5c, 14.5d, 14.7a, 14.7d, 14.7e, 14.7f, 17.25;
Air Fotos: 15.19;
Ken Andrew: 6.44, 8.11, 15.18i;
Heather Angel: 3.43, 4.7, 11.2b, 11.8, 11.9, 11.12, 11.14a, 11.14c, 12.5, 16.38;
Aspect/Derek Bayes: 16.42;
J. Allen Cash: 4.17a, 4.17b, 6.38, 8.12, 9.21(9), 10.2a, 11.26, 15.28, 17.18;
Celtic Picture Agency: 4.17c/M.J. Thomas, 6.28, 16.58, 16.59;
Bruce Coleman: 1.20/Wedigio Ferchland, 2.8/Bruce Doré, 3.33/ Ray Bryant, 7.4/P.R. Wilkinson, 8.6/L.C. Marigo, 9.21(5), 9.21(8)/R.K. Pilsbury, 12.21/Jen & Des Bartlett, 16.54;
Colour Library International: 6.19;
Susan Griggs Agency: 3.35, 15.18k/Nick Holland;
GSF Picture Library: 2.12/R.J.B. Goodale, 12.30, 12.37, 12.42;
Robert Harding: 1.21, 2.21, 3.36a/Michael Short, 6.17/G.M. Wilkins, 6.24/Christopher Nicholson, 7.16, 8.5, 11.27/Philip Craven, 12.18, 13.27, 15.20, 15.27, 15.31/Robert Francis, 18.7/Paul Van Riel, 17.36/G. & P. Corrigan, 18.8, 18.9;
David Horwell: 7.3;
Hutchison: 14.7c, 16.27, 16.26/Brian Moser, 15.30/P. Moszinski, 16.25, 16.51/Melanie Friend;
Image Bank/Don Landwehrle: 7.19;
Intermediate Technology Development Group (ITDG): 15.33, 17.33a, b, c, d;
Japanese Information Centre: 17.16;
A.F. Kersting: 6.15, 8.9, 12.39, 14.5e, 15.24;
Keystone: 2.18;
Landform: 4.23b, 5.7, 5.8, 6.31, 6.42, 7.7, 7.10, 7.11, 7.13, 7.14, 10.21;
Frank Lane Picture Agency: 9.21(1), 9.21(6), 12.50/Steve McCutcheon;
NASA: 6.26, 9.45;
Novosti: 13.32/Walter Holt, 16.50/Jeremy Hartley;
Oxford Scientific Films: 4.14/Pete O'Toole, 7.15/J.A.L. Cooke, 10.32b/Tony Romford, 12.17;
The Photo Source: 16.28;
Planet Earth: 7.12/Alex Williams, 12.22/Hans Christian Heap, 16.24/Shawn Avery;
Popperfoto: 3.27;
Port of Felixstowe: 18.6;
Stuart Powles: 15.18b, 15.18c, 15.18d, 15.18e, 15.18h;
Nigel Press Associates/Meteosat: 9.35, 16.41;
Walter Rawlings: 15.25;
David Simson: 11.2a, 12.7;
South American Pictures/Tony Morrison: 10.30, 12.6;
Daily Telegraph: 9.48, 16.36;
A.C. Waltham: 2.13, 2.17, 3.46, 4.12, 4.23a, 6.40, 8.2c, 9.26, 10.27, 11.14b, 17.23;
David Waugh: 1.22, 1.24, 1.29, 1.30, 2.1, 2.2, 2.6, 2.7, 3.19, 3.23, 3.31, 3.38, 3.40, 3.47, 4.5, 4.17d, 4.18, 4.23c, 4.23d, 5.10, 6.8, 6.13, 6.14, 6.23, 6.29, 6.30, 6.36, 6.39, 8.3, 8.4, 8.10, 9.24, 10.23a, 11.5a, 11.5b, 11.11, 11.28, 12.27, 12.44, 14.5a, 15.18a, 15.18g, 15.29, 15.30, 16.39;
Worldwide/Tony Stone: 12.28, 9.21(10);
ZEFA: 2.5, 9.21(2), 9.21(3), 9.21(4), 9.21(7), 12.33, 12.43, 13.30, 14.5b, 14.7b, 16.53, 16.56, 17.35, 18.5

The author wishes to thank the following for their ideas and contributions to a number of illustrations:
Marian Green: 1.14, 2.3a, 6.48, 9.15, 9.37a, 9.53, 9.54, 9.61, 10.2;
Brian Rigby: 10.16, 11.3;
Sue Rigby: 1.24, 1.26, 1.28, 1.29.

All illustrations produced by Dataset Marlborough Design Limited.
Front cover illustration by Martin Lloyd.

Contents

Contents

Matrix for use of quantitative techniques

Where to find
FRAMEWORKS FOR THINKING

Introduction for the reader

This book has been written as much for those students who have an interest in Geography, an enquiring mind and a concern for the future of the planet upon which they live, as for those specialising in the subject. The text has been written as concisely as seemed practical so as to minimise the time needed to be spent by the student in traditional note-taking and making, therefore maximising the time available for discussion, individual enquiry and wider reading. Colour photographs and sketches have been used wherever possible to show the wide range of natural and artificial environments. Annotated diagrams are included to illustrate interrelationships and the more difficult concepts and theories; a wide range of graphical skills are used to portray geographical data.

It is because Geography is concerned with interrelationships that this book has included, and aims to integrate, several fields of study. These involve the study of physical environments (atmosphere, lithosphere and hydrosphere) and the living world (biosphere); economic development (or lack of it) of the world's peoples; the frequent misuse of the environment and long-overdue concern about the consequences; and the application of a modern scientific approach using statistical methods in investigations. It is intended that a single book may satisfy the requirements made in the Geographical Association's booklet, *Geography in the National Curriculum* (1989), that:

> 'Geography is concerned with people as well as landscapes, with economic as well as ecological systems. The character of places — the subject's central focus — derives from the interaction of people and environments.'

By coincidence, the initial letters of the title of this book form the word GAIA. In ancient Greece, Gaia was the goddess of the earth: today the term has been reintroduced to mean 'a new look at life on earth', and the approach looks at the earth in its entirety as a living organism.

The reader should be aware that there is no rigid sequence proscribed in the order either of the chapters themselves nor their content — each is open to several routes of enquiry. Terminology is a major problem because geographers may use several terms, some borrowed from other disciplines, to describe the same phenomenon. When a term is introduced for the first time it is shown by emboldened print, has a list of alternatives (one of which is subsequently retained for consistency) and is defined. These definitions may be as margin notes where it is inappropriate for them to be incorporated into the main text.

Geography: An Integrated Approach sets out to provide a background store of information which will help you, the reader, to understand basic concepts, to be able to enter discussions and to develop your own informed, rather than prejudiced, values and attitudes. Due to the limited space, Case Studies are often brief and, building upon earlier knowledge, relatively well-known. They are intended to stimulate the reader to read more widely, in particular to make complementary use of the two *Readers*. References given after each chapter are those to which the author has himself referred and are not intended as a comprehensive list. The reader should realise that Geography is a dynamic subject with data, views, policies and terms all constantly changing. Consequently, the student's research cannot be limited to textbooks, which in any case are out-of-date even before their publication, but should be widened to include newspapers, journals, television, radio, and many 'non-academic' media.

The book also includes eleven 'Frameworks for thinking'. The purpose of these sections is to stimulate discussion on methodological and theoretical issues. They illustrate the skills required and the problems involved in geographical enquiry referring, for example, to the uses, limitations and reliability of models; quantitative techniques; the collection of data; making classifications; the dangers of stereotyping and of making broad generalisations. Geography is also concerned with the development of graphical skills. The media show an increasing amount of data in a graphical form, and this is likely to grow further as geographical information systems develop. It is assumed that the reader already understands those skills covered by present 16+ examination syllabuses and therefore only *new* skills are explained in this book. Quantitative and statistical techniques are incorporated at appropriate points, though, as shown by the matrix which follows, they may be used in examples elsewhere in the physical or the human-economic chapters. After the explanation of each technique there is a worked example and follow up exercises. Opportunities for self-assessment are provided throughout the book with examination-style questions.

Index of case studies

CHAPTER 1

Plate tectonics and vulcanicity

'... how does a supercontinent begin to rift and how do the pieces move apart? What effects do such movements have on the shaping of the continental land-scapes, on hot climates and ice ages, on the evolution of life in general and on humanity's relationship with the upper crust of the Earth in particular?'
The Making of a Continent, Ron Redfern, BBC, 1983

Our unstable earth

It is estimated that the earth was formed 4 600 000 000 years ago. Even if this figure is simplified to 4600 million years it still presents a timescale far beyond our understanding. Nigel Calder, in his book *Restless Earth* made a more comprehensible analogy by reducing the time span to 46 years by ignoring the eight noughts and comparing it to a human lifetime.

Geologists have been able to study rocks and fossils formed during the last 'six years of

the lady's life' and have produced a time-chart, or **geological timescale**. Not only have they been able to add dates with increasing confidence but they have made progress in describing and accounting for the major changes in the earth's surface, e.g. sea level fluctuations and landform development, and in its climate. This table, shown in Figure 1.1, should be a useful reference for later parts of this book.

▼ Figure 1.1

The geological timescale

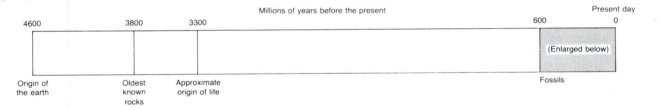

	Era	Geological period	Millions of years before present	Conditions and rocks in Britain	Major world events
CENOZOIC	QUATERNARY	Pleistocene	2	Ice Age in Britain with warm periods	First human civilisations
CENOZOIC	TERTIARY	Pliocene	7	Warm climate; Crag rocks in East Anglia	
CENOZOIC	TERTIARY	Miocene	26	No deposits in Britain	Formation of the Alps
CENOZOIC	TERTIARY	Oligocene	38	Warm shallow seas in south of England	Rockies and Himalayas begin to form
CENOZOIC	TERTIARY	Eocene	54	Nearly tropical; London clay	Volcanic activity in Scotland / End of the dinosaurs
MESOZOIC		Cretaceous	136	Chalk deposited; Atlantic ridge opens	Age of the dinosaurs / Pangaea breaks up
MESOZOIC		Jurassic	195	Oxford clays and limestones; warm	
MESOZOIC		Triassic	225	Desert; sandstones.	First mammals / Formation of Pangaea
PALAEOZOIC		Permian	280	Desert; new red sandstones, limestones	
PALAEOZOIC		Carboniferous	345	Tropical coast with swamps; coal	First amphibians and insects
PALAEOZOIC		Devonian	395	Warm desert coastline; sandstones	First land animals
PALAEOZOIC		Silurian	440	Warm seas with coral; limestones	First land plants
PALAEOZOIC		Ordovician	500	Warm seas; volcanoes (Snowdonia) sandstones, shales	First vertebrates
PALAEOZOIC		Cambrian	570	Cold at times; sea conditions	Abundant fossils begin
		Pre-Cambrian		Igneous and sedimentary rocks	

CASE STUDY

A simplified history of the earth, Nigel Calder

... Or we can depict Mother Earth as a lady of 46, if her 'years' are megacenturies. The first seven of those years are wholly lost to the biographer, but the deeds of her later childhood are to be seen in old rocks in Greenland and South Africa. Like the human memory, the surface of our planet distorts the record, emphasising more recent events and letting the rest pass into vagueness — or at least into unimpressive joints in worn down mountain chains.

Most of what we recognise on Earth, including all substantial animal life, is the product of the past six years of the lady's life. She flowered, literally, in her middle age. Her continents were quite bare of life until she was getting on for 42 and flowering plants did not appear until she was 45 — just one year ago. At that time, the great reptiles, including the dinosaurs, were her pets and the break-up of the last supercontinent was in progress.

The dinosaurs passed away eight months ago and the upstart mammals replaced them. In the middle of last week, in Africa, some man-like apes turned into ape-like men and at the weekend, Mother Earth began shivering with the latest series of ice ages. Just over four hours have elapsed since a new species calling itself *Homo sapiens* started chasing the other animals and in the last hour it has invented agriculture and settled down. A quarter of an hour ago, Moses led his people to safety across a crack in the Earth's shell, and about five minutes later Jesus was preaching on a hill farther along the fault line. Just one minute has passed, out of Mother Earth's 46 'years', since man began his industrial revolution, three human lifetimes ago. During that time he has multiplied his numbers and skills prodigiously and ransacked the planet for metal and fuel.

Earthquakes

Even the earliest civilisations were aware that the crust of the earth was not rigid and immobile. The first European civilisation, the Minoan, based in Crete, constructed buildings such as the Royal Palace at Knossus to be capable of withstanding earthquakes and the civilisation may have been destroyed following the huge volcanic eruption on the nearby island of Thera (Santorini). Later, inhabitants of places as far apart as Pompeii, Lisbon, Tokyo, San Francisco and Mexico City were to suffer from major earth movements.

It was by studying earthquakes that geologists were first able to determine the structure of the earth (Figure 1.2). At the **Mohorovičić** or **'Moho' discontinuity**, it was found that shock waves begin to travel faster indicating a change of structure, in this case the junction of the **crust** and **mantle** (Figure 1.2). Earthquakes result from a slow build up of pressure within crustal rocks. If this pressure is suddenly released then parts of the surface may experience a jerking movement. The point at which the release in pressure occurs is known as the **focus**. Above this, on the surface and usually receiving the worst of the shock or **seismic** waves, is the **epicentre**. Unfortunately, it is not only the immediate or primary effects of the earthquake which may cause loss of life and property, as often the secondary or after effects are even more serious.

'Moho' discontinuity is a zone between the earth's crust and the mantle where seismic waves are modified. The moho is at 30–40 km beneath continents and 10 km below the oceans.

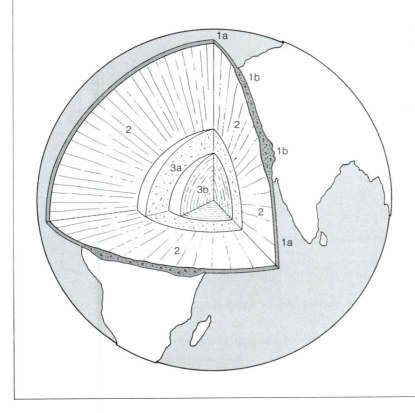

1 Crust Relatively speaking, this is as thin as the skin of an apple is to its flesh

(a) Oceanic crust (sima) is a layer mainly consisting of basalt averaging 6 to 10 km in thickness. At its deepest it has a temperature of 1200 °C
(b) Continental crust (sial) can be up to 65 km thick and approximately corresponds to the continents. The crust is separated from the mantle by the Moho Discontinuity

2 Mantle This is composed mainly of silicate rocks rich in iron and magnesium which are kept in a semi-molten state. It extends to a depth of 2900 km where temperatures may reach 5000°C, thus generating convection currents

3 Core This consists of iron and nickel. The outer core (3b) is kept in a semi-molten state but the inner core is solid. The temperature at the centre of the Earth (6371 km below the surface) is about 5500°C

▲ Figure 1.2
The internal structure of the earth

C A S E   S T U D Y

The San Francisco earthquakes

'At 0512 hours on the morning of 18 April 1906, the ground began to shake. There were three tremors, each one increasingly more severe. The ground moved by over six metres in an earthquake which measured 7.9 on the Richter scale. Many apartment buildings collapsed, bridges were destroyed — the Golden Gate had not then been built — and water pipes fractured. The worst damage was 'downtown' where the housing density was greatest. Although many people were trapped within collapsed buildings there were relatively few deaths.

Then came the fire! It started in numerous places resulting from overturned stoves or sparked by electricity or the ignition of gas escaping from the broken mains. As the water pipes had been fractured it hardly mattered that there were only 38 horse-drawn fire engines to cope with 52 fires. As the fire spread, houses were blown up with dynamite to try to create gaps to thwart the flames, but the explosions only caused further fires. It took over three days to put out the fires by which time over 450 people (mainly those pre-viously trapped) had died, 28 000 buildings in 500 blocks had been destroyed and an area six times greater than that destroyed by the Great Fire of London had been ravaged. Fortunately the threat of disease and starvation was averted.'

During rush-hour on 18 October 1989 an earthquake measuring 6.9 on the Richter scale shook the city for 15 seconds. The early warning system had given no clues. Skyscrapers swayed 3m, fractured gas pipes caused fires in one residential area and a 1.5km stretch of the two-tier highway in nearby Oakland collapsed, killing people in their vehicles. The figures of 67 dead and 2000 homeless were low, however, compared to the quake of similar magnitude in Armenia, USSR, 11 months earlier — in California, precautions had been taken and fully-equipped and trained emergency services were on hand.

Q Read the eye-witness account of the San Francisco earthquake of 1906 on page 11 and then answer the questions below.

1 What is an earthquake?

2 Refer to the diagram of the Richter scale (Figure 1.3).

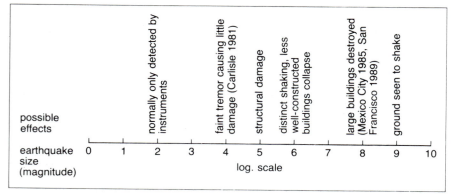

◄ Figure 1.3
The Richter scale

a Why is an earthquake of 7.0 on the Richter scale 100 times more severe than one which measures 5.0?

b How many times greater was the Mexican earthquake of 1985 than the one which affected Carlisle on Boxing Day 1981?

3 Describe **a** the primary and **b** the secondary effects of the San Francisco earthquake.

4 What types of damage other than that experienced at San Francisco may result from earthquakes?

Distribution of recent major earthquakes, volcanoes (active, dormant or exinct) and young fold mountains

■ Recent earthquakes:

1903	Mexico, Java, Aleutians, Greece	1943	Philippines, Java
1904	Tokyo, Kamchatka, Costa Rica	1944	Japan
1906	Japan, San Francisco, Aleutians, Chile	1946	West Indies, Japan
1907	Mexico, Afghanistan	1949	Alaska
1908	Mexico, Sicily	1950	Japan, Assam
1909	Fiji, Japan	1953	Turkey, Japan
1916	New Guinea	1956	California
1917	Java	1957	Mexico
1918	Tonga	1958	Alaska
1920	Taiwan, Fiji	1960	Chile, Morocco
1921	Peru	1962	Iran
1922	Chile	1963	Yugoslavia
1923	Japan	1964	Alaska, Turkey, Mexico, Japan, Taiwan
1924	Philippines	1965	El Salvador, Greece
1925	California	1966	Chile, Peru, Turkey
1926	Rhodes	1967	Colombia, Yugoslavia, Java, Japan
1927	Japan	1968	Iran
1928	Chile	1970	Peru
1929	Aleutians, Japan	1971	New Guinea, California
1931	New Zealand	1972	Nicaragua
1932	Mexico	1976	Guatemala, Italy, China, Philippines, Turkey
1933	California		
1935	Sumatra	1978	Japan
1938	Java	1980	Italy
1939	Chile, Turkey	1985	Mexico, Colombia
1940	Burma, Peru	1988	USSR
1940	Ecuador, Guatemala	1989	San Francisco

- Volcanoes:
  Aconcagua, Chimborazo, Cotopaxi, Nevado del Ruiz, Paricutin, Popocatepetl, Mount St Helens, Fujiyama, Mayon, Krakatoa, Tarawera, Erebus; Krafla, Helgafell, Surtsey, Azores, Ascension, St Helena, Tristan da Cunha; Vesuvius, Etna, Pelee; Mauna Loa, Kilauea

- Young fold mountains:
  Andes, Rockies, Atlas, Pyrenees, Alps, Caucasus, Hindu Kush, Himalayas, Southern Alps

5   On an outline map of the world mark on by means of a dot — no need to name the places — the location of each earthquake in the above list.

6   **a**   On a tracing overlay mark and name all the volcanoes given above.
     b   On a separate overlay mark and name the young fold mountains.

7   Describe the distribution of earthquakes, volcanoes and fold mountains on your two maps.

Continental drift and plate tectonics

As early as 1620 Francis Bacon noted the jigsaw-like fit between the east coast of South America and the west coast of Africa. Others were later to point out similarities between the shapes of coastlines of several adjacent continents.

In 1912, a German meteorologist, Alfred Wegener, published his theory that long ago all the continents were joined together in one large supercontinent which he named **Pangaea**. Later, this landmass somehow split up and the various continents, as we know them, drifted apart. Wegener collated evidence from several sciences:

- **Biology** Mesosaurus was a small reptile living in Permian times (Figure 1.1) which appeared to have been limited to South Africa and Brazil. A plant of similar age associated with coalfields had been located only in India and Antarctica.

- **Geology** Rocks of similar type, age, formation and structure had been found in southeast Brazil and South Africa, while the Appalachian Mountains of eastern USA corresponded with mountains in northwest Europe.

▼ Figure 1.4

The wandering continents

a Pangaea - The supercontinent of 200 million years ago

b Sub-oceanic forces send the landmasses wandering

c Tomorrow's world - 50 million years hence

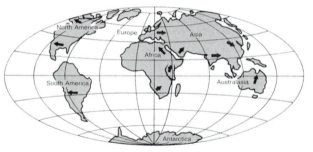

■ **Climatology** Coal, formed only under warm, wet conditions, was found beneath the Antarctic ice cap; evidence of glaciation had been noted in tropical Brazil and peninsular India; and how could coal, sandstone and limestone have formed in Britain with its present climate?

Wegener's theory of **continental drift** combined information from several subject areas, but his ideas were rejected by specialists in those disciplines partly because he was not regarded as an expert himself but perhaps mainly because he could not explain how solid continents could change their positions and could suggest no mechanism by which they might drift.

Figure 1.4 shows Wegener's Pangaea, how it began to divide up into two large continents which he named **Laurasia** and **Gondwanaland** and suggests how the world may look in the future if movements continue as at present.

1 Which two present day continents formed Laurasia?
2 Which present day continents formed Gondwanaland?
3 How may the world of 50 million years time be different from that of today?

Since Wegener first put forward his theory, new evidence has become available to support his ideas. Such evidence falls into three main groups of studies.

1 Investigating islands in the Atlantic in 1948, Ewing noted that there was a continuous mountain range extending the whole length of the ocean bed and that the rocks of this range were volcanic and recent in origin — not ancient rocks as previously assumed.

2 Studies of **palaeomagnetism** in the 1950s. During underwater volcanic eruptions, basaltic lava reaches the surface of the crust and cools. During the cooling process, individual minerals, especially iron, align themselves in the direction of the earth's magnetic field, i.e. in the direction of the magnetic pole. Recent refinements in dating techniques have enabled the time at which rocks were formed to be calculated very accurately. It was known before the 1950s that the earth's magnetic pole varied a little from year to year, but it was then discovered that periodically the magnetic field had been reversed, meaning that for a while the magnetic pole was in the south. It is claimed that there have been 171 reversals over 76 million years. So beginning at a given time, new basalt would be aligned towards a north magnetic pole. After a reversal in the magnetic poles, newer lava would be orientated to the south where the magnetic pole now was. After a further reversal, the alignment would again be to the north. Subsequent investigations have shown that this reversal in the rocks was almost symmetrical either side of the Mid-Atlantic Ridge which is the name given to the mountain range identified by Ewing (Figure 1.5).

3 **Sea floor spreading** In 1962, Hess studied the age of rocks from the middle of the Atlantic to the coast of North America. He confirmed that the newest rocks were in the centre of the ocean still being formed in Iceland and that the oldest ones were nearest to the USA and the Caribbean. He also suggested that the Atlantic could be widening by up to 5 cm a year.

◄ Figure 1.5
The repeated reversal of the earth's magnetic field

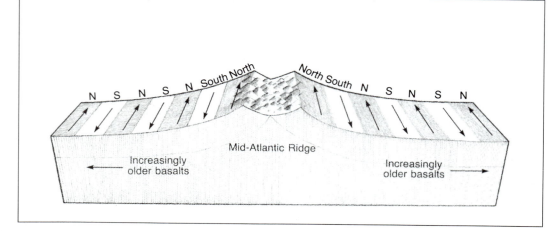

Mid-Atlantic Ridge

Increasingly older basalts

Increasingly older basalts

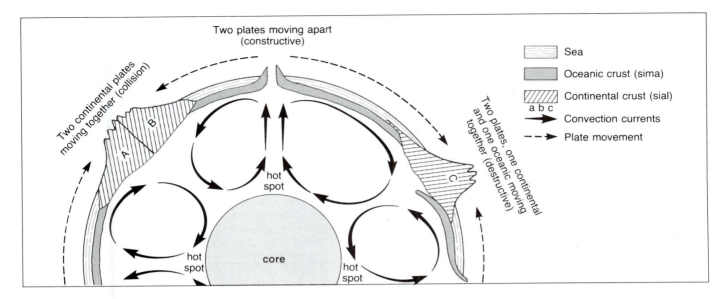

▲ Figure 1.6

How plates move

One major difficulty resulting from this concept of sea floor spreading was the implication that the earth must be increasing in size, so further investigations were needed to show that elsewhere parts of the crust were being destroyed. Such areas were found to correspond to the fringes of the Pacific Ocean — the regions where you plotted some major earthquakes and volcanic eruptions (page 13). These discoveries led to the development of the theory of **plate tectonics** which is now almost universally accepted.

The theory of plate tectonics

The earth's crust is divided into seven large and several smaller rigid **plates** which float like rafts on the underlying semi-molten mantle. These plates are moved by **convection currents** (Figure 1.6). Plate tectonics is the study of the movement of these plates and the landforms which result from those movements. There are two types of plate: **continental** and **oceanic**. However, these terms do not refer to actual continents and oceans but to different types of crust or rock.

Continental crust, or **sial**, is composed of older, lighter rock of granitic type. It is dominated by minerals rich in silicon (Si) and aluminium (Al), from which the term is derived. Oceanic crust, or **sima**, consists of

much younger and denser rock of basaltic composition. Its dominant minerals are silicon and magnesium (Ma). The major differences between the two types of crust are summarised in Figure 1.7.·

Plate movement

As a result of convection currents generated by heat from the centre of the earth, plates may move towards, away from or sideways along adjacent plates. It is at plate boundaries that most of the world's major landforms occur, and where earthquake, volcanic and mountain building zones are located (Figure 1.8). However, before trying to account for the formation of these landforms several 'rules' and generalisations should be noted.

1 Continental crust does not sink because of its relatively low density and so is permanent, whereas oceanic crust, being denser, does. Oceanic crust is continually being formed and destroyed.

2 Continental plates, such as the Eurasian Plate, may consist of both continental and oceanic crust.

3 Continental crust may extend far beyond the margins of the landmass.

4 Plates cannot overlap. This means that either they must be pushed upwards on impact to form mountains or one plate must be forced downwards into the mantle and destroyed.

► Figure 1.7

Differences between continental and oceanic crust

	Continental crust (sial)	Oceanic crust (sima)
Thickness	35 to 70 km on average	6 to 10 km on average
Age of rocks	very old, mainly over 1500 million years	very young, mainly under 200 million years
Weight of rocks	lighter with an average density of 2.6	heavier with an average density of 3.0
Nature of rocks	light in colour; numerous types, many contain silica and oxygen, granite is the most common	dark in colour; few types, mainly basalt

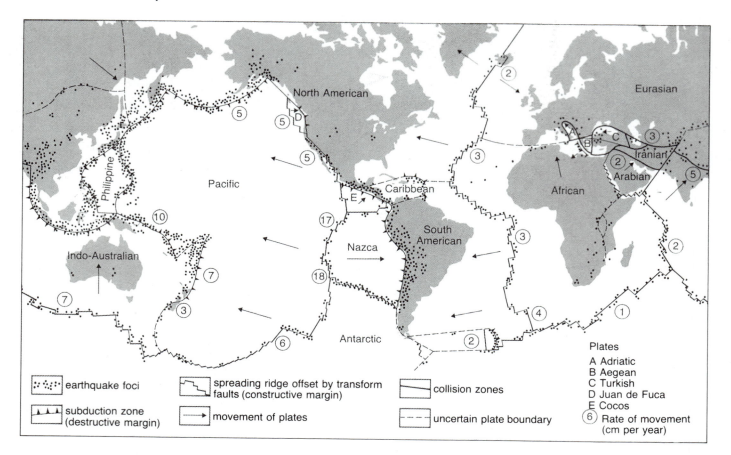

▲ Figure 1.8

Plate boundaries and active

zones of the earth's crust

Type of plate boundary	Description of changes	Examples
A Constructive margins (spreading or divergent plates)	two plates move away from each other new oceanic crust appears forming mid-ocean ridges with volcanoes	Mid-Atlantic Ridge (Americas moving away from Eurasia, Africa) East Pacific Rise (Nazca and Pacific Plates moving apart)
B Destructive margins (subduction zones)	oceanic crust moves towards continental crust but being heavier sinks and is destroyed forming deep sea trenches and island arcs with volcanoes	Nazca sinks under South America Plate (Andes) Juan de Fuca sinks under North America Plate (Rockies) island arcs of the West Indies and Aleutians
Collision zones	two continental crusts collide and as neither can sink, are forced up into fold mountains	Indian collided with Eurasian forming Himalayas African collided with Eurasian forming Alps
C Conservative or passive margins (transform faults)	two plates move sideways past each other — land is neither formed nor destroyed	San Andreas fault in California
Note: centres of plates are rigid . . .	rigid plate centres form	
	a shield lands of ancient worn down rocks	Canadian (Laurentian) Shield, Brazilian Shield
	b depressions on edges of the shield developing into large river basins	Mississippi-Missouri, Amazon
(. . . With one main exception)	(Africa dividing to form a rift valley and possibly a new sea)	(African Rift Valley and the Red Sea)

◄ Figure 1.9

The major landforms resulting

from plate movements

5 No 'gaps' may occur on the earth's surface so, if two plates are moving apart, new oceanic crust originating from the mantle must be formed.

6 As the earth is neither expanding nor shrinking in size then as new oceanic crust is being formed at one place in the world, elsewhere older oceanic crust must be being destroyed.

7 Plate movement is slow (though not in geological terms) but is usually continuous. Sudden movements are detected as earthquakes.

8 Most significant landforms (fold mountains, volcanoes, island arcs, deep sea trenches, and batholith intrusions) are found at plate boundaries. Very little change occurs in the centre of plates. Figure 1.9 shows the major landforms resulting from different types of plate movement.

Landforms at constructive plate margins

A constructive margin is when two plates diverge or move away from each other and new crust is created at the boundary. Sea floor spreading occurs in the Mid-Atlantic where the North and South American and the Eurasian–African plates are being pulled apart by convection currents. Initially this may cause huge **rift valleys** to form on the sea floor, but molten rock or **magma** from the mantle rises to fill any possible gap between the two plates. This magma produces **submarine volcanoes** which in time may grow above sea level, e.g. Surtsey, south of Iceland on the Mid-Atlantic Ridge and Easter Island, on the East Pacific Rise.

As the basaltic magma cools, it adds new land to the separating plates. The Atlantic Ocean did not exist some 150 million years ago (Figure 1.4, Pangaea) and is still widening by some 2 to 5 cm a year. When the magma cools, large cracks called **transform faults** are produced at right angles to the plate boundary. The largest visible product of constructive divergent plates is Iceland, where one-third of the lava emitted on to the earth's surface in the last 500 years can be found (Figures 1.10 and 1.20).

CASE STUDY

Iceland

On 14 November 1963, the crew of an Icelandic fishing boat reported an explosion under the sea southwest of the Westman Islands. This was followed by smoke, steam and emissions of pumice stone. Having built up an ash cone of 130 m from the sea bed, the island of Surtsey emerged above the waves. On 4 April 1964 a lava flow covered the unconsolidated ash and guaranteed the island's survival.

Just before 0200 hours on 23 January 1973 an earth tremor stopped the clock in the main street of Heimaey, Iceland's main fishing port (Figure 1.20). Once again the North American and Eurasian Plates were moving apart (Figure 1.10). Fishermen at sea witnessed the crust of the earth break open and lava and ash pouring out of a fissure 2 km in length. Eventually the activity became concentrated on the volcanic cone of Helgafell and the inhabitants of Heimaey were evacuated to safety. By the time volcanic activity had ceased six months later, many homes nearby had been burned, others farther afield had been buried under five metres of ash and the entrance to the harbour was all but blocked.

▼ Figure 1.10

A constructive plate margin: Iceland

a Location of Iceland on the Mid-Atlantic Ridge

b Cross-section of the Mid-Atlantic Ridge

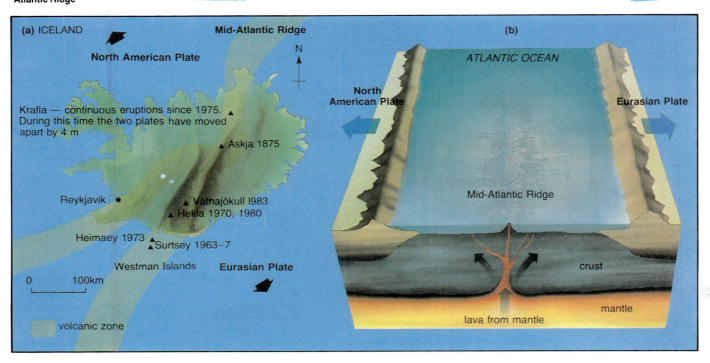

The formation of fold mountains is sometimes extremely complex. They usually occur when oceanic crust is subducted by continental crust. Figure 1.14 shows how a volcanic island arc migrates towards the main landmass and the intervening sediments are uplifted. A second, less frequent occurrence, is when two continental plates converge. In Figure 1.13 the Indian sub-continent is shown to have moved north-eastwards to collide with the Eurasian Plate and because continental crust does not sink, the intervening sediments containing sea shells have been raised into the Himalayas — an uplift which is still continuing. It is where these continental collisions have taken place that the earth's crust is at its thickest (Figure 1.7).

Landforms at destructive plate margins

When two plates converge or collide they form a destructive margin. The Pacific Ocean overlies five plates, all of which consist of oceanic crust, and is surrounded by continental plates (Figure 1.8). The large Pacific Plate moves northwest to collide with Eastern Asia, while the smaller but no less important Nazca, Cocos and Juan de Fuca plates travel eastwards towards South America, Central America and North America respectively. Figure 1.16 shows how the Nazca Plate, made of oceanic crust which cannot over-ride continental crust, is forced to dip down-wards at an angle to form a **subduction zone** and its associated deep sea **trench**. As crust descends, the increase in pressure can trigger off major earthquakes and the heat produced by friction helps to convert the disappearing crust back into magma. Being less dense than the mantle, the newly formed magma will try to rise to the earth's surface. Should it reach the surface, volcanoes may erupt either, as in the case of the Andes, forming a long chain of **fold mountains** or, if the eruption takes place offshore (e.g. Japan, West Indies, Figure 1.15), as an **island arc**. As this rising magma is more acidic than the basic lava of constructive margins, it is more viscous, flows less easily and may solidify within the mountains to form large **intrusive** features called **batholiths** (page 27).

The Atlantic Ocean was formed as the continent of Laurasia split into two. The process which caused the continental crust to become arched, weakened and stretched may be repeating itself in East Africa. Here the brittle crust has fractured and as sections moved apart the central portion dropped to form the Great African Rift Valley (Figure 1.11) with its associated volcanic activity. The valley extends for 4000 km from Mozambique to the Red Sea; in places the sides are over 600 m high and the width varies between 10 and 50 km. Where the land has been pulled apart and dropped sufficiently it has been invaded by the sea and it is suggested that the Red Sea is a newly forming ocean. Looking into the future, in 50 million years time will the east of Africa have detached itself from the remainder of the continent (Figure 1.4c)?

Conservative margins

This is when two plates are forced to slide past one another so that crust is neither created nor destroyed. The boundary between the two plates is characterised by pronounced transform faults of which the San Andreas in California (Figure 1.12) is the most notorious.

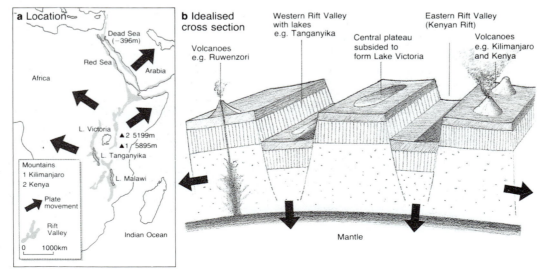

◄ Figure 1.11

The African Rift Valley

a Location

b Idealised cross-section

CASE STUDY

The San Andreas Fault

The San Andreas Fault forms a junction between the North American and Pacific Plates. Although both plates are moving northwest, the Pacific Plate moves faster giving the illusion that they are moving in opposite directions. The Pacific Plate moves about 6 cm a year but sometimes it sticks (like a machine without oil) until pressure builds up enabling it to jerk forwards. The last major movement resulted in the San Francisco earthquake of 1989 (page 12). Should these plates continue to slide past each other it is likely that Los Angeles will eventually be on an island off the Canadian coast.

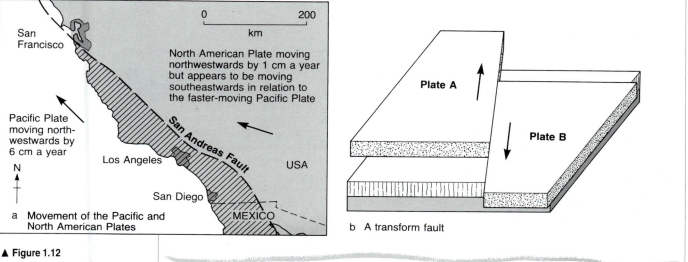

▲ Figure 1.12

A conservative plate margin: the San Andreas Fault

a Movement of the Pacific and North American Plates

b Diagram of a transform fault

Plate tectonics and the British Isles

During Cambrian times (Figure 1.1), Northern Scotland lay on the American Plate while the rest of Britain was on the Eurasian Plate, as it is today. Both plates are believed to have been in the latitude of present day South Africa. In Ordovician and Silurian times the two plates began to converge causing volcanic activity and the formation of mountains in Snowdonia and the Lake District. Being continental plates, sediment between them was pushed up to form the Caledonian Mountains, and NW Scotland became linked to the rest of Britain. During the Devonian period, the locked plates drifted northwards through a desert environment (of the present Kalahari) during which the old red sandstones were deposited. This northward movement continued in Carboniferous times and was accompanied by a sinking of the land to allow the limestones of that period to form in warm, clear seas.

As the land began to emerge from these seas, first millstone grit was formed from sediments in a shallow sea, and then coal measures were laid down under hot, wet, swampy conditions on the Equator. It was during Permian and Triassic times that the continents collided to form Pangaea; Africa moved towards Europe, and Britain's new red sandstones were laid down under dry, hot desert conditions in the position of the present Sahara. A further submergence during Jurassic–Cretaceous times enabled the Cotswold limestones and then the chalk of the Downs to form, again in warm, clear seas.

During the Tertiary, the North American and Eurasian Plates split apart forming a constructive boundary and the volcanoes of NW Scotland, while the African Plate moved further north pushing up the Alps and the hills of southern England. Since then, Britain, firmly settled away from the volcanoes and severe earthquakes of plate margins, has experienced landscape modifications during ice ages, but these have been a consequence of worldwide climatic change rather than of plate movement.

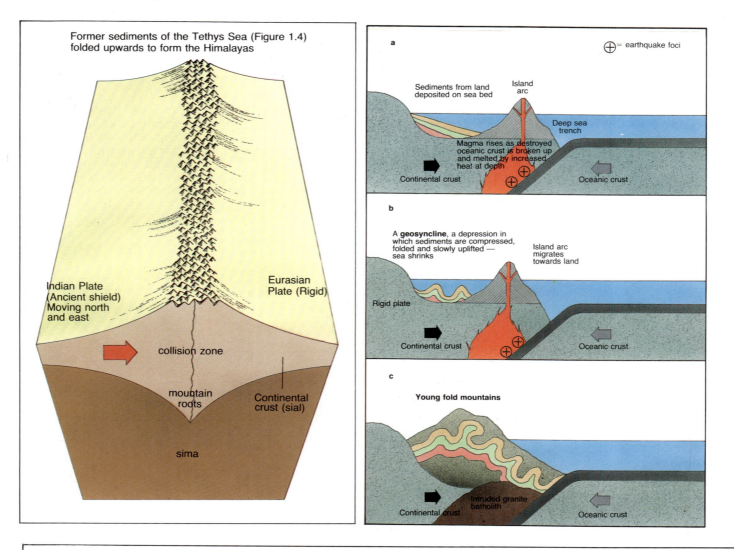

Former sediments of the Tethys Sea (Figure 1.4) folded upwards to form the Himalayas

Indian Plate (Ancient shield) Moving north and east

Eurasian Plate (Rigid)

collision zone

mountain roots

Continental crust (sial)

sima

a

⊕ = earthquake foci

Sediments from land deposited on sea bed

Island arc

Deep sea trench

Magma rises as destroyed oceanic crust is broken up and melted by increased heat at depth

Continental crust

Oceanic crust

b

A **geosyncline**, a depression in which sediments are compressed, folded and slowly uplifted — sea shrinks

Island arc migrates towards land

Rigid plate

Continental crust

Oceanic crust

c

Young fold mountains

Continental crust

Intruded granite batholith

Oceanic crust

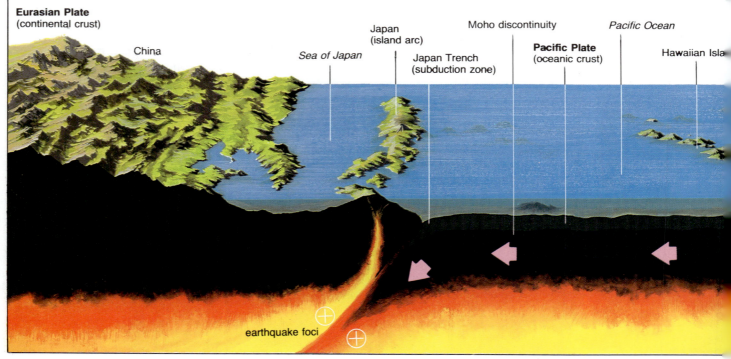

Eurasian Plate (continental crust)

China

Japan (island arc)

Moho discontinuity

Pacific Ocean

Sea of Japan

Japan Trench (subduction zone)

Pacific Plate (oceanic crust)

Hawaiian Isla

earthquake foci

◄ **Figure 1.13,** *far left*
Mountain building (orogenesis):
the Indian and Eurasian Plates

◄ **Figure 1.14**
The formation of fold mountains

▶ **Figure 1.16**
A destructive plate margin: the
Nazca and South American Plate
boundary

▼ **Figure 1.15**
Landforms resulting from plate
tectonics in the Pacific Ocean

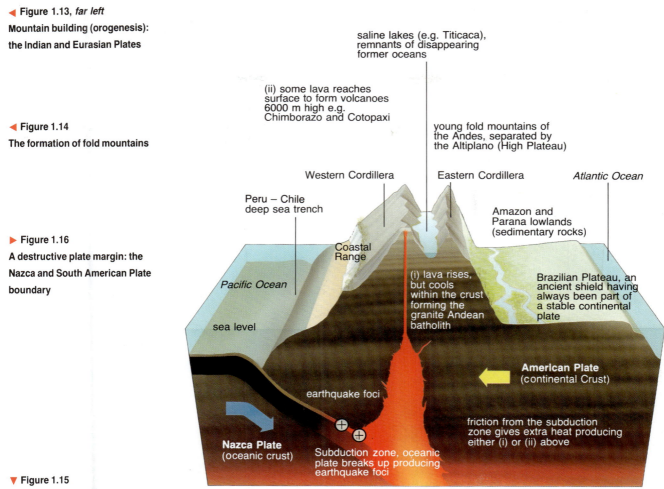

saline lakes (e.g. Titicaca),
remnants of disappearing
former oceans

(ii) some lava reaches
surface to form volcanoes
6000 m high e.g.
Chimborazo and Cotopaxi

young fold mountains of
the Andes, separated by
the Altiplano (High Plateau)

Western Cordillera

Eastern Cordillera

Atlantic Ocean

Peru – Chile
deep sea trench

Coastal
Range

Amazon and
Parana lowlands
(sedimentary rocks)

Pacific Ocean

(i) lava rises,
but cools
within the crust
forming the
granite Andean
batholith

Brazilian Plateau, an
ancient shield having
always been part of
a stable continental
plate

sea level

American Plate
(continental Crust)

earthquake foci

friction from the subduction
zone gives extra heat producing
either (i) or (ii) above

Nazca Plate
(oceanic crust)

Subduction zone, oceanic
plate breaks up producing
earthquake foci

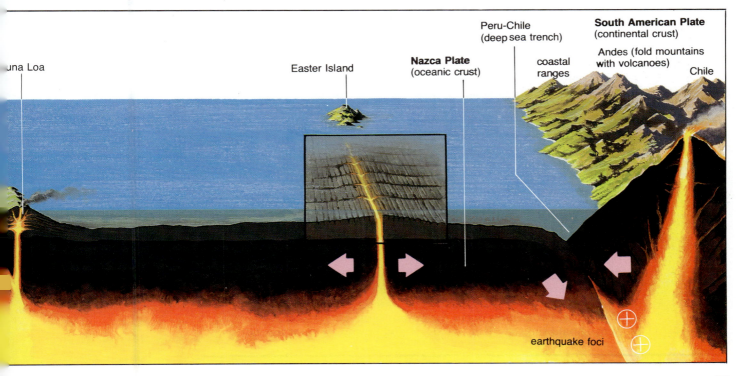

una Loa

Easter Island

Nazca Plate
(oceanic crust)

Peru-Chile
(deep sea trench)

South American Plate
(continental crust)

coastal
ranges

Andes (fold mountains
with volcanoes)

Chile

earthquake foci

21

Q

1 Refer to the geological timescale (Figure 1.1). Copy out the table and add notes to it describing mountain building and volcanic activity which has taken place in the British Isles.

2 Outline the evidence that supports the theory of plate tectonics.

3 How does an understanding of plate tectonics help to explain the distribution of **a** earthquake zones **b** volcanoes and **c** fold mountains?

4 With reference to any continent which you have studied, explain how an understanding of plate tectonics accounts for its major relief features.

▼ Figure 1.17

Landforms and major relief features resulting from plate tectonics in the Americas

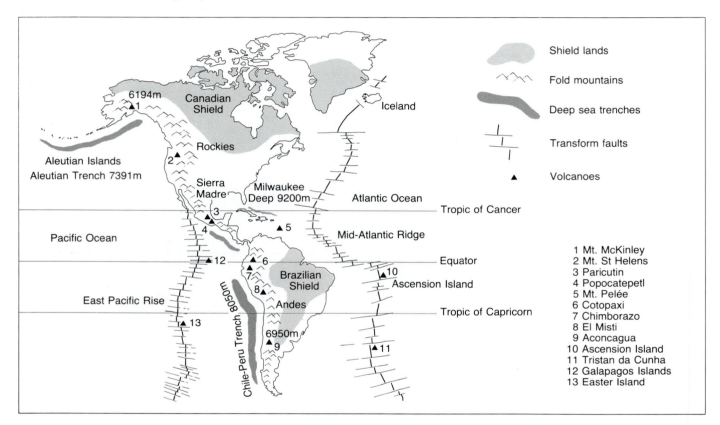

Shield lands
Fold mountains
Deep sea trenches
Transform faults
Volcanoes

1 Mt. McKinley
2 Mt. St Helens
3 Paricutin
4 Popocatepetl
5 Mt. Pelée
6 Cotopaxi
7 Chimborazo
8 El Misti
9 Aconcagua
10 Ascension Island
11 Tristan da Cunha
12 Galapagos Islands
13 Easter Island

5 With reference to the map of North and South America (Figure 1.7), explain how an understanding of plate tectonics helps to explain:
   the presence of the two landmasses and the Pacific Ocean,
   the presence of the Atlantic Ocean,
   the Canadian and Brazilian Shields,
   Ascension Island, Tristan da Cuhna and Easter Island,
   the three deep sea trenches,
   the Andes and Rockies,
   the West Indies and Aleutian Islands,
   Mount St Helens, Paricutin and Cotopaxi,
   the two different areas where transform faults occur.

6 Why do earthquakes on the Mid-Atlantic Ridge have a shallow focus whereas those in the Andes and Central America have a deep focus?

7 The Hawaiian Islands are an anomaly in that they are the summits of large volcanoes and yet they do not occur on a plate boundary. One theory suggests that they lie over a hot spot (Figure 1.6) in the Pacific Ocean.
   a Why should volcanoes occur over a hot spot?
   b Why should the Hawaiian Islands extend northwest from this hot spot?

Vulcanicity

The term **vulcanicity** includes all the processes by which solid, liquid or gaseous materials are forced into the earth's crust or are ejected on to the surface. Although material in the mantle has a high temperature it is kept in a semi-solid state because of the great pressure exerted upon it. However, if this pressure is locally released by folding, faulting or other movements at plate boundaries, some of this material becomes molten and rises, forcing its way into weaknesses in the crust, or on to the surface, where it cools, crystallises and solidifies. The molten rock is called **magma** when it is below the surface and **lava** on the surface. When lava and other materials reach the surface they are called **extrusive**: the resulting landforms vary in size from tiny cones to widespread lava flows. Materials injected into the crust are referred to as **intrusive**: they may later be exposed at the surface by erosion of the overlying rocks. Both extrusive and intrusive masses cooled from magma are known as **igneous** rocks.

Extrusive landforms

There are several types of extrusive landforms whose nature depends on how gaseous or viscous the lava is when it reaches the earth's surface (Figure 1.18).

- Lava produced by the upward movement of material from the mantle is **basaltic** and tends to be located along mid-ocean ridges, over hot spots and alongside rift valleys.
- Lava that results from the process of subduction is described as **andesitic** (after the Andes) and occurs as island arcs or at destructive plate boundaries where oceanic crust is being destroyed.

▶ Figure 1.18

Basic and acid lava

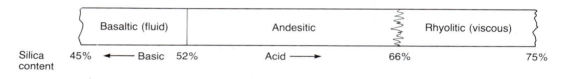

Comparing characteristics of lava types

Basaltic or basic lava	Andesitic or acid lava
Has low viscosity, is hot (1200°C) and runny, like warm treacle	Viscous, less hot (800°C), flows more slowly and less far
Has a lower silica content	Has a higher silica content
Takes a longer time to cool and solidify so flows considerable distances as rivers of molten rock	Soon cools and solidifies, flowing very short distances
Retains its gas content which makes it mobile	Quickly loses gases and becomes viscous
Produces extensive but gently sloping landforms	Produces steep-sided, more localised features
Eruptions are frequent but relatively gentle	Eruptions are less frequent but violent due to the build up of gases
Lava and steam ejected	Ash, rocks, gases, steam and lava ejected
Found at constructive plate margins where magma rises from the mantle e.g. fissures along the Mid-Atlantic Ridge (Heimaey) e.g. over hot spots (Mauna Loa, Hawaii)	At destructive margins where oceanic crust is destroyed (subducted) melts and rises e.g. subduction zones (Mount St Helens) e.g. as island arcs (Mt Pelée, Martinique)

How can we classify volcanoes?

Because of the large number of volcanoes and wide variety of eruptions it is convenient to group together those with similar characteristics. (Purpose of classification, page 130.) Unfortunately there is no universally accepted method of classification. The two most quoted groupings are according to the **shape** of the volcano and its vent which, because it describes landforms, is arguably of more value to the geographer; and the nature of the **eruption** which has traditionally been the method used by vulcanologists.

The shape of the volcano and its vent

1 **Fissure eruptions** When two plates move apart, lava may be ejected through fissures rather than via a central vent. The Heimaey eruption of 1973 (Figures 1.10, 1.20) began with a fissure over 3 km in length. This was small in comparison to that at Laki, also in Iceland, where in 1783 a fissure exceeding 30 km opened up. The basalt may form large plateaux, filling in hollows rather than building up into the more typical cone-shaped volcanic peak. The remains of one such lava flow, formed when the Eurasian and North American Plates began to divide, can be seen in Northern Ireland, NW Scotland, Iceland and Greenland. The columnar jointing produced by the slow cooling of the lava provides tourist attractions at the Giant's Causeway (Northern Ireland see Figure 1.21), and Fingal's Cave (Isle of Staffa).

2 **Basic** or **shield volcanoes** In volcanoes such as Mauna Loa, Hawaii, lava flows out of a central vent and can spread over wide areas before solidifying. The result is a 'cone' with long, gentle sides made up of many layers of lava from repeated flows.

3 **Acid** or **dome volcanoes** Acid lava quickly solidifies the air. This produces a steep-sided, convex cone as most lava builds up near to the vent. In the case of Mt Pelée, the lava actually solidified as it came up the vent and produced a spine rather than flowing down the sides.

4 **Composite cones** Many of the larger, classically shaped volcanoes result from alternating types of eruption in which first ash and then lava (usually acidic) is ejected. Mt Etna is a result of a series of both violent and more gentle eruptions.

▼ Figure 1.19

Classification of volcanoes based on their shape (not to scale)

a Fissure b Basic-shield
c Acid-dome d Ash-cinder
e Composite f Caldera
g Minor landforms

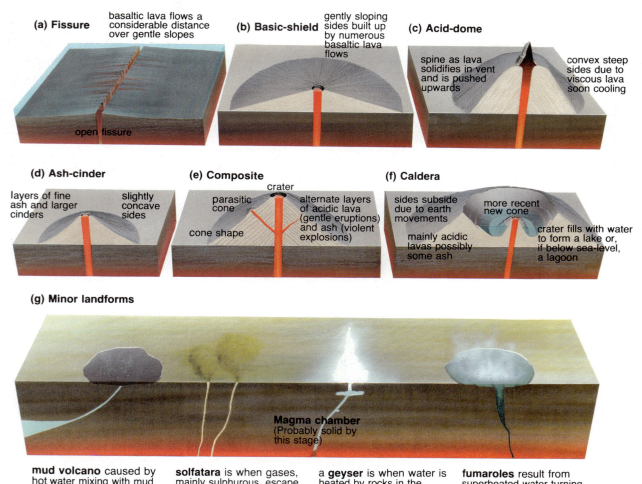

(a) Fissure basaltic lava flows a considerable distance over gentle slopes

open fissure

(b) Basic-shield gently sloping sides built up by numerous basaltic lava flows

(c) Acid-dome spine as lava solidifies in vent and is pushed upwards — convex steep sides due to viscous lava soon cooling

(d) Ash-cinder layers of fine ash and larger cinders — slightly concave sides

(e) Composite crater — parasitic cone — cone shape — alternate layers of acidic lava (gentle eruptions) and ash (violent explosions)

(f) Caldera sides subside due to earth movements — mainly acidic lavas possibly some ash — more recent new cone — crater fills with water to form a lake or, if below sea-level, a lagoon

(g) Minor landforms

Magma chamber (Probably solid by this stage)

mud volcano caused by hot water mixing with mud and surface deposits

solfatara is when gases, mainly sulphurous, escape on to the surface

a **geyser** is when water is heated by rocks in the lower crust, turns to steam, pressure increases and the steam and water explode onto the surface

fumaroles result from superheated water turning to steam as pressure drops as it emerges from the ground

▲ Figure 1.20
The Heimaey eruption, Iceland 1973

► Figure 1.21
The Giant's Causeway, Northern Ireland

5 **Ash** and **cinder cones** Paricutin, for example, was formed in the 1940s by ash and cinders building up into a symmetrical cone with a large crater.

6 **Calderas** When the build up of gases becomes extreme, huge explosions may clear the magma chamber beneath the volcano and remove the summit of the cone. This causes the sides of the crater to subside, thus widening the opening to several kilometres in diameter. In the case of both Thira (Santorini) and Krakatoa, the enlarged crater or caldera has been flooded by the sea and within the resultant lagoons, later eruptions have formed smaller cones (Figure 1.24).

Minor extrusive landforms
These are often associated with, but are not exclusive to, areas of declining volcanic activity. They include solfatara, fumaroles, geysers and mud volcanoes, illustrated in Figure 1.19.

CASE STUDY

Solfatara

Solfatara is a small volcano on the outskirts of Naples. Its crater is 2 km in diameter, making it larger than that of nearby Vesuvius, but there is no volcanic cone. **Solfatara** takes its name from the gases which escape to the surface; they are mainly sulphurous and can be smelled from a considerable distance. Many rocks are coated with sulphur. **Fumaroles**, resulting from superheated water being turned to steam as it cools on its ejection through the thin crust, are numerous in the area (Figure 1.22). Evidence of the thin crust (magma is only 3 m below the surface) is demonstrated by a guide who throws a boulder on to the surface and makes groups of tourist jump in harmony to hear the hollowness of the ground. A guide, who is needed to keep visitors safely away from bubbling mud volcanoes and areas too hot to walk upon, may also show volcanic activity by lighting twigs and stirring loose material to cause a miniature eruption.

The only minor landform missing is the geyser, an intermittent fountain of hot water (e.g. Old Faithful, Yellowstone National Park, USA).

During the mid 1980s the temperature (160°C), pressure and surface of Solfatara all rose, giving rise to fears of a new eruption — the last was in 1198. Despite the appearance of a small fissure near to the observatory, which has led to its abandonment, activity appears (in 1989) to have stabilised.

◄ **Figure 1.22**
Inside the Solfatara Crater, near Naples, Italy

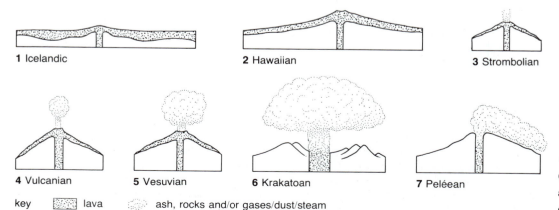

1 Icelandic

2 Hawaiian

3 Strombolian

4 Vulcanian

5 Vesuvian

6 Krakatoan

7 Peléean

key ▨ lava ⣿ ash, rocks and/or gases/dust/steam

◄ **Figure 1.23**
Classification of volcanoes according to the nature of the explosion

The nature of the eruption

This classification is based on a progressively more violent explosion which is a consequence of the pressure and amount of gas in the magma. Its categories may be summarised as follows:

- Icelandic type where lava flows gently from a fissure,
- Hawaiian type where lava is emitted gently but from a vent,
- Strombolian type where small but very frequent eruptions occur,
- Vulcanian type is more violent and less frequent,
- Vesuvian type (Figure 1.24) has a violent explosion after a long period of inactivity,
- Krakatoan or Plinian has an exceptionally violent explosion,
- Peléean type is where a violent eruption is accompanied by a cloud of gas, ash and pumice, called a **nuée ardente** ('glowing cloud'), which rolls at great speed down the mountain side.

▲ Figure 1.24
Vesuvius — notice the new cone within the old crater of Monte Somma

▶ Figure 1.25
Diagrammatic model showing intrusive landforms: batholith, dyke and sill

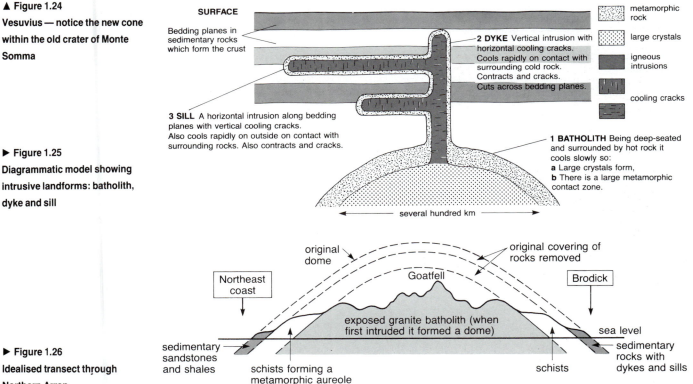

▶ Figure 1.26
Idealised transect through Northern Arran

Intrusive landforms

Geomorphology is the study of the structure, origin and development of the topographical features of the earth's crust

A relatively small amount of magma actually reaches the surface as most is **intruded** into the crust, where it solidifies. Such intrusions may initially have little impact upon the surface geomorphology but should the overlying rocks later be worn away then distinctive landforms may develop (Figure 1.25).

During Tertiary times an upthrust of magma was intruded into the sedimentary rocks of Arran to form the Northern Granite. As the magma slowly cooled it formed large crystals (unlike on the surface where rapid cooling forms fine crystals), contracted and cracked resulting in a series of joints. As the magma solidified it also produced a large, deep-seated, dome-shaped **batholith**.

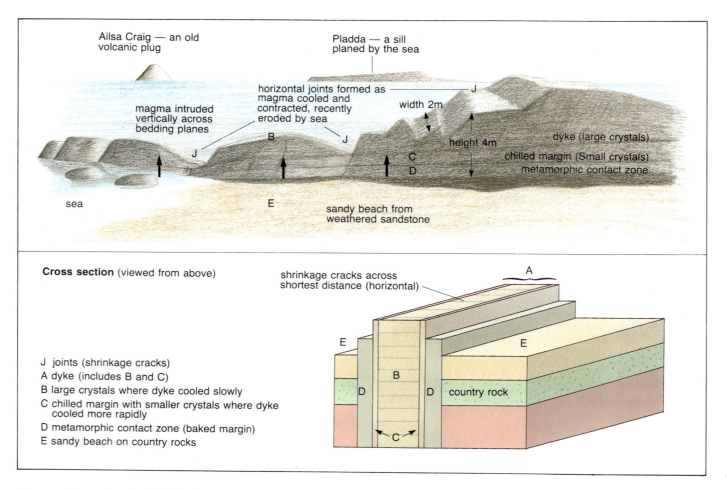

Ailsa Craig — an old volcanic plug

Pladda — a sill planed by the sea

horizontal joints formed as magma cooled and contracted, recently eroded by sea

magma intruded vertically across bedding planes

width 2m

height 4m

dyke (large crystals)
chilled margin (Small crystals)
metamorphic contact zone

sea

sandy beach from weathered sandstone

Cross section (viewed from above)

shrinkage cracks across shortest distance (horizontal)

country rock

J joints (shrinkage cracks)
A dyke (includes B and C)
B large crystals where dyke cooled slowly
C chilled margin with smaller crystals where dyke cooled more rapidly
D metamorphic contact zone (baked margin)
E sandy beach on country rocks

Surrounding the batholith is a **meta-morphic aureole** where the original sedimentary rocks have been changed (metamorphosed) by the heat and pressure of the intrusion from sandstones into **schists**. Since then the overlying rocks have been removed by water, ice and even the sea to leave the granite batholith with its jointing exposed (Figure 1.26). These joints have been widened by chemical weathering (Chapter 2) to form the large granite slabs and tors surrounding Goatfell (Figure 8.13).

If, in trying to rise to the surface, magma cuts across the bedding planes of the sedimentary rock it is called a **dyke** (Figure 1.27). The material which forms the dyke cools slowly although those parts which come into contact with the surrounding rock cool more rapidly to produce a chilled margin (Figure 1.28). Most dykes on Arran were formed after, and radiate from, the batholith intrusion and are so numerous that they have been termed a 'dyke swarm'. Most dykes are more resistant to erosion than the surrounding sandstones and where they cross the island's beaches they stand up like groynes (Figure 1.29). Although averaging 3 m, these dykes vary from 1 to 15 m in width.

A **sill** is formed when the igneous rock is intruded along the bedding planes between the existing sedimentary rocks. The magma cools and contracts but this time the resultant joints will be vertical and their hexagonal shapes can be seen when the landform is later exposed as on headlands such as that at Drumadoon on the west coast of Arran (Figures 1.30, 1.31). The sill at Drumadoon is 50 m thick.

▲ Figure 1.27, *top*
Fieldsketch of a dyke at Kildonan, Arran

▲ Figure 1.28, *above*
Diagrammatic cross-section of a dyke, Arran

▼ Figure 1.29
Dyke at Kildonan, Arran

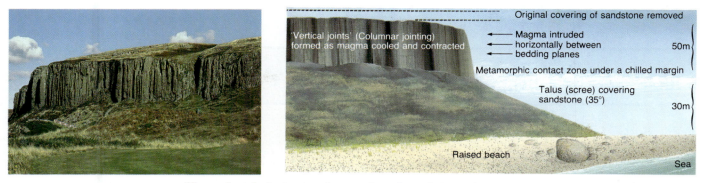

'Vertical joints' (Columnar jointing) formed as magma cooled and contracted

Original covering of sandstone removed

← Magma intruded
← horizontally between
← bedding planes

50m

Metamorphic contact zone under a chilled margin

Talus (scree) covering sandstone (35°)

30m

Raised beach

Sea

How do plate tectonics and vulcanicity affect human activity?

Benefits	Hazards
Lava and ash weather rapidly into fertile soils ideal for farming, e.g. the region surrounding Etna	Earthquakes destroy buildings and result in loss of life
Igneous rock contains minerals such as gold, copper, lead and silver	Violent eruptions with blast waves and gas may destroy life and property, e.g. Mt Pelée, Mt St Helens
Igneous rock is used for building purposes, e.g. Naples, Aberdeen	Mudflows may be caused by heavy rain and melting snow, e.g. Armero, Colombia
Extinct volcanoes may provide sites for defensive settlement, e.g. Edinburgh	Tidal waves/tsunamis, e.g. Krakatoa
Geothermal power is being developed, e.g. Iceland, New Zealand	Ejection of ash and lava ruins crops and kills animals
Geysers and volcanoes are tourist attractions, e.g. Yellowstone National Park, generating revenue for local communities	Interrupts communications
Volcanic eruptions may produce spectacular sunsets, e.g. Krakatoa	Short term climatic change as volcanic dust absorbs solar energy lowering temperatures and increasing rainfall

▲ **Figure 1.30,** *top left*
Sill at Drumadoon, Arran

▲ **Figure 1.31,** *top right*
Fieldsketch of a sill exposed at Drumadoon, Arran

Q

1 In Figure 1.32, which of the diagrams show sills and which dykes?

2 Using specific examples, discuss the benefits and dangers faced by people living near to plate boundaries.

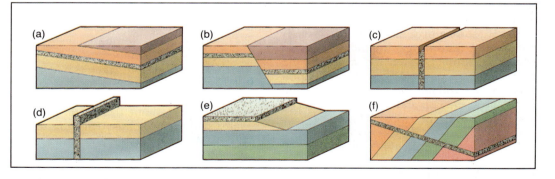

▶ **Figure 1.32**

References

Europe, David Waugh, Thomas Nelson, 1985.

Inside Volcanoes, Open University television programme.

North and South America, David Waugh, Thomas Nelson, 1984.

Planet Earth: Continents in Collison, Time–Life Books, 1983.

Planet Earth: Earthquakes, Time–Life Books, 1982.

Planet Earth: Volcanoes, Time–Life Books, 1982.

The Nature of the Environment, Andrew Goudie, Basil Blackwell, 1984.

The Restless Earth, Nigel Calder, BBC Publications, 1972.

The Unquiet Landscape, Denys Brunsden, John Doornkamp (editors), David & Charles, 1977.

The World, David Waugh, Thomas Nelson, 1987.

This Shaking Earth, John Gribben, Sidgwick & Jackson, 1978.

Volcanoes, Geological Museum Publications, 1974.

Volcanoes, Peter Francis, Penguin Books, 1984.

CHAPTER 2

Weathering and mass movement

'Every valley shall be exalted, and every mountain and hill shall be made low: and the crooked shall be made straight, and the rough places plain.'

The Bible, Isaiah, 40:4

Weathering processes

The majority of rocks have been formed at high temperature (**igneous** and many **metamorphic** rocks) and/or under great pressure (igneous, metamorphic and **sedimentary** rocks) but in the absence of oxygen and water. If, later, these rocks become exposed on the earth's surface they will experience a release of pressure, be subjected to fluctuating temperatures and exposure to oxygen in the air and to water. They are therefore vulnerable to **weathering** which is the disintegration and decomposition of rock *in situ* i.e. in its original position. Weathering is hence the natural breakdown of rock and can be distinguished from erosion because weathering need not involve any movement of material. Weathering is the first stage in the **denudation** or wearing down of the landscape: it loosens material which can subsequently be transported by such agents of erosion as running water, ice, wind and the sea. The degree of weathering depends upon the structure and mineral composition of the rocks, local climate and vegetation and the length of time the weathering processes operate.

There are two main types of weathering and some authorities add a third category.

1 **Physical** or **mechanical** weathering is the disintegration of rock into smaller particles without any change in the chemical composition of that rock. It is more likely to occur in areas void of vegetation such as deserts, high mountains and arctic regions. Physical weathering usually produces sands.

2 **Chemical** weathering is the decomposition of rock resulting from a chemical change. This is more likely to take place in warmer, more moist climates, with an associated vegetation cover, to form clays.

3 **Biological** weathering is the result of the activities of plants and animals but, as will be seen later, such actions may also be included under physical and chemical weathering.

It should also be appreciated that though in any given area either physical or chemical weathering may be locally dominant, both processes usually operate together rather than in isolation.

Physical weathering

Frost shattering

This is the most widespread form of physical weathering. It occurs in rocks which contain crevices and joints, (e.g. joints formed in granite as it cooled, bedding planes found in sedimentary rocks and pore spaces in porous rocks), where there is limited vegetation cover and where temperatures oscillate around 0°C. During the daytime when it is warmer, water enters the joints but during the cold nights it freezes. As ice occupies 9 per cent more volume than water it exerts pressure within the joints. This alternating **freeze–thaw** process slowly widens the joints and pieces of rock shatter from the main body. Where this **block disintegration** occurs on steep slopes, large angular rocks collect at the foot of the slope as **scree** or **talus** (Figure 2.1) whereas if the slopes are gentle, large **blockfields** or **felsenmeer** tend to develop. Frost shattering is more common in regions such as upland Britain where temperatures fluctuate around freezing point for several months in winter than in Polar areas where temperatures rarely rise above 0°C.

▲ Figure 2.1
The formation of screes resulting from frost shattering at Goredale Scar in the Yorkshire Dales National Park

▲ Figure 2.2, *top right*
Salt crystallisation on the coast of Arran

Salt crystallisation

If water entering pore spaces is slightly saline, and the water then evaporates, salt crystals form. As these become larger they will break up the rock. This process occurs in hot deserts when capillary action (page 214) draws water to the surface — where the rock is sandstone, individual grains are broken off by **granular disintegration**. Salt crystallisation occurs on coasts (Figure 2.2) where the supply of salt is constant throughout the year.

Pressure release

As stated earlier, many rocks, especially intrusive jointed granites, are developed under considerable pressure. When these rocks are exposed to the atmosphere, the slight decrease in pressure allows the rocks to expand. Consequently, cracks may develop parallel to the surface causing **sheeting**, a process which results in the peeling away of the outer layers. It is now believed that pressure release is responsible for the formation of large rounded rocks called **exfoliation domes** (Figures 2.3 and 2.5), and in part for the granite tors found on Dartmoor and the Isle of Arran (Chapter 8). The characteristic cirque shape found in glaciated areas (Chapter 4, Figure 4.13) is accentuated by pressure release cracking.

Thermal expansion or insolation weathering

Like all solids, rocks expand when heated and contract when cooled. In deserts where cloud and vegetation cover is minimal, the diurnal range of temperature is tremendous. It was believed that because the outer layers of rock warm up faster and cool more rapidly than the inner ones, stresses were set up which would cause the outer thickness to peel off like the layers of an onion — the process of **exfoliation**. It was thought, initially, that it was this expansion-contraction process which produced exfoliation domes. Different minerals expand and contract at different rates with changes in temperature. It has been suggested that this causes granular disintegration in rocks composed of several minerals, e.g. granite consisting of quartz, feldspar and mica, whereas thermal expansion in homogeneous rocks tends to produce block disintegration.

It was also assumed that this process was responsible for the granular disintegration of sandstones in desert areas to produce sand, but recent laboratory experiments, described on page 148, have cast doubt upon the effectiveness of thermal expansion without the presence of water.

▶ Figure 2.3
The process of pressure release tends to perpetuate landforms — as new surfaces are exposed, the reduction in pressure causes further jointing parallel to the surface

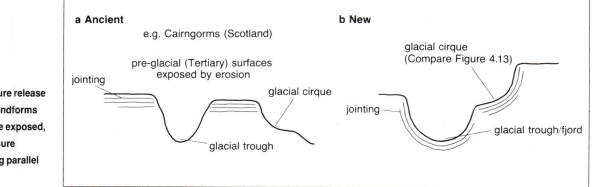

◀ Figure 2.4

The effect of raindrop impact

and the formation of earth pillars

Diagram labels:

once the capstone falls, the earth pillar will collapse

raindrop action removes softer material leaving a capstone

boulder clay or till (Chapter 4) with little vegetation cover — exposed to rain

◀ Figure 2.5

Sugarloaf mountain in Rio de Janeiro, Brazil, is one of the world's best known exfoliation domes

Raindrop impact

In areas of limited vegetation cover, the weathering effect of raindrops may be to remove soft, fine material from between larger boulders. This produces columns of softer material capped by a protective boulder and is most common in glacial deposits (Figure 2.4).

Biological

Tree roots may grow along bedding planes or extend into joints widening them until blocks of rock become detached (Figure 2.6). It is also claimed that burrowing creatures, such as worms and rabbits, may play a minor role in the excavation of partially weathered rocks.

Chemical weathering

Chemical weathering tends to:

■ attack certain minerals selectively

■ be common in zones of alternate wetting and drying, e.g. where the level of the water table fluctuates

■ be more common at the base of slopes where it is likely to be wetter and warmer.

This type of weathering involves a number of specific processes which may operate in isolation but are more likely to be found in conjunction with one another. Formulae for the various chemical reactions are listed at the end of this chapter.

◀ Figure 2.6

Biological weathering caused by expanding tree roots in Geltsdale, Cumbria

Oxidation

This occurs when rocks are exposed to the oxygen in air or water. The simplest and most easily recognised example is when iron in a **ferrous** state is changed by the addition of oxygen into a **ferric** state. The rock or soil, which may have been blue or grey in colour, characteristic of a lack of oxygen, is discoloured into a reddish brown — a process better known as rusting. Following oxidation, rocks are likely to crumble more easily.

In waterlogged areas oxidation may operate in reverse and is known as **reduction**. Here, the amount of oxygen is reduced and the soils take on a blue/green/grey tinge (see **gleying**, page 226).

Hydration

Certain rocks are capable of absorbing water into their structure which causes them to swell and to become vulnerable to future breakdown. For example, gypsum is the result of water having been added to anhydrite ($CaSO_4$). This process appears to be most active following successive periods of wet and dry weather and is important in forming clay particles. Hydration is in fact a physio-chemical process as the rocks may swell and exert pressure as well as changing their chemical structure.

Quartz	Mica	Feldspar
not affected by water, remains unchanged as sand (Figure 2.7)	may be affected by water under more acid conditions releasing aluminium and iron	readily attracts water producing a chemical change which turns the feldspar into clay (kaolin or china clay)

Hydrolysis

This is possibly the most significant chemical process in the decomposition of rocks and formation of clays. Hydrogen in water reacts with minerals in the rock (the hydrogen ions in water replace cations in the mineral — a process in soil development called **cation exchange**, page 222). An example of hydrolysis is the breakdown of feldspar, a mineral found in igneous rocks such as granite. Granite consists of three minerals: quartz, mica and feldspar (see table).

Carbonation

Rainwater contains carbon dioxide in solution which produces carbonic acid (H_2CO_3). This weak acid reacts with rocks, such as limestone, which are composed of calcium carbonate. The limestone dissolves and is removed in solution (calcium bicarbonate) by running water. Limestone is well-jointed and bedded (Chapter 8) which results in the development of a distinctive group of landforms (Figure 2.8).

Solution

Some minerals, e.g. rock salt, are soluble in water and simply dissolve *in situ*. The rate of solution may be determined by acidity because several minerals become more soluble when the pH of the solvent increases (page 223).

Biological

Humic acid is derived from the decomposition of vegetation (humus) and contains important elements such as calcium, magnesium and iron. These are released by a process called **chelation**. The presence of organic life (plants, animals, bacteria) increases the concentration of carbon dioxide in a soil and thus the level of carbonation. Lichens can extract iron from certain rocks and concentrate it at the surface. However, it should be remembered that the presence of a vegetation cover dramatically reduces the extent of physical weathering.

Acid rain

Human economic activities release increasingly more carbon dioxide, sulphur dioxide, and nitrogen oxide (power stations, cars and lorries) into the atmosphere. These gases then form acids in solution in rainwater. Acid rain readily attacks limestones and, to a lesser extent, sandstones, as shown by crumbling buildings and statues. The increased level of acidity in water passing through the soil tends to release more hydrogen and so speeds up the process of hydrolysis.

▼ Figure 2.7, *below*
Decomposition of granite by hydrolysis on Goatfell, Arran

▶ Figure 2.8, *right*
Carbonation of limestone near Malham, Yorkshire Dales National Park

Climatic controls on weathering
Physical weathering

Frost shattering is important if temperatures fluctuate around 0°C, but will not operate if the climate is too cold (permanently frozen), too warm (no freezing), too dry (no moisture to freeze) or too wet (covered by vegetation). Physical weathering will not take place at **X** (Figure 2.9a) where it is too warm and there is insufficient moisture, while at **Y**, the high temperature and heavy rainfall will give a thick protective vegetation cover against the insolation effects of the sun (page 168).

Chemical weathering

(Figure 2.9b) This increases as temperatures and rainfall totals increase. Recent views suggest that in tropical areas, direct removal by solution may be the major factor in the lowering of the landscape, as water flowing through the soil is a continuous weathering process. The rich vegetation at **S** will increase the amount of humic acid present. Chemical weathering will not operate at **P** because temperatures are too low, nor at **R**, because moisture is essential in the chemical decomposition of rocks.

Peltier, an American physicist/climatologist, attempted to show how it is possible to predict the type and rate of weathering at any given place in the world from a knowledge of its mean annual temperature and its mean annual rainfall (Figure 2.9c).

▼ Figure 2.9

Climatic controls on weathering (*after* Peltier)

a Mechanical weathering

b Chemical weathering

c Weathering regions

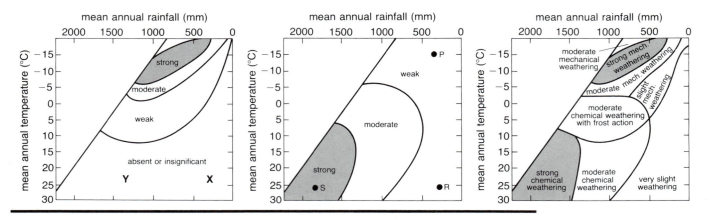

Using figure 2.9c:

1 What is the minimum mean annual rainfall which allows **a** strong chemical weathering and **b** strong mechanical (physical) weathering to occur?

2 What is the minimum mean annual temperature which enables strong chemical weathering to take place?

3 What are the temperature extremes for significant mechanical weathering?

4 Why is chemical weathering more important than mechanical weathering at location **X**?
The diagram below (Figure 2.10) is a model different from that of Peltier.

◄ Figure 2.10

A model showing the relationship between climate and weathering

5 Describe how the processes of mechanical weathering will differ between locations **Y** and **Z**.

6 Without referring to precipitation, describe three other factors which affect the rate and type of weathering in the British Isles.

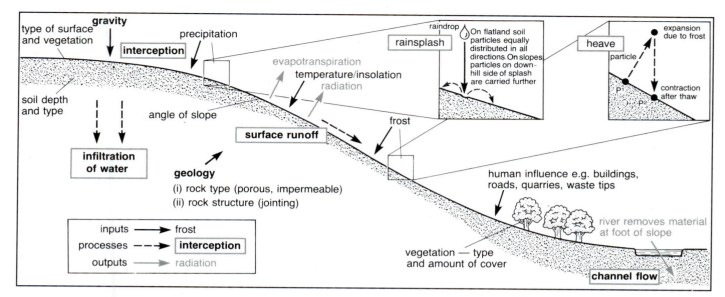

▲ Figure 2.11

The slope as a dynamic open system

Mass movement

The term **mass movement** decribes all downhill movements of soil, loose stones and rocks (**regolith**) in response to gravity but excludes movements when the material is carried by ice, water or wind. When gravitational forces exceed forces of resistance, slope failure occurs and material starts to move downwards. A slope is a **dynamic open system** affected by biotic, climatic, gravitational, groundwater and tectonic inputs which vary in scale and time, while the amount, rate and type of movement depends upon the degree of slope failure (Figure 2.11).

Processes of mass movement and resultant landforms

Although by definition mass movement refers only to the movement downhill of material under the force of gravity, in reality water is usually present and assists the process. When M.A. Carson and M.J. Kirkby (1972)

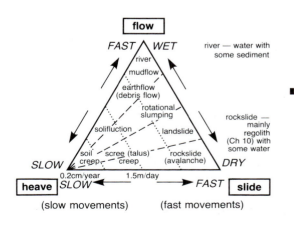

▶ Figure 2.12

A classification of mass movement processes (*after Carson and Kirkby, 1972*)

attempted to classify the main types of movement by means of a graph, they used as their basis the mechanism of movement (heave, flow and slide) and moisture content (Figure 2.12).

A   Slow movements

Soil creep

This is the slowest of downhill movements and is difficult to measure as it takes place at a rate of less than 1 cm a year. However, unlike faster movements, it is an almost continuous process. Soil creep occurs mainly in humid climates where there is a vegetation cover. There are two major causes of creep, both resulting from repeated expansion and contraction.

■ **Wet–dry periods**   During times of heavy rainfall, moisture increases the volume of the soil and adds weight, causing expansion and allowing the regolith to move down-hill under gravity. In a subsequent dry period the soil will dry out and then contract, especially in clay. An extreme case of contraction in clays was in SE England during the 1976 drought when buildings sited on almost imperceptible slopes suffered major structural damage.

■ **Freeze–thaw**   When the regolith freezes, the presence of ice crystals increases the volume of the soil by 9 per cent. As the soil expands, particles are lifted at right angles to the slope (Figure 2.11) in a process called **heave**. When the ground later thaws and the regolith contracts, these particles fall back vertically under the influence of gravity and so move downslope.

telegraph
pole tilted

base of tree
turned downslope

tension
gashes in road

fences
broken

terracettes

soil piled up
behind wall
forcing it to
bulge and break

◄ Figure 2.14
The effects of soil creep

Soil creep usually occurs on slopes of about 5°, and produces **terracettes** (Figure 2.13). These are step-like features, often 20–50 cm in height, which develop as the vegetation is stretched and torn: they are often accentuated by grazing animals, especially sheep. The effects of soil creep are shown in Figure 2.14.

Solifluction

This process, meaning 'soil flow', is a slightly faster movement averaging between 5 and 20 cm a year. It usually takes place under periglacial conditions (Chapter 5) where vegetation cover is limited. During the winter season, bedrock and regolith are both frozen. In summer, the surface layer thaws but the underlying layers remain frozen and acts like an impermeable rock. The surface meltwater cannot infiltrate downwards and temperatures are too low for much effective evaporation, so the meltwater soon saturates any topsoil, causing it to flow as an **active layer** over the frozen subsoil and rock. This process produces **solifluction lobes** (Figure 5.10) which are rounded, tongue-like features reaching up to 50 m in width and **head**, a mixture of sand and clay formed in valleys and at the foot of sea cliffs (page 106). Solifluction was widespread in southern Britain during the Pleistocene Ice Age, over most of Britain following the Pleistocene and continues to take place in the Scottish Highlands today.

B Flow movements

Earthflows

When the regolith on slopes of 5° to 15° becomes saturated with water, it begins to flow downhill at speeds of less than 15 km per hour. The movement of material may produce short **flow tracks** and small bulging lobes or tongues, yet may not be fast enough actually to break the vegetation.

Mudflows

These are more rapid movements, occurring on steeper slopes and exceeding 10 m/hr — when Nevado del Ruiz erupted in Colombia in 1985, the resultant mudflow reached the town of Armero at an estimated speed of 80 km/hr. Mudflows are most likely to happen following periods of intensive rainfall when both mass and volume are added to the soil giving it a higher water content than an earthflow. Mudflows may result from the combination of several factors as shown in Figure 2.15.

▼ Figure 2.15
Fieldsketch showing the causes of a mudflow, Glen Rosa, Arran

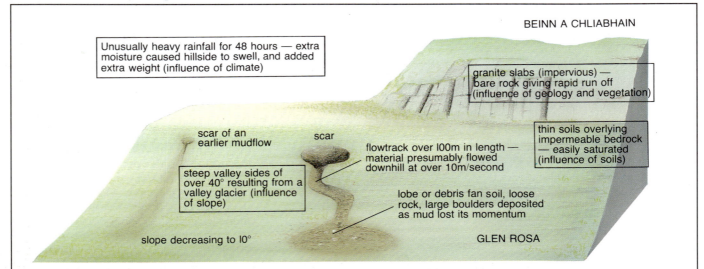

BEINN A CHLIABHAIN

Unusually heavy rainfall for 48 hours — extra moisture caused hillside to swell, and added extra weight (influence of climate)

granite slabs (impervious) — bare rock giving rapid run off (influence of geology and vegetation)

thin soils overlying impermeable bedrock — easily saturated (influence of soils)

scar of an earlier mudflow

scar

flowtrack over 100m in length — material presumably flowed downhill at over 10m/second

steep valley sides of over 40° resulting from a valley glacier (influence of slope)

lobe or debris fan soil, loose rock, large boulders deposited as mud lost its momentum

slope decreasing to 10°

GLEN ROSA

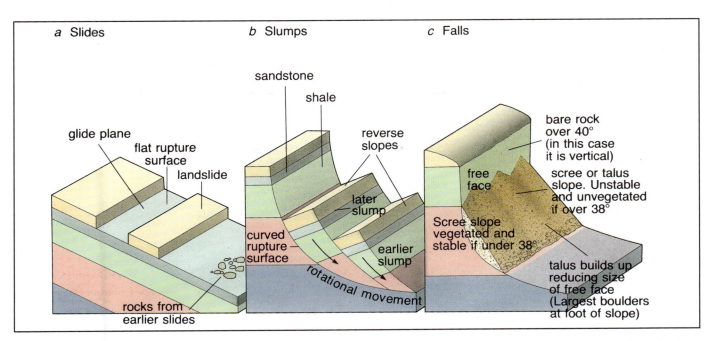

▲ Figure 2.16

Slides, slumps and falls

C   Rapid movements

Slides

Rocks which have jointing or bedding planes roughly parallel to the angle of slope are particularly susceptible to landslides. The rock, once weathered, may slide downhill leaving a flat rupture surface (Figure 2.16a).

Slumps

Rotational movement which produces a curved rupture surface is known as **slumping**. Slope failure often penetrates deep into a hillside which may be of homogeneous rock but is more likely to consist of softer materials — clay or sandstone — overlying more resistant or impermeable rocks — limestone or shale (Figure 2.16b). Slumping is common in many coastal areas of south and eastern England. The photograph in Figure 2.17 was taken near Lulworth Cove on the Dorset coast. Here the cliffs are composed mainly of soft glacial deposits and slumping is causing the land to retreat rapidly. In the foreground a mudflow with its lobe can be seen extending from the slumped material.

D   Very rapid movements

Rockfalls

These are spontaneous yet relatively rare debris movements on slopes in excess of 40°. They may result from extreme physical or chemical weathering in mountains, pressure release, storm wave action on sea cliffs, and earthquakes. Material, once broken from the surface, will either bounce or fall vertically to produce scree or talus at the foot of the slope (Figures 2.16c and 2.21).

► Figure 2.17

Slumping and mudflow on the coast at Lulworth Cove, Dorset

CASE STUDY

Mudflow at Aberfan

Aberfan, like many other settlements in the South Wales valleys, grew up around its colliery. However, the valley floors were rarely wide enough to store the coal waste and so it became common practice to tip it high on the steep valley sides. At Aberfan, the spoil tips were on slopes of 25° over 200 m above the village and, unknowingly, on a line of springs. Water from these springs added weight to the heaps and following a wet October in 1966 and a night of heavy rain, the material suddenly and rapidly flowed downhill. The time was just after 0900 on 21 October, soon after lessons had begun at the local junior school. Of the 147 deaths that morning in Aberfan, 116 were children in the school.

◄ Figure 2.18
The village of Aberfan, South Wales, where in 1966 a massive mud- or debris-flow caused the loss of over 100 lives

The Vaiont Dam

The Vaiont Dam was opened in northern Italy in 1960. The third highest concrete dam in the world, it had been built in a narrow, steep-sided valley whose sides consisted of alternating layers of clays and limestones. Three years later, following a period of heavy rain which saturated the clays, a mass of rock, mud, earth and vegetation slid over the limestone and into the reservoir. The dam itself withstood the landslide but a wave of water spilled over the lip creating a towering wall of water which swept down the valley below, destroying villages and causing almost 1200 deaths.

Peruvian earthquake

In May 1970 an earthquake recording 7.7 on the Richter scale shook parts of Peru. The shock waves loosened a mass of ice and snow near the summit of Huascarán, the country's highest peak at over 6500 m. As the ice and snow fell 3000 m it picked up rocks and boulders and the avalanche hit the town of Yungay at an estimated 480 km/hr. When rescue workers eventually reached the area three days later they found very few survivors out of a population of 20 000 and only the tops of several 30 m palm trees marking the spot where the town had previously stood.

The types of mass movement can be classified according to their speed of movement and the amount of water necessary to assist this movement.

1 Redraw the graph (Figure 2.19a) and add the following labels in the appropriate places: earth/mudflow, solifluction, rockfall, slumping, soil creep.

a

Extremely slow	Very slow	Slow	Moderate	Rapid	Very rapid	Extremely rapid
1 cm/year	1 m/year	1km/year	1m/hour	1 km/hour	1 m/sec	100m/sec

② ④

① ③ ⑤

◄ Figure 2.19a

2 Redraw the sketch shown in Figure 2.19b and add labels for the same five processes as listed in 1. Indicate the location of the following landforms: scree (talus), terracettes, lobe.

► Figure 2.19b, c

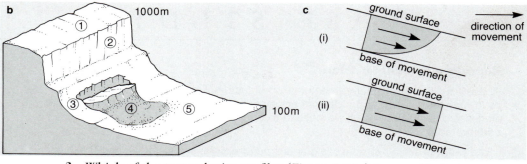

3   Which of the two velocity profiles (Figure 2.19c) is more likely to represent an earth/mudflow, and which a landslide? Explain your answer.

Development of slopes

Slope development is the result of the inter-action of several factors. Rock structure and lithology, soil, climate, vegetation and human activity are probably the most significant and all are influenced by the time over which processes operate.

The effect of rock structure and lithology

■ Areas of bare rock are vulnerable to mechanical weathering, e.g. frost shatter-ing, and certain types of chemical weathering processes.
■ Areas of alternating hard/resistant rocks and soft/less resistant rocks are more likely to experience movement, e.g. clays on limestones.
■ An impervious underlying rock will cause the topsoil to become saturated more quickly, e.g. glacial deposits overlying granite.
■ Steep gradients are more likely to suffer slope failure than gentler ones. In Britain most slopes are under 5° and few are over 40°.
■ Failure is also likely on slopes where the equilibrium (balance) of the system, Framework 1, has been disturbed, e.g. a glaciated valley.
■ The presence of joints, cracks and bed-ding planes can allow increased water content and so lead to sliding, (e.g. Vaiont Dam.
■ Earthquakes, e.g. Mt Husacaran in Peru.

Soil

■ Thin soils tend to be more unstable as they can support only limited vegetation with the result that there are fewer roots to bind the soil together.
■ Unconsolidated sands have lower internal cohesion than clays.
■ A porous soil is less likely to become saturated than one which is impermeable, e.g. sand and clay.

■ In a non-saturated soil, surface tension of the water tends to draw particles together. This increases cohesion and reduces soil movement. In a saturated soil, the pore water pressure (page 221) forces the particles apart, reducing friction and causing soil movement (Figure 2.20).

Climate

■ Heavy rain adds volume and mass to the soil.
■ Heavy rain increases the erosive power of any river at the base of the slope and so, by removing material, makes that slope less stable.
■ Areas with freeze–thaw or wet–dry periods are subjected to alternating expansion-contraction of the soil.
■ Heavy snowfall adds weight and is thus conducive to rapid movements.

Vegetation

■ A lack of vegetation means fewer roots to bind the soil together.
■ Sparse vegetation cover will encourage surface run off as precipitation is not intercepted (page 45).

Human influence

■ Deforestation increases (afforestation de-creases) the rate of slope movement.
■ Building roads or quarrying at the foot of slopes upsets the equilibrium, e.g. the building of the M5 in the Bristol area.
■ By adding weight to slopes by building on them or depositing industrial or mining waste, (Aberfan) slope development pro cesses may be accentuated.
■ The shaking action of heavy traffic (Mam Tor, Derbyshire).
■ The grazing of animals and ploughing loosen soil and remove protective veg-etation cover.

▼ Figure 2.20

The effect of pore-water pressure and capillary action on soil movement

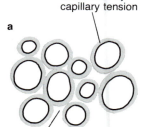

a

water held by capillary tension

air in pore spaces

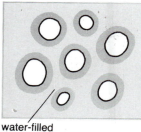

b

water-filled pore spaces friction is reduced

CASE STUDY

Slope development on Arran

Slopes usually develop from a combination of factors. Figure 2.23 shows the Glen Rosa Valley on the Isle of Arran. The steep sides of this valley are the result of several processes, some of which have ceased to operate and others which can still be seen working today.

The valley is a fine example of a glacial trough with the sides steepened by the action of a valley glacier. As the ice melted, the bare, jointed granite rocks provided ideal conditions for physical weathering in the form of freeze–thaw activity. During these periglacial times further fracturing of the rock may have occurred due to pressure release from the ice, and certainly scree (talus) slopes would have formed. Physical weathering still operates today as does chemical weathering with the heavy rainfall causing hydrolysis. The steep slopes and heavy rainfall contribute to a range of mass movement processes including creep, solifluction, mudflows (Figure 2.15) and rockfalls. The steepness of the slopes is partly maintained by the Rosa Water (which had formed the pre-glacial valley) as it removes material and keeps the sides unstable. Finally the slopes are modified by human activities (the tourist/walker/climber affecting vegetation, causing pollution and erosion and reducing wildlife) and the grazing of sheep and deer.

◀ Figure 2.21
Rockfalls within the crater of Vesuvius, Italy

Slope elements

Two models try to show the shape and form of a typical slope. The first, Figure 2.22a is more widely used, though in this author's view it is less conspicuous in the British landscape than the second (Figure 2.22b). Confusion arises, unfortunately, because of the variation in nomenclature used to describe the different facets of the slope, regardless of which model is used.

In reality, few slopes are likely to match up perfectly with either model, and each individual slope is likely to show more elements than those in Figure 2.22. In the field, one way to plot the distribution of these facets to see if they form any pattern is to use a **morphological map**. Morphological symbols can be used to show the **form** of the slope (**concave**, **convex** or **rectilinear**) and to show changes in slope, which may either be a sharp break or a gradual change. Figure 2.23 is the result of a transect taken down a valley side in Glen Rosa, Arran.

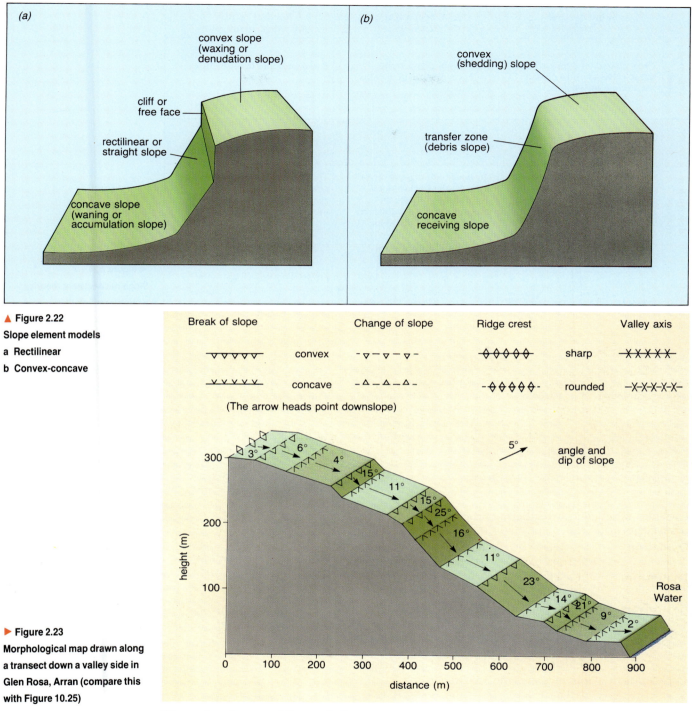

▲ Figure 2.22
Slope element models
a Rectilinear
b Convex-concave

▶ Figure 2.23
Morphological map drawn along a transect down a valley side in Glen Rosa, Arran (compare this with Figure 10.25)

Slope development through time

How slopes have developed over time is one of the more controversial topics in geomorphology. The contention arises partly because of the time needed for slopes to evolve and the variety of combinations of processes acting upon these slopes in various parts of the world. The differing environments have led to three divergent theories being proposed: **slope decline**, **slope replacement** and **parallel retreat**. Figure 2.24 is a summary of these hypotheses.

None of the models of slope development overleaf can be universally accepted, although each may have a local significance resulting from the climate and geology (structure) of the area under study. Two different climates and processes may produce the same type of slope, e.g. cliff retreat due to sea action in a humid climate or to weathering in a semi-arid climate.

41

	Slope decline (W.M. Davis, 1899)	Slope replacement (W. Penck, 1924)	Parallel retreat (L.C. King, 1948, 1957)
region of study	Theory based on slopes in what was to Davis a normal climate, NW Europe and NE USA.	Conclusions drawn from evidence of slopes in the Alps and Andes.	Based on slopes in South Africa.
climate	Humid climates.	Tectonic areas.	Semi-arid landscapes. Sea cliffs with wave-cut platforms.
description of slope	Steepest slopes at beginning of process with a progressively decreasing angle in time to give a convex upper slope and a concave lower slope.	The maximum angle decreases as the gentler lower slopes erode back to replace the steeper ones giving a concave central portion to the slope.	The maximum angle remains constant as do all slope facets apart from the lower one which increases in concavity.
	(a) **slope decline** 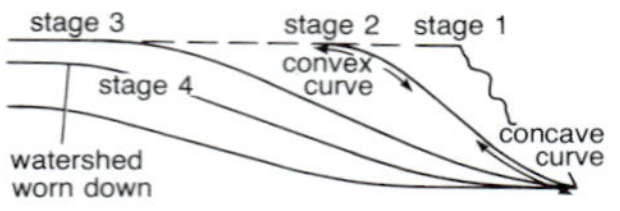 stage 3, stage 2, stage 1 convex curve stage 4 watershed worn down concave curve By stage 4 land has been worn down into a convex-concave slope	(b) **slope replacement** 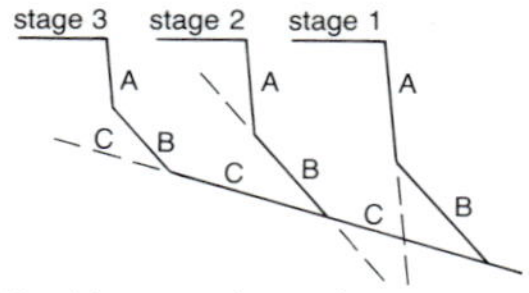 stage 3, stage 2, stage 1 A A A C B C B C B B = talus-scree slope and will replace slope A C will eventually replace slope B	(c) **slope retreat** 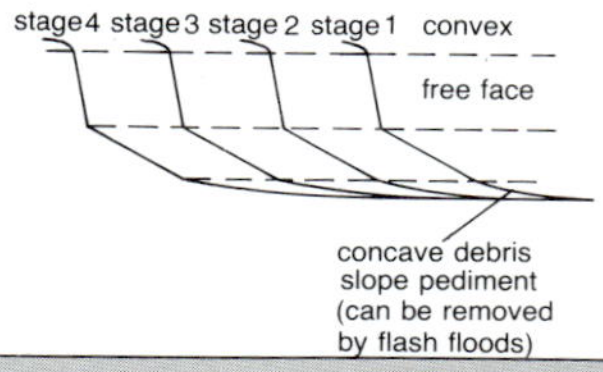 stage4 stage 3 stage 2 stage 1 convex free face concave debris slope pediment (can be removed by flash floods)
changes over time	Assumed a rapid uplift of land with an immediate onset of denudation. The uplifted land would undergo a cycle of erosion where slopes were initially made steeper by vertical erosion by rivers but later became less steep (slope decline) until the land was almost flat (peneplain).	Assumed landscape started with a straight rock slope with equal weathering overall. As scree (talus) collected at the foot of the cliff it gave a gentler slope which, as the scree grew, replaced the original one.	Assumed that slopes had two facets — a gently concave lower slope or pediment and a steeper upper slope (scarp). Weathering caused the parallel retreat of the scarp slope allowing the pediment to extend in size.

◄ Figure 2.24

Slope development theories

a Slope decline

b Slope replacement

c Parallel retreat

Formulae for chemical weathering processes referred to in text:

Oxidation

$$4FeO + O_2 \longrightarrow 2Fe_2O_3$$
(ferrous oxide + oxygen $\longrightarrow$ ferric oxide)

Hydration

$$CaSO_4 + 2H_2O \longrightarrow CaSO_4 2H_2O$$
(anhydrite + water $\longrightarrow$ gypsum)

Hydrolysis

The formula for hydrolysis differs depending on the rock type involved in the reaction, so no single formula is applicable. The following, for the hydrolysis of feldspar/granite to kaolin, is a common example.

$$K_2O, Al_2O_3, 6SiO_2 + H_2O \longrightarrow Al_2O_3, 2SiO_2, 2H_2O$$
(feldspar + water $\longrightarrow$ kaolin)

Carbonation

this process is in two stages:

$$H_2O + CO_2 \longrightarrow H_2CO_3$$
(water + carbon dioxide $\longrightarrow$ carbonic acid)

$$CaCO_3 + H_2CO_3 \longrightarrow Ca(HCO_3)_2$$
(calcium carbonate + carbonic acid $\longrightarrow$ calcium bicarbonate)

Acid rain

$$2SO_2 + O_2 + 2H_2O \longrightarrow 2H_2SO_4$$
(sulphur dioxide + oxygen + water $\longrightarrow$ weak sulphuric acid)

References

A Dictionary of the Natural Environment, F.J. Monkhouse and J. Small, E.J. Arnold, 1978.

Modern Concepts in Geomorphology, Patrick McCullagh, Oxford University Press, 1978.

Process and Landform, A. Clowes and P. Comfort, Oliver & Boyd, 1982.

Process and Pattern in Physical Geography, Keith Hilton, University Tutorial Press (UTP), 1979.

The Nature of the Environment, Andrew Goudie, Basil Blackwell, 1984.

The Unquiet Landscape, Brunsden and Doornkamp, David & Charles, 1977.

CHAPTER 3

Drainage basins and rivers

'All the rivers run into the sea; yet the sea is not full; unto the place from whence the rivers come, thither they return again.'

The Bible, Ecclesiastes, 1:7

A **drainage basin** is an area of land drained by a river and its tributaries. Its boundary is marked by a ridge of highland beyond which any precipitation will drain into adjacent basins. This boundary is called a **watershed** or waterparting.

The drainage basin as a system

A drainage basin may be described as an **open system** and it forms part of the hydrological or water cycle. If a drainage basin is viewed as a system then its characteristics are:
- **inputs** in the form of precipitation (rain and snow),
- **outputs** where the water is lost to the system either by the river carrying it into the sea or through evapotranspiration (the loss of water directly from the ground, water surfaces or vegetation).

Within this system some of the water
- is **stored** either in lakes or in the soil, or
- passes through a series of **transfers**, e.g. infiltration, percolation, throughflow.

Figure 3.1 shows the drainage basin system as it is likely to operate in a temperate humid region such as the British Isles.

Elements of the drainage basin system

Precipitation This forms the major input into the system, though amounts vary over time and space. As a rule, the greater the intensity of the storm, the shorter its duration. Convectional thunderstorms are short, heavy and may be confined to small areas whereas the passing of a warm sector of a depression (page 194) will give a longer period of more steady rainfall extending over all of the basin.

▼ Figure 3.1

The drainage basin as an open system

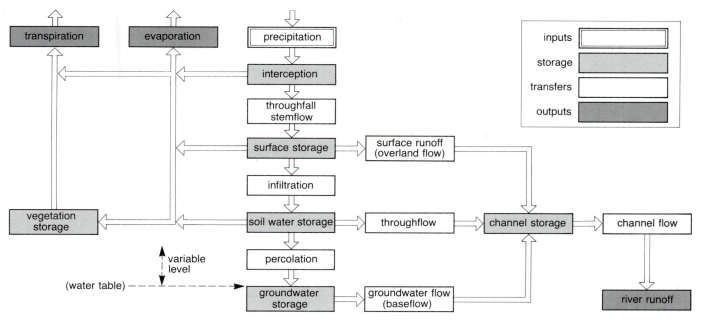

F R A M E W O R K   1

A systems approach

One type of model (page 293) widely adopted by geographers to help explain phenomena is the **system**. The system is a method of analysing relationships within a unit and consists of a number of components between which there are linkages. The model is usually illustrated schematically as a flow diagram.

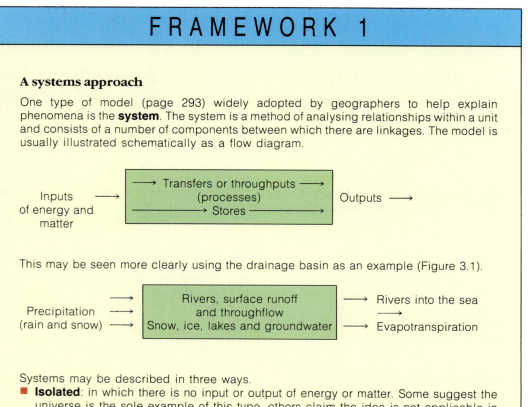

This may be seen more clearly using the drainage basin as an example (Figure 3.1).

Systems may be described in three ways.
- **Isolated**: in which there is no input or output of energy or matter. Some suggest the universe is the sole example of this type, others claim the idea is not applicable in geography.
- **Closed**: there is input, transfer and output of energy but not of matter (or mass).
- **Open**: most environmental systems are open and there are inputs and outputs of both energy and matter.

Examples of the systems approach used and referred to in this book (chapter number is given in brackets):

Geomorphological	Climate, soils and vegetation	Human and economic
Slopes (2)	Atmosphere energy budget (9)	Farming (15)
Drainage basin (3)	Hydrological cycle (9)	Industry (16)
Glaciers (4)	Soils (10)	Transport (17)
	Ecosystems (11)	
	Nutrient cycle (12)	

When opposing forces, or inputs and outputs, are balanced, the system is said to be in a state of **dynamic equilibrium**. If one element in the system changes because of some outside influence, then it upsets this equilibrium and affects the other components. This process is known as feedback which may be in two directions.
- **Negative feedback**, is when an increase in the activity of one component causes a decrease in activity in another, e.g. steep beaches generate large waves which in time comb down material causing the beach, and consequently the waves, to become less steep. Processes therefore work to restore the balance or equilibrium.
- **Positive feedback** refers to a series of progressive changes within the system, e.g. as the discharge of a river increases then so too does its velocity and capacity to carry more sediment and to erode its banks and bed. Processes therefore operate to move the system away from equilibrium.

Negative feedback is generally found to operate more commonly than positive.

Evapotranspiration The two components of evapotranspiration contribute to form an output from the system. **Evaporation** is the physical process by which moisture is directly lost into the atmosphere from various water surfaces and the soil due to the effect of air movement or the sun's heat. **Transpiration** is a biological process by which water is lost from a plant through the minute pores (stomata) in its leaves. Evaporation rates are affected by temperature, wind speed, humidity, hours of sunshine and other climatic factors. Transpiration rates depend on the time of year, the type and amount of vegetation and the length of the growing season. It is also possible to distinguish between the potential and the actual evapotranspiration of an area. For example, in deserts there is a high **potential evapotranspiration** because the amount of moisture that could be lost is greater than the amount of water which is actually available. On the other hand, in Britain the amount of water which is available for evapotranspiration exceeds the amount which actually takes place, hence the term **actual evapotranspiration**.

Interception The first raindrops of a storm will fall on trees or plants which shelter the underlying ground. This is called interception storage, and naturally will be greater in a woodland area than over grassland. If the precipitation is light and of short duration, much of the water may never reach the ground and may be quickly lost to the system through evaporation. Estimates suggest that in a woodland area up to 30 per cent of the precipitation may be lost because of interception, which helps to account for reduced soil erosion in forests. In an area of deciduous trees, both interception and evapotranspiration rates will be higher in summer.

If a storm persists, then water begins to reach the ground by three possible routes: dropping off the leaves, **throughfall**; flowing down the trunk, **stemflow**; or by undergoing **secondary interception** by any undergrowth. Following a warm/dry spell in summer, the ground may be hard, so at the onset of a storm water lies on the surface as soil **surface storage** until the upper layers become wet and soft enough to absorb the moisture. If precipitation is very heavy at the beginning of the storm then the ground may be incapable of absorbing all of the rain. As a result, excess water flows away over the surface, a transfer known as **surface runoff** or **overland flow**.

Infiltration In most environments overland flow is relatively rare except in urban areas, which have impermeable coverings of tarmac and concrete, or during exceptionally heavy storms. Usually the ground rapidly becomes soft and sufficiently absorbent for water, gradually, to infiltrate vertically through the pores in the soil. The speed at which water can pass through the soil is called its **infiltration capacity** and is expressed in mm/hour. The rate of infiltration depends upon the amount of water already in the soil, the **porosity** (Figure 8.2) and structure of the soil and the type, amount and seasonal changes in vegetation cover. Some of the water will flow more horizontally as **throughflow**. On reaching valley sides this forms springs which provide a constant supply of water to a river even in dry spells. During drier periods, some water may be drawn up towards the surface by **capillary action** while at all times plant roots are likely to take up moisture from the soil (vegetation storage) which may later be lost from the system by transpiration.

Percolation As the excess water reaches the underlying soil or rock layers, which tend to be more compact, its progress is slowed. This constant movement, called percolation, creates **groundwater storage**.

Groundwater Water eventually collects above an impermeable rock or fills all pore spaces, creating a **zone of saturation**. The upper level of saturated material, i.e. the upper surface of the groundwater layer, is known as the **water table**. Water may then be slowly transferred laterally as **groundwater flow** or **baseflow**. Groundwater levels usually respond slowly to surface storms or droughts (Figure 3.5). During a lengthy dry period some of the groundwater store will be utilised as river levels fall. In a subsequent wetter period, groundwater must be replaced before the level of the river can rise appreciably. If the water table reaches the surface it means that the ground will be saturated and excess water forms a marsh where the land is flat or becomes surface runoff if the ground is sloping.

Channel flow Although some rain does fall directly into the channel of a river, most water reaches it by a combination of three transfer processes: surface runoff (overland flow), throughflow, or groundwater (baseflow). Once in the river as channel storage, water flows towards the sea where it is lost to the drainage basin system.

Q Describe and account for the changes which take place in the rainfall-runoff process during the course of a storm as illustrated in Figure 3.2.

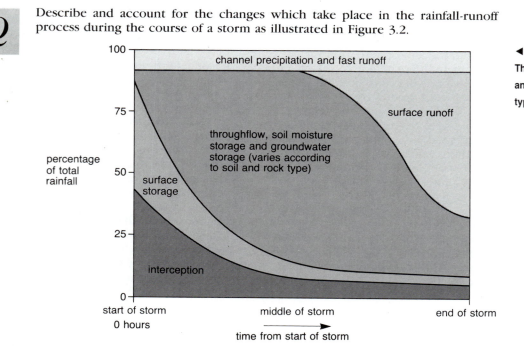

◄ Figure 3.2
The relationship between rainfall and runoff in the course of a typical storm

The water balance

This shows the state of equilibrium in the drainage basin between the inputs and outputs. It can be expressed as:

$$P = Q + E \pm \text{changes in storage}$$

where:

P = precipitation (measured using rain-gauges)

Q = runoff (measured by discharge flumes in the river channel)

E = evapotranspiration (this is far more difficult to measure — how can you accurately measure transpiration from a forest?)

In Britain the annual precipitation almost always exceeds evapotranspiration — though sometimes the situation may be reversed as in the dry summers of 1975, 1976 and 1984. (This reversal is more likely in the south and east of England than in the north and west of Scotland.) Should evapotranspiration exceed precipitation, any surplus soil moisture will be utilised to leave a **soil moisture deficiency**. Later, in the cooler, wetter autumn, there must be a period of **soil moisture recharge** until the water in the soil is replenished to its field capacity.

Figure 3.3 is a model based on an area of SE England. During the winter, precipitation exceeds evapotranspiration, producing a water surplus and considerable runoff — the soils will be wet and river levels high. In summer, however, evapotranspiration now

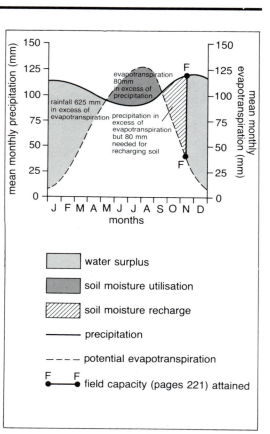

◄ Figure 3.3
A model illustrating the water balance

exceeds precipitation and so plants and humans utilise water from the soil store, leaving it depleted and causing river levels to fall. By autumn, precipitation again exceeds evapotranspiration although the first of the surplus water has to be used to recharge the soil store.

Use Figure 3.3 to answer the following questions.

1 Explain why precipitation exceeds evapotranspiration from September to April.

2 Define what is meant by the term *potential evapotranspiration*.

3 Why does evapotranspiration exceed precipitation between May and August?

4 What is meant by *soil moisture utilisation*?

5 Explain why the model does not show a water deficit in summer.

6 Why was 80 mm of precipitation needed in September and October to recharge soil moisture?

7 Why would a river in the area shown by the graph:
a not dry up in August? b not flood in September? c be liable to flood in January?

8 Do you think that the natural vegetation of the area is likely to have been deciduous forest, grassland or semi-desert? Give your reasons.

9 What might happen to the natural vegetation of the area if the climate changed to one with long, dry summers?

10 In what ways might a graph showing the water balance for the northwest of Scotland differ from that of Figure 3.3?

Figure 3.4 shows the water balance for two towns in the USA.

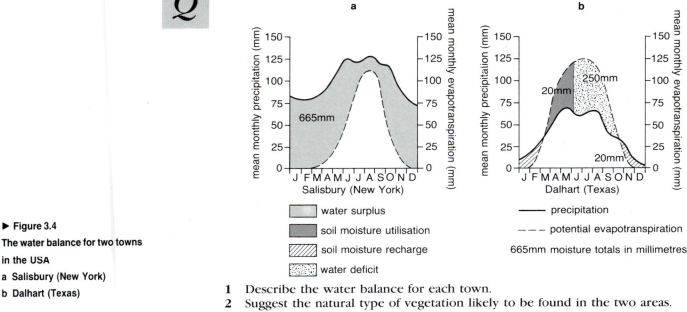

▶ Figure 3.4
The water balance for two towns in the USA
a Salisbury (New York)
b Dalhart (Texas)

1 Describe the water balance for each town.
2 Suggest the natural type of vegetation likely to be found in the two areas.

The storm hydrograph

The hydrograph is a means of showing the discharge of a river at a given point over a short period of time. The **discharge** is the amount of water originating as precipitation which reaches the channel by surface runoff, throughflow and baseflow. Discharge is therefore the water *not* stored in the drainage basin by interception, as surface storage, soil moisture storage or groundwater storage or lost through evapotranspiration (Figure 3.1). The model of a storm hydrograph, Figure 3.5, shows how the discharge of a river responds to an individual storm.

Measuring discharge

Discharge is the velocity (speed) of the river, measured in metres per second, multiplied by the cross-sectional area of the river, measured in square metres. This gives the volume in cubic metres per second or **cumecs** and can be expressed as:

$$Q = A \times V$$

where: Q = discharge
A = cross-sectional area
V = velocity (Figure 3.6)

Interpreting the hydrograph

Refer to the hydrograph in Figure 3.5. The graph includes the approach segment which shows the discharge of the river before the storm. At the time when the storm begins, the river's response is negligible for although some of the rain does fall directly into the channel, most falls elsewhere in the basin and takes time to reach the channel. However, when the initial surface runoff and later the throughflow eventually reach the river there is an increase in discharge. This is indicated by the **rising limb**. The period between maximum precipitation and peak discharge is referred to as **lag time**. Lag time varies according to conditions within the drainage basin, e.g. soil and rock type, slope and size of the basin, drainage density, type and amount of vegetation and water already in storage. The **falling** or **receding** limb is the segment of the graph where discharge is decreasing and the level of the river is falling. This segment is usually less steep than the rising limb because throughflow is still being released into the channel. By the time all the water from the storm has passed through a given point in the channel, the river will have returned to its baseflow level — unless there has been another storm within the basin. Baseflow is very slow to respond to a storm, but by continually releasing water from the lower ground it maintains the river's flow during periods of low precipitation. Indeed, baseflow is more significant

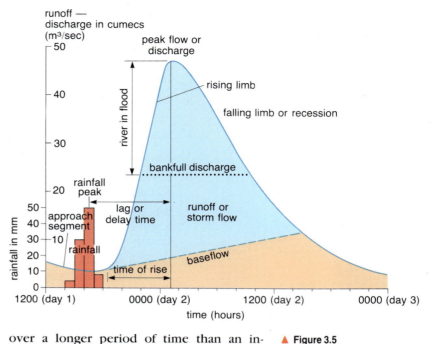

▲ Figure 3.5

The storm hydrograph

over a longer period of time than an individual storm and reflects seasonal changes in precipitation, snow melt, vegetation and evapotranspiration. Finally, on the graph, **bankfull discharge** is the point when the level of water has reached the top of its channel and any further increase in discharge will result in flooding of the surrounding land.

Controls in the drainage basin and on the storm hydrograph

In some drainage basins, discharge rises very quickly after a storm and may cause frequent and occasionally catastrophic flooding. Following a storm, the level of such rivers falls almost as rapidly and after dry spells can become very low. Rivers in other basins seem neither to flood nor to fall to very low levels. There are several factors which contribute to regulate the way in which a river responds to precipitation.

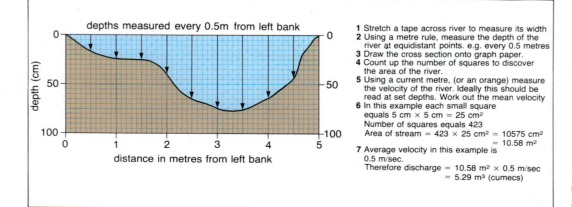

1 Stretch a tape across river to measure its width
2 Using a metre rule, measure the depth of the river at equidistant points. e.g. every 0.5 metres
3 Draw the cross section onto graph paper.
4 Count up the number of squares to discover the area of the river.
5 Using a current metre, (or an orange) measure the velocity of the river. Ideally this should be read at set depths. Work out the mean velocity
6 In this example each small square equals 5 cm × 5 cm = 25 cm²
Number of squares equals 423
Area of stream = 423 × 25 cm² = 10575 cm²
                    = 10.58 m²
7 Average velocity in this example is 0.5 m/sec.
Therefore discharge = 10.58 m² × 0.5 m/sec
                    = 5.29 m³ (cumecs)

◄ Figure 3.6

Measuring the discharge of a river

1 Basin size, shape and relief

Size If a basin is small it is likely that rainfall will reach the main channel more rapidly than in a larger basin where the water has much further to travel. Lag time will therefore be shorter in the smaller basin.

Shape A more circular basin will have a shorter lag time and a higher peak flow than an elongated basin (Figure 3.7a). All the points on the watershed of the former are approximately equidistant from the gauging station whereas in the latter it takes longer for water from the extremities of its basin to reach the gauging station.

Relief The slope of the basin and its valley sides also affects the hydrograph. In steep-sided upland valleys water is likely to reach the river more quickly than in gently sloping lowland areas (Figure 3.7c).

▼ Figure 3.7
Drainage basin shape
a Two basins, A and B, with widely different shapes
b Storm hydrographs for three drainage basins of differing shapes
c Basin relief and associated storm hydrographs showing the relationship between the long profile and the storm hydrographs

Q

1 Using Figure 3.7a, how many hours will pass before all the floodwater passes the gauging stations at **A** and **B**?

2 What do you think will be the shape of the storm hydrograph for basin *A* and basin **B**? Draw the storm hydrograph for these basins and explain the shape you have predicted.

3 Fit the three hydrographs *P*, *Q* and *R* in Figure 3.7b with the drainage basins **X**, **Y** and **Z**.

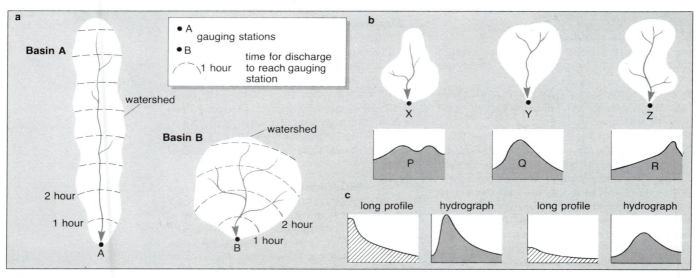

2 Types of precipitation

Prolonged rainfall The most frequent cause of flooding is after a long period of heavy rainfall when the ground has become saturated and infiltration is replaced by surface runoff (overland flow).

Intense storms, (e.g. convectional thunderstorms) When heavy rain occurs (e.g. in summer in Britain, when the ground may be harder), the intensity may be greater than the infiltration capacity of the soil. The resulting surface runoff is likely to produce rapid rises in river levels.

Snowfall Heavy snowfalls mean that water is held in storage and river levels drop. However, as temperature rises (in Britain this may be as a warm front passes with its associated rainfall, page 194) then the meltwater soon reaches the main river. It is possible that the ground will remain frozen for some time and infiltration is impeded.

3 Temperature

Extremes of temperature, as already seen, can restrict infiltration (very cold in winter, very hot and dry in summer) and thus increase surface runoff. If evapotranspiration rates are high then there will be less water available to flow into the main river.

4 Land use

Vegetation may help to prevent flooding by intercepting rainfall and storing moisture on its leaves before it evaporates back into the atmosphere. Estimates suggest that tropical rainforests intercept up to 80 per cent of rainfall (30 per cent of which may later evaporate) whereas arable land may intercept only 10 per cent. Interception is less during the winter in Britain when the deciduous trees have shed their leaves and crops have been harvested leaving bare earth. Vegetation, especially trees, also takes up water from the soil through roots and so reduces throughflow. Flooding is more likely to occur in deforested areas, e.g. the increasingly frequent and serious flooding in Bangladesh is attributed to the removal of trees in Nepal and other Himalayan areas. Flooding is less likely to occur in areas of afforestation. Figure 3.8 contrasts the storm hydrograph for two rivers. Although they rise very close together, one, the Wye, flows over moors and grassland, whereas the other, the Severn, flows through an area of coniferous forest.

Urbanisation has increased flood hazard. Water cannot infiltrate through tarmac and concrete — gutters and drains carry water more quickly to the nearest river. Small streams may be canalised so that, with friction reduced, the water flows away more quickly, while others may be culverted which allows only a limited amount of water to pass through at one time.

5 Rock type (geology)

Rocks which allow water to pass through them are said to be **permeable**. There are two types:

- **Porous rocks**, e.g. sandstone and chalk, contain numerous pores which can fill with and store water;
- **Pervious rocks**, e.g. carboniferous limestone, which allow water to flow along bedding planes and down joints though the rock itself is impervious (page 159).

Both types permit rapid infiltration of water so there is little surface runoff and a limited number of streams. In contrast, **impermeable rocks**, such as granite, do not allow water to pass through them and so they produce more surface runoff and many streams.

6 Soil type

This controls the speed of infiltration, the amount of storage and the rate of throughflow. Sandy soils, with large pore spaces, allow rapid infiltration and do not encourage flooding. Clays have much smaller pore spaces thus reducing infiltration and throughflow but encouraging surface runoff and increasing the risk of flooding.

7 Drainage density

This refers to the number of surface streams in a given area. The density is highest on impermeable rocks and clays and lowest on permeable rocks and sands. The higher the density the greater the chance of **flash** (sometimes known as 'flashy') **floods**. A flash flood is a sudden rise of water in the river, shown on the hydrograph as a short lag time and a high peak flow in relation to normal discharge.

8 Tides and storm surges

High spring tides tend to prevent flood water from rivers escaping into the sea. Flood water therefore builds up (or 'ponds') in the lower part of the valley. If high tides coincide with gale force winds blowing onshore and a narrowing estuary, the result may be a storm surge (page 116). This happened in southeast England and the Netherlands in 1953 and prompted the construction of the Thames Barrier and the Dutch Delta Plan to prevent the recurrence of such a flood.

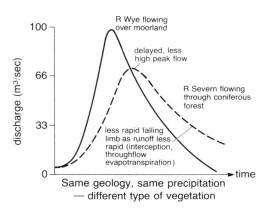

◄ Figure 3.8

The effect of vegetation on storm hydrographs for the River Wye and the River Severn; geology and precipitation are the same in both basins

Q

1 How will urbanisation alter the drainage basin system shown in Figure 3.1?

2 Complete the boxes in Figure 3.9 to show the effects of urbanisation on a drainage basin (see Figure 3.1).

3 How and why does this urban system differ from the system shown in Figure 3.1?

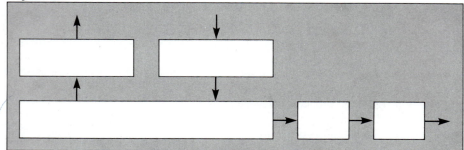

◄ Figure 3.9

The effect of urbanisation on the drainage basin system

Q Figure 3.10 shows two hydrographs. Which one, **a** or **b**, is more likely to correspond to each of the pairs in the following situations?

1 a long period of steady rain and a short, torrential downpour
2 a gently sloping and a steep-sided valley
3 an area with a low drainage density and one with a high density
4 an area of chalk and an area of granite
5 an area of mature deciduous trees and an area of heathland in summer
6 a rural area and an urban area
7 an area where the soil is saturated and an area which is not saturated, i.e. is below its field capacity
8 following a very cold period with some snow and a mild period in autumn
9 a mature deciduous woodland in winter and in summer
10 an elongated drainage basin and a round-shaped drainage basin

▶ Figure 3.10

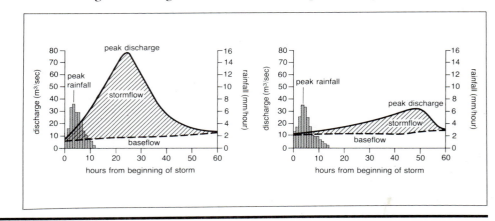

River régimes

The régime of a river is a term which allows us to describe the annual variation in discharge. The average régime, which can be shown by either the mean daily or the mean monthly figures, is primarily determined by the climate of an area, e.g. the amount and distribution of rainfall together with the rates of evapotranspiration and snowmelt. Local geology may also be significant. There are few rivers flowing today under wholly natural conditions, especially in Britain. They are managed, regulated systems resulting from human activity.

Régimes of rivers, which are used to demonstrate any seasonal variations, may be either simple, with one peak period of flow, or complex with several peaks.

CASE STUDY

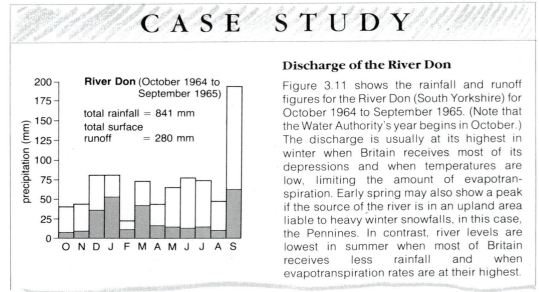

▶ Figure 3.11

Rainfall and runoff for the River Don, Yorkshire (*after* Riley, Briggs and Tolley)

Discharge of the River Don

Figure 3.11 shows the rainfall and runoff figures for the River Don (South Yorkshire) for October 1964 to September 1965. (Note that the Water Authority's year begins in October.) The discharge is usually at its highest in winter when Britain receives most of its depressions and when temperatures are low, limiting the amount of evapotranspiration. Early spring may also show a peak if the source of the river is in an upland area liable to heavy winter snowfalls, in this case, the Pennines. In contrast, river levels are lowest in summer when most of Britain receives less rainfall and when evapotranspiration rates are at their highest.

Q Figure 3.12 shows the régimes for four rivers in different parts of the world.

Explain how each régime, **a**, **b**, **c**, **d**, is determined by the climate of the area in which it is found. (You may find it useful to consult Chapter 12 and an atlas.)

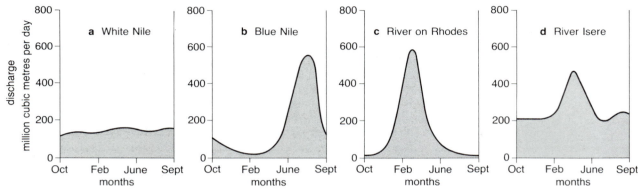

Morphometry of drainage basins

The development of morphometric techniques was a major advancement in the quantification (as opposed to the qualitative description) of drainage basins. It meant that instead of studies being purely subjective, it became possible to compare and contrast different basins with precision. Much of the early work in this field was by R.E. Horton. In the mid-1940s he suggested the 'Laws of drainage composition' which established a hierarchy of streams ranked according to 'order'. One of these laws, the law of **stream number**, showed that within a drainage basin a constant geometric relationship could be seen between the occurrence of one stream order in the hierarchy and the numbers of streams in the next highest order.

Figure 3.13 shows how one of his successors, A.N. Strahler, defined streams of different order. All the initial, unbranched source tributaries he called **first order** streams. When two first order streams join they form a **second order**; when two second order streams merge they form a third order, and so on. Notice that it needs two streams of equal order to join to produce a stream of a higher order, while the order remains unchanged if a lower order stream joins a higher order. For example, a second order plus a second order gives a third order but if a second order stream joins a third order, the resultant stream remains as a third order. A basin may therefore be described in terms of the highest order stream within it, e.g. a 'third order basin' or a 'fourth order basin'.

If the number of segments in a stream order is plotted on a semi-log graph against the stream order then the resultant best-fit line will be approximately straight (Figure 3.14a). On a semi-log graph, the vertical scale, showing the dependent variable (page 327), is divided into cycles, each of which begins and ends ten times greater than the previous cycle, e.g. a range of 1 to 10, 10 to 100, 100 to 1000, and so on. (If the horizontal scale, showing the independent variable, had also been divided into cycles instead of the arithmetic or constant scale then Figure 3.14 would have been referred to as a log-log graph.) Logarithmic graphs are valuable when:

■ The rate of change is of more interest than the amount of change: the steeper the line the greater the rate of change.
■ There is a greater range in the data than there is space to express on an arithmetic scale (a log scale compresses values).
■ There is considerably more data at one end of the range than the other.

▲ Figure 3.12
Selected river régimes – the influence of climate on régimes in various parts of the world
a The White Nile has its source in an area of equatorial climate (Figure 12.4)
b The Blue Nile rises in the Ethiopian Highlands in a region experiencing a savanna climate (Figure 12.13)
c A small seasonal river on the Mediterranean island of Rhodes (Figure 12.25)
d The River Isere, a tributary of the River Rhône, rises in the French Alps

Morphometry means the 'measurement of shape'

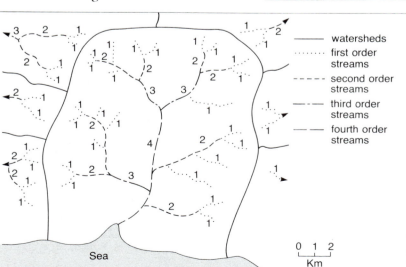

Figure 3.14a shows an almost perfect negative correlation (page 327) in that as the independent variable (in this case the stream order) increases, then the dependent variable (the number of streams) decreases. Studies of stream ordering for most rivers in the world produce a similar straight-line relationship. For any exceptions to Horton's law of stream ordering, further studies can be made to determine which local factors alter the relationship.

Q 1 Describe the relationships between the following:
   a mean stream length and stream order (Figure 3.14b),
   b mean drainage basin area and stream order (Figure 3.14c).
2 In each case is the correlation (the relationship between the independent and the dependent variables) positive or negative?

▶ Figure 3.14

Relationships between stream order and

a the number of streams
b stream length
c area of drainage basin

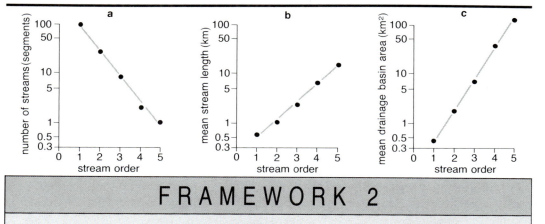

FRAMEWORK 2

Quantitative techniques and statistical methods of data interpretation

As geography adjusted to a more scientific approach in the 1960s, a series of statistical techniques were adopted which could be used to quantify field data and add objectivity to the testing of hypotheses and theories. This period is often referred to as the 'Quantitative Revolution'.

At first it seemed to many, the author included, that mathematics had taken over the subject but it is now accepted that these techniques are a useful aid provided they are not seen as an end in themselves. They provide a tool which, if carefully handled and understood, gives greater precision to arguments, helps in the identification of patterns and may contribute to the discovery of relationships and possible cause–effect links. In short, by providing greater accuracy in handling data they reduce the reliance upon subjective conclusions.

It is essential to select the most appropriate techniques for the data and for the job in hand. Therefore some understanding of the statistical methods involved is important.

Statistical methods may be most profitably employed in three areas.

1 **Sampling** (page 138) Rapid collection of data is made possible.
2 **Correlation and regression** (page 328) This not only shows possible relationships between two variables but quantifies or measures the strength of those relationships.
3 **Spatial distributions** Not only may this approach be used to identify patterns, but it may demonstrate how likely it is that the resultant distributions occurred by chance (page 329).

When these new techniques first appeared in schools in the 1970s, they appeared extremely daunting until it was realised that often the difficulty of the worked examples detracted from the usefulness of the technique itself. Where such techniques appear in this book, the mathematics has been simplified to show more clearly how methods may be used and to what effect. With the wider availability of calculators and computers it has become easier to take advantage of more complex calculations to test geographical hypotheses (page 259). Much of the 'number crunching' has now been removed by the increasing availability of statistical packages for micro-computers.

◀ Figure 3.13

Strahler's method of stream ordering

Comparing drainage basins

Horton's work on drainage basins has made it possible to compare different basins scientifically (quantitatively) rather than relying on subjective (qualitative) descriptions by individuals. It means that different people studying drainage basin morphometry use the same standards, measurements and 'language'. Figure 3.15 shows two imaginary and adjacent basins. The answers for basin **A**, which have already been completed, include two fundamental methods which can be used to compare it with basin **B**: the bifurcation ratio and drainage density.

The bifurcation ratio

This is the relationship between the number of streams of one order of magnitude with those of the next highest order. It is obtained, as shown in Figure 3.15, by dividing

the number of streams in one order by the number in the next highest order, i.e.

$$\frac{N1}{N2}\frac{(\text{number of first order streams})}{(\text{number of second order streams})} = \frac{26}{6} = 4.33$$

and then finding the mean of all the ratios in the basin to be studied, i.e.

$$\frac{4.33 + 3.00 + 2.00}{3} = 3.11 = \begin{array}{l}\text{bifurcation}\\ \text{ratio for}\\ \text{basin } \mathbf{A}\end{array}$$

The human significance of the bifurcation ratio is that as the ratio is reduced so the risk of flooding within the basin increases. Most British rivers have a bifurcation ratio of between 3 and 5.

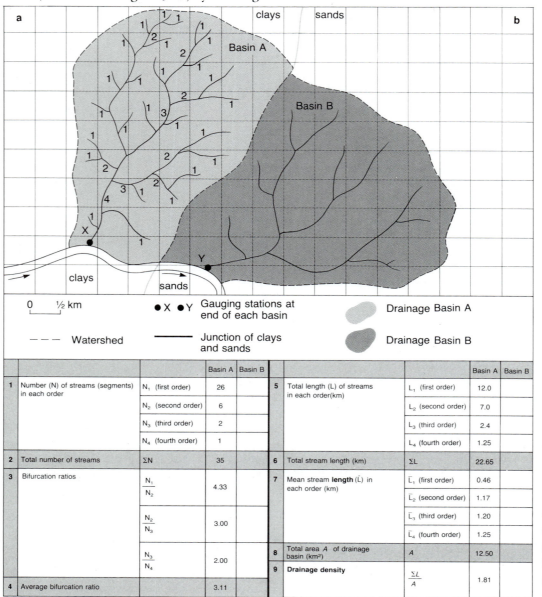

			Basin A	Basin B
1	Number (N) of streams (segments) in each order	N₁ (first order)	26	
		N₂ (second order)	6	
		N₃ (third order)	2	
		N₄ (fourth order)	1	
2	Total number of streams	ΣN	35	
3	Bifurcation ratios	$\frac{N_1}{N_2}$	4.33	
		$\frac{N_2}{N_3}$	3.00	
		$\frac{N_3}{N_4}$	2.00	
4	Average bifurcation ratio		3.11	

			Basin A	Basin B
5	Total length (L) of streams in each order(km)	L₁ (first order)	12.0	
		L₂ (second order)	7.0	
		L₃ (third order)	2.4	
		L₄ (fourth order)	1.25	
6	Total stream length (km)	ΣL	22.65	
7	Mean stream **length** (L̄) in each order (km)	L̄₁ (first order)	0.46	
		L̄₂ (second order)	1.17	
		L̄₃ (third order)	1.20	
		L̄₄ (fourth order)	1.25	
8	Total area A of drainage basin (km²)	A	12.50	
9	**Drainage density**	$\frac{\Sigma L}{A}$	1.81	

◄ **Figure 3.15**

A comparison between two drainage basins on

a clay

b sand

and data from drainage basins A and B

Drainage density

This is found by measuring the total length of all of the streams within the basin (L) and dividing by the area of the whole basin (A). It is therefore the average length of stream within each unit area. For basin **A** in Figure 3.15 this will be:

$$\frac{L}{A} = \frac{22.65}{12.50} = 1.81 \text{ km/km}^2$$

In Britain most drainage densities lie between 2 and 4 km/km² but this varies considerably according to local conditions.

A number of factors influence drainage density which tends to be highest in areas where the land surface is impermeable, rainfall is heavy and prolonged and vegetation cover is lacking.

1 Geology and soils On very permeable rocks or soils (e.g. chalk, sands) drainage densities may be under 1 km/km², whereas this increases to over 5 km/km² on highly impermeable surfaces (e.g. granite, clays). In Figure 3.15 with two adjacent drainage basins of approximately equal size, shape, and probably rainfall, the difference in drainage density is presumably due to basin **A** being on clay and basin **B** on sands.

2 Land use The drainage density, especially of first order streams, is much greater in areas with little vegetation cover. The density decreases as does the number of first order streams, if the area becomes afforested.

3 Time As a river pattern develops over a period of time the number of tributaries will decrease and with it the drainage density.

4 Precipitation

Q

1 Complete the table for drainage basin **B** in Figure 3.15.

2 Figure 3.16 illustrates the influence of rainfall on the density of surface rivers in a drainage basin.
Describe and give reasons for the changes in density following periods of dry weather (drought) and very wet conditions (flood).

3 Figure 3.17 shows the relationship between drainage density and discharge and between the number of segments and stream order.
Describe and give reasons for the changes in each graph following a period of drought and under flood conditions.

▶ Figure 3.16

The response of a British drainage basin to extremes of rainfall

a Normal flow

b Flood

c Drought

▶ Figure 3.17

The relationship between

a Density and discharge

b Stream order and the number of segments, under extremes of rainfall

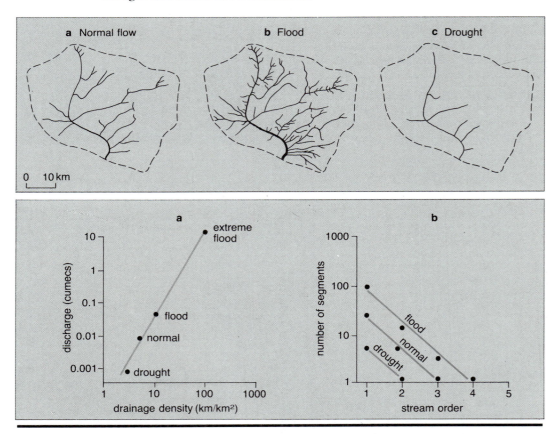

River form

A river will try to adopt a channel shape that best fulfils its two main functions: transporting water and sediment. It is important to understand the significance of channel shape in order to identify the controls on the flow of a river (Framework 1, page 44).

Types of flow

As water flows downhill under gravity it seeks the path of least resistance, i.e. a river possesses potential energy and follows a route which will maximise the rate of flow (velocity) and minimise the loss of this energy caused by friction. Most friction occurs along the banks and bed of the river but the internal friction of the water and air resistance on the surface are also significant.

There are two patterns of flow, **laminar** and **turbulent**. Laminar flow (Figure 3.19a) is a horizontal movement of water so rarely experienced in rivers that it is usually discounted. Such a method of flow, if it existed, would travel over sediment on the river bed without disturbing it. Turbulent flow, which is the dominant method, consists of a series of erratic eddies, both vertical and horizontal, in a downstream direction (Figures 3.19b and 3.18). Turbulence varies with the velocity of the river which, in turn, depends upon the amount of energy available after friction has been overcome. It is estimated that under 'normal' conditions about 95 per cent of a river's energy is expended in order to overcome friction.

Influence of velocity on turbulence

■ If the velocity is high, the amount of energy still available after friction has been overcome will be greater and so turbulence increases. This results in sediment on the bed being disturbed and carried downstream. The faster the flow of the river, the larger the quantity and size of particles which can be transported. This transported material is referred to as the river's **load**.

■ When the velocity is low then there is less energy to overcome friction. Turbulence decreases and may not be visible to the human eye and sediment on the river bed remains undisturbed. Indeed, as turbulence maintains the transport of the load, a reduction in turbulence may lead to deposition of sediment.

The velocity of a river is influenced by three main factors:

1 Channel shape in cross-section.
2 Roughness of the channel's bed and banks.
3 Channel slope.

◄ Figure 3.18
Turbulence in a river — the confluence of the Rio Amazon (red with silt, *left*) and Rio Negro (black with plant acids, *right*). The waters remain unmixed for 20–30 km

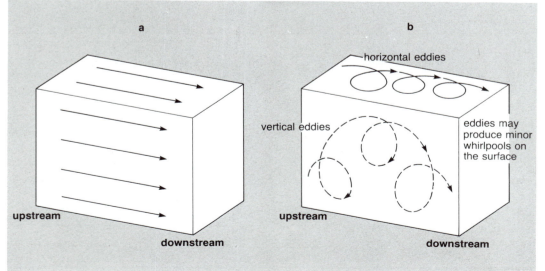

◄ Figure 3.19

Types of flow in a river
a Laminar flow
b Turbulent flow

1 Channel shape

This is best described by the term **hydraulic radius**, i.e. the ratio between the area of the cross-section of a river channel and the length of its wetted perimeter. The cross-section area is obtained by measuring the width and the mean depth of the channel (Figure 3.6). The **wetted perimeter** is the surface of the bed and banks which is in contact with the water in the channel. Figure 3.20 shows two channels with the same cross-section area but with different shapes and hydraulic radii.

Stream **A** has a larger hydraulic radius, meaning that it has a smaller amount of water in its cross-section in contact with the wetted perimeter. This creates less friction, and allows greater velocity. Stream **A** is said to be the more efficient of the two rivers.

Stream **B** has a smaller hydraulic radius, meaning that a relatively large amount of water is in contact with its wetted perimeter. This results in greater friction and a reduced velocity. Stream **B** is less efficient than stream **A**.

The point of maximum velocity is different in a river with a straight course where the channel is likely to be approximately symmetrical (Figure 3.21) compared to a meandering channel where the shape is asymmetrical (Figure 3.22).

▶ **Figure 3.20**

The wetted perimeter, hydraulic radius and efficiency of two differently shaped channels with equal area

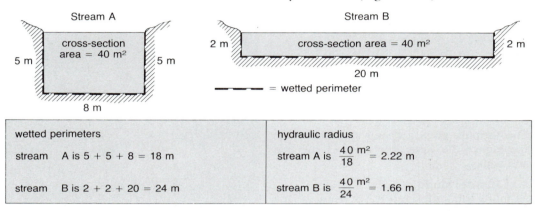

wetted perimeters	hydraulic radius
stream A is 5 + 5 + 8 = 18 m	stream A is $\frac{40 \text{ m}^2}{18}$ = 2.22 m
stream B is 2 + 2 + 20 = 24 m	stream B is $\frac{40 \text{ m}^2}{24}$ = 1.66 m

▼ **Figure 3.21**

A symmetrical channel — velocities in a straight stretch of river

▶ **Figure 3.22**, *below right*

An asymmetrical channel showing velocities through the cross-section of a typically meandering river

1 Account for the differences in velocity within the symmetrically shaped channel in Figure 3.21.

2 Figure 3.22 shows an asymmetrical channel.
   a Using the information given in the table, complete the data plotted on the cross-section diagram.
   b Draw **isovels** (lines joining places with equal velocity) at intervals of 0.1 m per second.
   c Shade in the part of the channel with the maximum velocity.
   d Label the area where you consider that deposition is most likely to occur.

3 Account for the differences in isovel patterns between Figures 3.21 and 3.22.

Depth (metres)	Distance from left bank (X) (metres)				
	0.2	0.4	0.6	0.8	1.0
0.1	0.47	0.29	0.23	0.11	0.05
0.2	0.53	0.32	0.18	0.08	
0.3	0.44	0.22	0.07		
0.4	0.32	0.10	0.01		
0.5	0.10				

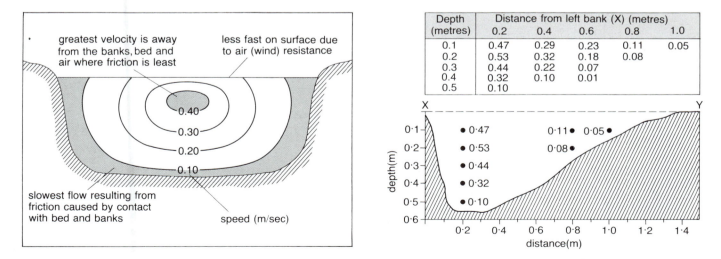

2 Roughness of channel bed and banks

A river flowing between banks composed of coarse material with numerous protrusions and over a bed of large, angular rocks meets with more resistance than a river with cohesive clays and silts forming its bed and banks (Figure 3.23).

Figure 3.24 explains why the velocity in a mountain stream is less than that of a lowland river. As bank and bed roughness increase, so does turbulence. Therefore a mountain stream is likely to pick up loose material and carry it downstream.

Roughness is difficult to measure but Manning, an engineer, calculated a **roughness coefficient** by which he inter-related the three factors affecting the velocity of the river. In his formula, known as 'Manning's N':

$$v = \frac{R^{0.67} S^{0.5}}{n}$$

where: v = mean velocity of flow
R = hydraulic radius
S = channel slope
n = boundary roughness

The formula gives a useful approximation: the higher the value, the rougher the bed and banks.

3 Channel slope

As more tributaries and water from surface runoff, throughflow and groundwater flow join the main river, the discharge, the channel cross-section area and the hydraulic radius will all increase. At the same time less energy will be lost through friction and the role of bedload material will decrease. As a result the river flows over a gradually decreasing gradient — the characteristic concave **long profile** or **thalweg** as shown in Figure 3.25.

▲ Figure 3.23
A boulder-strewn river bed in the upper Afon Glaslyn (the Aber Glaslyn Pass), Snowdonia — notice the scree slopes in the background

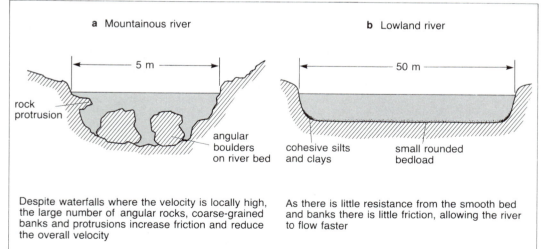

a Mountainous river

b Lowland river

rock protrusion

angular boulders on river bed

cohesive silts and clays

small rounded bedload

Despite waterfalls where the velocity is locally high, the large number of angular rocks, coarse-grained banks and protrusions increase friction and reduce the overall velocity

As there is little resistance from the smooth bed and banks there is little friction, allowing the river to flow faster

◄ Figure 3.24
Why a river increases in velocity towards its mouth
a Mountainous or upper course of a river
b Lowland or lower course river

▶ Figure 3.25

The characteristic long profile of a river

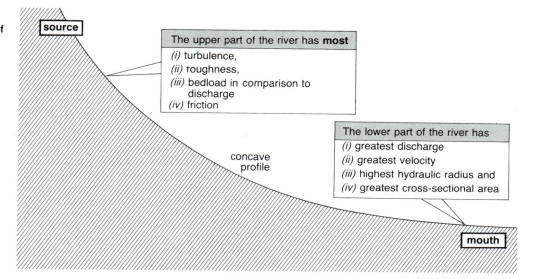

source

The upper part of the river has **most**
(i) turbulence,
(ii) roughness,
(iii) bedload in comparison to
       discharge
(iv) friction

The lower part of the river has
(i) greatest discharge
(ii) greatest velocity
(iii) highest hydraulic radius and
(iv) greatest cross-sectional area

concave profile

mouth

In summarising this section it should be noted that:

■ A river in a deep, broad channel, often with a gentle gradient and a small bedload will have a greater velocity than a river in a shallow, narrow, rock filled channel even if the gradient of the latter is steeper.

■ The velocity of a river increases as it nears the sea (unless like the Colorado and Nile it flows through deserts where water is lost through evaporation or human extraction for water supply).

■ The velocity increases as the depth, width and discharge of a river all increase.

■ As roughness increases, so too does turbulence and the ability of the river to pick up and transport sediment.

Transportation

Any energy remaining after the river has overcome friction can be used to transport sediment. The amount of energy available increases rapidly as the discharge, velocity and turbulence increase, until the river reaches flood levels. A river in flood has a large

wetted perimeter and the extra friction is likely to cause deposition on the flood plain. A river at bankfull stage can carry large quantities of soil and rock, its load, along the channel. The load is transported by three main processes: **suspension**, **solution** and as **bedload** (Figure 3.26).

Suspended load

Very fine particles of clay and silt are dislodged and carried by turbulence in a fast flowing river. The greater the turbulence and velocity the larger the quantity and size of particles which can be picked up. The material held in suspension usually forms the greatest part of the total load and the amount increases towards the river's mouth and gives the water its brown or black colour.

Dissolved or solution load

Water flowing within a river channel contains acids (e.g. carbonic acid from precipitation). If the bedrock is readily soluble, like limestone, it is constantly dissolved in the running water and removed in solution. Except in limestone areas the material in solution forms only a relatively small proportion of the total load.

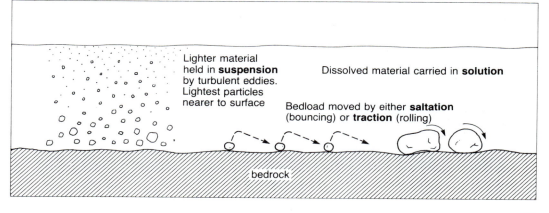

Lighter material held in **suspension** by turbulent eddies. Lightest particles nearer to surface

Dissolved material carried in **solution**

Bedload moved by either **saltation** (bouncing) or **traction** (rolling)

bedrock

▶ Figure 3.26

Processes of transportation in a river or stream

Bedload

Larger particles which cannot be picked up by the current may be moved along the bed of the river in one of two ways. **Saltation** is when pebbles, sand and gravel are temporarily lifted up by the current and bounced along the bed in a hopping motion (saltation in deserts, page 150). **Traction** is when the largest cobbles and boulders roll or slide along the bed. Some of these may be moved only during times of extreme flood.

It is much more difficult to measure the bedload than the suspended or dissolved load. Its contribution to the total load may be small unless the river is in flood (Figure 3.27). It has been suggested that the proportion of material carried in one year by the River Tyne is 57 per cent in suspension, 35 per cent in solution and 8 per cent as bedload and is the equivalent of a ten-tonne lorry tipping its load into the river every 20 minutes throughout the year. In comparison, the Amazon's load is equivalent to four ten-tonne lorries tipping every minute of the year!

C A S E   S T U D Y

The River Lyn and the Lynmouth flood

One of the worst floods in living memory in Britain was that which devastated the North Devon town of Lynmouth in 1952. The West Lyn River flows through a narrow, steep-sided valley and has a steep gradient. It forms a small drainage basin on the northern flanks of Exmoor. August of that year had been very wet and left the ground saturated. An estimated 230 mm of rain fell in a freak storm lasting 14 hours, and there was immediate surface runoff into the river causing a flash flood. It is claimed that only twice in the last century has the River Thames, with a drainage basin 100 times larger than the West Lyn, had a discharge greater than the Devon river had on that occasion.

The channel of the West Lyn had been narrowed at Lynmouth by the building of hotels and other tourist amenities and the already narrow arch of the road bridge had become blocked. The river, too swollen to be contained within its channel, flowed down the main street of the town following the path of least resistance. The size of the bedload was enormous. Over 100 000 tonnes of boulders were left in the main street after the waters receded, and one bounder found in the basement of a hotel weighed 7.5 tonnes. Elsewhere, rocks of up to 10 m^3 had been moved by traction. Measurement of the suspended load was impossible but it did, on deposition, form an offshore delta. The flood claimed 34 lives, destroyed 90 houses and hotels, 130 cars and 19 boats.

◀ Figure 3.27
Bedload left in the main street of Lynmouth in the aftermath of the 1952 flood

▶ Figure 3.28
The relationship between velocity and particle size. This shows the velocities necessary ('critical') for the initiation of movement (erosion), for deposition (sedimentation) and the area where transportation will continue to occur once movement has been initiated (*after* Hjulström, 1935)

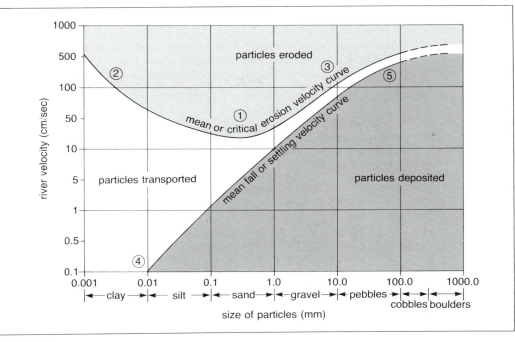

▼ Figure 3.29
Velocities at which particles of different sizes are picked up and deposited

Two further terms should be noted at this point: the competence and the capacity of a river. The **competence** of a river refers to the maximum size of material the river is capable of transporting. The **capacity** is the total load actually transported. When the velocity is low, only small particles such as clay, silt and fine sand can be picked up (Figure 3.28). As the velocity increases then larger material can be moved. Because the maximum particle mass which can be moved increases with the sixth power of the velocity it means that rivers in flood can move considerable amounts of material. For example, if the stream velocity increased by a factor of four, then the mass of boulder which could be moved would increase by 4^6 or 4096 times; if by a factor of five then the maximum mass it could transport would be multiplied 15 625 times.

The relationship between particle size (competence) and water velocity is shown in Figure 3.28. The **mean**, or **critical**, **erosion velocity** curve gives the approximate velocity needed to pick up and transport, in suspension, particles of various sizes from clay to boulders. The material carried by the river (capacity) is responsible for most of the subsequent erosion. The **mean fall**, or **settling**, **velocity** curve shows the velocities at which particles of a given size become too heavy to be transported and so will fall out of suspension and be deposited. The graph shows two important points:

- Sand can be transported at lower velocities than either finer or coarser particles. Particles of about 0.2 mm diameter can be picked up by a velocity of 20 cm/sec (labelled *1* on the graph, Figure 3.28) whereas finer clay particles (*2*), because of their cohesive properties, need a similar velocity to pebbles (*3*) to be dislodged.
- The velocity required to maintain particles in suspension is less than the velocity needed to pick them up. Indeed, for very fine clays (*4*) the velocity required to maintain them is virtually nil — at which point the river has presumably stopped flowing! This means that material picked up by turbulent tributaries and lower order streams can be kept in suspension by a less turbulent, higher order main river. For coarser particles (*5*) the boundary between transportation and deposition is narrow, indicating that only a relatively small drop in velocity is needed to cause sedimentation.

Size of particles in river	Velocity at which particles may be picked up (cm/sec) (*mean erosion velocity curve*)	Velocity at which particles may be deposited (cm/sec) (*mean fall velocity curve*)
clay	200	
silt		0.5
sand	20	5
gravel		
pebbles		
cobbles		
boulders		

Q

Use Figure 3.28 to answer the following questions.

1 a Complete the table in Figure 3.29 to show the velocity at which particles of different sizes will be picked up and deposited.

b Does this graph show the competence, capacity or load of a river?

2 Which process of transportation is likely to be most significant in a river flowing:

a over limestone

b on glacial sands and gravels

c on boulder clay

d through a narrow gorge and then on to a flat, wide plain?

3 The data given in Figure 3.30 and tables **a**, **b**, **c** refers to two adjacent drainage basins in northern Britain. The readings were taken on a day of average flow in late spring.

a Using evidence from the map give one reason why the range in Table **a** between the extreme maximum and extreme minimum discharge at site **A** is greater than at site **B**.

b Describe and give two reasons for the contrasts shown in table **b** between the particle size of the bedload found at sites **A** and **B**.

c Using the map, give one possible reason for the differences in the pattern of suspended load as shown in Table **b**.

d What is the importance of the downstream change in the hydraulic radius from measuring points **1** to **4** as given in Table **c**?

e Give three reasons why the average velocity increases downstream from measuring point **1**, despite a reduction in gradient.

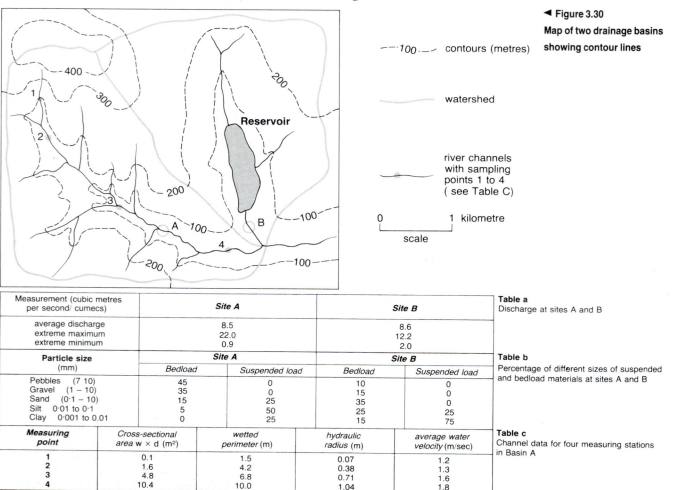

◄ **Figure 3.30**

Map of two drainage basins showing contour lines

---‧100‧--‧ contours (metres)

——— watershed

river channels with sampling points 1 to 4 (see Table C)

0 —————— 1 kilometre

scale

Table a
Discharge at sites A and B

Measurement (cubic metres per second/ cumecs)	Site A	Site B
average discharge	8.5	8.6
extreme maximum	22.0	12.2
extreme minimum	0.9	2.0

Table b
Percentage of different sizes of suspended and bedload materials at sites A and B

Particle size (mm)	Site A		Site B	
	Bedload	Suspended load	Bedload	Suspended load
Pebbles (7 10)	45	0	10	0
Gravel (1 – 10)	35	0	15	0
Sand (0·1 – 10)	15	25	35	0
Silt 0·01 to 0·1	5	50	25	25
Clay 0·001 to 0.01	0	25	15	75

Table c
Channel data for four measuring stations in Basin A

Measuring point	Cross-sectional area w × d (m²)	wetted perimeter (m)	hydraulic radius (m)	average water velocity (m/sec)
1	0.1	1.5	0.07	1.2
2	1.6	4.2	0.38	1.3
3	4.8	6.8	0.71	1.6
4	10.4	10.0	1.04	1.8

Erosion

The material carried by a river can contribute to the wearing away of its banks and bed. There are four main processes of erosion.

Corrasion

This is when the river picks up material and rubs it along its bed and banks, wearing them away by **abrasion**, rather like sandpaper. This process is most effective during times of flood and is the major method by which the river erodes both vertically and horizontally. If there are hollows in the river bed, pebbles are likely to become trapped. As the current produces turbulent eddies the pebbles will be swirled around to form **potholes** (Figure 3.31).

Attrition

As the bedload is moved downstream boulders collide with other material and the impact may break the rock into smaller pieces. In time these angular rocks become increasingly rounded in appearance.

Hydraulic action

The sheer force of the water as the turbulent current hits river banks, e.g. on the outside of a meander bend, means that water is forced into cracks. The air in the cracks is compressed, pressure is increased and, in time, the bank may collapse. **Cavitation** is a form of hydraulic action caused by bubbles of air collapsing. The resultant shock waves hit and slowly weaken the river banks. This is the slowest, least effective erosion process.

Solution or corrosion

This occurs continuously and is independent of river discharge or velocity. It is related to the chemical composition of the water, e.g. the concentration of carbonic and humic acid.

Deposition

When the velocity of a river begins to fall the stream no longer has the competence or capacity to carry all of its load. So, starting with the largest particles, material begins to be deposited (Figure 3.28).

Deposition occurs where:

- a river broadens out and therefore has a larger wetted perimeter which, assuming the volume of water remains constant, results in increased friction and a reduction of velocity,
- a river enters the sea or a lake and therefore velocity is lessened,
- discharge is reduced following a period of low precipitation,
- the river is shallower on the inside of a meander,
- the load is suddenly increased, e.g. by debris from a landslide.

As the river loses energy the following changes are likely:

- The heaviest or bedload material is deposited first. It is for this reason that the channels of mountainous streams are often filled with large boulders. These increase the size of the wetted perimeter.
- Gravel, sand and silt, transported either as bedload or in suspension, will be carried further to be deposited over flood plains or in the channel of rivers as they near their mouth.
- The finest particles of silt and clay, which are carried in suspension, may be deposited where rivers meet the sea either to infill estuaries or to form deltas.
- The dissolved load will not be deposited but will be carried out to sea where it will help to maintain the saltiness of the oceans.

▶ Figure 3.31
Potholes in the river bed of the Afon Glaslyn, Snowdonia (compare Figure 3.23)

Figure 3.32 is based upon an Open University television programme on the Afon Glaslyn in North Wales and aims to show the relationships between transportation, erosion and deposition.

1 Using evidence given in Figure 3.32, say whether you agree with the hypotheses (Framework 5 on page 208) that, as the river loses energy:
   a the largest material, carried as bedload, will be deposited first,
   b material carried in suspension will be deposited over floodplains or in the channel of the river as it nears its mouth,
   c as the competence of the river decreases, material is likely to be carried greater distances,
   d material transported as dissolved load will be carried out to sea.

2 How would you further test these hypotheses in the field?

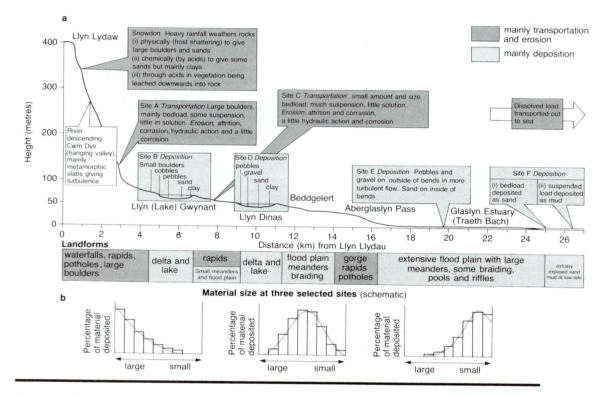

Fluvial landforms

As the velocity of a river increases surplus energy becomes available which may be harnessed to transport material and cause erosion. Where the velocity decreases, an energy deficit is likely to result in depositional features.

Results of erosion

V-shaped valleys and interlocking spurs
As was seen in Figure 3.23, in its upper course the channel of a river is often choked with large, angular boulders. This bedload produces a large wetted perimeter which uses up much of the river's energy. Erosion is minimal because little energy is left to pick up and transport material. However, following periods of heavy rainfall or after snowmelt,

the discharge of the river may rise rapidly. As the water flows between the boulders turbulence increases and may result either in the bedload being taken up into suspension or, as is more usual because of its size, in its being rolled or bounced along the river bed. The result is intensive **vertical erosion** which enables the river to create a steep-sided valley with a characteristic 'V' shape (Figure 3.33). The steepness of the valley sides depends upon several factors. **Climate** – is there sufficient rain to instigate mass movement on the valley sides and to increase discharge sufficiently for the river to generate enough energy to move its bedload? **Rock structure** – e.g. Carboniferous limestone is a resistant rock which tends to produce almost vertical valley sides. **Vegetation** may

▲ Figure 3.32

a Long profile of the Afon Glaslyn Snowdonia showing features of selected sites (0.S. 1:50000, sheets 115,124, grid reference 630545 to 575370)

b Material size at three selected sites

▶ Figure 3.33
'V' shape valley with interlocking spurs, small rapids and no flood plain, South Grey Mare's Tail, N.E. of Moffat, Scotland

help to bind the soil together and keep the hillslope more stable.

Interlocking spurs form because the river is forced to follow a winding course around the protrusions from the surrounding highland. As these spurs interlock, the view up or down the valley is restricted (Figure 3.33).

A process characteristic at the source of a river is **headward erosion**, or **spring sapping**. Here, where throughflow reaches the surface, the river may erode back towards its watershed as it undercuts the overlying rock, soil or vegetation.

Waterfalls

A waterfall forms when a river, after flowing over a relatively hard rock, meets a band of less resistant rock or, as is common in South America and Africa, where it flows over the edge of a plateau. As the water approaches the brink of the falls, velocity increases as the water in front of it loses contact with its bed and so is unhampered by friction (Figure 3.34). The underlying softer rock is worn away as water falls on to it. This may lead to the harder rock becoming undercut, unstable and eventually to collapse. As this process is repeated, the waterfall retreats upstream leaving a deep, steep-sided gorge (Figure 3.35). At Niagara, the falls are retreating by 1 m a year. The rock, which collapses to the foot of the falls, is swirled around by the turbulence, usually in times of high discharge, and carves out a deep **plunge pool**.

▼ Figure 3.34
Fieldsketch of the Iguaçu Falls at the border of Argentina and Brazil, from the Brazilian side (see Figure 3.35)

▶ Figure 3.35, *below right*
The Iguaçu Falls

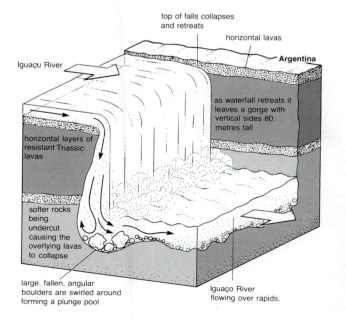

Rapids

Rapids develop where the gradient of the river bed increases without a sudden break of slope as in a waterfall, or where the stream flows over a series of gently dipping bands of harder rock. Rapids increase the turbulence of a river and hence its erosive power.

Effects of fluvial deposition

Deposition of sediment takes place when there is a decrease in energy or an increase in capacity which makes the river less competent to transport its load. This can occur anywhere from the upper course, where large boulders may be left, to the mouth, where fine clays may be deposited.

Flood plains

Rivers have most energy when at their bank-full stage. Should the river continue to rise, then the water will cover any adjacent flat land. (The land susceptible to flooding in this way is known as the **flood plain**.) At this point there will be a sudden increase in both the wetted perimeter and the hydraulic radius which in turn will produce an increase in friction and a corresponding decrease in velocity. The thin veneer of silt deposited over the flood plain often increases the fertility of the land (e.g. the Huang He in China, p. 419). Successive flooding means that the flood plain builds up in height (as yet nobody has managed to bore down to the bedrock in the lower Nile Valley). The flood plain may also be made up of material deposited as point bars on the inside of meanders (page 68) and can be widened by the **lateral erosion** of these meanders. The edge of the flood plain is often marked by a prominent slope known as the **bluff line** (Figure 3.37).

Levees

When a river overflows its banks the friction produced by the flood plain causes material to be deposited. The coarsest material is dropped first to form a small, natural embankment alongside the channel (Figure 3.37). If, later on, the river bed is raised by deposition of material, these embankments are sometimes artificially strengthened and heightened to try to contain the river. Some rivers, e.g. the Mississippi and the Huang He (Hwang Ho), flow above the level of their flood plains which means that if the levées collapse there can be serious danger to life and property.

▼ Figure 3.36

Rivers in flood

a The River Severn near Tirley, Gloucestershire

b The Mississippi River

a

b

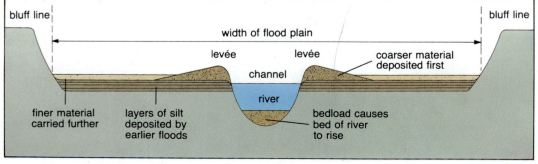

bluff line

bluff line

width of flood plain

levée

levée

coarser material deposited first

channel

river

finer material carried further

layers of silt deposited by earlier floods

bedload causes bed of river to rise

◀ Figure 3.37

Cross-section of a flood plain showing levées and bluffs

Braiding

For short periods of the year, some rivers carry a very high load in relation to their velocity, e.g. during snowmelt periods in Alpine or Arctic areas. When river levels fall rapidly, competence and capacity are reduced, and the channel may become choked with material which causes the river to divide into a series of diverging and converging segments. The 'islands' are known as **eyots** (Figure 3.38) and may also develop in semi-arid areas where rapid evaporation and infiltration of water reduces the volume, velocity and therefore the competence of the stream.

Deltas

A delta is usually composed of fine sediment which is deposited when a river loses energy and competence on flowing into an area of slow moving water such as a lake or into the sea. When rivers like the Mississippi or Nile reach the sea, the meeting of fresh and salt water produces an electric charge which causes clay particles to coagulate and to settle on the sea bed, a process called **flocculation**. Deposits are laid on the ocean bed in a threefold sequence (Figure 3.39). The finest materials are carried furthest and form the **bottomset beds**. These will be covered by slightly coarser materials which are deposited to form a slope and make up the **foreset beds**. The upper layers, nearest to the land and composed of still coarser deposits, are the horizontal **topset beds**.

Deltas were so-called because it was thought that their shape resembled that of the fourth letter of the Greek alphabet. In fact, deltas vary greatly in shape but geomorphologists have grouped them into three basic forms:

- **arcuate** which has a rounded, convex outer margin, e.g. the Nile;
- **cuspate** where the material brought down by a river is spread out evenly on either side of the channel, e.g. Tiber;
- **bird's foot** where the river has many distributaries bounded by sediment and which extend out to sea like the claws of a bird, e.g. Mississippi.

▶ **Figure 3.38**
A braided river, South Island, New Zealand

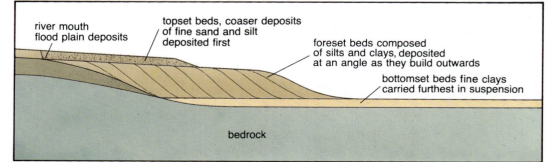

river mouth flood plain deposits

topset beds, coaser deposits of fine sand and silt deposited first

foreset beds composed of silts and clays, deposited at an angle as they build outwards

bottomset beds fine clays carried furthest in suspension

bedrock

▶ **Figure 3.39**
The structure of a delta

Effects of combined erosion and deposition

Pools, riffles and meanders

Rivers rarely flow in a straight line. Indeed, testing under laboratory conditions suggests that a straight course is abnormal and unstable. How meanders begin to form is uncertain, but they appear to have their origins in relatively straight sections where pools and riffles develop (Figure 3.41a, b, c). The usual spacing between **pools**, areas of deeper water, and **riffles**, areas of shallower water, is five to six times the bed width. The pool is an area of greater erosion where the available energy in the river builds up because of reduced friction. Hence velocity and erosive capacity increase. Across the riffle area a higher proportion of total energy is used in overcoming friction. Thus velocity and erosive capacity are reduced and further deposition may take place (Figure 3.40).

In order to avoid the riffles, the main current swings from side to side in a sinuous course. Consequently, the maximum discharge and velocity are directed towards one side of the channel, which will be eroded, while on the opposite, where volume and discharge are at a minimum, deposition occurs. In time, this process increases the sinuosity of the meander.

Meanders, point bars and oxbow lakes

A meander has an asymmetrical cross-section shape (Figure 3.22). It is thought that the material eroded from the outside of one bend is moved downstream in a corkscrew movement (known as **helicoidal flow**) and that much of this material is deposited on the inside of the next bend. The re-mainder is carried, mainly in suspension, towards the river mouth. Material deposited on the convex inside of the bend may take the form of a curving **point bar** (Figure 3.42), and its particles are usually graded in size with the largest material being found highest up the slope. As erosion continues on the concave outer rim of the meander the whole feature tends to migrate slowly downstream. The material which formed the point bar is a contributory factor in the formation of the flood plain. Over time, the sinuosity may become so pronounced that during a flood the river cuts through the narrow neck of land in order to shorten its course. Having achieved a temporary straightening of its channel, the main current flows in the centre of the channel and deposition occurs near to the banks. This means that the old curve of the river will be cut off. The remaining crescent-shaped feature is an **oxbow lake** or **cutoff** (Figure 3.43).

◄ Figure 3.40
A pool and riffles in the River Gelt, Cumbria

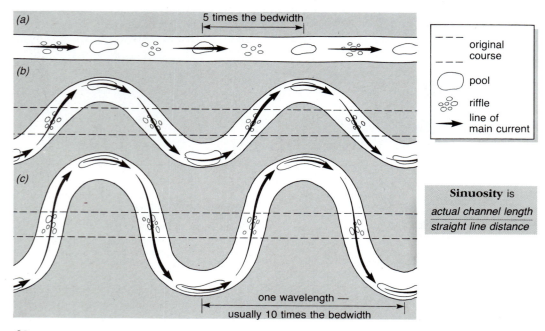

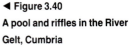

original course

pool

riffle

line of main current

Sinuosity is

actual channel length
straight line distance

◄ Figure 3.41
A possible sequence in the development of a meander through time

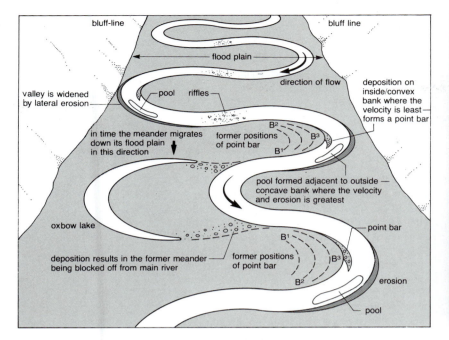

▲ **Figure 3.42**
Meanders, point bars and oxbow lakes showing changes in the position of the point bar over time

▶ **Figure 3.43,** *above right*
Meanders and oxbow lakes in rainforest in Peru

▼ **Figure 3.44**
a The graded profile
b Irregularities in the long profile

Base level and the graded river

Base level

This is the lowest point to which the erosion by running water can take place. In the case of rivers the ultimate base level is sea level. Exceptions are when the river flows into an inland sea (e.g. the River Jordan flows into the Dead Sea) or there happens to be a temporary local base level such as where a river flows into a lake, where a tributary joins the main river or where there is a resistant band of rock crossing a valley.

Grade

The concept of grade is that a river is capable of existing in a state of balance, or dynamic equilibrium, with the rate of erosion being equal to the rate of deposition. In its simplest interpretation, a graded river has a gently sloping long profile with the gradient decreasing towards its mouth (Figure 3.44a). This balance is always transitory as changes in volume, velocity and load increase either the rate of erosion or deposition until a state of equilibrium has again been reached (Framework 1, page 44). This may be illustrated by two examples.

■ The long profile of a river happens to contain a waterfall and a lake (Figure 3.44b). Erosion is likely to be greatest at the waterfall while deposition slowly fills in the lake so that in time both features are eliminated.

■ There is a lengthy period of heavy rainfall within the river basin. As the volume of water rises and consequently the velocity and load of the river increase, so too will the rate of erosion. Ultimately the extra load carried by the river leads to extra deposition further down the valley or out at sea.

In a wider interpretatation, grade is a balance not only in the long profile, but also in the river's cross profile and the roughness of its channel. In this sense, balance or grade is when all aspects of the river's channel (width, depth and gradient) are adjusted to the discharge and load of the river at a given point in time. If the volume and load change then the river's channel morphology must adjust accordingly.

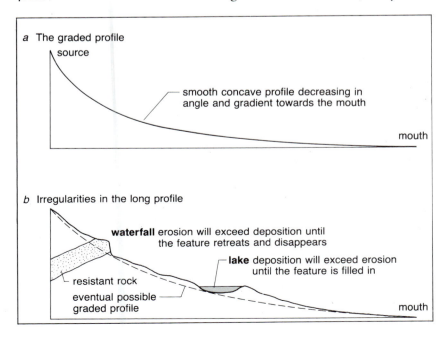

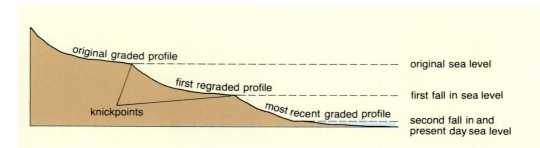

Changes in base level

Factors which influence changes in base level can be divided into two groups:

- **Climatic** the effect of glaciation and changes in rainfall (either an increase or drought).
- **Tectonic** crustal uplift following plate movement, and local volcanic activity.

As will be seen in Chapter 6, changes in base level affect coasts as well as rivers. There are two types of base level movement: positive and negative.

Positive change is when sea level rises in relation to the land (which also means that land can sink in relation to the sea). This results in a decrease in the gradient of the river with a corresponding increase in deposition and potential flooding of coastal areas.

Negative movement is when sea level falls in relation to the land (or the land rises in relation to the sea). This movement causes land to emerge from the sea, steepening the gradient of the river and therefore increasing the rate of fluvial erosion. This process is called **rejuvenation**.

Rejuvenation

A negative change in base level increases the potential energy of a river enabling it to revive its erosive activity and so upsetting any possible graded long profile. Beginning in its lowest reaches, next to the sea, the river will try to regrade itself. During the pleistocene glacial period, Britain was depressed by the weight of ice. Following deglaciation, the land slowly and intermittently rose again (**isostatic uplift**, page 100), so that rejuvenation took place on more than one occasion: many rivers today show several partly graded profiles (Figure 3.45). Should the rise in the land be rapid, the river does not have sufficient time to erode vertically to the new sea level, so rivers may descend as waterfalls over recently emerged sea cliffs (Figure 3.46). In time, the river cuts downwards and backwards and the waterfall, or **knickpoint**, retreats upstream and marks the maximum extent of the newly graded profile. Should a river become completely regraded, which is unlikely because of the

timescale involved, the knickpoint and all of the original graded profile will disappear.

A good example of a rejuvenated river is the Greta in northwest Yorkshire. Figure 3.47 is a construction of what the valley above the present-day town of Ingleton may have looked like before the change in base level and how it appears today. The Beezley Falls are a knickpoint, while below these the new profile forms a **valley in valley** feature. Above the present small flood plain are **river terraces**. A terrace is the area which was once the flood plain of the river but which, following vertical erosion, is now left high and dry above the maximum level of flooding.

River terraces offer excellent sites for the location of towns (e.g., London, Figures 3.48 and 14.3). Above the present flood plain of the Thames at London are two earlier ones forming the Taplow and Boyn terraces. If a river erodes rapidly into its flood plain, a pair of terraces of equal height may be seen flanking the river. However, more often than not, the river cuts down relatively slowly, enabling it to meander at the same time. The result is that the terrace to one side of the river may be removed as the meanders migrate. Figure 3.49 shows terraces, not paired, on a small stream crossing a beach on southern Arran. In this case rejuvenation takes place twice daily as the tide ebbs and sea level falls.

▼ Figure 3.46

A river recently rejuvenated —

this river has not had time to

readjust to the new sea level and

descends to the beach by means

of a waterfall where the land has

experienced uplift, Antalaya,

Mediterranean coast of Turkey

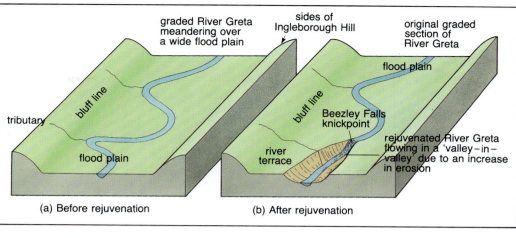

Figure 3.47
The River Greta, Yorkshire Dales
National Park
a Before and
b After rejuvenation (*after* D.S.
Walker)

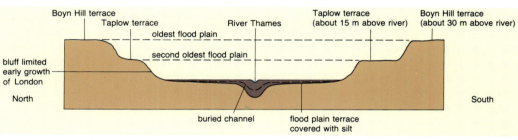

Figure 3.48
Cross-section illustrating the
paired river terraces of the
Thames at London

Q 1 Make a sketch of the photograph and label 'ingrown meanders', 'river terraces' and 'valley-in-valley' features.

2 What evidence suggests that this stream has already regraded itself?

Figure 3.49
Rejuvenation on a microscale —
a small stream crossing a beach
at Kildonan, Arran, has cut
downwards to the level of the
falling sea (tide)

If the uplift of the land, or fall in sea level, continues for a lengthy period, the river may cut downwards to form **incised meanders** (e.g. the Wear at Durham, Figure 14.5). There are two types of incised meanders (Figure 3.50). **Entrenched meanders** have a symmetrical cross-section and result from either a very rapid incision by the river or the valley sides being resistant to erosion. **Ingrown meanders** occur when the uplift of the land, or incision by the river, is less rapid, allowing the river to have time to shift laterally and to produce an asymmetrical cross valley shape (e.g. the River Wye at Tintern Abbey). As with meanders in the lower course of a normal river, incised meanders can also change their channels to leave an abandoned meander with a central meander core (Figure 3.50).

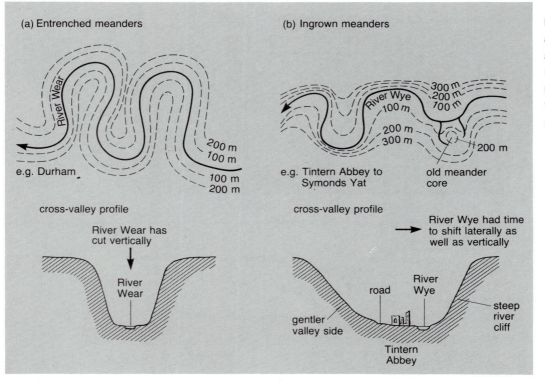

◀ Figure 3.50

Incised meanders and
associated cross-valley profiles
a Entrenched meanders, e.g.
River Wear at Durham
b Ingrown meanders, e.g. River
Wye between Tintern Abbey and
Symonds Yat

Drainage patterns

A **drainage pattern** is the way in which a river and its tributaries arrange themselves within their drainage basin (see Horton's Laws, page 52). Most patterns evolve over a lengthy period of time and usually become adjusted to the structure of the basin. There is no widely accepted classification partly because several patterns are descriptive.

Patterns independent of structure

Parallel   This, the simplest pattern, occurs on newly uplifted land when rivers and tributaries flow downhill more or less parallel with each other (Figure 3.51a).

Dendritic   Deriving its name from the Greek word *dendron*, meaning a tree, this is a tree-like pattern in which the many tributaries (branches) converge upon the main river (trunk). It is a common pattern and develops in basins having one rock type with no variations in structure (Figure 3.51b).

Patterns dependent on structure

Radial   In areas where the rocks have been lifted into a dome structure (e.g. the batholiths of Dartmoor or Arran) or where a conical volcanic cone has formed (e.g. Mount Etna), rivers radiate outwards from a central point like the spokes of a wheel (Figure 3.52).

Trellis or **rectangular**   In areas of alternating resistant and less resistant rocks, tributaries will form and will join the main river at right angles. It is possible that each individual segment may be of approximately equal length (Figure 3.53a). The main river, called a **consequent** because it is a consequence of the initial uplift or slope, flows in the same direction as the dip of the rocks (Figure 3.53b). The tributaries which develop, mainly by headward erosion, along areas of weaker rocks are called **subsequents** because they form at a later date than the consequents. In time these subsequents create wide valleys or vales (Figure 3.53c). An

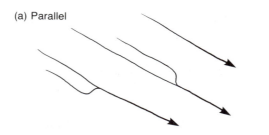

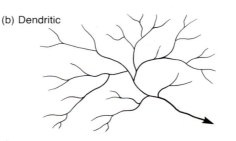

◀ Figure 3.51
Drainage patterns
a Parallel
b Dendritic

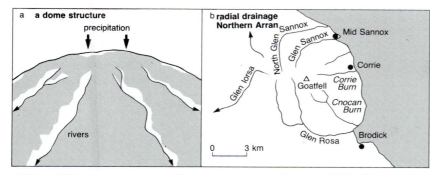

▲ Figure 3.52
Radial drainage pattern
a Dome structure
b Sketch map of a radial drainage system in northern Arran

▼ Figure 3.53
Development of a trellis or rectangular drainage pattern showing consequent, subsequent and obsequent streams
a plan
b before and
c after river capture

obsequent stream is one which flows in the opposite direction to the consequent, i.e. down the steep scarp slope of the escarpment. It is these obsequents which often provide the source of water for the scarp foot springline settlements (Figure 14.4). The development of this drainage pattern is also responsible for the formation of the **scarp and vale** topography of SE England.

Patterns apparently unrelated to structure

Antecedent Antecedence is when the drainage pattern had developed before structural movements, such as the uplift or folding of the land, and where the vertical erosion of the river was able to keep pace with any uplift. The Brahmaputra River rises in Tibet, yet it flows through the Himalayas in a series of deep gorges before reaching the Bay of Bengal (Figure 3.54 overleaf). It must at one stage have flowed southwards into the Tethys Sea (Figure 1.4) which had existed before the Indo-Australian Plate moved northwards forming the Himalayas on collision with the Eurasian Plate. The Brahmaputra,

with an increasing gradient and load, was able to cut downwards to maintain its original course.

Superimposed In several parts of the world, including the English Lake District, the drainage pattern seems to have no relationship to the present day surface rocks. When the Lake District was uplifted into a dome, the newly formed volcanic rocks were covered by sedimentary limestones and sandstones. The radial drainage which developed, together with later glacial processes, cut through and ultimately removed the surface layers to leave a pattern which bears no relation to the structure of the newly exposed rocks.

River capture

Rivers, in attempting to adjust to structure, may capture the headwaters of their neighbours. Figure 3.55a (overleaf) shows a case where there are two consequent rivers with one having a greater discharge and a higher erosional activity than the other. Each has a subsequent flowing along a valley of weaker rock, but the tributary of the master, or larger, consequent is likely to be the more vigorous. This subsequent will cut backwards by headward erosion until it first reaches the tributary of the weaker consequent and then, by a process known as **watershed migration**, begins to enlarge its own drainage basin at the expense of the smaller river. In time the headwaters of the minor consequent will be captured and diverted into the drainage basin of the major consequent.

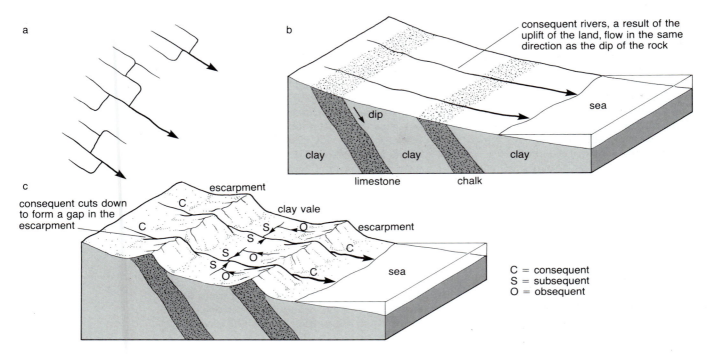

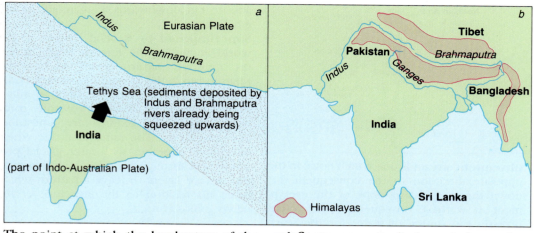

◀ Figure 3.54

Antecedent drainage related to
the tectonic movement of the
Eurasian and Indian Plates
a Before b After

The point at which the headwaters of the minor river change direction is known as the **elbow of capture**. Below this point a **wind gap** marks the former course of the now beheaded consequent. (A wind gap is a dry valley cut through hills by a former river.) This beheaded river is also known as a **misfit stream** as its discharge is far too low to account for the size of the valley through which it flows. Most eastward flowing English rivers between the Humber and the Wansbeck (Northumberland) have had their courses altered by river capture or piracy (Figure 3.56).

Q
1 Redraw the diagram in Figure 3.57 (page 75).
   a On it label: master consequent, a subsequent river, an obsequent river, elbow of capture, wind gap, misfit stream, beheaded consequent.
   b Mark on the watershed for river **A**.

2 What changes in the drainage pattern are likely to take place in the near future? Give reasons for your answer.

Human activity in the drainage basin

The drainage basin system can affect human activity and be modified greatly by it. Unfortunately this activity, often with short term objectives, frequently leads to longer term problems if it is not carefully planned. For most of history, humans have tended to extract fresh water from rivers and to replace it with their wastes. It has now been recognised that drainage basin management, from source to mouth, is essential — a difficult task because many major drainage basins straddle several countries which may have different priorities in the use of that basin and differing amounts of wealth to develop and/or conserve it.

The early settled civilisations (page 330) grew up along river banks. Water was needed for domestic use, for farm animals and crops, and to supply food in the form of fish. Later, the river was used to harness power (initially using water wheels and today hydro-electricity), to transport people and their goods and for recreation. Yet in trying to utilise and control the river, human society has created many difficulties both deliberate and unforeseen.

River régimes vary annually and seasonally.

Because some rivers experience extremes of discharge, dams have been built to hold back flood waters and to release them when levels would otherwise be very low. Such schemes reduce the risk of loss of human and animal life, and of property, as well as maintaining levels sufficient for transport and for fish. They enable crops to be grown in areas or at times of the year when previously the climate was too arid. However, the building of dams has often had an adverse effect because by altering discharge, sedimentation levels are affected. The release of sediment-free water can overdeepen the channel below the dam, while silt may no longer be spread over flood plains. Elsewhere, the reduced discharge has encouraged deposition, braiding and the raising of river beds. Sediment-free water has also, in some tropical areas, encouraged the rapid growth of floating plants, such as water hyacinth, which has blocked channels.

The creation of lakes behind dams can alter local climates, e.g. reduction in the discharge of the River Ob in the USSR might accelerate ice formation so that it freezes 19 days earlier.

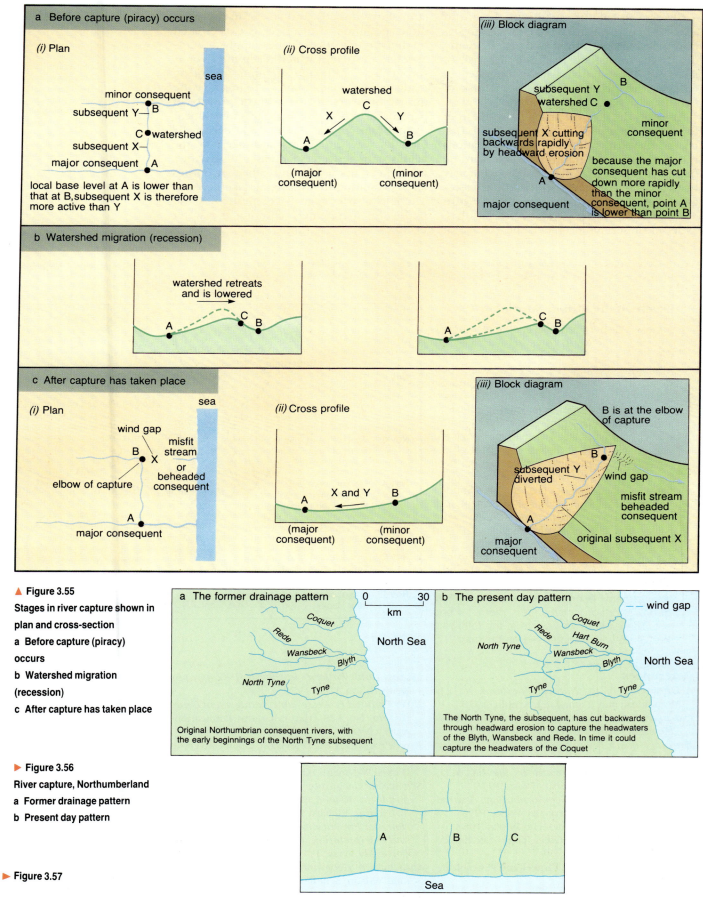

a Before capture (piracy) occurs

(i) Plan

sea

minor consequent
subsequent Y — B
C watershed
subsequent X —
major consequent A

local base level at A is lower than that at B, subsequent X is therefore more active than Y

(ii) Cross profile

watershed

X C Y

A (major consequent)   B (minor consequent)

(iii) Block diagram

B
subsequent Y watershed C
subsequent X cutting backwards rapidly by headward erosion
minor consequent
A
major consequent
because the major consequent has cut down more rapidly than the minor consequent, point A is lower than point B

b Watershed migration (recession)

watershed retreats and is lowered

A   C B

A   C B

c After capture has taken place

(i) Plan

sea

wind gap
B misfit stream
X or beheaded consequent
elbow of capture
A
major consequent

(ii) Cross profile

X and Y
A   B
(major consequent)   (minor consequent)

(iii) Block diagram

B is at the elbow of capture
B
subsequent Y diverted
wind gap
misfit stream beheaded consequent
A
major consequent
original subsequent X

▲ **Figure 3.55**

Stages in river capture shown in plan and cross-section

a Before capture (piracy) occurs

b Watershed migration (recession)

c After capture has taken place

a The former drainage pattern

0 ——— 30 km

Coquet
Rede
Wansbeck
Blyth
North Sea
North Tyne
Tyne

Original Northumbrian consequent rivers, with the early beginnings of the North Tyne subsequent

b The present day pattern

--- wind gap

Coquet
Rede   Hart Burn
North Tyne
Wansbeck
Blyth
North Sea
Tyne   Tyne

The North Tyne, the subsequent, has cut backwards through headward erosion to capture the headwaters of the Blyth, Wansbeck and Rede. In time it could capture the headwaters of the Coquet

▶ **Figure 3.56**

River capture, Northumberland

a Former drainage pattern

b Present day pattern

▶ **Figure 3.57**

A   B   C

Sea

Changes in land use within the basin can also change the natural system. Deforestation may increase runoff and consequently discharge, suspended load and the risk of flooding. Afforestation usually operates in reverse. Urbanisation increases the risk of flash floods (page 50).

Arguably the greatest problems result from the river being used as an open drain. Discharge of warm water from power stations raises water temperatures which can be harmful to animal and plant life though it may be beneficial to certain species. Toxic wastes are a major threat to river ecology (Figure 3.57). The Northwest Water Authority has, as its most serious problem, the release of

farm slurry directly into rivers. In this area, direct pollution of the river is exacerbated by the after-effects of farmers applying fertiliser and pesticides which reach the river as throughflow. Although attempts are being made in developed countries to improve the quality of rivers, this is a lower priority in parts of the world which are less developed.

Acid rain (pages 33 and 183) eventually finds its way into rivers and lakes. As the acidity of these waters increases it is having a fatal effect upon fish and other organisms and is resulting in a changing lake ecology. There is also concern over pesticides and herbicides increasingly entering the water cycle.

C A S E   S T U D Y

▼ Figure 3.58
'Death on the Rhine', November 1987

Death on the Rhine, November 1987

Almost 20 per cent of chemicals manufactured in the world come from factories on the banks of the Rhine and its major tributaries. Estimates claim that over 2000 different chemicals have, at one time or another, been released into the river. Untreated sewage and other industrial wastes have limited the use of the Rhine by plant and animal life. The river lacks oxygen and often smells unpleasant at times of low water. Recent attempts to clean up the river were shattered on 1 November 1987 when there was a major fire at the Sandoz chemical factory near Basel in Switzerland. Chemicals from exploding drums were released into the river and began a 1200 km journey to the North Sea in the form of a slick 80 km long. It was 36 hours before the real seriousness of the fire was realised. By then it was too late — part of the river below Basel was already ecologically dead, with fish, birds, micro-organisms and plants wiped out. The diary of the events of the following twelve days is summarised on the map.

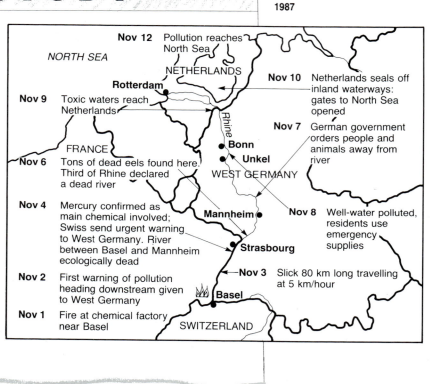

References

A Dictionary of the Natural Environment, F.J. Monkhouse and J. Small, E.J. Arnold, 1978.

Catchment Studies: Geography Today, BBC Television, Open University.

Drainage Basin Form and Process, K.J. Gregory and D.E. Walling, Arnold, 1973.

Flooding and Flood Hazard in the UK, M.D. Newson, Oxford University Press, 1975.

Modern Concepts in Geomorphology, Patrick McCullagh, Oxford University Press, 1978.

Planet Earth: Floods, Time–Life Books, 1983.

Planet Earth: Rivers and Lakes, Time–Life Books, 1985.

Process and Landform, A. Clowes and P. Comfort, Oliver & Boyd, 1982.

Process and Pattern in Physical Geography, Keith Hilton, UTP, 1979.

River Channel Processes and Management Issues, Geofile, No 128, Mary Glasgow Publications, April 1989.

River Processes, BBC Television, Open University.

Runoff Processes and Streamflow Modelling, Darrell Weyman, Oxford University Press, 1973.

Science in Geography: 1 Developments in Geographical Method, Brian P. Fitzgerald, *2 Data Collection*, Richard Daugherty, *3 Data Collection and Presentation*, Peter Davis, *4 Data Use and Interpretation*, Patrick McCullagh, Oxford University Press, 1974.

Statistics in Geography for 'A' level Students, John Wilson, Schofield & Sims, 1984.

CHAPTER 4

Glaciation

'Great God! this is an awful place.'
The South Pole, Robert Falcon Scott, Journal, 1912

Ice ages

It appears that roughly every 200 to 250 million years in the earth's history there have been major periods of ice activity (Figure 4.2). Of these, the most recent and significant occurred during the Pleistocene period of the Quaternary (Figure 1.1). In the two million years since the onset of the Quaternary there have been fluctuations of between 5° and 6°C in global temperature which have led to cold phases (**glacials**) and warm phases (**interglacials**). Until recently it was believed that there had been four main glacials but in the last decade evidence obtained from cores taken from ocean floor deposits and polar ice caps has led glaciologists to claim that there have been more than 20 (Figure 4.1).

When the ice reached its maximum extent it is estimated that 30 per cent of the earth's land surface was covered (compared with some 10 per cent today). However, the effect was felt not only in polar latitudes and mountainous areas, for each time the ice advanced there was a change in the global climatic belts (Figure 4.3). Only 18 000 years ago, at the time of the maximum advance within the last glacial, ice covered Britain as far as South Wales, the Midlands and Norfolk. The southern part of Britain experienced **tundra** conditions, as did most of France (page 281).

Climatic change

Although it is accepted that climatic fluctuations occur on a variety of timescales, as yet there is no single explanation for the onset of major ice ages or for fluctuations within each ice age. The most feasible of theories to date is that of Milutin Milankovitch, a Yugoslav mathematician/astronomer, who, between 1912 and 1941, performed exhaustive calculations to show that the earth's position in space, its tilt and its orbit

▼ Figure 4.1

Generalised trends in mean global temperatures during the past million years

► Figure 4.2, *right*

A chronology of ice ages

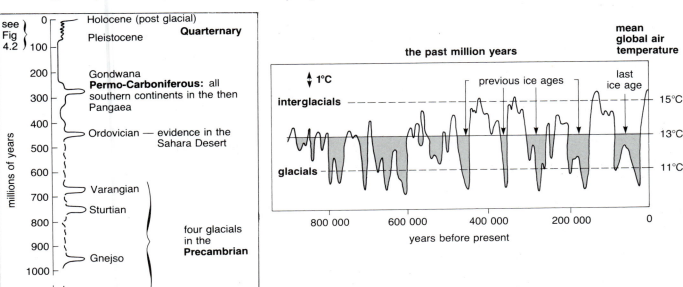

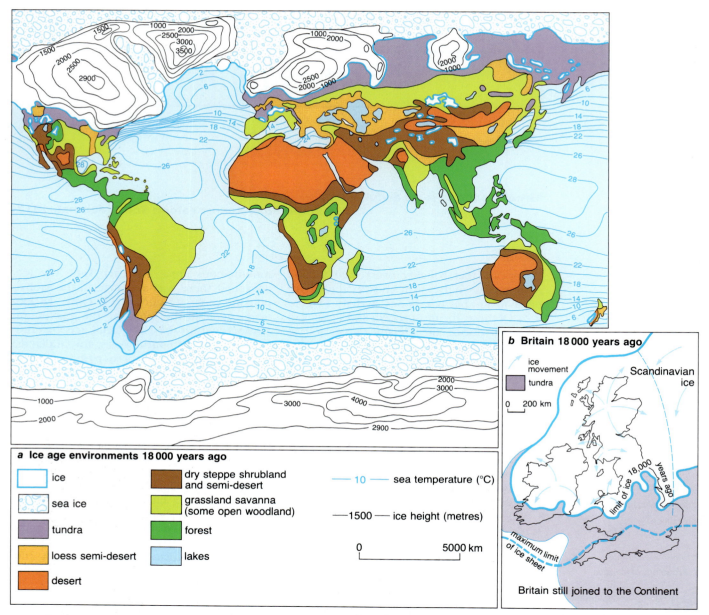

a Ice age environments 18 000 years ago

- ☐ ice
- ▨ sea ice
- ▨ tundra
- ▨ loess semi-desert
- ▨ desert
- ▨ dry steppe shrubland and semi-desert
- ▨ grassland savanna (some open woodland)
- ▨ forest
- ▨ lakes
- —— 10 —— sea temperature (°C)
- —— 1500 —— ice height (metres)
- 0 —————— 5000 km

b Britain 18 000 years ago

- ice movement
- ▨ tundra
- 0 —— 200 km

Scandinavian ice

18 000 years ago

limit of ice

maximum limit of ice sheet

Britain still joined to the Continent

▲ **Figure 4.3**

World climates and vegetation 18 000 years ago

a Major ice age environments

b Britain 18 000 years ago

around the sun all change. These changes, he claimed, affect incoming radiation from the sun and produce three main cycles of 95, 42 and 21 thousand years (Figure 4.4). His theory, and the timescale of each cycle, has been given considerable support by evidence gained, since the mid 1970s, from ocean floor cores. As yet, although the relationship appears to have been established, it is not known precisely how these celestial cycles are related to climatic change.

Other suggestions have been made as to the causes of ice ages, and it is possible that one or more of them may, at some later stage, be shown to be significant (case study p 79).

- Variations in sunspot activity may increase or decrease the amount of radiation received by earth.
- Injections of volcanic dust into the atmosphere can reflect and absorb radiation from the sun (page 168).
- Changes in atmospheric carbon dioxide gas which could accentuate the greenhouse effect (page 170). Initially extra CO_2 traps heat in the atmosphere, possibly raising world temperatures by an estimated 3°C. In time some of this CO_2 will be absorbed by the seas, reducing the amount remaining in the atmosphere and causing a drop in world temperatures and the onset of another ice age.
- The movement of plates either into colder latitudes or at constructive margins, where there is an increase in altitude, could lead to an overall drop in world land temperatures.
- Changes in ocean currents or jet streams (page 190).

CASE STUDY

Evidence of climatic change from Antarctica

The first results of a five year drilling experiment by the Russians in Antarctica were announced in January 1988. It took from 1980 to 1985 to drill and carefully to extract a core of ice extending downwards for 2 km. By dating the rings, which show the annual accumulation of snow it has been estimated that the team has reached ice 160 000 years old. This would provide information about the whole of the last interglacial and glacial. The researchers hope that eventually they will reach ice half-a-million years old. Three of several experiments being undertaken are:

1 measuring dust from volcanic activity,
2 extracting bubbles of air which will contain carbon dioxide to see if the levels of that gas have changed over periods of time,
3 comparing ratios between two types of hydrogen which help to indicate past temperatures — these two types of hydrogen occur in rain clouds which drift southwards and if the climate gets colder then there will be less moisture in the air giving less precipitation and hydrogen.

Early results seem to indicate that:
- there have been periods of considerable volcanic activity,
- the earth may well wobble on its axis causing the 21 000 year cycle suggested by Milankovitch,
- there have been earlier periods when carbon dioxide built up in the atmosphere and there are no plausible suggestions as to why this should occur. It does appear though that the last glacial began at a time when the carbon dioxide content was very low.

▶ **Figure 4.5**

Dirt bands (englacial moraine) in a glacier in Iceland

Evidence seems to support the view that the earth is nearing the end of a brief interglacial and that following a period of colder, drier climate (which would seriously affect food crop production), the next glacial is likely to start within the next few thousand years. This might be temporarily delayed by the increase in temperatures resulting from the extra carbon dioxide content of the atmosphere.

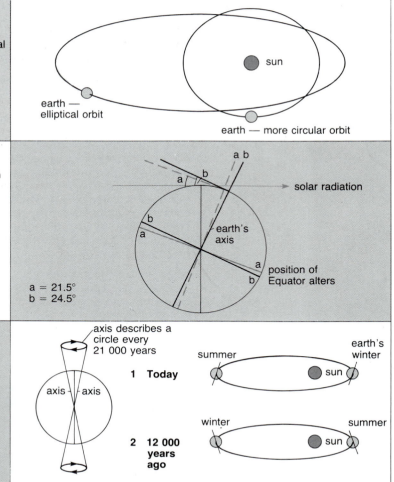

The 95 000 year stretch

The earth's orbit stretches from being nearly circular to an elliptical shape and back again in a cycle of about 95 000 years. During the *Quaternary*, the major glacial – interglacial cycle was almost 100 000 years. Glacials occur when the orbit is almost circular and interglacials when it is a more elliptical shape.

The 42 000 year tilt

Although the tropics are set at 23.5° N and S to equate with the angle of the earth's tilt, in reality the earth's axis varies from its plane of orbit by between 21.5° and 24.5°. When the tilt increases, summers will become hotter and winters colder, leading to conditions favouring interglacials.

The 21 000 year wobble

As the earth slowly wobbles in space its axis describes a circle once in every 21 000 years.
1 At present the orbit places the earth closest to the sun in the northern hemisphere's winter and furthest away in summer. This tends to make winters mild and summers cool. These are ideal conditions for glacials to develop.
2 12 000 years ago the position was in reverse and this has contributed to our present warm 'interglacial'.

Snow accumulation and ice formation

As the climate gets colder, more precipitation is likely to be in the form of snow in winter and there is less time in the shorter summer for that snow to melt. If the climate continues to deteriorate then snow will lie throughout the year forming a permanent **snow line**. The snow line, the level above which snow will lie all year, is at a lower altitude on north-facing slopes than on south-facing ones in the northern hemisphere as these receive less insolation. The snow line is also lower nearer the poles and higher at the Equator. It is at sea level in northern Greenland, about 1500 m in Southern Norway, 3000 m in the Alps and 6000 m on the Equator. It is estimated that the Cairngorms in Scotland would be snow-covered all year had they been 200 m higher.

When snowflakes fall they have an open feathery appearance, trap air, and have a low density. Where snow collects in hollows it becomes compressed by the weight of subsequent falls and gradually develops into a more compact, dense form called **firn** or **névé**. Firn is compacted snow which has experienced one winter's freezing and survived a summer's melting. In temperate latitudes, such as in the Alps, summer meltwater percolates into the firn only to freeze either at night or during the following winter and so forms an increasingly dense mass. Air is progressively squeezed out and after 20 to 40 years the firn will have turned into solid ice. This same process may take over 200 years in Antarctica and Greenland where there is no summer melting. Once ice has formed it may begin to flow downhill, under the force of gravity, as a **glacier**.

Classification of glaciers and ice masses

It is customary to classify glaciers according to size and shape as these characteristics are relatively easy to identify by field observation.
1 **Niche** glaciers are very small occupying hollows and gulleys on north-facing slopes in the northern hemisphere.

▲ Figure 4.4

Milankovitch's theories on climatic change

a The 95000 year stretch

b The 42000 year tilt

c The 21000 year wobble

▶ Figure 4.6

The glacial budget — inputs, outputs and storage elements of the glacier system

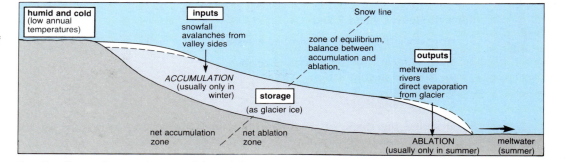

▼ Figure 4.7

Accumulation and ablation zones in Gigjökull Glacier, emerging from the crater at Fljötshlid, Iceland

▼ Figure 4.8, *below right*

Graph illustrating the glacial budget or net balance

2 **Corrie** or **cirque** glaciers, although larger than niche glaciers, are small masses of ice occupying armchair-shaped hollows in mountains. They may overspill to feed valley glaciers.

3 **Valley** glaciers are larger bodies of ice which move down from a mountain source. These glaciers usually follow former river courses and are bounded by steep valley sides.

4 **Piedmont** glaciers are formed when valley glaciers extend on to lowland areas, spread out and merge.

5 **Ice caps** and **ice-sheets** are huge areas of ice which spread outwards from central domes. Apart from the summits of high mountains, called **nunataks**, which poke up through the ice, the whole landscape is buried. Ice sheets, which once covered much of northern Europe and North America (Figure 4.3) are now confined to Antarctica and Greenland.

Glacial systems and budgets

A glacier behaves as a system, with inputs, stores and outputs (Figure 4.6). Inputs are derived from snow falling directly on to the glacier or from **avalanches** along valley sides. The glacier itself is water in storage. Outputs from the system include evaporation from the glacier or, more noticeably, in the form of meltwater streams which flow either on top of or under the ice during the summer months. The upper part of the glacier, where inputs exceed outputs, is known as the **zone of accumulation** while the lower part, where outputs exceed inputs, is called the **zone of ablation** or **melting** (Figures 4.6, 4.7).

The **zone of equilibrium** is where the rates of accumulation and ablation are equal and it corresponds with the snow line. The **glacier budget**, or **net balance** is the difference between the total accumulation and total ablation for one year. In temperate glaciers (page 83) there is likely to be a negative net balance in summer when ablation exceeds accumulation, and a positive net balance in winter when the reverse occurs (Figure 4.8). If the summer and winter budgets cancel each other out, the glacier appears to be stationary — it seems stationary because the snout is neither advancing nor retreating although ice from the accumulation zone is still moving down-valley into the ablation zone.

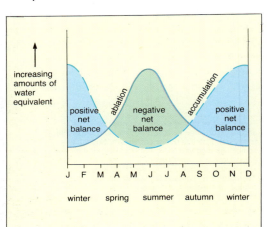

Q

Refer to Figures 4.6, 4.7 and 4.8.

1 How does the glacier operate as a system?

2 What is meant by the following terms?
   a zone of accumulation
   b storage
   c zone of ablation

3 Describe the shape of the two graphs in Figure 4.8. Give reasons for the differences between them.

4 What is meant by the following terms?
   a positive net balance in winter
   b negative net balance in summer
   c net annual balance (or glacier budget)

5 Under what conditions might a glacier **a** advance, **b** retreat?

6 How might the following affect the net balance of a glacier?
   a a decrease in mean annual temperature
   b a decrease in mean annual precipitation

▼ Figure 4.9

Processes of glacier movement

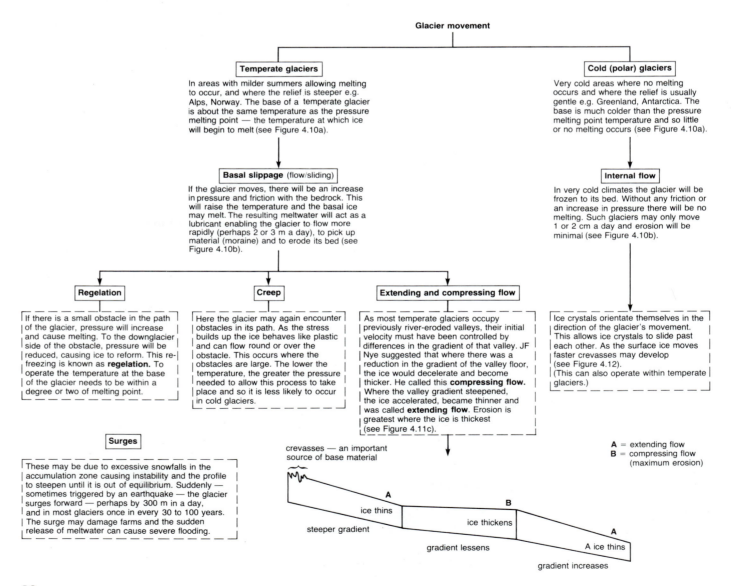

Glacier movement

Temperate glaciers

In areas with milder summers allowing melting to occur, and where the relief is steeper e.g. Alps, Norway. The base of a temperate glacier is about the same temperature as the pressure melting point — the temperature at which ice will begin to melt (see Figure 4.10a).

Cold (polar) glaciers

Very cold areas where no melting occurs and where the relief is usually gentle e.g. Greenland, Antarctica. The base is much colder than the pressure melting point temperature and so little or no melting occurs (see Figure 4.10a).

Basal slippage (flow/sliding)

If the glacier moves, there will be an increase in pressure and friction with the bedrock. This will raise the temperature and the basal ice may melt. The resulting meltwater will act as a lubricant enabling the glacier to flow more rapidly (perhaps 2 or 3 m a day), to pick up material (moraine) and to erode its bed (see Figure 4.10b).

Internal flow

In very cold climates the glacier will be frozen to its bed. Without any friction or an increase in pressure there will be no melting. Such glaciers may only move 1 or 2 cm a day and erosion will be minimal (see Figure 4.10b).

Regelation

If there is a small obstacle in the path of the glacier, pressure will increase and cause melting. To the downglacier side of the obstacle, pressure is reduced, causing ice to reform. This re-freezing is known as **regelation.** To operate the temperature at the base of the glacier needs to be within a degree or two of melting point.

Creep

Here the glacier may again encounter obstacles in its path. As the stress builds up the ice behaves like plastic and can flow round or over the obstacle. This occurs where the obstacles are large. The lower the temperature, the greater the pressure needed to allow this process to take place and so it is less likely to occur in cold glaciers.

Extending and compressing flow

As most temperate glaciers occupy previously river-eroded valleys, their initial velocity must have been controlled by differences in the gradient of that valley. JF Nye suggested that where there was a reduction in the gradient of the valley floor, the ice would decelerate and become thicker. He called this **compressing flow.** Where the valley gradient steepened, the ice accelerated, became thinner and was called **extending flow.** Erosion is greatest where the ice is thickest (see Figure 4.11c).

Ice crystals orientate themselves in the direction of the glacier's movement. This allows ice crystals to slide past each other. As the surface ice moves faster crevasses may develop (see Figure 4.12). (This can also operate within temperate glaciers.)

Surges

These may be due to excessive snowfalls in the accumulation zone causing instability and the profile to steepen until it is out of equilibrium. Suddenly — sometimes triggered by an earthquake — the glacier surges forward — perhaps by 300 m in a day, and in most glaciers once in every 30 to 100 years. The surge may damage farms and the sudden release of meltwater can cause severe flooding.

crevasses — an important source of base material

A = extending flow
B = compressing flow
   (maximum erosion)

A

B

ice thins

ice thickens

A

steeper gradient

gradient lessens

A ice thins

gradient increases

Glacier movement and temperature

Warm and cold glaciers

Temperature is an alternative to size or shape as a criterion when categorising glaciers — they may be either **temperate** or **cold** (Figure 4.9). Movement is much faster in the temperate glacier and takes place by **basal flow or slipping**. Figure 4.9 summarises the three types of basal flow: **regelation**, **creep** and **extending–compressing flow**. Cold glaciers move less quickly by a process called **internal flow**. The difference in velocity is related to the **pressure melting point** temperature (Figure 4.10).

Both glaciers move more rapidly on the surface and away from their valley sides (Figure 4.11a) but it is the temperate one which is more likely to erode its bed and to carry and deposit most material or **moraine** (Figure 4.11b). Recent research suggests that any single glacier may exhibit, at different points along its profile, the characteristics of both cold and temperate glaciers.

Movement is greatest:
- ☐ at the point of equilibrium as this is where the greatest volume of ice passes and consequently there is most energy available,
- ☐ in areas with high precipitation and ablation,
- ☐ in small glaciers which respond more readily to short term climatic fluctuations,
- ☐ in temperate glaciers where there is more meltwater available,
- ☐ in areas with steep gradients.

▶ Figure 4.10

Temperature and velocity profiles in cold and temperate glaciers

a Comparison of temperature change with depth of ice

b Speed of movement and depth of ice

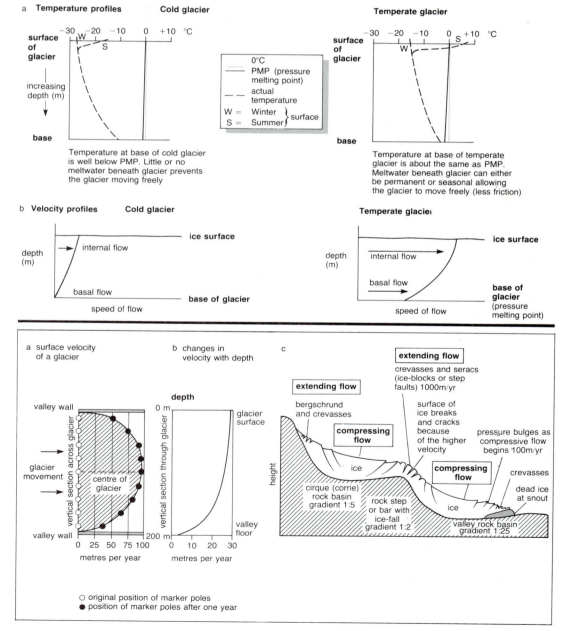

▶ Figure 4.11

a Surface velocity of a glacier

b Changes in velocity with depth

c Extending and compressing flow

Q

1   For *both* diagrams (Figures 4.11a, 4.11b on previous page) describe where the velocity is greatest. Give reasons for your answer.

2   Account for the location of the areas of extending and compressing flow.

3   Account for the formation of crevasses and seracs on the diagram.

4   What glacial features might be found in the area shown on the sketch once the glacier melts?

Processes of glacial erosion

Ice that is stationery or contains little debris has limited erosive power whereas moving ice carrying with it much morainic material can drastically alter the landscape. Although ice lacks the turbulence and velocity of the water in a river, it has the 'advantage' of being able to melt and to refreeze in order to overcome obstacles in its path. Virtually all of the glacial processes of erosion are physical as the climate tends to be too cold for chemical reactions to operate (Figure 2.9b).

Frost shattering This process (page 30) produces much loose material which may fall from the valley sides on to the edges of the glacier to form **lateral moraine**, be covered by later snowfall or plunge down crevasses to form **englacial** (within the glacier) **moraine**. Some of this material may be added to rock loosened by frost action as the climate deteriorated (but before glaciers formed) to form **basal** or **ground moraine**.
Abrasion This is the sandpapering effect of angular material embedded in the glacier as it rubs against the valley sides and floor. It usually produces smoothened gently sloping landforms.

Plucking At its simplest, this process involves the glacier freezing on to the valley sides and subsequent ice movement pulling away masses of rock. In reality, as the strength of the bedrock is greater than that of the ice, it would seem that only previously loosened material can be removed. Material may be continually loosened by one of three processes.

1   The relationship between local pressure and temperature (the pressure melting point) produces sufficient meltwater for freeze–thaw activity to break up the ice-contact rock.
2   Water flowing down a **bergschrund** (Figure 4.13) or smaller crevasses later freezes on to rock surfaces.
3   Removal of layers of bedrock by the glacier causes a release in pressure and an enlarging of joints in the underlying rocks (pressure release, page 31).
Plucking generally creates a jagged-featured landscape.
Rotational movement This is a downhill movement of ice which, like a land-slump, pivots about a point (Figure 2.17).

▲ Figure 4.12
Crevasses on an icefall, Skafta Glacier, Iceland

Bergschrund: a large crevasse-like feature found near the head of a glacier

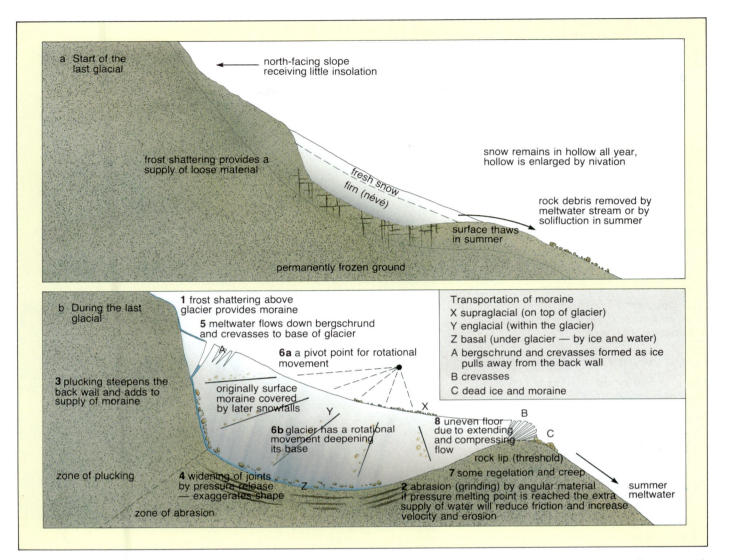

The following labels appear within the figure:

a Start of the last glacial

north-facing slope receiving little insolation

frost shattering provides a supply of loose material

fresh snow
firn (névé)

snow remains in hollow all year, hollow is enlarged by nivation

rock debris removed by meltwater stream or by solifluction in summer

surface thaws in summer

permanently frozen ground

b During the last glacial

1 frost shattering above glacier provides moraine

5 meltwater flows down bergschrund and crevasses to base of glacier

6a a pivot point for rotational movement

3 plucking steepens the back wall and adds to supply of moraine

originally surface moraine covered by later snowfalls

6b glacier has a rotational movement deepening its base

8 uneven floor due to extending and compressing flow

zone of plucking

4 widening of joints by pressure release — exaggerates shape

7 some regelation and creep

rock lip (threshold)

summer meltwater

zone of abrasion

2 abrasion (grinding) by angular material if pressure melting point is reached the extra supply of water will reduce friction and increase velocity and erosion

Transportation of moraine
X supraglacial (on top of glacier)
Y englacial (within the glacier)
Z basal (under glacier — by ice and water)
A bergschrund and crevasses formed as ice pulls away from the back wall
B crevasses
C dead ice and moraine

▲ Figure 4.13

Processes in the formation of a cirque

a Early in the last glacial

b During the last glacial

The increase in pressure is responsible for the overdeepening of a cirque floor.

Extending and **compressing flow** Figure 4.9 shows how this process causes differences in the rate of erosion at the base of a glacier.

Maximum erosion occurs where temperatures fluctuate around 0°C, allowing frequent freeze–thaw to operate; in areas of jointed rocks which can be more easily frost shattered; where two tributary glaciers join or the valley narrows giving an increased depth of ice; and in steep mountainous regions in temperate latitudes where the velocity of the glacier is greatest.

Landforms produced by glacial erosion

Cirques

These are amphitheatre or armchair-shaped hollows with a steep backwall and a rock basin. They are also known as **corries** (Scotland) and **cwms** (Wales).

During the periglacial times (Chapter 5) which preceded the last glacial, snow collected in hollows especially on north-facing slopes. Freeze–thaw action beneath the snow caused the underlying rocks to disintegrate, a process called **nivation** (Figure 4.13a). Debris was removed by summer meltwater streams leaving, in the enlarged hollow, an embryo cirque. As the snow patch grew, its layers would become compressed to form firn and eventually ice.

It is accepted that several processes interact to form a fully developed cirque (Figure 4.13b). Plucking is the process mainly responsible for steepening the back wall but this partly relies upon a supply of water for freeze–thaw and partly upon pressure-release in well-jointed rocks. A rotational movement, aided by water from pressure point melting and angular moraine from frost shattering, enables abrasion to over-deepen the floor of the cirque. A **rock lip** develops where erosion decreases. This may be increased in height by the deposition of moraine at the glacier's snout. When the climate begins to get warmer the ice remaining in the hollow may melt to leave a deep, rounded lake or tarn (Figure 4.14).

In Britain, as elsewhere in the northern hemisphere, cirques are nearly always orientated between the northwest, through the northeast (where the frequency peaks) to the southeast. This is largely attributable to two facts:

- Northern slopes receive least **insolation** and so glaciers remained here much longer than those facing in more southerly directions (less melting on north-facing slopes).
- Western slopes faced the sea and, although still cold, the relatively warmer winds which blew from that direction were more likely to melt the snow and ice (more snow accumulated on east-facing slopes). Of 56 cirques identified in the Snowdon area, 51 have a lip orientation of between 310° and 120°.

Mean, median and mode

These are all types of average (measures of dispersion, page 208).

1 The **mean** (or arithmetic average) is obtained by totalling the values in a set of data and dividing by the number of values in that set. It is expressed by the formula:

$$\bar{x} = \frac{\Sigma x}{n}$$

where:
$\bar{x}$ = mean
Σ = the sum of
x = the value of the variable
n = the number of values in the set

The mean is reliable when the number of values in the sample is high and their range, i.e. the difference between the highest and lowest values, is low, but it becomes less reliable as the number in the sample decreases as it then becomes influenced by extreme values.

2 The **median** is the mid-point value of a set of data. In the sample of 15 cirques, values must first be ranked in descending order. The mid-point is the orientation of the eighth cirque because there are seven values above and seven below. Had there been an even number of values, then the median would have been the mean of the two middle values. The median is a less accurate measure of dispersion than the mean because widely differing sets of data can return the same median but it is less distorted by extreme values.

3 The **mode** is the value or class which occurs most frequently in the data. In the set of values 4, 6, 4, 2, 4 the mode would be 4. Although it is the easiest of the three 'averages' to obtain it has limited value. Some data may not have two values in the same class (e.g. 1, 2, 3, 4, 5), while others may have more than one modal value (e.g. 1, 1, 2, 4, 4).

Relationships between mean, median and mode

When data is plotted on a graph we can often make useful observations about the shape of the curve. For example, we would expect 'A'-level results nationally to show a few top grades, a small number of 'unclassifieds' and a large number of average passes. Graphically this would show a **normal distribution**, with all three averages at the peak. If the distribution is **skewed**, then by definition only the mode will lie at the peak (Figure 4.15).

▲ Figure 4.14

A cirque, Llyn Cau, Cader Idris, West Wales

The steep back wall maintains its shape as frost shattering is still taking place. The loosened material forms scree which is beginning to fill in the lake which itself has been dammed behind a natural rock lip

Q The table shows the orientation of 15 cirques (the minimum needed for an acceptable sample) in the Glyders of Snowdonia, and 15 on the Isle of Arran.

1 What is meant by bearings of 45° and 350°?

2 For both groups of cirques give the mean, median and mode of their orientation.

3 What do your answers show regarding the orientation of cirques in those two areas?

▼ Figure 4.15

Normal and skewed distributions

a Normal distribution

b Positive skew

c Negative skew

Cirque number	1	2	3	4	5	6	7	8	9	10	11	12	13	14	15
Cirque orientation (degrees)															
Glyders	30	60	45	55	75	50	80	50	10	15	10	35	45	50	85
Arran	05	05	10	55	15	30	95	05	185	70	120	40	30	115	110

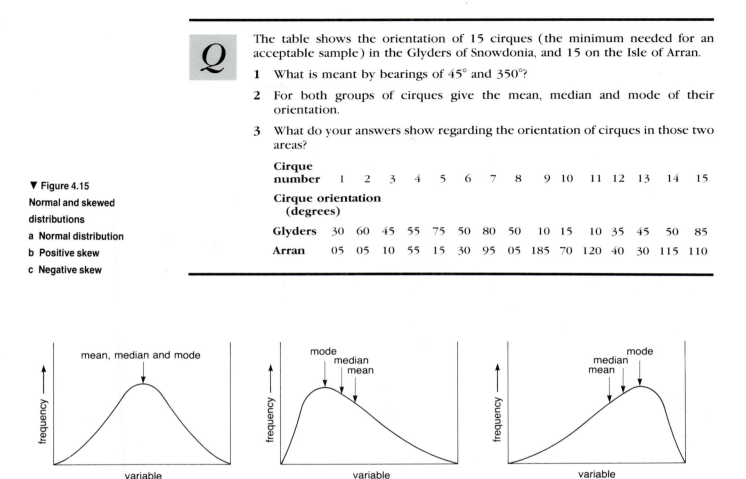

normal distribution

positive skew distribution

negative skew distribution

Lip orientation is the direction of an imaginary line from the centre of the back wall of the cirque to the lip

Arêtes and pyramidal peaks

When two adjacent cirques erode backwards or sideways towards each other, the previously rounded landscape is transformed into a narrow, rocky, steep-sided ridge called an **arête**, e.g. Grib Goch on Snowdon (land-sketch in Figure 4.16) and Striding Edge on Helvellyn in the Lake District (Figure 4.17a). If three or more cirques develop on all sides of a mountain a pyramidal peak, or horn, may be formed. This feature has steep sides and several arêtes radiating from its peak, e.g. Matterhorn, Figure 4.17b.

Glacial troughs, rock steps, truncated spurs and hanging valleys

These features are inter-related in their formation. Valley glaciers straighten, widen and deepen pre-glacial valleys turning the original 'V'-shape river-formed feature into the characteristic 'U'-shape typical of glacial erosion (e.g. Nant Francon Figure 4.17c). These steep-sided, flat-floored valleys are known as **glacial troughs**. The over-deepening of the valleys is credited to the movement of ice which, aided by large volumes of meltwater and moraine has a greater erosive power than that of rivers. Extending and compressing flow may over-deepen parts of the trough floor, which later may be occupied by long, narrow **ribbon lakes**, such as Windermere in the Lake District, or leave less eroded, more resistant **rock steps**.

Theories to explain pronounced over-deepening of valley floors are debated amongst glaciologists and geomorphologists. Suggested causes include: extra erosion following the confluence of two glaciers; the presence of weaker rocks; an area of rock deeply weathered in pre-glacial times; or a zone of well-jointed rock. Should the deepening of the trough continue below the former sea level, then during deglaciation and concurrent rises in sea level, the valley may become submerged to form a **fjord** (page 132).

Abrasion by englacial moraine and plucking along the valley sides removes the tips of pre-glacial interlocking spurs leaving cliff-like **truncated spurs** (Figures 4.16, 4.17c).

Snowdon
pyramidal peak

Crib y Ddysgl

Crib Goch
arête

Y Lliwedd
arête

Glaslyn
corrie

hanging valley

Llyn Llydaw
corrie

truncated spur

Cwm Dyli
hanging valley

Llyn Gwynant

Nant Gwynant
glacial trough

ribbon lake

▲ **Figure 4.16**
Landsketch of features in a glaciated upland, Snowdonia

▼ **Figure 4.17**
Features of glaciated upland areas

a, *below left*, An arête, Striding Edge, Cumbria, English Lake District

b *below* A pyramidal peak, the Matterhorn, Switzerland

▲ Figure 4.17, *continued*

c Glacial trough, Nant Francon, Snowdonia

d *above right* Hanging valley, Glen Rosa, Arran

Hanging valleys are the result of differential erosion between glaciers in the main and tributary valleys. The floor of the tributary is deepened at a slower rate so that when the glaciers melt it is left hanging high above the main valley and its river has to descend by a single or a series of waterfalls, e.g. Cym Dyli, Snowdonia (Figure 4.16), Glen Rosa, Arran (Figure 4.17d).

Striations, roche moutonnées and crag and tail

These are smaller erosion features, which nonetheless help to indicate the direction of ice movement. As a glacier moves across areas of exposed rock, larger fragments of angular moraine embedded in the ice tend to leave a series of parallel scratches and grooves or **striations** (e.g. Central Park in New York). A **roche moutonnée** is a mass of more resistant rock. It has a smooth, rounded upvalley or stoss slope, facing the direction of ice flow, formed by abrasion, and a steep, jagged downvalley or lee slope resulting from plucking (Figures 4.18, 4.19, overleaf).

A **crag and tail** consists of a larger mass of resistant rock (e.g. the basaltic crag upon which Edinburgh Castle has been built) which protected the lee side rocks from erosion (e.g. the tail down which the sloping Royal Mile extends).

▶ Figure 4.18

A roche moutonnée, Nant Francon Valley, Snowdonia

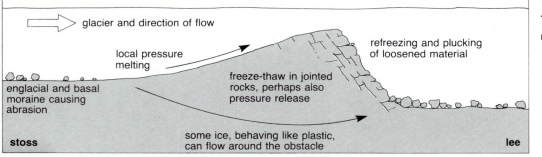

Q The left-hand column in the table is a list of features produced by glacial erosion. The second column describes these features and the third summarises their formation. At present neither the descriptive nor the formation points match the features. Re-write the list using the appropriate descriptions and formation processes.

Feature	Description	Formation
Cirque	pointed peak with radiating arêtes	sub- or englacial moraine dragged along exposed rock
Arête	tributary glacier left high above the main valley	extending and compressing flow overdeepens parts of valley floor
Pyramidal peak	stepped long profile in a glacial trough (valley)	resistant rock remains after ice abrasion on the ice-direction-facing side and plucking on the lee
Glacial trough	ice-smoothened rocks with a steeper side facing down-valley	frost shattering, abrasion, plucking and rotational ice movement
Hanging valley	steep-sided U-shaped valley	three or more cirques cut backwards
Truncated spur	small, deep, circular lake	two cirques cut back towards each other
Rock steps	an amphitheatre-shaped depression in mountain side with steep back wall and a rock lip	overdeepening by abrasion by moraine moving in a rotational movement. Fills with water after deglaciation
Ribbon lake	steep, cliff-like valley sides	formed by extending and compressing flow in main valley or where two glaciers met
Cirque lake	narrow, knife-edged ridge	widened and deepened by valley glacier
Roche moutonnée	rocks scarred with thin parallel scratches	valley glaciers have removed the ends of interlocking spurs by abrasion
Striations	long, narrow lake in a glacial trough	ice in main valley eroded more rapidly than ice in the tributary valleys often producing a waterfall

Transportation by ice

Glaciers are capable of moving large quantities of moraine. This rock debris may be carried in three ways.

Supraglacial debris is carried on the surface of the glacier as lateral and medial moraines. In summer, surface meltwater streams carry only a small load and this often disappears down crevasses.

Englacial debris is moraine carried within the body of the glacier.

Subglacial debris is moved along the floor of the valley either by the ice as ground moraine or by meltwater streams formed by pressure melting.

Glacial deposition

The collective name for all of the boulders, gravel, sand and clay deposited under glacial conditions is **drift** (Figure 4.20). These deposits may be sub-divided into **till**, which includes all material deposited directly by the ice and **fluvioglacial** material which is the debris deposited by meltwater streams. This includes material which may have initially been deposited by the ice and later picked up and redeposited by water.

Till consists of unsorted material while fluvioglacial deposits have been sorted (Figure 3.28). Deposition occurs in upland

valleys as well as across lowland areas. A study of drift deposits helps to explain:
- the nature and extent of the ice advance,
- the frequency of ice advances,
- the source and direction of ice movement,
- post-glacial chronology (including climatic changes, page 249).

Till deposits

Although the term **till** is often applied today to all materials deposited by ice, it is more accurately used to mean an unsorted mixture of rocks, clays and sands. This material was largely transported as ground moraine and deposited as the glacier retreated. In Britain it has commonly been called **boulder clay** but since some deposits may contain neither boulders nor clay this term is now less fashionable. Individual stones are sub-angular, i.e. they are not rounded like river or beach material but neither do they possess the sharp edges of rocks which have recently been broken up by frost shattering. The composition of till reflects the character of the rocks over which it has passed, e.g. East Anglia is covered by chalky till because the ice passed over a chalk escarpment.

Technically, there are two types of till. **Lodgement till**, synonymous with ground moraine, is material smeared on to the valley floor when its weight becomes too great to be moved by the glacier. **Ablation till**, a combination of englacial and supraglacial moraine, is released as a stationary glacier begins to melt and material is dropped *in situ*.

Till fabric analysis is a fieldwork technique used to determine the direction and source of the deposits. Stones and pebbles carried by a glacier tend to become aligned. Their long axes move parallel to the direction of ice flow as this offers least resistance to the ice. For example, a small sample of 50 stones was taken from a moraine in Glen Rosa, Arran. As each stone was removed, its orientation was carefully measured using a compass and its angle of dip from the horizontal was measured with a clinometer.

Three conclusions could then be drawn from the results.

1 The pebbles were grouped into classes of 20° and plotted on to a star diagram (Figure 4.21). The classes were plotted as respective radii from the midpoint and then the ends of the radii were joined up to form a star-like polygonal graph. As each stone has two orientations which must be opposites (e.g. 10° and 190°) then the graph will be symmetrical. The results showed that the ice must have come from the NNW or the SSE.

▶ Figure 4.20
Landforms associated with glacial deposition

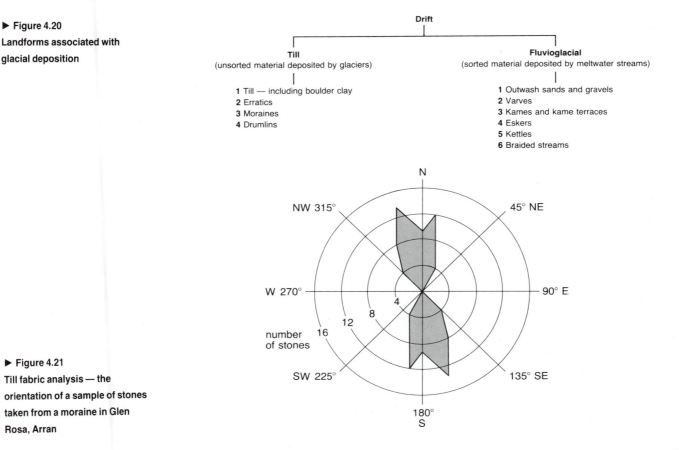

▶ Figure 4.21
Till fabric analysis — the orientation of a sample of stones taken from a moraine in Glen Rosa, Arran

2 Stones will normally be deposited horizontally but if there has been a subsequent advance of the ice then they may have been pushed upwards. The average dip of stones taken from the moraine was 25°, indicating a re-advance.

3 Although most of the pebbles taken in the sample were composed of local rock, some were of material not found on the island. This suggests that some of the ice must have come from the Scottish mainland.

Landforms characteristic of glacial deposition

Erratics

These are boulders picked up and carried by ice, often for many kilometres, and deposited in areas of completely different lithology (page 158) (Figure 4.23a). By determining where the boulders originally came from it is possible to track ice movements. For example, volcanic material from Ailsa Craig in the Firth of Clyde has been found 250 km to the south on the Lancashire Plain, while some deposits on the north Norfolk coast have their origins in southern Norway.

Moraine

This is the material carried and later deposited by the glacier. It is possible to recognise six types of morainic landforms.

■ **Lateral** moraine is debris derived from frost shattering of valley sides and carried along the edges of the glacier (Figure 4.23b). When the glacier melts it leaves an embankment of material along the valley side.

■ **Medial** moraine is found in the centre of a valley and results from the merging of two lateral moraines at the confluence of two glaciers (Figure 4.23b).

■ **Ground** moraine is featureless till deposited over the valley floor.

■ **Terminal** or end moraine is often a high mound of material extending across a valley, or lowland area, at right angles to and marking the maximum advance of the glacier or ice sheet (Figure 4.23c).

■ **Recessional** moraines mark interruptions in the retreat of the ice when the glacier or ice sheet remained stationary long enough for a mound of material to build up. Recessional moraines are usually parallel to the terminal moraine.

■ **Push** moraines may develop if the climate deteriorates sufficiently for the ice temporarily to advance again. Previously deposited moraine may be shunted up into a mound. It can be recognised by individual stones which have been pushed upwards from their original horizontal position.

▼ Figure 4.22

Types of moraine

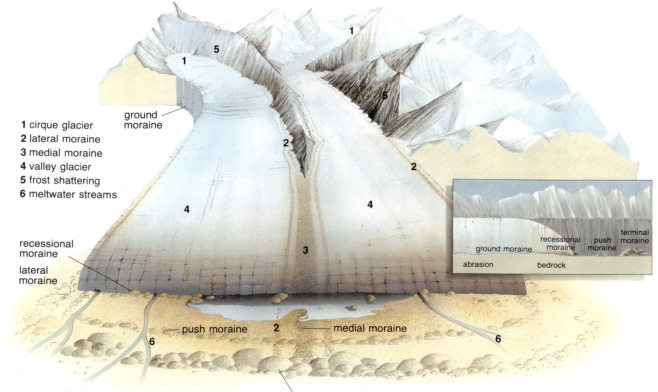

1 cirque glacier
2 lateral moraine
3 medial moraine
4 valley glacier
5 frost shattering
6 meltwater streams

ground moraine

recessional moraine

lateral moraine

push moraine

medial moraine

terminal moraine

recessional moraine
push moraine
terminal moraine
ground moraine
abrasion
bedrock

▲ Figure 4.23

a Erratics near Ingleborough, Yorkshire Dales

b Lateral and medial moraine near Juneau, Alaska where the glacier flows into the sea

c Morainic mounds at the head of Haweswater, Cumbria

d A drumlin in the Eden Valley, Cumbria

Drumlins

These are smooth, elongated mounds of till with their long axis parallel to the direction of ice movement. Drumlins may be over 50 m in height, over 1 km in length and nearly 0.5 km in width. The steep stoss end faces the direction from which the ice came while the lee side has a more gentle, streamlined appearance (Figure 4.23d). The highest point of the feature is near to the stoss end (Figure 4.24). The shape of drumlins can be described by using the **elongation ratio**. This is calculated by dividing the maximum length by the maximum width — drumlins are always longer than they are wide. They are usually found in **swarms** or *en echelon*.

There is much disagreement as to how drumlins are formed. Theories include their being erosion features, or formed by deposition around a central rock, or even by meltwater. However, none of these accounts for the fact that the majority of drumlins are

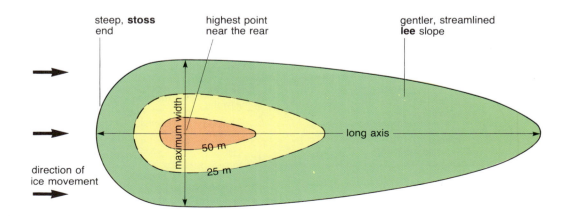

steep, **stoss** end

highest point near the rear

gentler, streamlined **lee** slope

maximum width

50 m

25 m

long axis

direction of ice movement

▶ Figure 4.24

Plan of a drumlin showing typical dimensions

composed of till which, lacking a central core of rock and consisting of unsorted material, would be totally eroded by moving ice. The most widely accepted view is that they were formed when the ice became overloaded with material and so the competence of the glacier was reduced. The reduced competence may have been due to the melting of the glacier or to changes in velocity related to the pattern of extending–compressing flow. Once the material had been deposited it may then have been moulded and streamlined by later ice movement.

Fluvioglacial deposition

Fluvioglacial (or glaciofluvial) landforms are those moulded by glacial meltwater. They tend to be predominantly depositional.

Outwash plains

These are comprised of gravels, sands and, uppermost and furthest from the snout, clays. They are deposited by meltwater streams issuing from the ice during summer or as the glacier melts. The material may originally have been deposited by the glacier and later picked up, sorted and dropped by running water beyond the maximum extent of the ice sheets. In parts of the North German Plain, deposits are up to 75 m deep. Outwash material may also be deposited on top of till following the retreat of the ice (Figure 4.26).

Varves

A varve is a distinct layer of silt lying on top of a layer of sand, deposited annually in lakes found near to glacial margins. The coarser, lighter coloured sand is deposited during late spring when meltwater streams are at their peak discharge and carrying their maximum load. As discharge decreases towards autumn when temperatures begin to drop, the finer, darker coloured silt will settle. Each band of light and dark material represents one year's accumulation (Figure 4.25). By counting the number of varves it is possible to date the origin of the lake

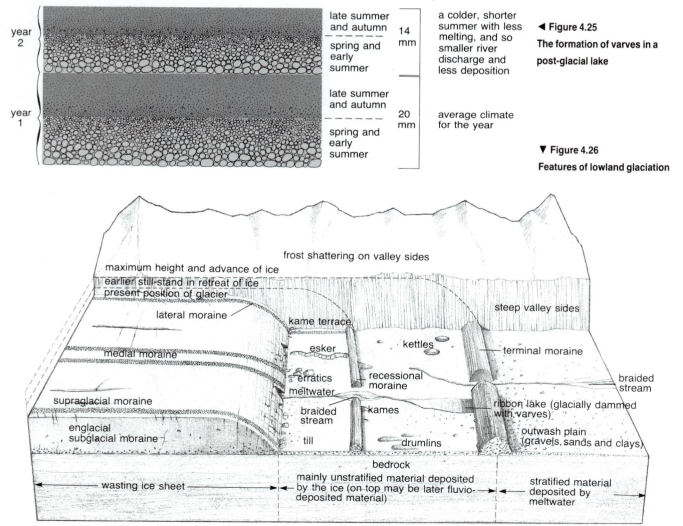

◄ Figure 4.25
The formation of varves in a post-glacial lake

▼ Figure 4.26
Features of lowland glaciation

and variations in the thickness of each varve will indicate warmer and colder periods (e.g. greater melting causing increased deposition).

Kames and kame terraces

Kames are undulating mounds of sand and gravel deposited unevenly by meltwater, similar to a series of deltas, along the front of a stationary or slowly melting ice-sheet. As the ice retreats further the unsupported kame often collapses. Kame terraces are ridges, also of sand and gravel, found along the sides of valleys. They are deposited by meltwater streams flowing between the glacier and the valley wall in a trough. Troughs form in this position because the valley side heats up faster in summer than the glacier ice and so the ice in contact with it melts. They are distinguishable from lateral moraines by their sorted deposits.

Eskers

These are very long, narrow, sinuous ridges composed of sorted coarse sands and gravel. It is thought that eskers are the fossilised courses of subglacial meltwater streams. As the channel is restricted by ice walls the hydrostatic pressure and load carried are both considerable. As the bed of the channel is built up (there is no flood plain) material is left above the surrounding land following the retreat of the ice. Like kames, eskers can be formed only during a period of deglaciation.

Kettles

These form when blocks of ice are left detached from the glacier as it retreats. They may then be partially buried by fluvioglacial deposits left by the meltwater streams. When these ice blocks melt, they leave enclosed depressions which often fill with water to form kettle-hole lakes and 'kame and kettle' topography.

Braided streams

Channels of meltwater rivers may become choked with coarse material resulting from the marked seasonal variations in discharge. Figure 4.26 attempts to show the location and inter-relationships of the various features of glacial deposition.

 The left-hand column in the table below is a list of features formed by glacial deposition, the second column describes these features and the third summarises their formation. At present neither the descriptive nor the formation points match the features. Rewrite the list using the appropriate description and formation processes.

Feature	Description	Formation
Till	mounds of sorted material deposited by meltwater along valley sides or as a delta at an ice front	material deposited by ice when englacial material is too heavy to be carried by a melting glacier
Terminal moraine	small, shallow lake containing stratified material	rocks carried by glacier from their source to an area of different rock
Recessional moraine	sorted deposits of gravel, sand and clay spread over a lowland area	formed by the meeting of two lateral moraines
Lateral moraine	a long, narrow, winding ridge of sorted material	material deposited where meltwater streams are in contact with the ice
Medial moraine	unsorted, angular deposits of rock, sand and clay	meltwater spreads out over low-lying areas, decreases in velocity and deposits its load of gravels, then sands and clays
Push moraine	narrow ridge of unsorted material extending across a valley	moraine deposited during a stillstand in the ice retreat
Drumlins	unsorted material found along sides of glaciers and glaciated valleys	deposited by subglacial streams
Erratics	series of narrow ridges extending across a valley	mainly unsorted subglacial material deposited by melting glaciers
Kettles	mounds of material found in centres of glaciers	mounds of boulders deposited at maximum advance of ice
Esker	rocks which are not native to the area in which they are found	material already deposited on valley floor pushed upwards by a temporary ice re-advance
Kame	small, elongated mounds with a steep end facing upvalley, a streamlined shape and found in swarms	blocks of detached, dead ice left by a melting ice-sheet, later surrounded by fluvioglacial material, melts and forms a small lake
Outwash plain	unsorted material found across a valley with the stones tilted at an angle	material found along the sides of a glacier or a valley as a result of frost shattering

The star graph shows the orientation between 0° and 210° of the long axes of stones obtained from a sample of till. The table beneath it shows some of the remaining orientations.

1   a   Complete the remainder of the table.
     b   Complete the star graph.

2   From which direction is the ice likely to have come? Give a reason for your answer.

3   Describe three other methods that you could use in the field to determine either the direction of flow or the source of the ice.

4   In Chapters 2, 3 and 4 mention has been made of solifluction material, a river terrace, scree, a lateral moraine and a kame terrace — all of which may be found along the foot of a valley side. How would you tackle the problem, in the field, of trying to identify the differences between these five named features?

5   Describe appropriate outline fieldwork methods which might be useful in answering the following questions.
     a   Describe the landforms found near to the snout of a glacier.
     b   Describe and account for the main characteristics of the depositional landforms which result from the downwasting and retreat of an ice-sheet. Consider the nature of the deposits as well as their surface forms.

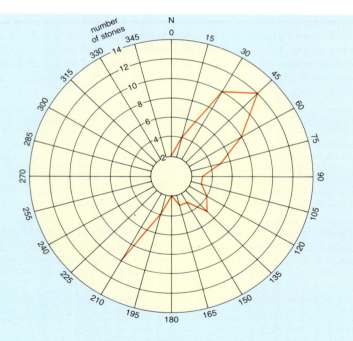

The orientation of the long axes of stones in the till sample

◀ Figure 4.27
A rose diagram — a technique used to show orientation of stones in a sample of till

Direction (°)	Number of stones
225	
240	
255	
270	3
285	3
300	
315	
330	
345	

CASE STUDY

Using fieldwork to answer an 'A' level question
Glacial landforms found near the snout of a glacier: the Isle of Arran

Figure 4.20 listed the types of features formed by glacial deposition, sub-dividing them into those composed of unsorted material left by the glacier itself and those which had been sorted, indicating fluvioglacial action. If the snout of a glacier had remained stationary for some time, indicating a balance between accumulation and ablation, and had then slowly retreated, it might be expected that several of these landforms would be visible in that area following deglaciation. One such area studied by a sixth form was the Lower Glen Rosa Valley on the Isle of Arran.

The sketch map (Figure 4.28) shows a 1 km stretch of the valley, and locates several stopping points. The dominant feature was a mound (labelled *A*), 14 m high, into which the Rosa Water had cut to give a fine exposure of the deposited material. As the mound was a long, narrow ridge-like feature extending across the valley it was suggested that it might be either a terminal or a recessional moraine. It was concluded that the feature was ice-deposited because the material was unsorted. Many of the largest boulders were high up in the exposure and most stones were sub-angular (not more rounded as might be expected in fluvioglacial deposits).

However, an observation downstream at point *B* revealed that material there was also unsorted and this, together with some large granite erratics seen earlier, nearer the coast, seemed to indicate that the mound must be a recessional rather than a terminal moraine as it did not mark the maximum advance of the ice. When a till fabric analysis was carried out it was noted that the average dip of the stones was about 25°, suggesting that in fact the feature was more likely to be a push moraine resulting from a minor re-advance during deglaciation. The orientation of 50 sample stones showed that the ice must have come either from the NNW (probable as this was the highland) or the SSE (unlikely as the lower ground would not be the source of a glacier). An examination of the geology of the stones showed that 80 per cent were granite, and therefore erratics carried from the Upper Rosa Valley, 15 per cent were schists (the local rock) and 5 per cent were other igneous rocks not found on the island. It was inferred from the presence of these other rocks that some of the ice must have originated on the Scottish mainland. Also at point *B*, an investigation of river banks showed a mass of sand and gravel with some level of sorting as might be expected in an outwash area.

▼ **Figure 4.28**

Sketch to show features of glacial deposition in the Lower Glen Rosa Valley, Arran

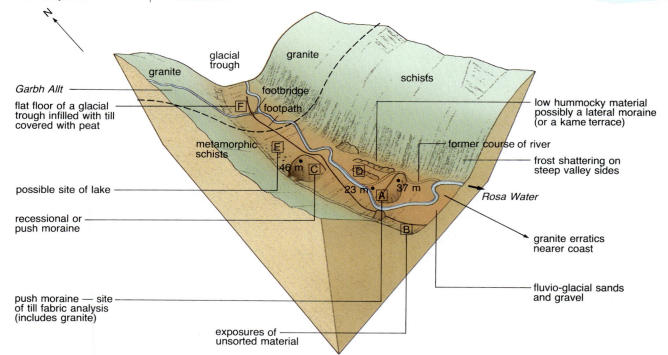

Looking upstream from *A* was a second mound filling much of the valley floor (labelled *C* on main sketch map). Once a detailed field sketch had been drawn (Figure 4.29) suggestions as to the nature of this feature included its being a drumlin, a lateral, a medial, a recessional or even another push moraine. On reaching the feature its length and width were measured. Although the length was slightly greater than the width (an elongation ratio of 1.25 : 1) and the highest point was nearest the upvalley end, it had neither the streamlined shape nor a sufficiently high elongation ratio to be a drumlin (and there were no signs of a swarm!). It appeared to be too far from the valley side to be a lateral moraine and as two glaciers could not have met here, neither could it have been a medial moraine. In the absence of exposures to determine the angle of the stones it was concluded that the mound was likely to be either a recessional or another push moraine. This moraine could have been formed during a stillstand in the glacier's retreat, possibly as the climate became cooler for a time, or it may have been due to the glacier losing momentum and therefore dropping material, having negotiated a bend in the glacial trough.

Across the river (site *D*) was an area of low hummocky material winding along the foot of the valley side as far as the push moraine. As the river was too difficult to cross, there were no obvious exposures, and with insufficient time and equipment to make one, it could only be speculated that the feature may have been formed in one of three ways: meltwater depositing sands and gravel between the valley side and the former glacier as a kame terrace; a lateral moraine from frost shattering on the valley sides; or solifluction deposits formed as the climate grew milder and the glacier retreated. (The feature was not flat enough for a river terrace to be seriously considered).

Upstream, the valley floor was extremely flat (labelled *E*). This could have been the remains of a former glacial lake, formed when the meltwater from the retreating glacier had been trapped behind the recessional moraine at *C* and before it had had time to cut through the deposits. With a soil auger only able to penetrate one metre, it was impossible to gain a profile to prove or disprove the existence of a lake. Had there been a lake, the profile might have shown a chronological sequence since the last glacial as depicted in Figure 4.30.

After crossing the Garbh Allt (a hanging valley), the steep sided, flat floored 'U'-shape of the glacial trough through which the Rosa Water flows was visible. The flatness of the floor is probably due to the deposition of ground moraine although the till has since been covered by peat — a symptom of the cold, wet conditions in the valley.

Although not every feature of glacial deposition was present and there was no evidence of such features as eskers or kettles, this small area did exhibit several landforms and the nature of the deposits expected at, or near to, the snout of a former glacier.

◄ **Figure 4.29,** *below left*
Fieldsketch of landform at C in Figure 4.28, Glen Rosa Valley, Arran

▼ **Figure 4.30**
Idealised profile at point E in Figure 4.28, Glen Rosa Valley, showing layers of deposited material

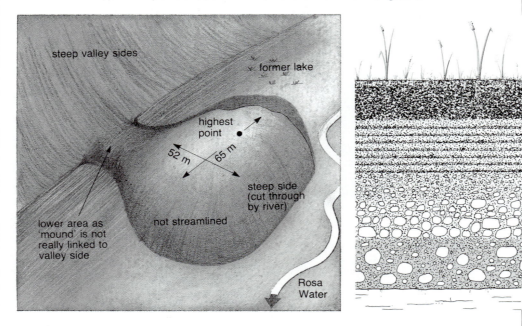

Juncus (reeds) cotton grass heather

thick layer of peat formed under cold, wet conditions

varves and silt deposited on bed of post glacial lake

stratified sands and gravels deposited by meltwater stream from the retreating glacier

unsorted till — ground moraine, deposited by glacier

bedrock (schists)

Other effects of glaciation

Proglacial lakes and drainage diversion

Where ice-sheets expand and dam rivers, **proglacial** lakes are created, e.g. Lakes Lapworth and Harrison (Figure 4.31). Before the ice age the River Severn flowed northwards into the River Dee, but this route became blocked during the Pleistocene by the Irish Sea ice. A large lake, Lapworth, was impounded against the edge of the ice until the waters rose high enough to breach the lowest point in the southern watershed. As the water overflowed through an overspill channel there was rapid vertical erosion which formed what is now Ironbridge Gorge.

When the ice had completely melted, the level of this new route was lower than the original course (which was also blocked by drift), forcing the present day River Severn to flow southwards.

Other rivers, e.g. the Thames, the Warwickshire Avon and the Yorkshire Derwent, have all been diverted as a consequence of glacial activity. Sometimes the glacial overspill channels have been abandoned e.g. at Fenny Compton the Warwickshire Avon temporarily flowed southeast into the Thames (O¹ in Figure 4.31). Proglacial lakes are also found behind eskers and recessional moraines.

▼ Figure 4.31
Glacial diversion of drainage and proglacial lakes in England and Wales

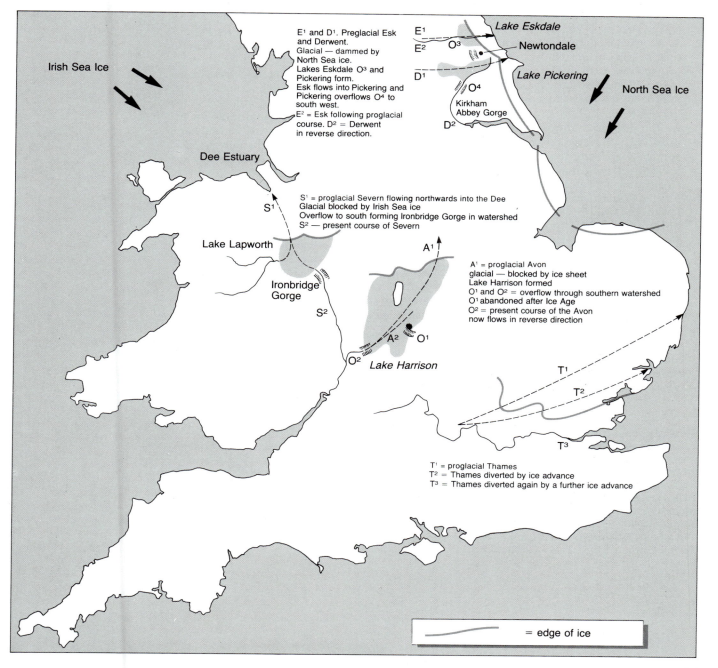

Changes in sea level

The expansion and contraction of ice sheets affected sea level in two different ways. **Eustatic** change refers to a worldwide fall or rise in sea level due to changes in the hydrological cycle. **Isostatic** adjustment is a local change in sea level resulting from differences in the weight imposed upon the land by a growing or declining ice-sheet.

The sequence of events caused by eustatic and isostatic changes during and after the last glacial can be summarised as follows.

1 At the beginning of the glacial, water in the hydrological cycle was stored as ice on the land instead of returning to the sea. There was a universal (eustatic) fall in sea level giving a negative change in base level (page 70).

2 As the glacial continued towards its peak, the weight of ice increased and depressed the earth's crust beneath it. This led to a local (isostatic) rise in sea level relative to the land and a positive change in base level.

3 As the ice-sheets began to melt, large quantities of water, previously held in storage, were returned to the sea causing a worldwide (eustatic) rise in sea level (a positive change in base level). This formed fjords, rias and drowned estuaries (page 132).

4 Finally, and still continuing in several places today, there was a local (isostatic) uplift of the land as the weight of the ice-sheets decreased (a negative change in base level). This change created raised beaches (Figure 6.39) and caused rejuvenation of rivers (page 70).

Looking into the future:

■ If the ice-sheets all continue to melt because of the greenhouse effect or a milder climate, sea levels could rise by another 60–100 m.

■ If isostatic uplift continues in Britain it will increase the tilt which has already resulted in NW Scotland rising by an estimated 10 m in the last 9000 years and SE England sinking — tides in London are over 4 m higher than in Roman times, hence the need for the Thames Barrier.

Human activity and glaciation

Agriculture Glacial till usually improves the fertility of the soil although this depends on the type of rock over which the ice-sheet has passed. In comparison, outwash sands and gravels have limited agricultural value, although they may be used for afforestation. Soils tend to be thin on glacially oversteepened slopes and deeper on valley floors. Some clay deposits may be prone to waterlogging.

Communications Where valleys have been straightened and have flat floors, communications are facilitated. Abandoned spillways may provide gaps through hills while on the coast fjords often provide natural harbours. It may, however, be harder to pass between valleys where their sides have been steepened.

Energy supplies Waterfalls which flow from hanging valleys may be harnessed to provide hydro-electricity, while various lakes provide natural reservoirs to maintain water supplies throughout the year.

Tourism and leisure In addition to winter ski-ing, the jagged mountain landscape attracts climbers and sightseers in summer and glacial lakes provide opportunities for water sports.

References

Fieldwork in Geography, Janet Boyce and Jane Ferretti, Cambridge University Press, 1984.

Glacial Deposits; Geography Today, A Search for Order, BBC Television.

Glaciers, BBC Television, Open University.

Modern Concepts in Geomorphology, Patrick McCullagh, Oxford University Press, 1978.

Planet Earth: Glaciers, Time–Life Books.

Planet Earth: Ice Ages, Time–Life Books.

Process and Landform, A. Clowes and P. Comfort, Oliver & Boyd, 1982.

Process and Pattern in Physical Geography, Keith Hilton, UTP, 1979.

The Nature of the Environment, Andrew Goudie, Blackwell, 1984.

Valley Glaciers; Geography Today, A Search for Order, BBC Television.

CHAPTER 5

Periglaciation

'Computer simulation models are now being used to predict ... soil instability due to frost where buildings, roads, airfields, and other utilities are to be built in regions of frost climate.'

The Unquiet Landscape, D. Brunsden and J. Doornkamp, 1977

The term **periglacial**, strictly speaking, means 'near to or at the fringe of an ice-sheet' where frost and snow have a major impact upon the landscape. However, the term is often more widely used to include any area which has a cold climate, e.g mountains in temperate latitudes, or which has experienced severe frost action in the past, e.g. southern England during the Quaternary ice age, Figure 4.3. Today the most extensive periglacial areas lie in the Arctic regions of Canada, the USA and the USSR. These areas exhibit characteristic landforms and tundra climate, soils and vegetation.

▼ Figure 5.1

Permafrost zones of the Arctic

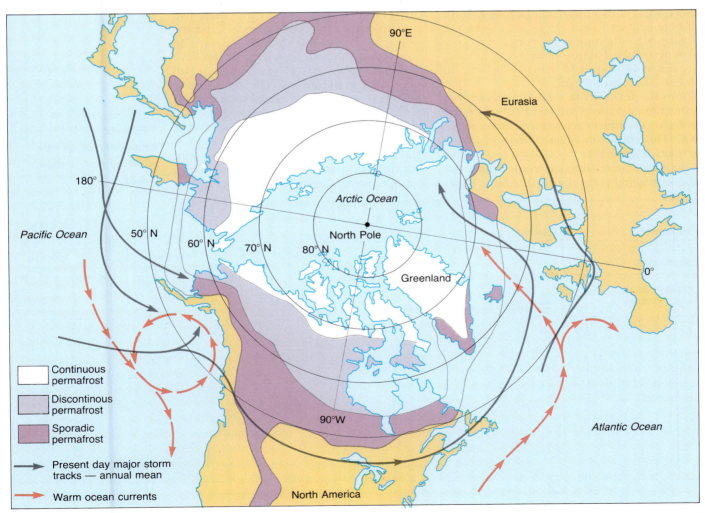

Continuous permafrost

Discontinous permafrost

Sporadic permafrost

Present day major storm tracks — annual mean

Warm ocean currents

Permafrost

Permafrost is permanently frozen ground. It occurs where soil temperatures remain below 0°C for at least two consecutive summers. Permafrost covers almost 25 per cent of the earth's land surface although its extent changes over a period of time. The depth and continuity of permafrost varies (Figures 5.1 and 5.2).

Continuous permafrost is found mainly within the Arctic Circle where the mean annual temperature is below −5°C. Here winter temperatures may fall to −50°C and summers are too cold and short to allow anything but a superficial melting of the ground. The permafrost has been estimated to reach a depth of 700 m in Northern Canada and 1500 m in Siberia. As Figure 5.1 shows, continuous permafrost extends further south in the centre of continents than in coastal areas which are subject to the warming influence of the sea, e.g. the North Atlantic Drift in NW Europe.

Discontinuous permafrost lies further south in the northern hemisphere, reaching 50°N in central Russia, and corresponds to those areas with a mean annual temperature of between −1°C and −5°C. As is shown in Figure 5.2, discontinuous permafrost consists of islands of permanently frozen ground, separated by less cold areas lying near to rivers, lakes and the sea.

Sporadic permafrost is found where mean annual temperatures are just below freezing point and summers are a few degrees above 0°C. This results in isolated areas of frozen ground (Figure 5.2).

In areas where summer temperatures rise above freezing point the surface layer thaws to form the **active layer**. This zone, which under some local conditions can become very mobile for a few months before freezing again, can vary in depth from a few centimetres (where peat or vegetation cover protects the ground from insolation) to 5 m. The active layer is often saturated because meltwater cannot infiltrate downwards through the impermeable permafrost (Figure 12.52). Meltwater is unlikely to be evaporated in the low summer temperatures or to drain downhill since most of the slopes are very gentle. The result is that permafrost regions provide some of the few wetland environments still left in the world.

The unfrozen layer beneath, or indeed any unfrozen material within, the permafrost is known as **talik**. The lower limit of the permafrost is determined by geothermal heat which causes temperatures to rise (Figure 5.3).

Periglacial processes and landforms

Most periglacial regions are sparsely populated and underdeveloped and until the search for oil and gas in the 1960s there had been little need to study or understand the geographical processes which operate in those areas. Although significant strides have been made in the last 30 years, there is still uncertainty as to how specific features have developed and whether indeed such features are still being formed today or are a legacy of a previous, even colder climate, i.e. a fossil or relict landscape. The table on page 104 is a classification of the range of processes with the resultant landforms and terminology found in modern texts.

▼ Figure 5.2
Transect through part of the permafrost zone in northern Canada

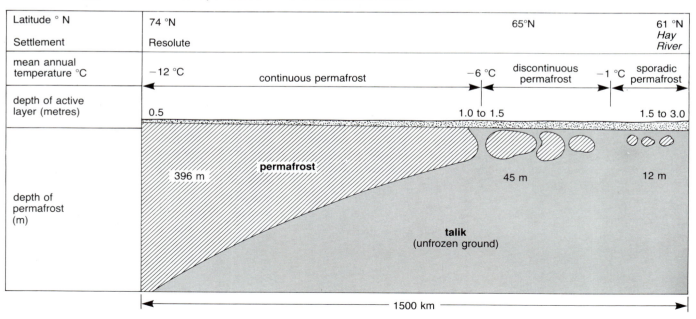

Latitude ° N	74 °N		65°N	61 °N
Settlement	Resolute			*Hay River*
mean annual temperature °C	−12 °C	continuous permafrost	−6 °C discontinuous permafrost	−1 °C sporadic permafrost
depth of active layer (metres)	0.5		1.0 to 1.5	1.5 to 3.0
depth of permafrost (m)	396 m **permafrost**		45 m **talik** (unfrozen ground)	12 m

1500 km

▶ Figure 5.3

Soil temperatures in permafrost
at Yakutsk, Siberia

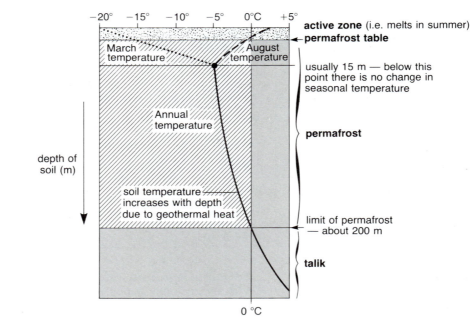

Ground ice

Frost-heave: ice crystals and lenses

Frost-heave includes several processes which cause fine-grained soils such as silts and clays to expand, forming small domes or lifting individual stones to the surface. It results from the direct formation of ice — either as crystals or lenses. The **thermal conductivity** of stones is greater than that of soil. As a result the area under a stone becomes colder than the surrounding soil and ice crystals form. Further expansion by the ice widens the capillaries in the soil allowing more moisture to rise and to freeze. The crystals, or the larger ice lenses which form at a greater depth, force the stones above them upwards until they eventually reach the surface. (Ask a gardener in northern Britain to explain why a plot which was left stoneless in autumn has become stone covered by the spring following a cold winter.)

During periods of thaw, meltwater leaves fine material under the uplifted stone preventing it from falling back into its original position. In areas of repeated freezing (ideally where temperatures fall to between −4°C and −6°C) and thawing, frost heave both lifts and sorts material to form **patterned ground** on the surface (Figure 5.4). The larger stones, with their extra weight, move outwards to form, on almost flat areas, stone circles or, more accurately, stone polygons. Where this process occurs on slopes with a gradient in excess of 6°, the stones will slowly move downhill under gravity to form elongated **stone stripes** or **garlands** (Figure 5.5).

▼ Figure 5.4

Frost-heave and stone sorting
a Doming occurs when the ground freezes in winter and may disappear in summer when the ground thaws — the ground is chilled from above
b Stones roll down into the hollows between mounds and material becomes sorted in size with the finest deposits left in the centre of the polygon and on top of the mound

▶ Figure 5.5, *below right*

Frost-heave — the formation of polygons and stone stripes

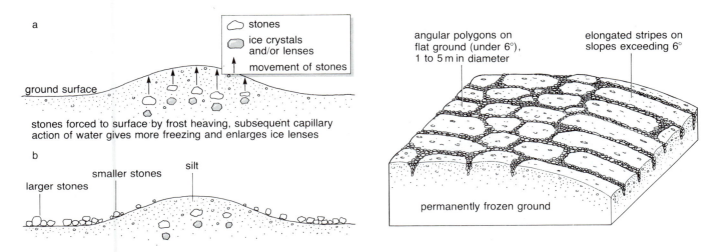

Classification of periglacial processes and landforms

	Processes	Landforms
Ground ice	Ice crystals and lenses (frost-heave)	Sorted stone polygons (stone circles and stripes − patterned ground)
	Ground contraction	Ice wedges with unsorted polygons − patterned ground
Frost weathering	Freezing of ground water	Pingos
	Frost shattering	Blockfields, talus (scree), tors (Chapter 8)
Snow	Nivation	Hollows
Meltwater	Solifluction	Lobes, rock streams
	Streams	Braiding, dry valleys in chalk (Chapter 8)
Wind	Windblown	Loess (limon)

Ground contraction

When the active layer refreezes during severe winter cold, the reduction in the volume of water means that the soil contracts. Cracks open up which are similar in appearance to the irregularly shaped polygons found on the bed of a dried-up lake. These cracks fill with meltwater and probably deposits from both water and wind during the following summer. When this water re-freezes, either the following winter or during cold summer nights, the cracks widen and deepen to form **ice wedges** (Figure 5.6). This process is repeated from year to year until the wedges, which underlie the perimeters of the polygons, grow to as much as 3 m in thickness and 30 m in depth. The diameter of an individual polygon can reach over 30 m (Figure 5.7). **Fossil wedges**, i.e. cracks filled with sands and silt left by meltwater, are a sign of earlier periglacial conditions (e.g. southern England).

Processes of frost-heaving and ground contraction both produce **patterned ground**. However, frost-heaving results in small dome-shaped polygons with larger stones found to the outside of the circles, whereas ice contraction produces larger polygons with the centre of the circles depressed in height and containing the bigger stones.

Freezing of ground water

Pingos are dome-shaped, isolated hills which interrupt the flat tundra plains. They can have a diameter of up to 500 m and may rise 50 m in height to a summit which is often ruptured to expose an icy core. As they occur mainly in sand they are not susceptible to frost-heaving. American geographers believe pingos form in one of two ways (Figure 5.9).

Closed-system pingos form where groundwater is trapped between advancing underlying permafrost and a newly frozen surface. This water eventually freezes, expands and forces its way upwards in the same way as frozen milk lifts the cap off a bottle. In this case the pingo grows from below. These pingos, sometimes known as the **Mackenzie** type, are usually found in areas of thin or discontinuous permafrost.

Open-system pingos result from the downward extension of permafrost into the talik. They may be formed by the freezing of water entrapped within a lake which, together with any sediment on the lake floor, will expand and lift up the surface. These are sometimes referred to as the **East Greenland** type and are more likely to be located where the permafrost is continuous.

▼ Figure 5.6

The formation of ice wedges

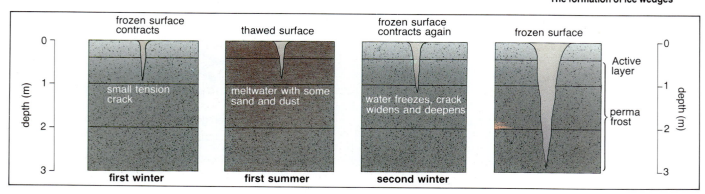

▲ Figure 5.7
Low centre ice wedge polygons — near Barrow, Alaska. This is patterned ground formed by unsorted polygons, each up to 30m in diameter. The boundaries of the polygons mark the positions of the ice wedges. They tend to be slightly raised and any stone lifted to the surface will roll to the centre of the polygon

▲ Figure 5.8, *above right*

A pingo at Prudhoe Bay, Alaska

▶ Figure 5.9

Formation of pingos

a Closed system
(Mackenzie type)

b Open system
(East Greenland type)

c Ruptured pingo

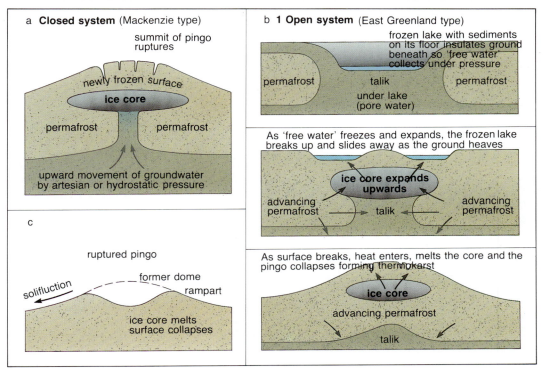

a **Closed system** (Mackenzie type)

summit of pingo ruptures

newly frozen surface

ice core

permafrost   permafrost

upward movement of groundwater by artesian or hydrostatic pressure

c

ruptured pingo

former dome

solifluction   rampart

ice core melts
surface collapses

b **1 Open system** (East Greenland type)

frozen lake with sediments on its floor insulates ground beneath so 'free water' collects under pressure

permafrost   talik   permafrost

under lake
(pore water)

As 'free water' freezes and expands, the frozen lake breaks up and slides away as the ground heaves

ice core expands upwards

advancing
permafrost   talik   advancing
permafrost

As surface breaks, heat enters, melts the core and the pingo collapses forming thermokarst

ice core

advancing permafrost

talik

This blistering of the surface sometimes causes the summit of the hill to crack while subsequent melting may result in the mound collapsing, leaving a meltwater-filled hollow.

Frost weathering

Mechanical weathering is far more significant in periglacial areas than chemical weathering, with frost shattering being the dominant process. On relatively flat upland surfaces, e.g. Scafell Range in the Lake District and the Glyders in Snowdonia, extensive spreads of large angular boulders, formed *in situ* by frost action, are known as **blockfields** or **felsenmeer** (literally a 'rock sea').

Scree, or talus, develops at the foot of steep slopes, especially those composed of well-jointed rocks prone to frost action. Frost shattering may also turn well-jointed rocks, such as granite, into tors (Figure 8.10). There are diverse schools of thought on tor formation: one suggests that these landforms are a result of frost shattering with the weathered debris having been removed by solifluction. If this is the case, tors are therefore a relict of periglacial times.

Snow

The process of nivation (page 85) is assumed to be responsible for forming hollows in hillsides (embryo cirque, Figure 4.13a).

Meltwater

During periods of thaw, the upper zone or active layer melts, becomes saturated and, if on a slope, begins to move downhill under gravity by the process of solifluction (page 36). Solifluction leads to the infilling of valleys and hollows by sands and clays to form **lobes** (Figure 5.10) or, if the source of the flow was a nivation hollow, a rock stream (page 107). Solifluction deposits, either those filling valleys or having flowed over cliffs, e.g. in southern England, are also known as **head** or, in chalky areas, **coombe** (Figure 5.11).

The chalklands of southern England are characterised by numerous dry valleys (Figure 8.13). The most favoured of several theories put forward to explain their origin suggests that the valleys were carved out under periglacial conditions. Any water in the porous chalk would freeze to produce permafrost, leaving the surface impermeable. Meltwater rivers would flow over this frozen ground to form textbook 'V' shaped valleys.

Rivers in periglacial areas have a different régime from that of rivers flowing in warmer climates. Many may stop flowing altogether when winters are sufficiently long and cold (Figure 5.12) and reach a peak in late spring or early summer when melting is at its maximum. With their high velocity, rivers are capable of transporting large amounts of material. Many north-flowing rivers in periglacial regions of North America and the USSR flood frequently as their outlets into the Arctic Ocean remain frozen.

◄ **Figure 5.10**
Solifluction lobes in the Cheviot Hills, Northumberland

▼ **Figure 5.11**
Solifluction lobes and head
a At the foot of a chalk escarpment in SE England
b Solifluction head — sketch of part of a cliff in SW England

a Lobes **b** Head

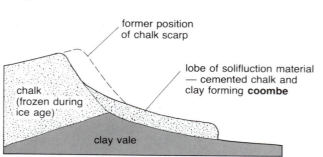
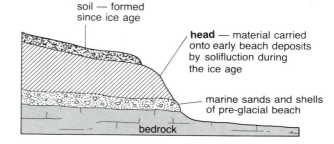

▶ Figure 5.12

Seasonal runoff régime for a river in a periglacial area

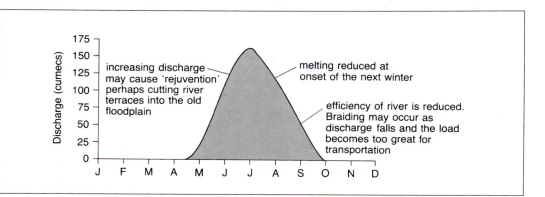

increasing discharge may cause 'rejuvention' perhaps cutting river terraces into the old floodplain

melting reduced at onset of the next winter

efficiency of river is reduced. Braiding may occur as discharge falls and the load becomes too great for transportation

Wind

In glacial environments a lack of vegetation and the plentiful supply of fine, loose material (i.e. silt) found in drift deposits enabled strong, cold, outblowing winds to pick up large amounts of debris and to redeposit it as **loess** in areas far beyond its source. Loess is found covering large areas in a discontinous zone extending from the Mississippi–Missouri Valley in the USA, across France (where it is called **limon**) and the North European Plain into NW China (where in places it exceeds 300 m in depth and forms the yellow soils of the Huang He Valley). In all areas it gives an agriculturally productive, fine-textured, deep, well-drained and easily worked soil which is, however, susceptible to further erosion by wind.

Q

1 What is meant by *periglacial*?

2 Describe the processes which lead to the formation of the landforms shown in Figure 5.13.

KEY

A blockfield
B stone polygons, garlands and stripes
C solifluction lobes/benches
D nivation hollow with snow patch
E rock stream

F debris fan
G braided stream
H ice-wedge polygons
K pingo
L tor

M talus (scree)
N cliffs with head deposits

horizontal scale
0   metres   500

upper limit of bedrock
upper limit of permafrost

▶ Figure 5.13

Landsketch showing typical landforms found in a periglacial area

3 Why do gentle slopes, such as between **X** and **Y** on Figure 5.13, develop under periglacial conditions?

4 Under what conditions might the depth of permafrost alter?

5 Why are many periglacial landforms in Britain said to be 'relict'?

6 In which parts of Britain may periglacial processes still be operating today?

Thermokarst

When ice melts, the pores in the active layer cannot hold the additional water. This results in the development of a group of subsidence landforms known as **thermokarst**. Thermokarst is the general name given to irregular, hummocky terrain with hollows and basins created by the disruption of the thermal equilibrium of the permafrost and an increase in the depth of the active layer. Although thermokarst can develop through climatic change, the major agent in the last few years has been human economic development. The delicate thermal equilibrium may be upset by a number of factors on which human activity has a direct influence.

- Removal of the tundra vegetation for construction purposes means that in summer more heat reaches the soil and so the depth of thaw increases.
- Construction of centrally heated buildings has warmed the ground underneath them causing the buildings to subside.
- Siting of oil, sewerage and water pipes in the active zone has increased the rate of thaw, sometimes causing fracturing of the pipes as the ground moves. Similar earth movements have caused road and railways to lose alignment and dams and bridges to crack.
- Removal of mosses and other tundra plants increases the likelihood of melting and thus of more extensive flooding.
- Drilling for oil and gas poses problems because the heat from the drill melts the permafrost. Having enlarged the drilling hole the machinery vibrates and is less effective.
- Road construction in Britain, especially motorways in southern England, has upset the equilibrium of many slopes produced by solifluction under different climatic conditions.

Human attempts to exploit commercially periglacial regions have been relatively recent. This is one reason why so far insufficient attention has been paid to the conflict between economic gain and environmental loss with the result that the thermal equilibrium has been severely disturbed in some places.

Development may also upset the thermal balance in the opposite direction. The construction of unheated buildings reduces the already low amounts of heat received during the short summer. This causes the upper surface of permafrost to rise and buildings to tilt. Similarly, early road construction increased the permafrost and several Arctic highways now run nearly a metre above the surrounding land. Consequently, large modern unheated buildings and airstrips are now constructed on thick gravel pads to try to maintain the level of the frost table.

Some innovations which try to prevent the extension of thermokarst and to maintain the environment of present day periglacial areas are illustrated in Figure 5.14.

▼ Figure 5.14

Attempts to reduce problems of construction when developing periglacial areas

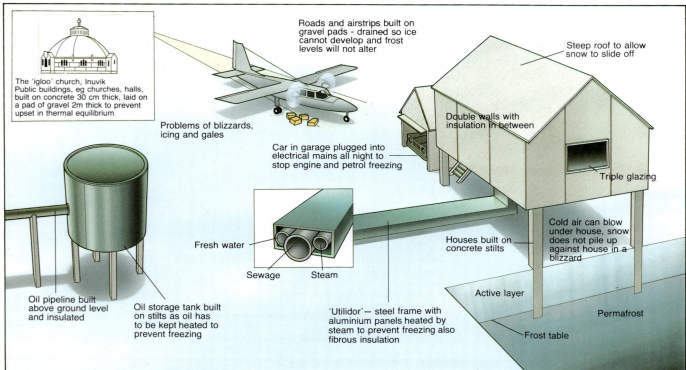

The 'igloo' church, Inuvik
Public buildings, eg churches, halls, built on concrete 30 cm thick, laid on a pad of gravel 2m thick to prevent upset in thermal equilibrium

Roads and airstrips built on gravel pads - drained so ice cannot develop and frost levels will not alter

Steep roof to allow snow to slide off

Double walls with insulation in between

Triple glazing

Problems of blizzards, icing and gales

Car in garage plugged into electrical mains all night to stop engine and petrol freezing

Fresh water

Sewage   Steam

Cold air can blow under house, snow does not pile up against house in a blizzard

Houses built on concrete stilts

Active layer

Permafrost

Oil pipeline built above ground level and insulated

Oil storage tank built on stilts as oil has to be kept heated to prevent freezing

'Utilidor'— steel frame with aluminium panels heated by steam to prevent freezing also fibrous insulation

Frost table

References

Modern Concepts in Geomorphology, Patrick McCullagh, Oxford University Press, 1978.
Planet Earth: Glaciers, Time–Life Books, 1982.
Planet Earth: Ice Ages, Time–Life Books, 1983.

Process and Landform, A. Clowes and P. Comfort, Oliver & Boyd, 1982.
The Nature of the Environment, Andrew Goudie, Blackwell, 1984.

CHAPTER 6

Coasts

'A recent estimate of the coastline of England and Wales is 2750 miles and it is very rare to find the same kind of coastal scenery for more than 10 or 15 miles together.'

The Coastline of England and Wales, J.A. Steers, CUP, 1960

'I do not know what I may appear to the world; but to myself I seem to have been only a boy playing on the sea-shore, and diverting myself in now and then finding a smoother pebble or a prettier shell than ordinary, while the great ocean of truth lay all undiscovered before me.'

Philosophiae Naturalis Principia Mathematica, Isaac Newton, 1687

The coast is a narrow zone where the land and the sea overlap and directly interact. It is affected by terrestrial, atmospheric and marine processes (Figure 6.1) and represents the most varied and rapidly changing of all landforms and ecosystems.

Waves

Waves are created by the transfer of energy from the wind blowing over the surface of the sea. (An exception to this definition is those waves, **tsunamis**, which result from submarine shock waves generated by earthquake or volcanic activity.) As the strength of the wind increases, so too does **frictional drag** and the size of the waves. Waves which result from local winds and travel only short distances are known as **sea** whereas those waves formed by distant storms and travelling large distances are referred to as **swell**. The energy acquired by the waves depends upon three factors: the velocity, the period of time during which the wind has blown, and the length of the fetch. The **fetch** is the maximum distance of open water over which the wind can blow and the greatest fetch potentially creates the highest energy waves. Parts of SW England are exposed to the Atlantic Ocean and when the south-westerly winds blow it is possible that some waves may have originated several thousand kilometres away. Dover, by comparision, has less than 40 km of open water between it and France and consequently receives lower energy waves.

▼ Figure 6.1

Factors affecting coasts — the interface of land and sea

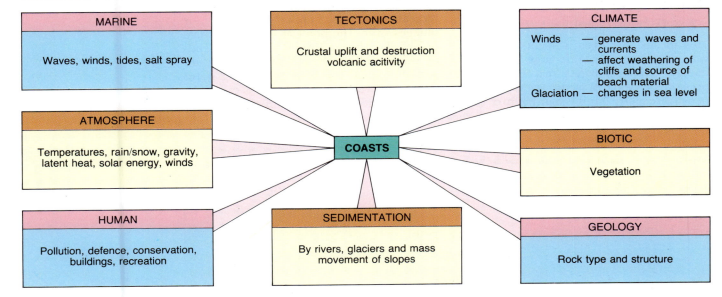

Q Locate Cromer, Lowestoft and Felixstowe on a map which shows East Anglia and the North Sea. Which is the direction of the greatest fetch for each place? Which location is likely to receive the largest waves, and which the smallest? Could the direction of greatest fetch be one factor in determining the size of the three ports?

Wave terminology

The **crest** and the **trough** are respectively the highest and lowest points of a wave (Figure 6.2).

Wave height (*H*) is the distance between the crest and the trough. The height has to be estimated when in deep water. Wave height rarely exceeds 6 m although freak, unexpected waves can be a hazard to life, property and shipping.

Wave period (*T*) is the time taken for a wave to travel through one wave length. This can be timed either by counting the number of crests per minute or by timing eleven waves and dividing by ten, i.e. the number of intervals.

Wave length (*L*) is the distance between two successive crests. It can be determined by either of two formulae:

$$L = 1.56 \, T^2$$
or $$L = CT$$

Wave velocity (*C*) is the speed of movement of a crest in a given period of time.

Wave steepness $\left(\dfrac{H}{L}\right)$ is the ratio of the wave height to the wave length. This ratio cannot exceed 1:7 (0.14) because at that point the wave will break. Steepness determines whether waves will build up or degrade beaches. Most waves have a steepness of between 0.005 and 0.05.

The **energy** (*E*) of a wave in deep water is expressed by the formula:

$$E \propto \text{(is proportional to) } L \, H^2$$

This means that even a slight increase in wave height can generate large increases in energy. It is estimated that the average pressure of a wave in winter is eleven tonnes per square metre but this may be three times greater during a storm — it is little wonder

that under such conditions sea defences may be destroyed and that wave power is a potential source of renewable energy for the future.

Swell is characterised by waves of low height, gentle steepness, long wavelength and a long period. **Sea**, with opposite characteristics, will usually have higher energy waves.

What happens to waves in deep water?
Deep water is when the depth of water is greater than half the wavelength $\left(D = >\dfrac{L}{2}\right)$. The drag of the wind over the sea surface causes water and floating objects to move in an **orbital motion** (Figure 6.3). Waves are surface features (submerged submarines are unaffected by storms) and therefore the size of the orbits decrease rapidly with depth. Any floating object in the sea has a small net horizontal movement but a much larger vertical motion (Figure 6.3).

Waves in shallow water
As waves approach shallow water, i.e. when their depth is less than half the wavelength $\left(D = <\dfrac{L}{2}\right)$, then friction with the sea bed increases. As the base of the wave begins to slow down then the circular oscillation becomes more elliptical (Figure 6.4). As the water depth continues to decrease, so does the wavelength. Meanwhile the height and steepness of the waves increase until the upper part spills or plunges over. The point at which the wave breaks, the **plunge line**, is where the depth of water and the height of the wave are virtually equal. The body of foaming water which then rushes up the beach is called the **swash**, while any water returning down to the sea is the **backwash**.

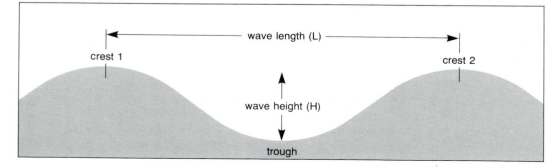

◄ **Figure 6.2**

Wave terminology

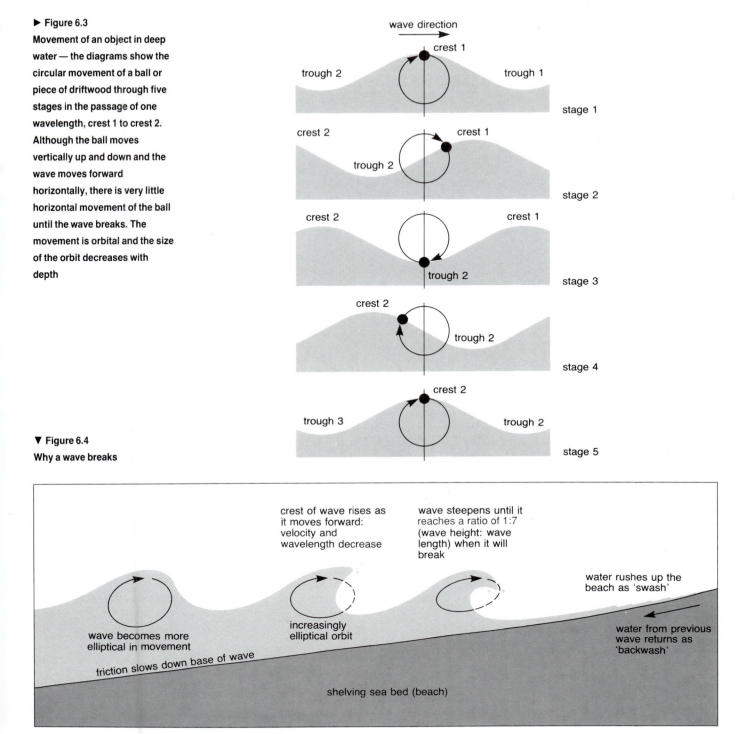

▶ Figure 6.3
Movement of an object in deep water — the diagrams show the circular movement of a ball or piece of driftwood through five stages in the passage of one wavelength, crest 1 to crest 2. Although the ball moves vertically up and down and the wave moves forward horizontally, there is very little horizontal movement of the ball until the wave breaks. The movement is orbital and the size of the orbit decreases with depth

▼ Figure 6.4
Why a wave breaks

Wave refraction

Where waves approach an irregular coastline they are refracted, i.e. they become increasingly parallel to that coastline (Figure 6.5). This is best illustrated where a headland separates two bays. It has already been seen that a wave approaching the shore loses velocity as the depth of water decreases. As the sea bed usually shelves more rapidly off a headland than in a bay, the wave loses velocity more quickly there. The

orthogonals (lines drawn at right angles to wave crests) in Figure 6.5 represent four stages in the advance of a particular wave. It is apparent from the convergence of lines S^1, S^2, S^3, S^4, that the wave energy becomes concentrated upon, and so accentuates erosion at, the headland. The diagram also shows the formation of **longshore currents**, which carry sediment away from the headland.

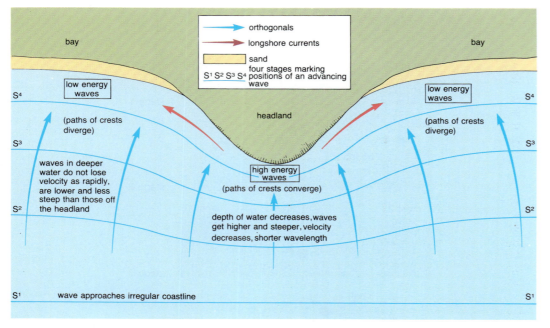

Beaches

Beaches may be divided into three sections — **backshore**, **foreshore** and **nearshore** — based upon the influence of waves (Figure 6.6). A beach forms a buffer zone between the waves and the coast and if it is an effective buffer then it will dissipate wave energy without experiencing any net change itself. Because it is composed of loose material a beach can rapidly adapt its shape to changes in wave energy and it is therefore in dynamic equilibrium with its environment (Framework 1).

Beach profiles fall between two extremes — those which are wide and relatively flat and those which are narrow and steep. The gradient of natural beaches is dependent upon the interrelationship between two main variables.

■ **Wave energy** Field studies have shown a close relationship between the profile of a beach and the action of two types of wave: constructive and destructive. However, the effect of wave steepness on beach profiles is complicated by the second variable.

■ **Particle size** While it is recognised that sand produces a different profile from that of shingle it is not yet fully understood how variations in beach material may modify the gradient between the two extremes mentioned above (i.e. wide and flat, narrow and steep).

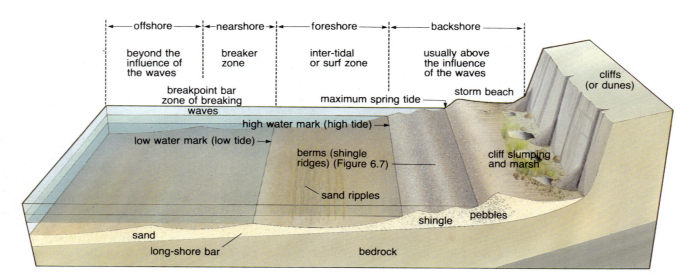

Wave energy

The steepness of waves determines whether they are likely to build up or degrade a beach. There are two extreme forms.

Constructive (surging or spilling) These waves result from swell and, being flat and gentle, will move material up the beach. They are low in height, usually less than a metre; have long wave lengths with up to 100 m between crests; and a long wave period with only six to eight breaking in each minute. Because the waves are gentle they tend to spill over and are therefore low in energy. As constructive waves commonly occur on beaches with a low angle they have a wide area to cross and so the energy in the swash is soon dissipated, leaving a weak backwash. Consequently, sand and shingle are slowly but constantly moved up the beach (Figure 6.7a). This will gradually increase the gradient of the beach and form a **berm** at its crest (Figure 6.6). Waves which approach the shore at an angle other than 90° have lost energy by the time they reach it because they have to overcome friction as they travel over the beach.

Destructive (surfing or plunging) These waves are steeper; have a height greater than 1 m; have a shorter wave length of about 20 m between crests; and have a more frequent wave period with ten to 14 breaking each minute. They are often associated with the increased energy available during storms and with steeply sloping shingle beaches. These steep waves plunge over when breaking and so their energy is concentrated on a small area of beach. Where the beach material does not allow rapid percolation the backwash (undertow) will be at least as strong as the swash. Although some stones may be thrown up above the high tide mark by very large waves, forming a **storm beach** (Figure 6.6), most material is carried downwards by the backwash to form a **longshore (break point) bar** (Figures 6.7b and 6.6). Waves are more likely to be destructive on beaches with a steep gradient and where they aproach the coastline at right angles. However, as material is constantly combed downwards, this profile becomes increasingly more gentle in its lower section.

▼ Figure 6.7

Differences between constructive and destructive waves

a Constructive (spilling) waves

b Destructive (plunging) waves

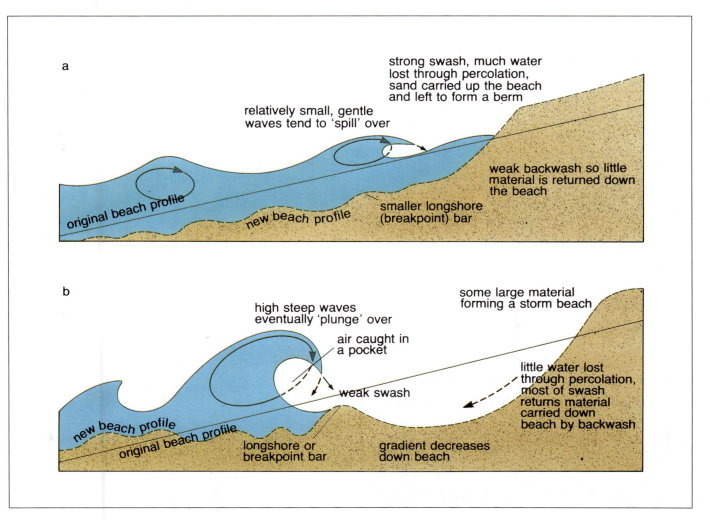

A beach is a system in equilibrium. If the gradient is gentle then constructive waves are dominant. These move material up the beach which will steepen the gradient, shorten the wave length and increase the wave steepness until they are eventually replaced by destructive waves. These in turn will degrade the beach producing a gentler gradient, which will increase the wave length and reduce wave steepness until the destructive waves give way to constructive waves again. This is an example of negative feedback (Framework 1), where the original process or landform is reversed. Britain tends to have more storms in winter and more settled weather in summer, therefore beaches are likely to be degraded in winter and built up again the following summer.

Q Complete the following table.

	Constructive or spilling waves	Destructive or plunging waves
Wave height		
Wave steepness		
Wave length		
Wave period		
Frequency per minute		
High or low energy?		
Beach gradient		
Stronger swash or backwash?		

Particle size

This factor complicates the influence of wave steepness on the morphology of a beach. Many studies have suggested that the explanation of the link between beach material and gradient lies in the rate of percolation which the material allows, i.e. water will pass through coarse-grained shingle more rapidly than through fine-grained sand.

Shingle beaches Shingle may make up the whole or just the upper part of the beach and like sand it will have been sorted by wave action. Usually, the larger the size of the shingle the steeper the gradient of the beach, i.e. the gradient is in direct proportion to shingle size. This is an interesting hypothesis to test by experiment in the field (page 259).

Regardless of whether the waves are constructive or destructive, most of the swash rapidly percolates downwards leaving limited surface backwash. This, together with the loss of energy resulting from friction caused by the uneven surface of the shingle (compare this with the effects of bed roughness of a stream, page 58), means that under normal conditions, very little shingle is moved back down the beach. Indeed, the strong swash will probably transport material up the beach forming a berm at the spring high tide level. Above the berm there is often a storm beach, comprised of even bigger boulders thrown there by the largest of waves, while below may be several smaller ridges, each marking the height of successively lower high tides which follow the maximum spring tide (Figures 6.8 and 6.9).

◄ Figure 6.8

Berms and storm beaches in northeast Anglesey, Wales

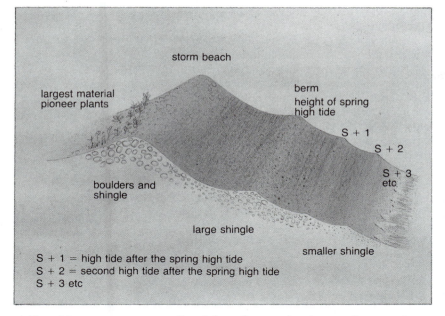

S + 1 = high tide after the spring high tide
S + 2 = second high tide after the spring high tide
S + 3 etc

▲ Figure 6.9
Storm beaches and berms —
berms mark the limits of
successively lower tides

The interrelationship between wave energy, beach material and beach profiles may be summarised by the following generalisations which refer to net movements.

■ Destructive waves carry material down the beach.
■ Constructive waves carry material up the beach.
■ Material is carried upwards on shingle beaches.
■ Material is carried downwards on sandy beaches.

Therefore as destructive waves usually occur on shingle beaches and constructive waves on sandy beaches, the movement of beach material will, under normal conditions, be evened out. The beach system will thus be in long term equilibium.

Tides

The position and range where waves break over the beach is determined by the state of the tide. You have already seen that the levels of high tides vary (berms are formed at the time of spring tides above progressively lower shingle ridges, Figure 6.9). Tides are controlled mainly by the gravitational effect of the moon and partly by the similar effect of the sun, together with the rotation of the earth and, more locally, the geomorphology of sea basins.

The moon has the greatest influence. Although its mass is much smaller than that of the sun, this is more than compensated for by its closer proximity to the earth. The moon attracts or pulls water to the side of the earth nearest to it. This creates a bulge or **high tide** (Figure 6.10a) with a complimentary bulge on the opposite side of the earth. This bulge is compensated for by the intervening areas where water is repelled and which experience a **low tide**. As the moon orbits the earth, the high tides follow it.

Sand beaches Sand usually produces beaches with a gentle gradient. This is because the small particle size allows the sand to become compact when wet which severely restricts the rate of percolation. Percolation is also hindered by the storage of water in pore spaces in sand which enables most of the swash from both constructive and destructive waves to return as backwash. Little energy is lost by friction (sand presents a smoother surface than shingle) so material will be carried down the beach. The material will build up to form a longshore bar at the low tide mark (Figure 6.6). This will cause waves to break further from the shore giving them a wider beach over which their energy is dissipated.

As the tide ebbs, the sand may dry out allowing any onshore winds to carry material up the beach. Should these winds be strong (**dominant**) or frequent (**prevailing**) then sand dunes may develop beyond the backshore zone.

▼ Figure 6.10
Causes of tides
a The gravitational pull of the moon
b Spring tides
c Neap tides

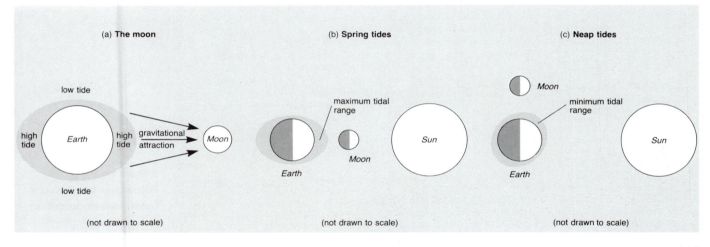

A lunar month — the time it takes the moon to orbit the earth — is 28 days and the tidal cycle (the time between two successive high tides) is 12 hours and 25 minutes, meaning there are two high tides, near enough, per day. The sun, with its smaller gravitational attraction, is the cause of the difference in tidal range rather than of the tides themselves. Once every 14 days (i.e. twice in a lunar month) the moon and sun are in alignment on the same side of the earth (Figure 6.10b). This increase in gravitational attraction generates the **spring tide** which produces the highest high tide, the lowest low tide and the maximum tidal range.

Midway between the spring tides are the **neap tides**. At this time the sun, earth and moon form a right angle with the earth at the apex (Figure 6.10c). As the sun's attraction partly counterbalances that of the moon, this is when the tidal range is at a minimum with the lowest of high tides and the highest of low tides. Spring and neap tides vary by approximately 20 per cent above and below the mean high and low tide levels.

So far we have seen how tides might change on a uniform or totally sea-covered earth. However, the earth is not uniform and often the explanation of local tide patterns is very complex.

The **Coriolis force** (page 186) is the effect of the earth's rotation on moving air and water. In the northern hemisphere this effect is deflection to the right. The rising tide behaves like a wave and as it moves towards Britain, part of the water will travel northwards up the Irish Sea. As it is deflected to the right there will be higher tides on the Welsh and English coasts rather than in Ireland. Later, this wave will travel southwards through the North Sea producing

higher tides on the English coast than on Danish and Dutch coasts (Figure 6.11).

The morphology of the sea bed and coastline also affects tidal range. In the example of the North Sea, as the tidal wave travels south it moves into an area where both the width and depth of the sea area decrease. This results in a rapid accumulation or funnelling of water, giving an increasingly higher tidal range — the range at Dover is about 5 m greater than that in the north of Scotland (Figure 6.11). Estuaries where incoming tides are forced into rapidly narrowing valleys will also have considerable tidal ranges. Examples are: the Severn Estuary, 13 m; the Rance (Brittany), 11.6 m; the Bay of Fundy (Canada), 15 m. It is due to such a range that the Rance has the world's first tidal power station, while the Bay of Fundy (Canada) and the Severn have, respectively, experimental and proposed schemes for electricity generation. Extreme narrowing of estuaries can concentrate the tidal rise so rapidly that an advancing wall of water, or **tidal bore**, will travel upriver. This happens, for example, in the rivers Severn and Amazon.

In contrast, small enclosed seas have only minimal tidal ranges e.g. the Mediterranean, 0.01 m.

Storm surges

These are rapid rises in sea level where water is piled up against a coastline far in excess of normal conditions, and where loss of life and property is often considerable. Two areas particularly prone to storm surges, the southern North Sea and the Bay of Bengal, are described in the case studies opposite.

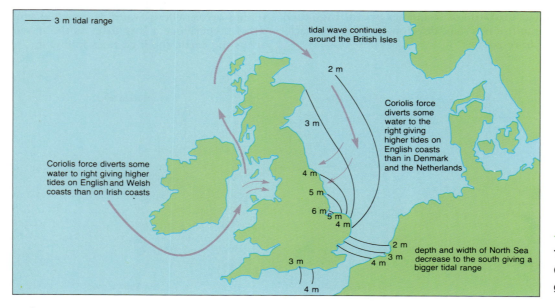

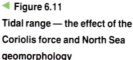

Figure 6.11
Tidal range — the effect of the Coriolis force and North Sea geomorphology

CASE STUDY

North Sea (31 January – 1 February 1953)

A deep depression to the north of Scotland, instead of following the usual track which would have taken it over Scandinavia, turned southwards into the North Sea (Figure 6.12). As air is forced to rise in a depression (page 193) the reduced pressure tends to pull up the surface of the sea area underneath it. If pressure falls by 56 millibars, as it did on this occasion, the level of the sea may rise by up to 0.5 m. The gale force winds, travelling over the maximum fetch, produced storm waves over 6 m high which caused water to pile up in the southern part of the North Sea. This coincided with spring tides and with rivers discharging into the sea at flood levels. The result was a high tide, excluding the extra height of the waves, of over 2 m in Lincolnshire, over 2.5 m in the Thames Estuary and over 3 m in the Netherlands. The immediate result was the drowning of 264 people in SE England and over 1800 in the Netherlands. To prevent such devastation by future surges, the Thames Barrier and the Dutch Delta Scheme have since been constructed.

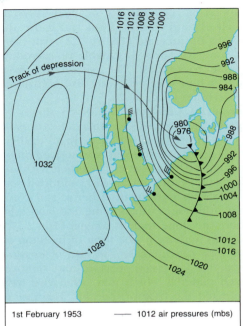

1st February 1953   —— 1012 air pressures (mbs)

Bay of Bengal

The south of Bangladesh includes may flat islands formed by deposition from the rivers Ganges and Brahmaputra. This delta region is ideal for rice growing and is home to an estimated 30 million people. However during the autumn, typhoon winds (tropical storms) funnel water northwards up the Bay of Bengal which becomes increasingly narrower and shallower towards Bangladesh. These waters sometimes build up into a surge which may exceed 4 m in height and which may be capped by waves reaching a further 4 m. A wall of water occasionally sweeps over the defenceless islands. Three days after one such surge in 1985, the Red Cross suggested that over 40 000 people had probably been drowned, many having been washed out to sea as they slept, the only survivors being those who had climbed to the tops of palm trees and managed to cling on despite the 180 km/hr winds. The Red Cross feared outbreaks of typhoid and cholera in the area because fresh water had been contaminated. Famine was a serious threat as the rice harvest had been lost under the salty waters.

A similar surge, a few years earlier, claimed the lives of over 300 000 people in the Indian state of Andhra Pradesh.

▶ Figure 6.12

The North Sea storm surge of 1 February 1953

Which erosion processes operate at the coast?

Wave pounding Steep waves have considerable energy. If these break as they hit the foot of cliffs or sea walls they may generate shock-waves of up to 30 tonnes per m². Some sea walls in parts of eastern England need replacing within 25 years of being built as a result of wave pounding.

Hydraulic pressure When a parcel of air is trapped and compressed, either in a joint in a cliff or between a breaking wave and a cliff, then the resultant increase in pressure may, over a period of time, weaken and break off pieces of rock or damage sea defences.

Abrasion/corrasion This is the wearing away of the cliffs by sand, shingle and boulders (i.e. load) hurled against them by the waves. It is the most effective method of erosion and is most rapid on coasts exposed to storm waves.

Attrition Rocks and boulders already eroded from the cliffs are broken down into smaller and more rounded particles.

Corrosion/solution This includes the dissolving of limestones by the carbonic acid in sea water (compare Figure 2.8); and the evaporation of salts to produce crystals which expand as they form and cause the rock to disintegrate (Figure 2.2). Salt from sea water or spray is capable of corroding several rock-types. The secretions from pioneer blue-green bacteria also contribute to corrosion (page 242).

Sub-aerial Cliffs may be worn down by non-marine processes. Rain may affect erosion of cliffs either by falling directly on to them or as a result of throughflow or surface runoff from inland. This water, together with the effects of weathering by wind and frost, can lead to mass movement either as soil creep on gentle slopes or slumping and landslides where the slope angle is steeper (Fig. 2.17).

Human activity in the form of building on cliff tops can increase the rate of erosion, since this increases pressure and removes beach material which may otherwise have protected the base of the cliff. The rate of erosion may be reduced by the construction of sea defences. Human activity therefore has the effect of disturbing the equilibrium of the coast system.

What affects the rate of erosion?

1 **Breaking point of the wave** A wave which breaks as it hits the foot of a cliff releases most energy and causes maximum erosion. If the wave hits the cliff before it breaks, then much less energy is transmitted whereas a wave breaking further offshore will have had its energy dissipated as it travelled across the beach (Figure 6.13).

2 **Wave steepness** Very steep destructive waves formed locally (sea) have more energy and erosive power than gentle constructive waves formed many kilometres away (swell).

3 **Depth of sea, length and direction of fetch, configuration of coastline** A steeply shelving beach creates higher and steeper waves than one with a gentle gradient. The longer the fetch, the greater the time available for waves to collect energy from the wind. The existence of headlands with vertical cliffs tends to concentrate energy by wave refraction (page 111).

4 **Supply of beach material** Although a continual supply of material is needed to erode the cliffs, a surfeit protects the coast by absorbing wave energy.

5 **Beach width** Cliffs containing a readily available supply of sand (e.g. northern Norfolk) will form wider beaches than those with limited amounts (e.g. Holderness, South Yorkshire). The wider the beach the greater will be the loss of energy (waves take longer to pass over it) and the slower the rate of cliff erosion.

6 **Rock resistance, structure and dip** The strength of coastal rocks influences rates of erosion with unconsolidated volcanic ash offering the least resistance to wave attack, as shown in the table below. In Britain it is coastal areas where glacial till was deposited which are being worn back most rapidly (Case Study, Holderness). When Surtsey first arose out of the sea in 1963, off the southwest coast of Iceland, it was composed of unconsolidated volcanic ash. It was only when it was covered and protected by a lava flow the following year that its survival was guaranteed, at least in the medium term.

▲ Figure 6.13

Waves breaking on Filey Brigg, Yorkshire — wave energy is absorbed by a band of rock and so the cliff behind is protected

Rock type and average rates of erosion		
Rock type	**Location**	**Rate of erosion** (metres/year)
Volcanic ash	Krakatoa	40
Glacial till	Holderness	2
Glacial till	Norfolk	1
Chalk	SE England	0.3
Shale	North Yorks.	0.09
Granite	SW England	0.001

Rocks which are well-jointed or have been subject to faulting have an increased vulnerability to erosion. The steepest cliffs are usually where the rock's structure is horizontal or vertical and the gentlest where the rock dips upwards away from the sea. In the latter case blocks may break off and slide downwards (Figure 2.16). Erosion is also rapid where rocks of different resistance overlie one another, e.g. chalk and Gault clay in Kent.

CASE STUDY

Coastal erosion on the coast of Holderness

This stretch of coastline is retreating by an average of 2 m a year. Since Roman times the sea has encroached by nearly 3 km, and some fifty villages mentioned in the *Domesday Book* of 1086 have disappeared. The storm surge of 1953 removed up to 4 m in one night. The coast is composed of glacial till which presents little resistance to storm waves. These waves can reach 4 m in height when the wind is in the northeast, i.e. the direction of maximum fetch. The local till contains much fine material but limited amounts of sand. The finer material is washed out to sea to leave a narrow beach which is incapable of absorbing the energy from the breaking waves. The destructive storm waves comb material down the beach leaving the cliff face unprotected. Waves attack the base of the cliff making it unstable and together with the weight added to the till by rainwater may cause the cliff to collapse. This cliff-foot rubble is then available for removal by wave energy, backwash and longshore drift.

Erosion landforms

Headlands and bays

These are most likely to be found in areas of alternating resistant and less resistant rock. Initially the less resistant rock experiences most erosion and develops into bays, leaving the more resistant outcrops as headlands. Later, (Figure 6.5) the headlands receive the highest energy waves and are thus more vulnerable to erosion than the sheltered bays. The latter now experience low energy breakers which will cause sand to accumulate and help to protect that part of the coastline.

Wave cut platforms

Wave energy is at its maximum when a high, steep wave breaks at the foot of a cliff. This results in undercutting of the cliff to form a **wave cut notch** (Figure 6.14, overleaf). The continual undercutting causes increased stress and tension in the cliff until it eventually collapses. As these processes are repeated the cliff retreats, leaving a gently sloping **wave cut platform** which has a slope angle less than 4° (Figure 6.15). The platform, which appears relatively even when viewed from a distance, cuts across rocks regardless of their type and structure. A closer inspection of this inter-tidal feature usually reveals that it is deeply dissected by abrasion, resulting from material carried across it by tidal movements, and corrosion. As the cliff continues to retreat then the widening of the platform means that waves break further out to sea and incoming waves have to travel over a wider area. This will dissipate their energy, reduce the rate of erosion of the headland and limit the further extension of the platform. It has been hypothesised that wave cut platforms cannot exceed 0.5 km in width.

Where there has been negative change in sea level, former wave cut platforms remain as **raised beaches** above the present influence of the sea (page 133).

Caves, blowholes, arches and stacks

Where cliffs are of resistant rock, wave action attacks any lines of weakness such as joints and faults. Sometimes the sea cuts inland along a joint to form a narrow, steep-sided inlet called a **geo**, or to undercut part of the cliff to form a **cave**. As shown in Figure 6.16, caves are often enlarged by the combined processes of marine erosion. Erosion may be vertical, to form blowholes, but is more typically backwards through a headland to form arches and stacks (Figures 6.16 and 6.17).

These landforms, which often prove to be attractions to sightseers and mountaineers, can be found at the Needles (Isle of Wight), Old Harry (near Swanage) and Flamborough Head (Yorkshire), which are all cut into chalk, and at the Old Man of Hoy (Orkneys) which is Old Red Sandstone.

Transportation of beach material

Up and down the beach

As we have already seen, constructive waves tend to move sand and shingle up the beach whereas the net effect of destructive waves is to comb the material downwards.

Longshore drift

The angle at which waves approach the land is determined by wind direction, the local configuration of the coastline, and refraction at headlands. Rarely, however, do waves approach a beach at right angles. When a wave breaks the swash carries material up the beach at the same angle at which the wave approached the beach (Figure 6.18).

◄ Figure 6.14
Wave cut notch at Flamborough
Head, Yorkshire

◄ Figure 6.15
Wave cut platform at
Flamborough Head, Yorkshire

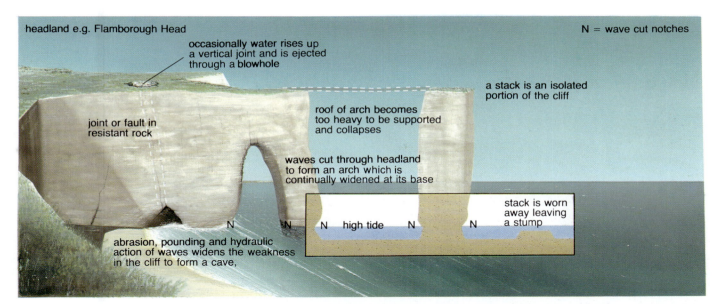

headland e.g. Flamborough Head

N = wave cut notches

occasionally water rises up
a vertical joint and is ejected
through a **blowhole**

a stack is an isolated
portion of the cliff

joint or fault in
resistant rock

roof of arch becomes
too heavy to be supported
and collapses

waves cut through headland
to form an arch which is
continually widened at its base

stack is worn
away leaving
a stump

N   high tide   N   N

abrasion, pounding and hydraulic
action of waves widens the weakness
in the cliff to form a cave,

▲ Figure 6.16
The formation of caves,
blowholes, arches and stacks at
a headland

► Figure 6.17
'Old Harry', near Swanage,
Dorset — this shows a small
wave cut platform (centre), a
cave (left foreground) and an
arch within a large stack ('Old
Harry'). A second stack ('Old
Harry's wife') lies to the right

▼ Figure 6.18
The process of longshore drift

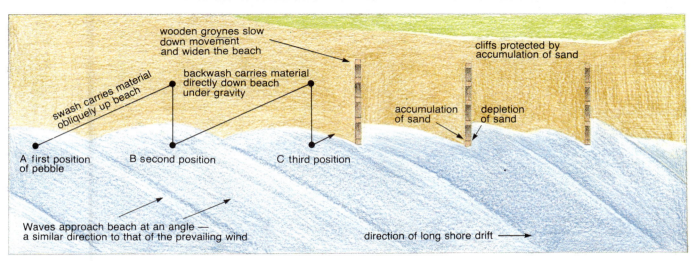

wooden groynes slow
down movement
and widen the beach

cliffs protected by
accumulation of sand

swash carries material
obliquely up beach

backwash carries material
directly down beach
under gravity

accumulation
of sand

depletion
of sand

A first position
of pebble

B second position

C third position

Waves approach beach at an angle —
a similar direction to that of the prevailing wind

direction of long shore drift ➞

◄ Figure 6.19

The effect of groynes on longshore drift, Southwold, Suffolk — this type of coastal management is usually undertaken on beaches at holiday resorts where the sand and sea are a major attraction to tourists

As the swash dies away, the backwash and any material carried by it returns straight down the beach at right angles to the water-line under the influence of gravity. In time, sand and shingle are slowly moved in a zig-zag course along the coast, a process known as **longshore drift**. If this material is carried a considerable distance it becomes smaller, more rounded and better sorted. Longshore drift is usually in the same direction as the prevailing wind, but it can alter if the wind changes its direction. On the south coast of England, where the maximum fetch and pre-vailing wind are both from the southwest, there is a predominantly eastward move-ment of beach material.

Where beach material is being lost through longshore drift then the coastline in that locality is likely to be worn back more quickly because the buffering effect of the beach is lessened. To counteract this pro-cess, wooden breakwaters or **groynes** are built (Figure 6.19). Groynes will encourage the local accumulation of sand (important in tourist resorts) but can result in a depletion of material further along the coast.

Coastal deposition

Deposition occurs where the accumulation of sand and shingle exceeds its depletion. This may take place in sheltered areas with low energy waves or where rapid coastal erosion further along the coast provides an abundant supply of material. In terms of the coastal system, deposition takes place as inputs exceed outputs and so the beach is a store of eroded material.

Beaches

Although beaches are not permanent fea-tures and may change after each high tide, they are zones where material has been deposited. They are likely to be more permanent in sheltered bays or as storm beaches where erosive wave activity is rare or absent.

Spits

Spits are long, narrow accumulations of sand and/or shingle with one end joined to the mainland and the other projecting out to sea or extending part way across a river estuary. Sandy spits are formed by constructive waves produced by swell; shingle spits occur where destructive waves have resulted from sea; composite spits occur when the larger sized shingle is deposited before the finer sands.

In Figure 6.21, the line **X–Y** marks the position of the original coastline. At point **A**, because the prevailing winds and maximum fetch are from the southwest, material is carried eastwards by longshore drift. When the orientation of the old coastline began to change at **B**, some of the larger shingle and pebbles were deposited in the slacker water in the lee of the headland. As the spit continued to grow, storm waves threw some larger material above the high water mark (**C**) making the feature more permanent, while under normal conditions the finer sand was carried towards the end of the spit at **D**. Many spits develop a hooked or curved end. This may be for two reasons: a change in the prevailing wind to coincide with the second most dominant wave direction and second longest fetch, or wave refraction at the end of the spit carrying some material around the end and into more sheltered water.

Eventually the seaward side of the spit will retreat while longshore drift continues to extend the feature eastwards. A series of recurved ends may form (**E**) each time there is a storm from the northeast or a lengthy period with an altered wind direction. Having reached its present day position (**F**), the spit is unlikely to grow any further partly as the faster current of the river will carry material out to sea and partly as the depth of water becomes too great for the spit to build upwards above sea level. Meanwhile, the prevailing winds from the southwest will pick up sand from the beach as it dries out at low tide and blow it inland to form **dunes** (**G**).

The stability of the spit may be increased by the anchoring qualities of marram grass. At the same time, gentle, low energy waves entering the sheltered area behind the spit deposit fine silt and mud, creating an area of **saltmarsh** (**H**).

Figure 6.25 on page 126 locates some of the larger spits around the coast of England and Wales. How do these relate to the direction of the prevailing or dominant winds?

Bars and tombolos
If a spit extends across a bay linking two headlands, straightening the coastline, it is called a **bar** (Figure 6.24). When a spit or bar joins the mainland to an island (Figure 6.23) it is a **tombolo**. Chesil Beach, off the Dorset coast, presents a gently smoothing face 30 km long and up to 14 m high to the prevailing winds in the English Channel and links the Isle of Portland to the mainland (Figure 6.24).

Cuspate forelands
These are more complex depositional features formed when longshore drift from two opposite directions meets to produce a series of ridges at right angles to each other (Case Study, Dungeness).

Barrier beaches and islands
Although relatively uncommon in Britain, these depositional coastal features are the most widespread globally. The most extensive series of barrier beaches extends from

CASE STUDY

Dungeness

Some 2000 years ago, when the sea level in SE England was 2 m lower than it is today, the coastline at Dungeness was 15 km inland. With predominant winds and the fetch of over 5000 km both from the southwest beach material is usually carried by longshore drift from west to east along the Channel coast. However, some of the fiercest storms are when the wind blows from the northeast, and, although the second greatest fetch is only 600 km, during these storms much shingle is carried from east to west. The two sets of storm waves have built up a series of shingle ridges, each newer ridge protecting the one behind it (Figure 6.20). The triangular shaped feature provides a site for Lydd airport and the Dungeness nuclear power stations.

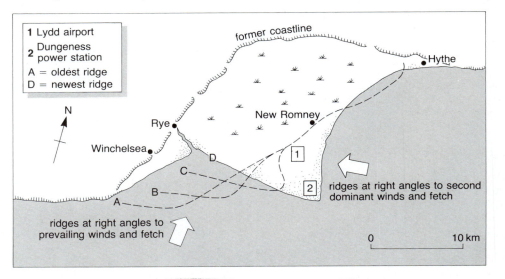

▶ Figure 6.20
The growth of Dungeness, a cuspate foreland (*after* Lewis)

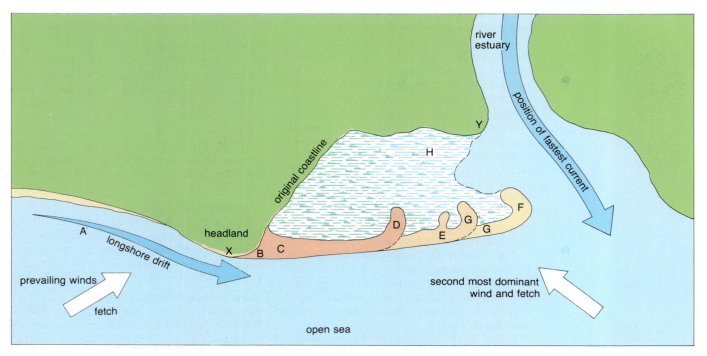

▲ Figure 6.21
Stages in the formation of a spit,
(lettering is explained in the text
on page 123)

◄ Figure 6.22

A spit — notice the vegetation
beginning to colonise the dunes
(see page 127)

▶ Figure 6.23
A tombolo, Llanddwyn Island, Anglesey

▶ Figure 6.24
Chesil beach — the shingle bar joins the Isle of Portland to the coast

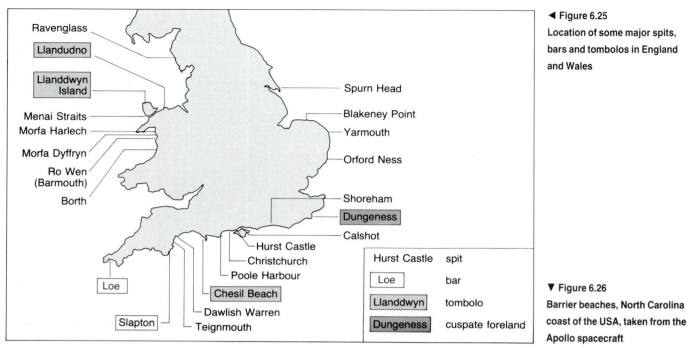

◀ Figure 6.25
Location of some major spits,
bars and tombolos in England
and Wales

Hurst Castle	spit
Loe	bar
Llanddwyn	tombolo
Dungeness	cuspate foreland

▼ Figure 6.26
Barrier beaches, North Carolina
coast of the USA, taken from the
Apollo spacecraft

New Jersey down the East Coast of the USA to include Cape Hatteras and Florida — the resort of Miami is built upon a barrier beach — and then along the southern coast into Mexico. This feature is easily identified using an Atlas.

These beaches are believed to form as offshore bars of sand, which accumulate below the low tide mark and move progressively inland. The relatively shallow water generates constructive waves which transport material landwards creating a long smooth coastal feature topped by sand dunes and separated from the mainland by a lagoon (Figure 6.26). Storm waves, especially those produced by hurricanes in the SE of the United States, usually expend their force harmlessly upon the long, low beaches of the barrier islands, but where human settlement has taken place on these islands, storm waves may cause considerable damage.

Sand dunes

Longshore drift may deposit sand in the inter-tidal zone. As the tide ebbs this sand dries out. When winds blow from the sea this dry sand will be moved up the beach by saltation (page 150). This is most likely to occur when the prevailing winds come from the sea and where there is a large tidal range which exposes large stretches of sand at low tide. Sand may become trapped by seaweed and driftwood on berms or at the point of the highest spring tides. Plants begin to colonise the area (Figure 11.7) stabilising

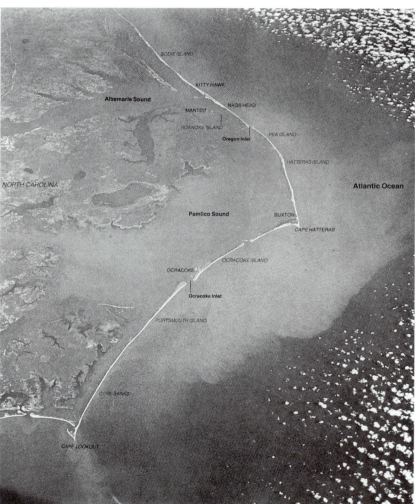

the sand and encouraging further accumulation. This regolith has a high pH value due to calcium from seashells.

Embryo dunes are then likely to develop (Figure 6.27) and the surface of the area becomes increasingly arid due to high rates of percolation into the sand. Marram grass, with its long roots to tap underground water supplies, and Spinifex, having low growth which helps to reduce transpiration in the strong winds, are often the first plants to colonise the dunes.

As more sand accumulates, embryo dunes join up to form **foredunes** which can attain a height of 5 m (Figure 6.28). Due to the lack of humus, their colour gives them the name **yellow dunes**. These dunes become increasingly grey as humus and bacteria from plants and animals are added and they gradually become more acidic and completely vegetation-covered. These **grey dunes** may reach a height of 10−30 m before the supply of fresh sand is cut off by their increasing distance from the beach. There may be several parallel ridges of old dunes (e.g. Morfa Harlech, Figure 6.29) separated by low-lying, damp slacks. Heath plants begin to dominate the area as acidity, humus and moisture content all increase (Figure 11.8). Paths cut by humans and animals expose areas of sand. As the wind funnels along these tracks, **blowouts** may form in the now **wasting** dunes. To combat further erosion at Morfa Harlech, parts of the dunes have been fenced off and marram grass and other vegetation has been planted to try to re-stabilise the area. Elsewhere, e.g. Les Landes on the west coast of France, attempts have had to be made to try to control the inland migration of dunes because of the potential damage to settlement or agricultural land.

Saltmarshes

Where there is shelter in river estuaries or behind spits, silt and mud will be deposited by either the gently rising and falling tide or the river. Initially the area will be uncovered for less than one hour in every 12 hour tidal cycle. Plants such as algae and Salicornia can tolerate this lengthy submergence and the high levels of salinity. They are able to trap more mud around them allowing this,

▼ Figure 6.27

Transect across sand dunes based on fieldwork at Morfa Harlech, North Wales

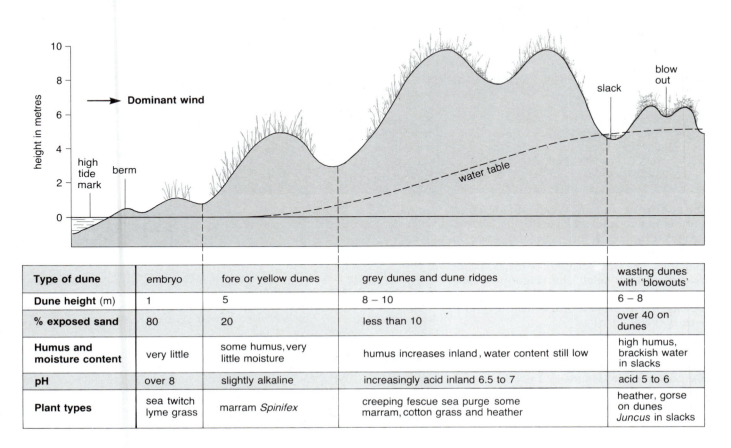

Type of dune	embryo	fore or yellow dunes	grey dunes and dune ridges	wasting dunes with 'blowouts'
Dune height (m)	1	5	8 – 10	6 – 8
% exposed sand	80	20	less than 10	over 40 on dunes
Humus and moisture content	very little	some humus, very little moisture	humus increases inland, water content still low	high humus, brackish water in slacks
pH	over 8	slightly alkaline	increasingly acid inland 6.5 to 7	acid 5 to 6
Plant types	sea twitch lyme grass	marram *Spinifex*	creeping fescue sea purge some marram, cotton grass and heather	heather, gorse on dunes *Juncus* in slacks

▲ Figure 6.28, *top left*
Embryo and foredunes at Morfa Harlech, North Wales (refer also to Figures 11.8 and 11.9)

▲ Figure 6.29, *top right*
Morfa Harlech from Harlech Castle showing foredunes, grey or wasting dunes, old cliff-line and reclaimed salt marsh

◄ Figure 6.30
Llanhridian Salt Marsh, Gower Penninsular, South Wales (refer also to Figures 11.13 and 11.14)

▼ Figure 6.31
Llanhridian Salt Marsh — notice the sward zone, creeks and salt pan

the **slob zone**, to remain exposed for longer periods between tides (Figure 6.30). Spartina grows throughout the year and since its introduction into Britain has colonised most estuaries. The landward side of the slob zone is marked by a small cliff (Figure 11.14b). Above this is the flat **sward zone** which may be covered by the sea for less than an hour in a tidal cycle.

Seawater collects in hollows which, as salinity increases following evaporation, enlarge into **saltpans**, void of vegetation except for certain algae (Figure 11.14c). As the tide retreats, water drains into **creeks** which are eroded rapidly both laterally and vertically. The upper sward zone will only be inundated by the highest of spring tides and eventually the land may be reclaimed for economic use.

Figure 6.32 shows Blakeney spit and the adjacent part of the coastline of north Norfolk as it was in circa AD 1600. Figure 16.33 shows the same coast in 1979.

1   a   State four coastal changes which have occurred in the area between 1600 and 1979.

      b   Explain each of the changes you have identified in **1a**.

2   a   With reference to 16.33 how might the shingle found at **A** differ from that found at **B**?

      b   Explain the difference.

3   a   With reference to Figure 16.33 describe the changes at Old Far Point which occurred between 1954 and 1979.

      b   Explain how these differences might have occurred.

4   Why have sand dunes formed at point **C**?

5   a   Where would you expect an area of saltmarsh to form?

      b   Give reasons for your answer.

▼ **Figure 6.32,** *below*

Blakeney Point, Norfolk in c.1600

▼ **Figure 6.33,** *bottom*

Blakeney Point in 1979, (compare with Figure 6.32)

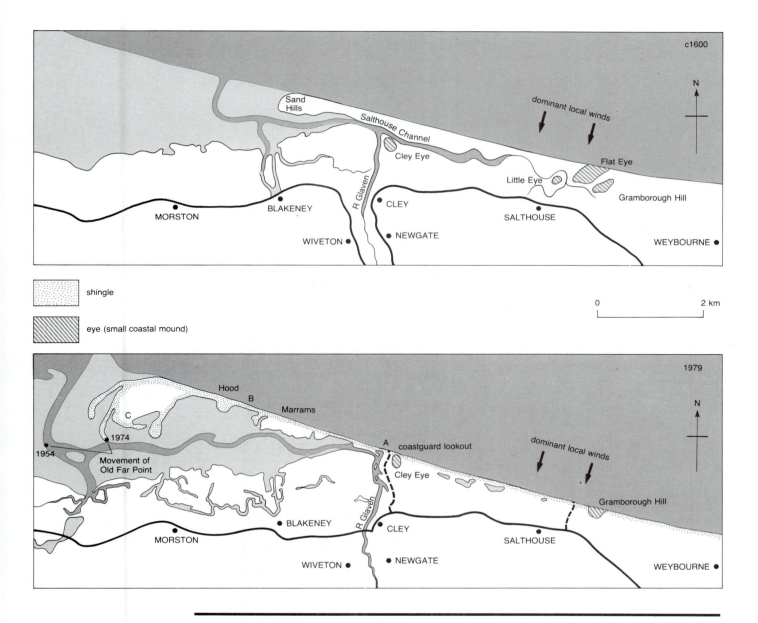

129

FRAMEWORK 3

Why classify?

Geographers frequently utilise classifications, e.g. types of climate, soils and vegetation, forms and hierarchy of settlement, landforms. This is done to try to create a sense of order by grouping together features which have similar, if not identical, characteristics into identifiable categories. For example, no two stretches of coastline will be exactly the same, yet by describing Blakeney Point as a spit it may be assumed that the processes leading to its formation and the description of its appearance are similar to those outlined in Figure 6.20, even if there are local differences in detail. When determining the basis for any classification care must be taken to ensure that:

- only meaningful data and measures are used,
- within each group or category there should be the maximum number of similarities,
- between each group there should be the maximum number of differences,
- there should be no exceptions, i.e. all of the features should fit into one group or another,
- as classifications are used for convenience and to assist understanding, each one should be easy to use,
- it should not be oversimplified, making it too generalised, nor too complex, rendering it unwieldy.

No classification is likely to be perfect and several approaches may be possible.

The following landforms have already been referred to in this book:
arch braided river corrie delta esker hanging valley knickpoint moraine raised beach rapids spit wave cut platform.

Can you think of at least three different ways in which they may be categorised?
- Perhaps the simplest is a division based upon whether they result from erosion or deposition.
- They could be rearranged into two different categories, those formed under a previous climate (i.e. relict features) and those still being formed today.
- The most obvious may be a threefold division into coastal, glacial and fluvial landforms.
- The most complex could result from combining the first two to give six groups.

Classification of coasts

Several attempts to classify coasts usefully have been made, based upon a range of criteria.

- **Changes in sea level**
  (D.W. Johnson, 1919)

This, the earliest method, divided coasts into four categories.

1 **Emergent:** those resulting from a fall in sea level and/or uplift of the land. Coastlines formed in this way are basically straight in appearance (e.g. barrier beaches).

2 **Submergent:** those showing a recent rise in sea level and/or a fall in the land surface. This type is characterised by an indented coastline (e.g. rias and fjords).

3 **Stable** or **neutral:** (a) those showing no signs of changes in sea level or in the land, (b) those produced by non-marine processes (e.g. a river delta, a volcanic island).

4 **Compound:** those with a mixture of at least two of the three previously listed groups.

This classification came to be regarded as too simplistic as new evidence accumulated to show that there had been several changes in sea level (local and global) and that many coasts exhibited signs both of submergence and emergence.

- **Relationship between coastal and other processes of erosion and deposition** (F.P. Shepard, 1963)

This has been shown to be over-generalised, having only two categories.

1 **Primary:** where the influence of the sea has been minimal, e.g. fjords (ice), rias and deltas (rivers), and islands (volcanic or coral).

2 **Secondary:** where marine processes have been dominant, e.g. stacks and spits.

- **Advancing and retreating coasts**
  (R. Valentin, 1952)

This method also suggested that all coasts could be fitted into a twofold classification.

1 **Advancing:** where marine deposition or the uplift of the land is dominant.

2 **Retreating:** where marine erosion or the submergence of the land is more significant.

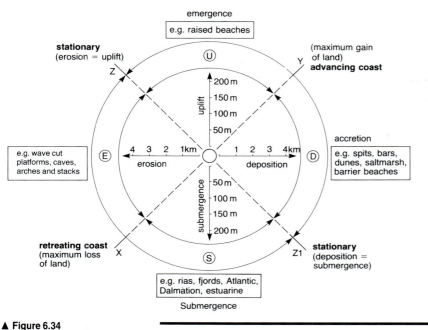

emergence
e.g. raised beaches

stationary
(erosion = uplift)

(maximum gain
of land)
advancing coast

e.g. wave cut
platforms, caves,
arches and stacks

accretion
e.g. spits, bars,
dunes, saltmarsh,
barrier beaches

retreating coast
(maximum loss
of land)

stationary
(deposition =
submergence)

e.g. rias, fjords, Atlantic,
Dalmation, estuarine

Submergence

▲ Figure 6.34
Valentin's classification of coasts — this shows the conditions under which the maximum advance or retreat of the coast will occur and how it is possible for a coast to be in equilibrium where there is neither advance or retreat

■ **Energy-produced coastlines**
(J.L. Davis, 1980)
This, the most recent attempt to categorise knowledge about coasts, is gaining credibility because it relates directly to the amount of energy expended by different types of wave upon a particular stretch of coastline.
1 **High energy** environments: where destructive storm waves breaking on shingle beaches are most typical.
2 **Low energy** environments: where constructive swell waves breaking upon sandy beaches are more frequent.
3 **Protected** environments: where wave action is limited in small, sheltered sea areas.

1 a Choose one of the above systems of coastal classification. Describe and explain the principles upon which the system is based.
b Describe some of the problems of applying your chosen classification to cover all coastal processes and/or landforms.

2 If you get an opportunity to study a stretch of coastline, discuss the problems you found in trying to fit it into any one coastal classification with which you are familiar.

Changes in sea level

Although the daily movement of the tide alters the level at which waves break on to the foreshore, the average position of sea level in relation to the land has remained relatively constant for nearly 8000 years (Figure 6.35). Before that time there had been several major changes in this mean level, the most dramatic being a result of the Quarternary ice age and plate movements.

During times of maximum glaciation, large volumes of water were stored on the land as ice, probably three times more than today.

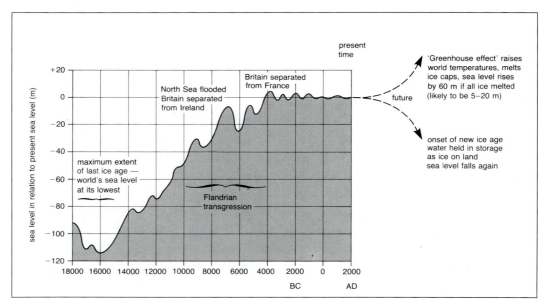

▶ Figure 6.35
Eustatic changes in sea level since 18 000 BC

This modification of the **hydrological cycle** meant that there was a worldwide, or eustatic, fall in sea level of an estimated 100–150 m.

As ice accumulated, its weight began to depress those parts of the crust lying beneath it. This caused a local, or **isostatic**, change in sea level.

The world's sea level was at its minimum 18 000 years ago when the ice was at its maximum (Figure 6.35). As the temperatures began to rise and ice caps melted there was firstly a eustatic rise in sea level followed by a slower isostatic uplift which is still operative in parts of the world today. This sequence of sea level changes may be summarised as follows:

1 Formation of glaciers and ice sheets. Eustatic fall in sea level giving rise to a negative change in base level (page 70).

2 Continued growth of ice sheets. Isostatic depression of the land under the ice producing a positive change in base level.

3 Ice sheets begin to melt. Eustatic rise in sea level with a positive change in base level.

4 Continued decline of ice sheets and glaciers. Isostatic uplift of the land under former ice sheets resulting in a negative change in base level. During this de-glaciation there may have been a continuing, albeit small, eustatic rise in sea level but it was less rapid than the isostatic uplift, so base level appeared to be falling. Measurements suggest that parts of NW Scotland are still rising by 4 mm a year and some northern areas of the Gulf of Bothnia (Scandinavia) by 20 mm a year.

Tectonic changes have resulted in the uplift (**orogeny**) of new mountain ranges: sea shells dating from over 18 000 BP (before present) have been found high in the rocks of the Himalayas, Alps and Andes. Local tilting (**epeirogeny**) of the land in the Mediterranean has led to the submergence of several ancient ports with others now stranded above the present day sea level. In SE England and the Netherlands epeirogeny has increased the risk of marine flooding. The emergence of the Mid-Atlantic Ridge (page 17) could have resulted in a eustatic rise in sea level of several hundred metres while the effects of ocean floor spreading could have led to a reversed eustatic change — a 1 per cent increase in the area of the oceans could produce a fall of 40 m in ocean levels. Local volcanic and earthquake activity have also altered the balance between land and sea levels.

Landforms created by sea level changes

Changes in sea level have affected:
- The shape of coastlines and the formation of new features by increased erosion or deposition.
- The balance between erosion and deposition by rivers (page 69) resulting in the drowning of lower sections of valleys or rejuvenation of rivers.
- Migration of plants, animals and humans.

Landforms resulting from submergence

Eustatic rises in sea level following the decay of the ice sheets led to the drowning of many low-lying coastal areas.

During the ice age those rivers still flowing were able to cut their valleys downwards to the lower base level. When the sea level subsequently rose, the lower parts of the main valley and its tributaries were drowned to produce sheltered, winding inlets, called **rias**, characteristic of southwest Britain (Figure 6.36). Whereas rias formed when river valleys flowed at right angles to the sea, **Dalmatian** coasts are found where river valleys had formed parallel to the coastline (Figure 6.37).

Fjords are found where glaciers were able to erode below sea level. When the ice melted, valleys were flooded by the sea leaving long, deep, narrow inlets with precipitous sides and hanging valleys, i.e. a drowned glacial trough, Figure 6.38. Glaciers appear to have followed lines of weakness, e.g. a pre-glacial river valley or a major fault line. Unlike a ria, which gets progressively deeper towards the sea, a fjord has a relatively shallow seaward entrance, known as a **threshold**, which is debatable in origin. Lowland areas which have been glaciated and drowned are called **fjards**, e.g. Strangford Lough, Northern Ireland.

▼ Figure 6.36
A ria — the Camel estuary, north Cornwall

▶ Figure 6.37
The Dalmation coast of
Yugoslavia

▶ Figure 6.38, *far right*
Geiranger Fjord, Norway

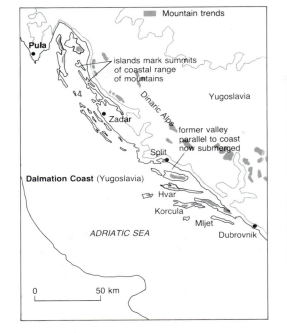

▼ Figure 6.39
Raised beaches on the Isle of
Arran — the lower relates to the
8m; the upper, older to the 30m
beach

▶ Figure 6.40, *below right*
The abandoned cliffline at
King's Cave, Arran — the
foreground is the 8m raised
beach with former wave cut
notches and caves visible

Landforms resulting from emergence

Following the global rise in sea level, and still occurring in several parts of the world today, was the isostatic uplift of land as the weight of the ice sheets decreased. Landforms created as a result of land rising relative to the sea are raised beaches and erosion surfaces.

Raised beaches As the land rose, former wave cut platforms and their beaches were raised above the reach of the waves. Raised beaches are characteristic of the west coast of Scotland. They are recognised by a line of degraded cliffs fronted by what was originally a wave cut platform (Figure 6.39). Within the old cliffline may be relict landforms such as wave cut notches, caves, arches and stacks (Figure 6.40). The presence of such features

indicates that isostatic uplift could not have been constant. It has been suggested that it would have taken an unchanging sea level up to 2000 years to cut each wave cut platform — this evidence has also been used to prove that the climate did ameliorate steadily following the ice age.

Early workers in the field claimed that there were three levels of raised beach found at 25, 50 and 100 ft above the present sea level. These are now referred to as the 8-, 15- and 30-m raised beaches. However, this description is now considered too simplistic for it has been accepted that places nearest to the centre of the ice depression have risen the most and the amount of uplift decreases with distance from that point — the much

quoted 8-m raised beach on Arran in fact lies between 4 and 6 m. Where the raised beach is extensive there is a considerable difference in height between the old cliff on its landward side and the more recent cliff to the seaward — the 30-m beach in southwest Arran rises from 24 to 38 m. It is now more acceptable to estimate the time at which a raised beach was formed by carbon-dating sea shells found in former beach deposits rather than by referring solely to its height above sea level (i.e. to indicate a 'late glacial raised beach' rather than a '100-ft/30-m beach'). Figure 6.41 is a labelled transect, not drawn to scale, showing raised beaches on the west coast of Arran.

Erosion surfaces In Dyfed, the Gower Peninsula (South Wales) and Cornwall, flat planation surfaces dominate the scenery. Where their general level is between 45 and 200 m, the surfaces are thought to have been cut during the Pleistocene when sea levels were higher: hence the name **marine platforms** (Figure 6.42).

People and the coastal environment

Why do you think human influence on the coastal ecosystem and environmental equilibrium is more noticeable on lower energy, protected environments that on higher energy coasts? It is a question which is becoming of increasing concern and leads to the issue of how far *should* people try to make an impact on the area where land and sea meet.

Land loss and pollution

Loss of land
Beaches dissipate wave energy. Where sand and shingle is removed, e.g. for use in the construction industry, less energy is absorbed by the beach thus permitting larger and steeper waves to attack and erode cliffs and dunes. During 1887, 660 000 tonnes of shingle were removed from the beach at Hallsands in Devon to help in the construction of the naval dockyard at Plymouth. As there was virtually no replenishment of material by longshore drift, the level of the beach fell by 4 m. Erosion of the cliffs behind the beach was accelerated — 6 m were lost between 1907 and 1957 — so that today the village of Hallsands is almost completely abandoned and left in ruins.

By trying to protect one part of the coast, engineers may increase the vulnerability of another area. Groynes are built to increase the width of a beach by limiting longshore drift (page 121). However, this action has repercussions because places further along that coast will lose their supply of beach material and become more susceptible to erosion, e.g. West Bay in Dorset. Added to this problem is the increasing evidence that the supply of sand and shingle to form beaches is a relict feature. Much was probably deposited around 18 000 years ago at the time of the maximum extent of the last glacial advance (Figure 4.3) when sea levels were at their lowest (Figure 6.41). Later the material was probably transported landwards during the eustatic Flandrian Transgression. Since then, sea levels having remained fairly constant, considerably less material has been added to the foreshore.

Many coastal areas have experienced a removal of their vegetation cover while others, especially sand dunes, have suffered from trampling by humans. In tropical areas the clearance of mangrove forest has removed a protective barrier. In many places vegetation has been destroyed for port and

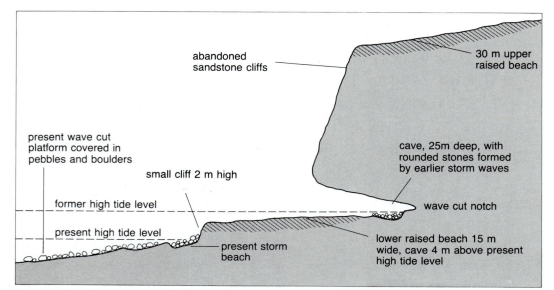

abandoned sandstone cliffs

30 m upper raised beach

present wave cut platform covered in pebbles and boulders

cave, 25m deep, with rounded stones formed by earlier storm waves

small cliff 2 m high

former high tide level

wave cut notch

present high tide level

present storm beach

lower raised beach 15 m wide, cave 4 m above present high tide level

◄ Figure 6.41
Diagrammatic transect across raised beaches in west Arran

urban development and, more recently, tourism. While the backshore may become unsightly with high density tower block hotels, the foreshore may become littered with rubbish such as broken glass, plastic bottles and tin cans. At the same time there is a loss of natural habitat for wildlife. In the Mediterranean area, the once quiet, sandy breeding grounds for the monk seal and the loggerhead turtle (SW Turkey and Crete) have been spoiled. Elsewhere, fragile wetlands, e.g. the Norfolk Broads and Florida Everglades, which provide a transition zone between land and sea are under threat. Although at times inhospitable to humans, these areas form an irreplaceable habitat for many plants, birds and animals.

Pollution

A major source of pollution on beaches is untreated sewage which at best is unsightly and at worst dangerously unhealthy. In Britain many outlets disgorging sewage into the sea are not long enough to prevent tides and currents returning the waste to beaches (Figure 6.43). In developing countries the problem may be accentuated by a lack of finance to provide treatment works and by pressure from a rapidly growing local population and possibly tourists as well. Over nine million tonnes of sewage are dumped annually into the sea outside New York harbour. Sewage decay utilises oxygen and so deprives entire marine communities of the oxygen they need to survive.

Industries often locate on estuaries not only for trading or extraction of water for cooling and washing but to use the sea as a dumping ground for their waste products. Nuclear reactors and oil refineries tend to be sited on coasts and both have been responsible for accidental leakages of harmful substances. Radioactive waste is dropped far out to sea, but is it safe there? — There are serious doubts about the long term implications of this nuclear waste disposal. Several beaches in County Durham are covered in coal waste (Figure 6.44).

During the 1950s mercury waste was released into Minamata Bay in Japan. It was converted into a substance which was absorbed by fish. The fish were later eaten by birds, cats and, at the top of the food chain, humans. The result was the death of many birds, cats and eventually, as levels of mercury accumulated, of over 100 people. Children were born with mental and physical defects such as blindness and deformed limbs.

Accidents to oil tankers, e.g. Torrey Canyon off southwest England, 1967, Amaco Cadiz off Brittany, 1978, destroy marine life and ruin beaches. Yet the amount of oil released into the sea by such headline-making spillages is quite small compared with the amount illegally discharged by ships washing out their tanks while at sea. Conflicts between nations may result in tankers and oil platforms being destroyed and oil spilled, as has happened in the Persian Gulf. Fertilisers and pesticides from farming eventually make their way into rivers, are caught up in the hydrological cycle, and finally enter the sea. Phosphates and nitrates encourage the growth of algae and other water plants which use up supplies of oxygen required by other marine life. Although the volume of agricultural runoff is small in comparison with the size of the ocean, it constitutes a serious problem when added to all of the other sources of pollution. The question being asked by governments as well as environmentalists is: 'How much waste can the oceans absorb?'.

Protection and conservation of coasts

The traditional way of protecting agricultural or populated land from wave attack is the erection of sea walls and other defences. However, this method is expensive, seems to increase the concentration of high energy waves and walls have to be replaced. A cheaper, more effective solution is to add more sand, shingle and boulders in order to widen the beach and so to dissipate wave energy. The construction of breakwaters is not aimed at halting waves but reducing their energy. Breakwaters may be built outwards at right angles from the land or parallel to the coast, as at the entrance to Plymouth Sound, beyond the low water mark. Sand dunes are stabilised by planting marram grass and erecting brushwood fences, as at Morfa Harlech, or by planting coniferous trees. The Dutch have protected those parts of the Netherlands which lie below sea level by constructing a series of dykes. The land behind these defences (polders) is often reclaimed for farming and new housing. The Dutch Delta scheme, similar to the Thames Barrage, is aimed at limiting the effect of future storm surges (page 116).

A more recently popular view is that future coastal development must be planned and controlled, with areas of scenic beauty and habitats for wildlife being protected and conserved. The Economic Community has issued standards for clean beaches and coastal waters in Europe which, by 1988, had not been met by a high proportion of British resorts. New resorts, like West Indian beach villages, are (theoretically at least) built to fit into rather than destroy the local environment. National Parks are one method

◄ Figure 6.42
Erosion surfaces (marine
peneplanation) at St David's,
Dyfed, South Wales

◄ Figure 6.43
Power station 'effluent outfall'
at Shoreham-by-Sea, Sussex

of protecting the countryside at the coast, while nature reserves and bird sanctuaries fulfil a similar role for wildlife.

An international agreement to prevent pollution from ships was signed in 1973. The discharge of toxic waste was banned in the Baltic and Black Seas, and the release of oil into the Black, Baltic, Red, and Mediterranean Seas was forbidden. This ban also operated in the Persian Gulf before the Iranian–Iraqui war. Elsewhere, treated sewage must not be discharged within 4 km of the shore, untreated sewage within 12 km, and oil within 50 km. Disturbingly, only 30 countries had ratified this agreement by the mid-1980s. The largest single scheme was the 'Mediterranean Blue Plan' (Case Study) which aimed to clean up the world's most polluted sea.

◄ Figure 6.44
Marine pollution by coal
dumping on a Durham beach

CASE STUDY

The Mediterranean Blue Plan

The Mediterranean is said to be 'the largest open air swimming pool' and 'the largest sewer' in the world. Increasingly in the last few decades it has been a dumping ground for untreated sewage, industrial and chemical waste, agricultural pesticides, fertilisers and oil. In 1976, the Barcelona Convention was signed by 18 countries which border the Mediterranean Sea. Setting aside their national rivalries, they agreed that the top priority was to build more sewage treatment plants and to control the release of industrial waste and pesticides. Two lists of contaminating chemical substances were drawn up. The 'black' list, which included highly toxic substances such as mercury and DDT, were completely banned while those on the 'grey' list, which were less toxic, needed a special dumping licence. A united plan to deal with possible oil spillages was accepted. Fifteen marine reserves were created, with additional safeguards to protect the endangered loggerhead turtles and monk seals. In addition, 83 marine laboratories were set up to monitor pollution and to pool resources and information.

Despite considerable success the continued operation of the plan faces many difficulties. Estimates suggest that over £2500 million is needed just to treat sewage — the poorer countries in the south and east cannot afford such schemes. It needs the goodwill of all of the member nations, yet several are not on good terms with their neighbours. The countries in the northwest, which are already industrialised and responsible for much of the pollution, wish to clean up their beaches for holidaymakers while those in the south and east see the protection of the environment as a secondary priority to the creation of new industries.

Sampling

Sampling techniques

Several different methods may be used according to the demands of the required sample and the nature of the parent population. There are two major types with one refinement.
Random This is the most accurate as it has no bias.
Systematic This is often quicker and easier to use although some bias or selection is involved.
Stratified This frequently proves to be a very useful refinement for geographers and can be used with either a random or a systematic technique.

Random sampling

Under normal circumstances this is the ideal type of sample as it shows no bias. Every value, or member of the population, has an equal chance of being selected and the selection of one member does not affect the probability of selection of another member. The ideal random sample may be obtained using **random numbers**. These are generated by a computer but if necessary they can be obtained by drawing numbers out of a hat. Random number tables usually consist of columns of pairs of digits which vary from 00 to 99. Numbers can be chosen by reading either along the rows or down the columns,

provided only one method is used. Similarly, any number of figures may be selected, e.g. six for a grid reference or four for a grid square.

Part of a random numbers table

5094	1323	7841	6058
6698	3796	4413	4505
6691	4283	6077	9091
3358	1218	0207	1940
1060	8846	3021	4598
3459	7585	4897	2719
6090	7962	5766	7228
2129	3945	9042	5884

Using this table, a six-figure grid reference could be 509413 (reading across), 509466 (reading down) or 669142 (as there is no need to begin with the first number). A four-figure grid square could be 5094 followed by either 1323 or 6698. If a random figure does not correspond to a grid reference on the map then it is disregarded. One feature of a genuine random sample is that the same number could be selected more than once — so remember that if you are pulling numbers from a hat, they should be replaced immediately after they have been read and recorded.

F R A M E W O R K   4

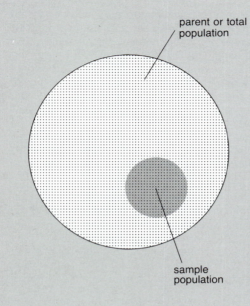

parent or total population

sample population

Why sample?

Geographers are part of a growing number of people who find it increasingly useful and/or necessary to use data to quantify the results of their research. The problem in this trend is that collection of data may be too expensive, too time-consuming or just impracticable, e.g. to investigate every-body's shopping patterns in a large city, to find the number of stones on a spit, or to map the land use of all the farms in Britain. Sampling is the method used to make statistically valid inferences when it is im-possible to measure the **total population** (Figure 6.45). It is essential to find the most accurate and practical method of obtaining a **representative sample**. If that sample can be made without, or at least with minimum bias or selection then statistically significant conclusions may be drawn. However, even if every effort is made to achieve precision, it must be remembered that any sample is only at best a close estimate.

◄ Figure 6.45
A sample population in relation to the parent population

Most sampling procedures assume that the total population would produce a **normal distribution** (Figure 4.15) which is a symmetrical curve on either side of the mean value, i.e. a large proportion of values are close to the average, with few extremes. Figure 6.46 shows a normal distribution curve as well as *one* measure of dispersion, the **standard deviation** from that mean (page 209). Where most of the values are clustered near to the mean then the standard deviation is said to be low. The larger the parent population, the greater the accuracy of the estimate and the more likely it is to conform to the normal dis-tribution. While the normally accepted minimum size for a sample is 30, there is no upper limit although there is a point beyond which the extra time and cost involved in increasing sample size does not give a significant improvement in accuracy — this is an example of the law of diminishing returns.

Figure 6.46 shows that in a normal distribution it is probable that 68.27 per cent of the values in the sample occur within + or − 1 standard deviation from the mean. Of the three probability levels, 68 per cent, 95 per cent and 99 per cent (page 329), geographers usually accept the middle level when sampling. This means that they accept the chance that in five times out of every 100 the true mean will lie outside two standard deviations of their sample mean.

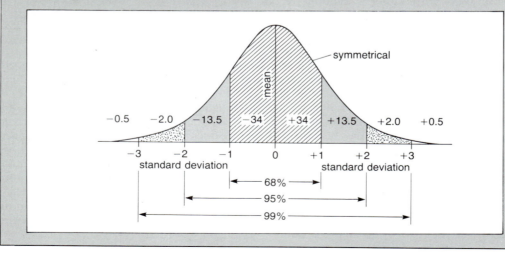

symmetrical

mean

−0.5   −2.0   −13.5   −34   +34   +13.5   +2.0   +0.5

−3   −2   −1   0   +1   +2   +3
standard deviation   standard deviation

68%
95%
99%

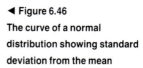

◄ Figure 6.46
The curve of a normal distribution showing standard deviation from the mean

There are three alternative ways of using random number figures to sample areal distributions (patterns over space) (Figure 6.47).

1 **Random point** A grid has to be superimposed over a map of the area to be sampled. Points, or map references, are then identified using random number tables. A large number of points may be needed to ensure coverage of the whole area.

2 **Random line** Random numbers are used to obtain two end points which are then joined by a line. Several random lines are needed to get a representative sample (e.g. lines across a city to show transects of variations in land use).

3 **Random square** or **area** Areas of constant size, e.g. grid squares, are obtained using random numbers. This method can be used to sample areas of rural land use or the distribution of plant communities over space.

The advantages of random sampling include its ability to be used with large numbers and its avoidance of bias. Its disadvantages include the problem that it is not always representative when testing small quantities of data (one plant species over a large area); it does not show changes over time or distance (it might not show a concentration of plants in one part of a large area); and when employed in the field it may involve considerable time and energy in visiting every point.

Systematic sampling

A systematic sample is one where values are selected in some regular way, e.g. choosing every tenth person on a list or every twentieth house in a street. This provides a much easier method in terms of time and effort than random sampling. Like random sampling it can be operated using individual points, lines or areas (Figure 6.48). In all cases points are pre-determined, e.g. grid intersections.

1 **Systematic point** This can show changes over distance, e.g. land use every 100 m, or time, e.g. size of Britain's population using ten year census figures.

2 **Systematic line** This may be used to choose a series of equally spaced transects across an area of land, e.g. up a shingle spit.

3 **Systematic area** In this case a series of quadrats (for measuring plant distributions) are selected at equal intervals.

The advantages of systematic sampling include its being straightforward and often quicker to use than random sampling, though it does not necessarily produce more accurate results. Systematic sampling is particularly facilitated by the use of micro computers. It is useful when sampling large areas as the points/lines can be more widely spaced; it is appropriate when selecting from a list because no time is lost disregarding unwanted random numbers; and it is more useful when looking for small data such as a

▼ Figure 6.47, *below*
Random sampling using point, line and square (area) techniques to sample an area

▼ Figure 6.48, *bottom*
Systematic sampling using point, line and square (area) techniques

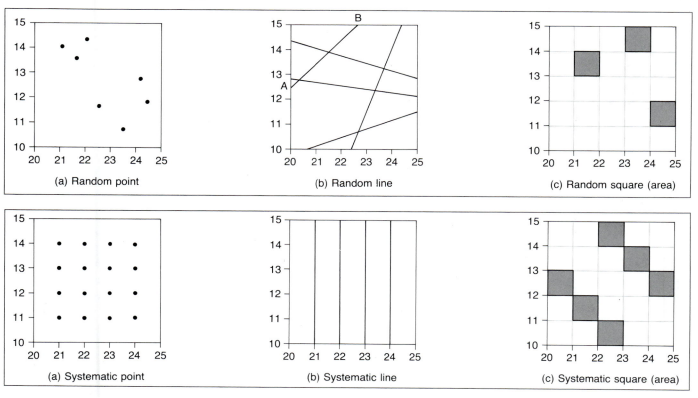

(a) Random point

(b) Random line

(c) Random square (area)

(a) Systematic point

(b) Systematic line

(c) Systematic square (area)

single plant species over a large area. However, it has disadvantages in that all points do not have an equal chance of selection, i.e. it shows bias, and it may either overstress or miss an underlying pattern (Figure 6.49).

Stratified sampling

When there are significant groups of known size within the parent population, in order to ensure adequate coverage of all the sub-groups it is sometimes necessary to stratify the sample. Although dividing populations into groups (layers or strata) may be a subjective decision, the practical application of this technique has considerable advantages to the geographer. Once the groups have been decided then the population can be sampled by a random or systematic method (Figure 6.50).

1 **Stratified random** This method can be used to cover a wide range of data such as in political opinion polls. Here the total population to be sampled is divided into equal age or socio-economic groups, e.g. 10–19, 20–29, 30–39, etc. If there are eight age groups then perhaps 800 people would be interviewed with 100 in each category. Strictly, the number interviewed in each age group should be in proportion to its known size relative to the parent population.
2 **Stratified area** This tries to overcome a weakness in systematic sampling (Figure 6.49). Figure 6.50c shows the pattern of moorland on two contrasting rock types.

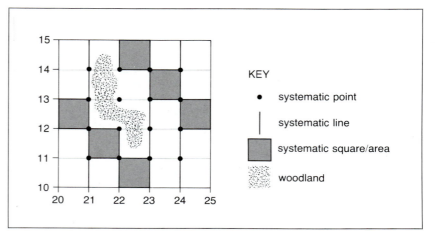

▲ Figure 6.49

A weakness of systematic sampling is shown where an area of woodland (shaded) is completely missed in this example

Although the previously described techniques could give the percentage of cover they would not show whether this proportion varied with rock type. As granite occupies an estimated 60 per cent of the map area and limestone 40 per cent, if 50 sample points were to be chosen by random numbers, then 30 of these (60 per cent) would have to be taken from the granite area and 20 points (40 per cent) from the limestone.

The advantages of stratified sampling include its potential to be used in conjunction with either random or systematic techniques and, as many populations have geographical sub-groups, this method helps to avoid these sub-groups being missed. The major disadvantage is the subjectivity (bias) of division into sub-groups.

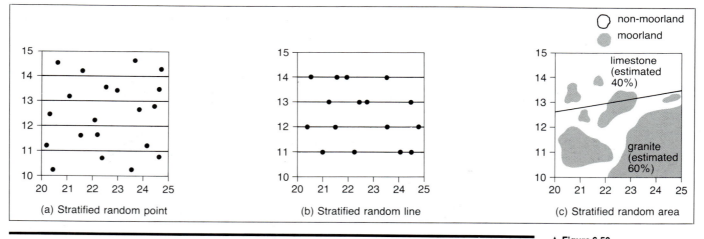

(a) Stratified random point

(b) Stratified random line

(c) Stratified random area

▲ Figure 6.50

Stratified sampling using random point, line and area methods

Q In an attempt to see the merits and problems of sampling in the field, four groups of sixth formers were taken to a spit on the east coast of the Isle of Arran. They were told to devise an appropriate sampling technique to test the following:
a to see if beach material decreased in size towards low water mark,
b to assess the percentage of vegetation cover over the spit at the time of their study.

Each group decided independently on a method different from the other three. These have been summarised in Figure 6.51.

Group	a Beach material decreases in size towards low water mark	b Assess the percentage of vegetation cover on the spit
1	Transect line chosen (subjective). Systematic sampling takes place along that transect line. 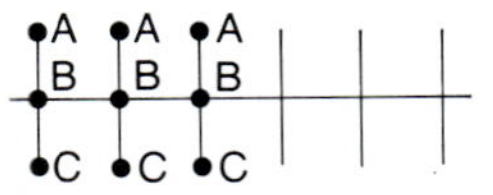 Every metre a rule is placed at right angles to transect. Three stones chosen at A B and C. Mean of the three taken.	Six transects chosen every 20 m. Systematic point sampling
2	Transect line chosen by random numbers. Systematic point sampling taken every 1 m. One stone (end of boot) chosen.	Random point sampling
3	Transect line chosen (subjective). 18 randomly chosen sites along line chosen by using random numbers. One stone chosen at each point.	Random area sampling. Having obtained each point by using random numbers to obtain a bearing — quadrats were used. Random number of 18 76 meant a bearing of 187° and 6 m had to be measured.
4	Transect line chosen by randon numbers. Systematic line sampling used every 5 m. 15 stones chosen and arranged in size. The one in the middle is chosen as mean.	Stratified sampling. It was estimated that 30% was vegetated and 70% was not. Random numbers were taken so that 30% of the sites were on vegetation and 70% on non-vegetated ground.

▲ Figure 6.51

Sampling techniques on a spit as selected by sixth formers conducting a fieldwork exercise

1 What do you consider to be the advantages and disadvantages of each of the techniques used by the four groups in terms of the objectives of their study?

2 Which group do you consider should have obtained the most accurate results? Why?

3 Can you suggest a more appropriate method for conducting sampling to test hypotheses **a** and **b**?

Standard error calculations

Having completed any sampling exercise it is important to remember that patterns exhibited may not necessarily reflect the parent population. In other words, the results may have been obtained purely by chance. Having determined the mean of the sample size it is possible to calculate the difference between it and the mean of the parent population by assuming that the parent population will conform to the normal distribution curve of Figure 6.46. However, while the sample mean must be liable to some error as it was based on a sample, it is possible to estimate this error by using a formula which calculates the **standard error of the mean** (SE).

$$SE\,\bar{x} = \frac{\sigma(\textit{standard deviation of parent population})}{\sqrt{n}\,(\textit{the square root of number of samples})}$$

where: $\bar{x}$ = mean of the parent population

We can then state the reliability of the relationship between the sample mean and the parent mean within the three confidence levels of 68, 95 and 99 per cent (Framework 4). Unfortunately, when sampling, the standard deviation of the parent population is not available and so to get the standard error we have to use the standard deviation of the sample, i.e. using s rather than σ. Although this introduces a margin of error, it will be

small if n is large — ideally n should be at least 30.

For example: a sample of 50 pebbles was taken from a spit off the coast of eastern England. The mean pebble diameter was found to be 2.7 cm and the standard deviation 0.4 cm. What would be the mean diameter of the total population (all the pebbles) at that point on the spit?

$$SE = \frac{0.4}{\sqrt{50}} = \frac{0.4}{7.07} = 0.06 \quad \text{(to two decimal places)}$$

This means we can say:

1 with 68 per cent confidence that the mean diameter will lie between 2.7 cm ± 0.06 cm, i.e. 2.64–2.76 cm

2 with 95 per cent confidence that the mean diameter will lie between 2.7 cm ± 2 × SE (2 × 0.06 = 0.12 cm), i.e. 2.58–2.82 cm

3 with 99 per cent confidence that the mean diameter will lie between 2.7 cm ± 3 × SE (3 × 0.06 = 0.18 cm), i.e. 2.52–2.88 cm.

If we wanted to be more accurate, or to reduce the range of error, then we would need to take a larger sample. Had we taken 100 values in the above example we would have had

$$SE = \frac{0.4}{\sqrt{100}} = \frac{0.4}{10.00} = 0.04 \text{ (to two decimal places)}$$

141

which means we can now say with 68 per cent confidence that the mean diameter size will lie between 2.7 cm ± 0.04 cm (i.e. 2.66 to 2.74 cm). Of course, this also means there is a 32 per cent chance that the mean of the parent population is not within these values. This is why most statistical techniques require answers at the 95 per cent confidence level at least in order to attain credibility as reliable indicators.

This standard error formula is applicable only when sampling actual values (**interval** or **measured data**). If we wish to make a count to discover the frequency of occurrence where the data is **binomial** (i.e. it could be placed into one of two categories), then we have to use the binomial standard error. For example, we may wish to determine how much of an area of sand dune is covered in vegetation and how much is *not* covered in vegetation. When using binary data, the estimates of the population for samples are given as percentages, not actual quantities, i.e. x per cent of points on the sand dune *were* covered by vegetation; x per cent of points on the sand dune were *not* covered by vegetation.

The formula for calculating standard error using binomial data is

$$SE = \sqrt{\frac{p \times q}{n}}$$

where: p = the percentage of occurrence of points in one category
q = the percentage of points not in that category
n = the number of points in the sample.

A random sample of 50 points was taken over an area of sand dunes similar to those found at Morfa Harlech (Figure 6.29). Of the 50 points, 32 lay on vegetation and 18 on non-vegetation (sand) which, expressed as a percentage, was 64 per cent and 36 per cent respectively.

How confident can you be about the accuracy of the sample?

$$SE = \sqrt{\frac{64 \times 36}{50}} = \sqrt{46.08} = 6.79 \text{ (to two decimal places)}$$

As the sample found 64 per cent of the sand dunes to be covered in vegetation and knowing the standard error to be ± 6.79, we can say:

1 with 68 per cent confidence that the vegetated area will lie between 64 per cent ± 6.79, i.e. between 57.21 and 70.79 per cent
2 with 95 per cent confidence that between 64 per cent ± 2 × SE (2 × 6.79 = 13.58), i.e. between 50.42 and 77.58 per cent will be vegetated

3 with 99 per cent confidence that it will lie between 64 per cent ± 3 × SE (3 × 6.79 = 20.37), i.e. between 43.63 and 84.37 per cent.

Minimum sample size

It seems obvious that the larger the size of the sample then the greater is the probability that it accurately reflects the distribution of the parent population. It is equally obvious that the larger the sample, the more costly and time consuming it is likely to be. There is, however, a method to determine the minimum sample size needed to get a satisfactory degree of accuracy. For example, this may be to find the mean diameter of pebbles on a spit, or the amount of vegetation cover on sand dunes. This is achieved by reversing the two standard error calculations.

For measured data Imagine you wish to know the diameter of pebble size at a given point on a spit to within ±0.1 cm at the 99 per cent confidence level.
The 99 per cent confidence level is 3 × SE.

$$3\,SE = \frac{3s}{\sqrt{n}}$$

i.e. $3s = 0.1 \sqrt{n}$

i.e. $\dfrac{3s}{0.1} = \sqrt{n}$

We determined earlier that s (standard deviation of the sample) for the pebble size was 0.4, and so by substitution we get:

$$\frac{1.2}{0.1} = \sqrt{n}$$

i.e. $12 = \sqrt{n}$

$$n = 12^2$$

$$n = 144$$

We would need, therefore, to measure the diameter of 144 pebbles to get an estimate of the parent population at the 99 per cent confidence level.

For binomial data How many sample values are needed to estimate the area of sand dunes which is vegetated, correct to the 5 per cent of the actual population (i.e. at the 95 per cent confidence level)?

$$n = \frac{p \times q}{(SE)^2}$$

Again by substitution we get:

$$n = \frac{64 \times 36}{(5)^2}$$

$$n = 92.16$$

We would therefore have to take a sample of 93 values to achieve results within 5 per cent of the parent population.

Q

1 Define the following terms used in sampling:
   a population,
   b random sample,
   c systematic sample,
   d stratified random sample,
   e sampling error.

2 Suggest an appropriate method of sampling which could be used to ascertain the following:
   a variations in industrial land use in a large town by fieldwork and using a map,
   b the mean width of a raised beach,
   c the relative importance of heather in a small area of moorland,
   d the origins of shoppers to the central shopping area of a town on a Saturday.

3 Study Figure 6.52 which shows the land use within a 40 km² sector of west Norfolk. The area is divided into three main soil types in which a cut-off channel separates the zone of drained fen peats to the west from the zones of free-draining brown earths and of podsolic soils to the east. The soil zones are defined on the map.

▼ Figure 6.52

Land use in a 40 km² sector in west Norfolk

land use	arable		market gardening		grassland		woodland		other land uses	
soil zone	tally	percentage	tally	percentage	tally	percentage	tally	percentage	tally	percentage
zone of drained fen peats										
zone of podsolic soils										

Standard error of the estimate formula: $SE = \sqrt{\dfrac{p \times q}{n}}$ where SE = standard error
p = percentage of land in a certain category
q = percentage of land *not* in that category
n = number of points in the sample
$\sqrt{}$ = square root

Standard error calculations
(a) (i) for the zone of drained fen peats

a) (ii) for the zone of podsolic soils

arable	market gardening	grassland	woodland
SE =	SE =	SE =	SE =

a Undertake a systematic sampling of the five land use categories defined in the table for the zone of drained fen peats and for the zone of podsolic soils in order to estimate the percentage of land occupied by various land uses. Use all the grid intersections within the respective areas as sampling points, including those on the margins of the map. Record your tallies in the spaces provided in the table and convert the tallies into percentages. *Note:* the four grid intersects falling within the cut-off channel and its enclosing banks are not to be considered).

b For the zone of the drained fen peats, calculate the standard error of the estimate for each of the arable land and the market gardening land.

c For the zone of podsolic soils, calculate the standard error of the estimate for each of the grassland and the woodland.

d Comment upon the usefulness of the standard error calculations undertaken in answers to **b** and **c**.

▲ Figure 6.52, *continued*

References

An Introduction to Coastal Geomorphology, John Pethick, Edward Arnold, 1984.

Coastal Dunes, Geography Today, ITV.

Coastal Management, Geography 16–19 Project, Longman, 1985.

Defence Against the Sea, Place and People, ITV.

Estuary Management and Planning, Geography 16–19 Project, Longman, 1985.

Modern Concepts in Geomorphology, Patrick McCullagh, Oxford University Press, 1978.

Pattern and Process in Human Geography, Vincent Tidswell, UTP, 1976.

Process and Landform, A. Clowes and P. Comfort, Oliver & Boyd, 1982.

Science in Geography, Books 1 to 4, Oxford University Press, 1974.

Statistics in Geography for 'A' Level Students, John Wilson, Schofield & Sims, 1984.

The Nature of the Environment, Andrew Goudie, Blackwell, 1984.

The Unquiet Landscape, D. Brunsden and J. Doornkamp (Editors), David & Charles, 1977.

Waves and Beaches, Geography Today, ITV.

CHAPTER 7

Deserts

'Little by little the sky was darkened by the mixing dust, and the wind felt over the earth, loosened the dust and carried it away'
The Grapes of Wrath, John Steinbeck, Penguin, 1939

What is a desert?

A desert environment has conventionally been described in terms of its deficiencies — water, soils, vegetation and population. Deserts include those parts of the world which produce the smallest amount of organic matter and have the lowest net primary production (NPP, page 260). More precisely, the term 'desert' refers to environments with a water deficit, the hot arid lands, rather than to those characterised by low temperatures, such as Arctic areas. In reality, many desert areas have potentially fertile soils, evidenced by successful irrigation schemes, all have some plant and animal life, even if special adaptations are necessary for their survival, and some are populated by humans, either seasonally by nomads or permanently by mineral companies.

The traditional definition of a desert is an area receiving less than 250 mm of rain per year. Very few areas receive no rain at all — the coast of the Atacama, Figure 7.2, allegedly the only truly rainless desert, in fact is often shrouded in fog. Amounts of precipitation are small and occurrences are infrequent, unreliable and often in the form of heavy downpours. Climatologists have sometimes tried to differentiate between cold deserts where for at least one month a year the mean temperature is below 6°C, and hot deserts. Several geomorphologists have used this to distinguish the landforms found in the hot subtropical deserts, i.e. our usual mental image of a desert, from those found in colder latitudes, e.g. the Gobi and the tundra.

Modern attempts to define deserts are more scientific and are specifically linked to the water balance. This approach is based on the relationship between the input of water as precipitation (P), the output of moisture resulting from evapotranspiration (E) and changes in water held in storage in the ground. In parts of the world where there is little precipitation annually or where there is a seasonal drought, the term potential evapotranspiration (PE) is used, i.e. where the amount of water available is less than the amount which could be evaporated. C.W. Thornthwaite (1931) was the first to define an **aridity index** using this relationship between precipitation and potential evapotranspiration (Figure 7.1).

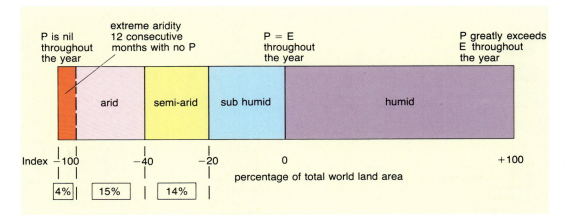

▶ Figure 7.1
The index of aridity

Deserts

Explanations for deserts

As shown in Figure 7.2, the majority of deserts lie in the centre or on the west coast of continents between 15° and 30° north or south of the Equator. This is the zone of subtropical high pressure where air is subsiding (the descending limb of the Hadley Cell, Figure 9.33). On page 188 there is an explanation as to how warm, tropical air is forced to rise at the Equator, producing convectional rain, and how later that air, once cooled and stripped of its moisture, descends at approximately 30° north and south of the Equator. As this air descends it is compressed, warmed and produces an area of permanent high pressure. As the air warms it can hold an increasing amount of water vapour which causes the lower atmosphere to become very dry. The cloudless skies are a result of low relative humidity combined with the fact that there is little surface water to evaporate.

A second explanation for deserts is the **rain shadow** effect produced by tall mountain ranges. As the prevailing winds in the subtropics are the trade winds, blowing from the NE in the northern hemisphere and the SE in the southern, then any continental barrier, such as the Andes or Rockies, prevents moisture from reaching the western slopes. Where plate movements have pushed up mountain ranges in the east of a continent then the rain shadow effect creates a much larger extent of desert (e.g. 82 per cent of

land area in Australia) than when the mountains are to the west, as in the Americas.

Aridity is increased as the trade winds blow towards the Equator, becoming warmer and therefore drier. Where the trades blow from the sea, any moisture which they might have held will be precipitated on eastern coasts leaving little moisture for mid-continental areas. The three major deserts in the northern hemisphere which lie beyond the subtropical high pressure zone (the Gobi and Turkestan in Asia and the Great Basins of the USA) are mid-continental regions far removed from any rain bearing winds while at the same time they are surrounded by protective mountains.

A third combination of circumstances giving rise to deserts is also shown in Figure 7.2: several deserts lie along western coasts where the ocean water is cold. In each case the prevailing winds blow parallel to the coastline and due to the earth's rotation these tend to push surface water seaward at right angles to the wind direction. The Coriolis force (page 186) pushes air and water coming from the south towards the left in the southern hemisphere and water from the north to the right in the northern hemisphere. Consequently, very cold water is drawn upwards to the ocean surface, a process called **upwelling**, to replace that driven out to sea. Any air which then crosses this cold water is cooled and its capacity to

▼ Figure 7.2

Arid lands of the world

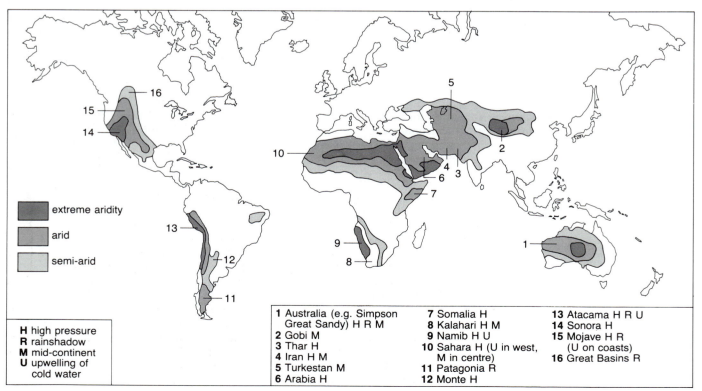

H high pressure
R rainshadow
M mid-continent
U upwelling of cold water

1 Australia (e.g. Simpson Great Sandy) H R M
2 Gobi M
3 Thar H
4 Iran H M
5 Turkestan M
6 Arabia H
7 Somalia H
8 Kalahari H M
9 Namib H U
10 Sahara H (U in west, M in centre)
11 Patagonia R
12 Monte H
13 Atacama H R U
14 Sonora H
15 Mojave H R (U on coasts)
16 Great Basins R

hold moisture is diminished. Where these cooled winds from the sea blow on to a warm land surface, advection fogs form (page 183).

Thornthwaite's aridity index indicates that 34.6 per cent of the world's land surface is already arid or semi-arid. Alarmingly, this is increasing by an estimated 102 km² every

day. The main cause of this may be a shift in the rainfall belts but it is certainly being abetted through deforestation and over-grazing which reduce the amounts of evapo-transpiration. With less vapour in the atmosphere, precipitation totals are there-fore likely to decrease.

CASE STUDY

The Atacama

Here the prevailing winds blow northwards along the South American coast. These winds and the northward-flowing Humboldt (Peruvian) current over which they blow, are pushed westwards (to the left) by the Coriolis force as they approach the Equator. It is the resulting upwellings of cold water from the deep Peru–Chile sea trench (Figure 1.12) that provide the rich nutrients to nourish the plankton which form the basis of Peru's fishing industry. The upwellings also cool the air which drifts inland over the warmer Atacama desert to produce advection fogs and some vegetation cover in coastal areas.

Desert landscapes — what does a desert look like?

Deserts provide a classic example of how easy it is to portray or to accept an in-accurate mental picture of different places (or people) in the world. What is your image of a desert? Is it a landscape of sand dunes with a camel or palm tree somewhere in the background? Large areas of dunes, known as **erg**, do exist but they cover only about 12 per cent of the world's deserts. The traditional 'Western' movie, as filmed in North America, does not show the cowboy–indian confrontation taking place amid dunes but in areas of bare rock, known as **hamada**, or across stone-covered plains, called **reg**. Figure 7.3 showing the Atacama Desert, shows an area of reg, with some vegetation cover, forming a foreground to barren, rocky volcanic peaks.

Arid processes and landforms

In their attempts to understand the develop-ment of arid landforms, geographers have come up against three main difficulties.

- How should the nature of the weathering processes be assessed? Desert weathering was initially assumed to be largely mechanical and to result from extreme diurnal ranges in temperature. More re-cently the realisation that water is present in all deserts in some form or other has led to the view that chemical weathering is far more significant than had previously been accepted. Latest opinions seem to suggest that the major processes, e.g. exfoliation and salt weathering, may in-

volve a combination of both mechanical and chemical weathering.
- What is the relative importance of wind and water as agents of erosion, transporta-tion and deposition in deserts?
- How important has been effect of climatic change on desert landforms? During the Quaternary ice age, and previously when continental plates were in different lati-tudes, the climate of the present arid areas was much wetter than it is today. How many of the landforms that we see now are therefore relict and how many are still in the process of being formed?

Weathering processes

Coastal deserts, especially those where the climate is modified by cold currents and upwelling, have limited annual and diurnal ranges in temperature — about 15°C and 10°C respectively in the Atacama. Inland areas are characterised by extremes. Because of the lack of cloud cover, day temperatures may exceed 40°C for several months while at night, rapid radiation often causes tempera-tures to fall to zero. Indeed, in some higher latitude deserts, frost shattering is a common process. Under the direct rays of the sun, the surface layers of rock, lacking any protective vegetation cover, may reach 80°C. The dif-ferent types and colours of minerals in most rocks, especially igneous, will heat up and cool down at different rates, causing internal stresses. **Insolation weathering** was thought to cause the surface layers of the rock to peel off—**exfoliation**—or individual grains to break away — **granular dis-integration**.

Deserts

◄ Figure 7.3
The Atacama, a stony and rocky
desert

Where surface layers do peel away, newly exposed surfaces experience pressure release (page 31) which is believed to be a contributory process in the formation of rounded exfoliation domes (e.g. Ayers Rock, Australia, Figure 7.4).

However, there was little evidence of exfoliation on the 4500 year old ancient monuments in Egypt. But it was realised that those found in Lower Egypt, where there is limited rainfall, were more noticeably weathered than those located in Upper Egypt, which experiences an extremely arid climate.

D.T. Griggs conducted a series of laboratory experiments in which he subjected granite blocks to extremes of temperature in excess of 100°C. After the equivalent of almost 250 years of diurnal temperature changes he found no discernible difference in the rock. Later he subjected the granite to the same temperature extremes while at the same time spraying it with water. Within the equivalent of two and a half years of diurnal temperature change he found the rock beginning to crack. His conclusion, which has been widely accepted, was that chemical weathering was far more important in forming desert landscapes than had previously been believed.

◄ Figure 7.4
Ayers Rock, Australia, an
exfoliation dome (compare
Figure 2.5)

Although precipitation in deserts may be limited, rain falls on an average of 17 days a year even in California's Death Valley. The rapid loss of temperature at night frequently produces dew (175 nights a year in Israel's Negev) and coastal deserts, e.g. Atacama and Namib, experience advection fogs. There is sufficient moisture, therefore, to combine with certain minerals and cause the rock to swell (hydration) and the outer layers to peel off (exfoliation).

The most recent experiments in the Mexican desert, using more advanced equipment and techniques than those of Griggs, have shown that temperature extremes, even without moisture, can actually cause hairline cracks in rock. At present it would appear that exfoliation is a consequence of both, or either, mechanical and chemical weathering.

The second major process, **salt weathering**, is quoted by some authorities as mechanical and by others as chemical. In fact this too is probably a combination of both types of process. Salts in rainwater, or salts brought to the surface by capillary action, form crystals because moisture is readily evaporated in the high temperatures and low relative humidities. Further evaporation causes the salt crystals to expand and mechanically to break off pieces of the rock in which they have formed. Subsequent rainfall or dew may be absorbed by salt minerals causing them to swell (hydration) and change chemically when water is added to their crystal structure. Where the salts accumulate near or on the surface, particles may become cemented together to form **duricrusts**.

These hard crusts are classified according to the nature of their chemical composition (students with a special interest in geology or chemistry may wish to research the meaning of the terms **calcretes**, **silcretes** and **gypcretes**). Another form of crust, **desert varnish**, is a hard, dark glazed surface found on rocks which have been coated by a film of iron or manganese oxide following the evaporation of capillary water.

The importance of wind and water

Aeolian (wind) processes

Wind appears to play a more important geomorphological role in deserts than in any other environment. This is partly because there is insufficient vegetation and ground moisture to bind the soil together, and partly because the effects of other processes, e.g. running water and ice, are at a minimum.

Transport

The movement of particles is determined by several factors. Aeolian movement is greatest where winds are strong, turbulent, come from a constant direction and blow steadily for a lengthy period of time. Of considerable importance, too, is the nature of the regolith. It is more likely to be moved if there is no vegetation to bind it together or to absorb some of the wind's energy, if it is dry and unconsolidated, if particles are small enough to be transported, and if material has been loosened by farming practices. While such conditions do occur locally in temperate latitudes, e.g. coastal dunes, summits of mountains and during dry summers in arable areas, the optimum conditions for transport by wind are in arid and semi-arid environments.

Wind can move material by three processes: suspension, saltation and surface creep. The effectiveness of each method is related to particle size (Figure 7.5).

▼ Figure 7.5
Processes of wind transportation

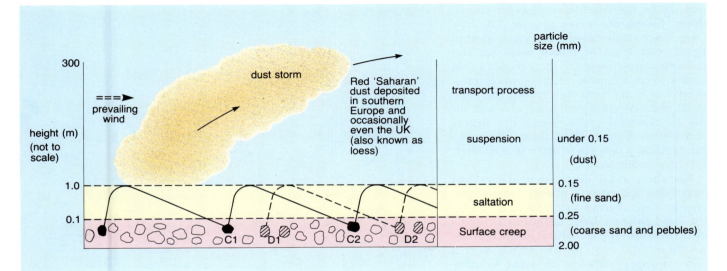

149

Suspension Where material is very fine, i.e. less than 0.15 mm in diameter, it can be picked up by the wind, raised to considerable heights and carried great distances. There have been occasions, though perhaps recorded only once a decade, when red dust from the Sahara has been carried northwards and deposited as 'red rain' over parts of Britain. Visibility in deserts is sometimes reduced to less than 1000 m and this is called a **dust storm**. The number of dust storms on the margins of the Sahara has increased rapidly in the last 20 years as the drought of that region has intensified. In Mauritania there was an average of only five days a year with dust storms during the early part of the 1960s compared with an average of 58 days of a year over a similar period in the early 1980s.

Saltation When wind speeds exceed the threshold velocity (the speed required to initiate grain movement), fine and coarse grained sand particles are lifted up a few centimetres almost vertically before returning to the ground in a relatively flat trajectory of less than 12° (Figure 7.5). As the wind continues to blow, the sand particles bounce along, leapfrogging over one another. Even in the worst storms sand grains are rarely lifted higher than 2 m above the ground.

Surface creep Every time a sand particle, transported by saltation, lands it may dislodge and push forward larger particles of more than 0.25 mm in diameter which are too heavy to be uplifted. This constant bombardment gradually moves small stones and pebbles over the desert surface.

Erosion

There are two main processes of wind erosion: deflation and abrasion.

Deflation is the progressive removal of fine material by the wind leaving pebble-strewn desert pavements or reg (Figure 7.6, 7.7). Over much of the Sahara, especially Sinai in Egypt, vast areas of monotonous flat and colourless pavement are the product of an earlier, wetter climate. Pebbles were transported by water from the surrounding highlands and deposited with sand, clay and silt on the lowland plains. Later the lighter particles were removed by the wind causing the remaining pebbles to settle and to interlock like cobblestones.

Elsewhere in the desert, dew may collect in hollows and material may be loosened by chemical weathering and then removed by the wind to leave large **deflation hollows** or **blow outs**. The Dust Bowl, formed in the American Mid-west in the 1930s, was a consequence of deflation and much valuable

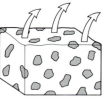
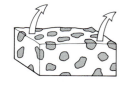

Deflation
silt and sand removed by wind leaving stones

land surface is lowered

Desert pavement — a coarse mosaic of stones resembling a cobbled street which acts as a protection against further erosion

▲ Figure 7.6

The process of deflation

topsoil was lost as a result of heavy ploughing followed by a series of drought years.

Abrasion is a sandblasting action effected by materials as they are moved by saltation. This process smooths, pits, polishes and wears away rock close to the ground (Figure 7.8). Since sand particles cannot be lifted very high, the zone of maximum erosion tends to be within 1 m of the earth's surface. At ground level, saltation is a more efficient method of erosion than abrasion by soil creep. Abrasion produces a number of distinctive landforms including **rock pedestals** (Figure 7.8), **ventifacts** (rocks with many smooth sides), **yardangs** (parallel troughs cut into softer rock, running in the direction of the wind and separated by ridges) and **zeugens** (tabular masses of resistant rock separated by trenches where the wind has cut vertically through the cap into underlying softer rock).

▼ Figure 7.7

A desert pavement created by deflation in the Algerian Sahara

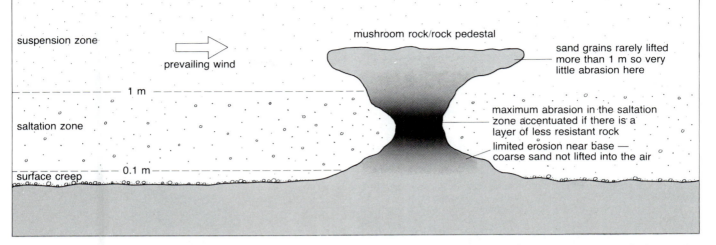

▲ Figure 7.8

Abrasion and the formation of a rock pedestal

Labels on figure:
suspension zone
prevailing wind
1 m
saltation zone
0.1 m
surface creep
mushroom rock/rock pedestal
sand grains rarely lifted more than 1 m so very little abrasion here
maximum abrasion in the saltation zone accentuated if there is a layer of less resistant rock
limited erosion near base — coarse sand not lifted into the air

Deposition

Wind has the competence to move only lightweight particles and much of the finer dust may be deposited outside of the actual desert environment. This leaves only the sand-sized particles to form depositional features, usually in the form of a dune. The main areas of ergs, or sand seas, which cover only 12 per cent of arid lands, occur in the Sahara and Arabian deserts.

Much of the early fieldwork on sand dunes was carried out by R.A. Bagnold. He was sent to Egypt with the British Army in the 1920s and was originally interested principally in trying to discover remains of that country's ancient civilisation. However, his desert expeditions saw this interest change to the studying of dunes. He noted that some dunes formed around an obstacle — a rock, a dead camel, a bush or a small hill — while others appeared to develop on even surfaces. He concentrated on two types of dune: the **barchan** and the **seif**. Textbooks often over-emphasise the barchan, perhaps because of its appealing crescent-shape and the relative ease of explaining its movement (Figures 7.9 and 7.10), yet it is a relatively uncommon feature. The seif dune, named after an Arab curved sword, is much larger and more common. Bagnold claimed that this linear dune, which may extend for over 100 km and reach a height of 200 m, is formed in the direction of the prevailing wind (Figure 7.1). Minor variations in wind direction account for its sinuous shape. It is now accepted that the origins of seifs are far more complicated than Bagnold had first suggested,

Bagnold travelled the desert by jeep. Modern geographers derive their picture of desert landforms using satellite photographs and Landsat images (Figure 15.37).

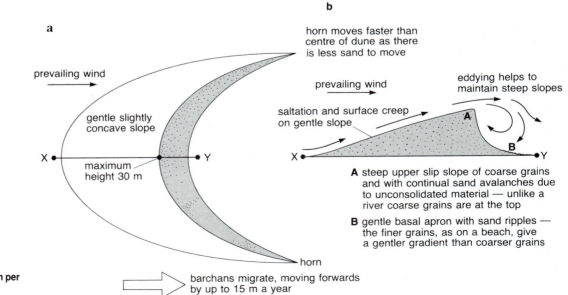

▶ Figure 7.9

The movement of crescent-shaped barchan dunes

a In plan

b Profile — barchans may migrate at a rate of up to 15m per year

Labels on figure a:
prevailing wind
gentle slightly concave slope
maximum height 30 m
X — Y
horn
barchans migrate, moving forwards by up to 15 m a year

Labels on figure b:
horn moves faster than centre of dune as there is less sand to move
prevailing wind
eddying helps to maintain steep slopes
saltation and surface creep on gentle slope
A
B
X — Y
A steep upper slip slope of coarse grains and with continual sand avalanches due to unconsolidated material — unlike a river coarse grains are at the top
B gentle basal apron with sand ripples — the finer grains, as on a beach, give a gentler gradient than coarser grains

▲ Figure 7.12, *centre left*
Longitudinal dunes,
Fuertaventura, Canary Islands
— winds are channelled up a
central trough, blowing sand to
the side and forming long,
symmetrical dunes most
common in the trade wind belt

▲ Figure 7.13, *centre right*
Transverse dunes near Djanet,
Algeria — these are similar to
barchans in their formation but
are straight, perpendicular to the
prevailing wind, rather than
crescent-shaped

◄ Figure 7.14
Star dunes near Arak, Algeria —
their extending arms are the
result of winds blowing from
several directions

This new technique has helped to identify and locate several other types of desert dune. **Longitudinal** dunes are formed by steady winds cutting troughs in the desert floor and piling the sand particles in symmetrical ridges on either side. **Parabolic** dunes are hairpin shaped with their noses pointing downwind. **Transverse** dunes are found where sand is plentiful and they resemble ocean waves with their crests at right angles to the prevailing wind direction. **Star** dunes develop where periodic changes in wind direction cause the dune to resemble a starfish, or arêtes radiating from a central peak.

Q Read the introduction of John Steinbeck's *The Grapes of Wrath*. Why was the area described prone to soil erosion?

◀ Figure 7.11, *top right*
Seif or linear dunes leading to star dunes, Sossusule, Namibia

▼ Figure 7.15
The Grand Canyon, Arizona, USA

The effects of water

It has already been noted that in arid areas, moisture must be present for processes of chemical weathering to operate. We have also seen that although rainfall totals may be low, irregular, infrequent and generally sudden, heavy downpours do occur. Indeed, there are records of individual desert rainstorms being equivalent to three months mean rainfall in London! The impact of rainfall is, therefore, very significant in forming desert landscapes.

Rivers in arid environments fall into three main categories.

Exogenous rivers are those like the Colorado, Nile, Indus and Tigris-Euphrates, which rise in mountains beyond the deserts. These rivers continue to flow throughout the year even if their discharge is reduced by evaporation as they cross the arid land. (The last three rivers mentioned above provided the location for the first urban settlements, page 330) For over 300 km of its course the Colorado has cut down vertically to form the Grand Canyon. The canyon, which in places is almost 2000 m (over one mile) deep, has steep sides because there is insufficient rainfall to degrade them (Figure 7.15).

Endoreic drainage is when rivers terminate in inland lakes. Examples are the River Jordan into the Dead Sea and the Bear into the Great Salt Lake.

Ephemeral streams, which are more typical of desert areas, flow intermittently, or seasonally, after rainstorms. Although often shortlived these streams can generate high levels of discharge because of several local characteristics. First, the torrential nature of the rain exceeds the infiltration capacity of the ground and so most of the water drains away as surface runoff (overland flow, page 45). Second, the high temperatures and the frequent presence of duricrust combine to give a hard, impermeable surface which inhibits infiltration. Third, the lack of vegetation means that no moisture is lost or delayed through interception and the rain is able to hit the ground with maximum force. Fine particles are displaced by rainsplash action and, by infilling surface pore spaces, further reduce the infiltration capacity of the soil. It is as a result of these minimal infiltration rates that slopes of less than 2° can experience extensive overland flow.

◀ Figure 7.16

A wadi, Qumran, Israel.
In the background is the Dead
Sea and beyond it, Jordan

Studies in Kenya, Israel and Arizona suggest that surface runoff is likely to occur within ten minutes of the start of a downpour. This may initially be in the form of a **sheet flood** where the water flows evenly over the land and is not confined to channels. Much of the sand, gravel and pebbles covering the desert floor are believed to have been deposited by this process yet, as the event has rarely been witnessed, it is thought that deposition by sheet floods occurred mainly during earlier wetter periods called **pluvials**.

Very soon the collective runoff becomes concentrated into deep, steep-sided ravines known as **wadis** (Figure 7.16). Normally dry, wadis or **arroyos** may be subjected to irregular flash floods. The average occurrence of these floods is once a year in the semi-arid margins of the Sahara and once a decade in the extremely arid interior. This infrequency of floods compared to the great number and size of wadis, suggests that they are a relict feature, created by processes no longer taking place.

Q The typical storm hydrograph for a wadi (Figure 7.17) contrasts with that of a British river (Figure 3.5).

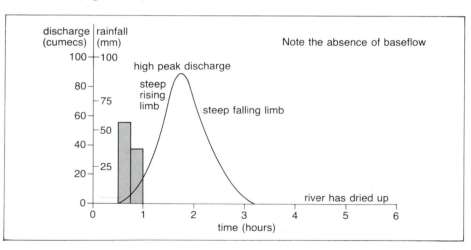

◀ Figure 7.17

Typical storm hydrograph for a
flash flood in a wadi

Can you account for the differences in lag time, the steep rising *and* falling limbs, the high peak discharge and the short duration of flow? The case study which follows may help with your answer.

CASE STUDY

Flash flood in a wadi

Camping in a wadi is something which experienced desert travellers avoid as it is possible to be swept away without any warning of rain, save the possible distant rumble of thunder.

The first warning may be the roar of an approaching wall of water. One minute the bed of the wadi is dry, baked hard under the sun and littered with weathered debris fallen from the steep sides or deposited by the previous flood, while the next minute it is a raging torrent. The energy of the flood enables enormous amounts of coarse material to be taken into suspension — some witnesses have claimed it is more like a mudflow — and large boulders to be moved by traction. Friction from the roughness of the bed, the large amounts of sediment and the high rates of evaporation soon cause a reduction of the stream's velocity and braiding occurs as the deposited material chokes the channel. Within hours the floor of the wadi is dry again. This is because the rapid runoff does not replenish groundwater supplies and without the groundwater contribution to base flow characteristic of humid climates, rivers cease to flow. At the mouth of the wadi, where the water can spread out and energy is dissipated, the river deposits its load as an **alluvial fan** or **cone**. If several wadis cut through a highland close to each other, their semi-circular shaped fans may merge to form a **bahada (bajada)** which is an almost continuous deposit of sand and gravel.

Pediments and playas

Stretching from the foot of the highlands is often a gently sloping area either of bare rock or covered in a thin veil of debris (Figure 7.18). This is the **pediment**. Two main theories suggest the origin of the pediment, one of which involves water. This claims that weathered material from the cliff faces or debris from alluvial fans was carried during pluvials by sheet floods. The sediment planed the lowlands before being deposited, leaving a gently concave slope of less than 7° (Figure 7.18). The alternative theory involves the parallel retreat of slopes resulting from weathering (King's hypothesis, Figure 2.19).

Playas are often found at the lowest point of the pediment. They are shallow, ephemeral, saline lakes formed after rainstorms. As the rainwater evaporates rapidly it leaves flat layers of either clay, silt or salt. Where the dried out surface consists of clay, large

▼ Figure 7.18

Pediments and playas

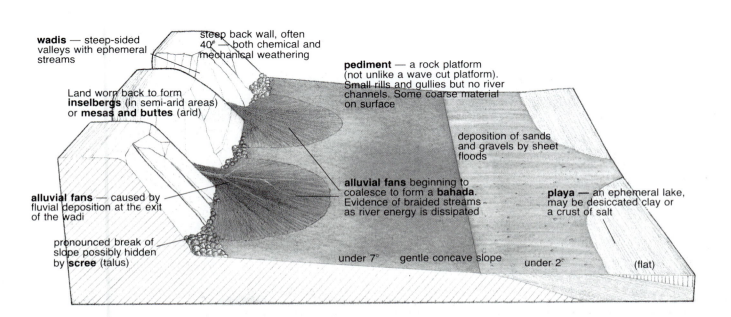

wadis — steep-sided valleys with ephemeral streams

steep back wall, often 40° — both chemical and mechanical weathering

pediment — a rock platform (not unlike a wave cut platform). Small rills and gullies but no river channels. Some coarse material on surface

Land worn back to form inselbergs (in semi-arid areas) or mesas and buttes (arid)

deposition of sands and gravels by sheet floods

alluvial fans — caused by fluvial deposition at the exit of the wadi

alluvial fans beginning to coalesce to form a bahada. Evidence of braided streams as river energy is dissipated

playa — an ephemeral lake, may be desiccated clay or a crust of salt

pronounced break of slope possibly hidden by scree (talus)

under 7°   gentle concave slope   under 2°   (flat)

desiccation cracks, up to 5 m deep, are formed. When the surface is salt-covered, it produces the 'flattest landform on land'. Rogers Lake, in the Mojave desert, California, has been used for spacecraft landings, while the Bonneville salt flats in Utah have been the location for land speed record attempts.

Occasionally, isolated, flat-topped remnants of former highlands rise sheer from the pediment. **Mesas** in Arizona have summits large enough for the Hopi Indians to have built their villages thereon. **Buttes** are smaller versions of mesas. The most spectacular mesas and buttes lie in Monument Valley National Park in Arizona (Figure 7.19).

Climatic change

There have already been several references to pluvials within the Sahara Desert (pages 147, 154). In pre-Quaternary times they may have been when the African plate lay further to the south and the Sahara was in a latitude equivalent to that of the present day savannas. In the Quaternary, the advance of the ice sheets may have pushed the Mediterranean winter rainfall patterns further south. Whatever the cause, there is sufficient evidence to suggest that many desert landforms are relict.

Herodotus, a historian living in ancient Greece, described the Garamantes civilisation which flourished in the Ahaggar Mountains 30 centuries ago. This people, whose exploits have been preserved as cave paintings at Tassili des Ajjers, hunted elephants, giraffes, rhinos and antelope. Twenty centuries ago, North Africa was the 'granary for the Roman Empire'. Wadis are too large and deep and alluvial cones too widespread to have been formed by today's occasional storms, while sheet floods are too infrequent to have recently moved so much material over pediment.

Satellite 'Landsat' photos (Figure 16.41) have revealed many dry valleys radiating from the Ahaggar and Tibesti Mountains which once must have held permanent rivers. Lakes were also once much larger and deeper. Around Lake Chad, shorelines 50 m above its present level are visible and research suggests that these levels might once have been twice that height. (Lake Bonneville in the USA is only one tenth of its former maximum size and like Lake Chad is drying up rapidly.) Small crocodiles found in the Tibesti must have been trapped in the slightly wetter uplands as the desert advanced, while pollen analysis has shown that oak and cedar forests abounded in the same region 20 000 years ago. Groundwater in the Nubian sandstone has been dated, by radio-isotope methods, as being over 25 000 years old and may have accumulated at about the same time as fossil laterite soils (page 268).

Is climate still changing?

Certainly the increased frequency of drought in the African Sahel and surrounding countries since the early 1970s suggest that the answer to this question is 'yes'. Computer forecasts using existing climatic data indicate that changes in the jet stream (page 190) will cause more areas of North Africa and the USA to turn into desert (**desertification**, (page 234). Yet it is undeniable that the problems of drought have been aggravated by human activity. Deforestation, to provide timber for fuel and land for farming, and overgrazing, which has removed the fragile vegetation cover, have bared the soil to the effects of wind and water. The reduction in evapotranspiration from plants has lowered the amount of atmospheric water vapour and reduced the chance of rain.

◄ Figure 7.19

Mesas and buttes in the Monument National Park, Arizona, USA

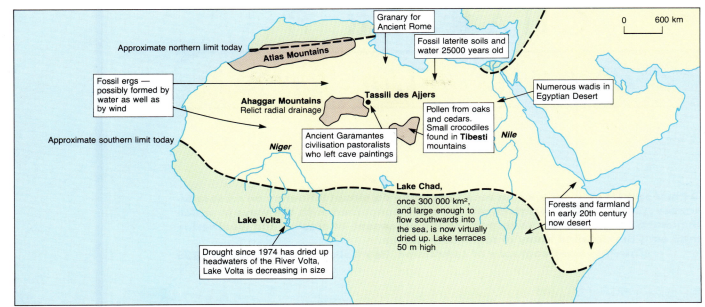

Labels on map:
- Granary for Ancient Rome
- Fossil laterite soils and water 25000 years old
- Approximate northern limit today
- Atlas Mountains
- 0   600 km
- Fossil ergs — possibly formed by water as well as by wind
- Numerous wadis in Egyptian Desert
- Ahaggar Mountains Relict radial drainage
- Tassili des Ajjers
- Pollen from oaks and cedars. Small crocodiles found in **Tibesti** mountains
- Nile
- Approximate southern limit today
- Ancient Garamantes civilisation pastoralists who left cave paintings
- Niger
- Lake Chad, once 300 000 km², and large enough to flow southwards into the sea, is now virtually dried up. Lake terraces 50 m high
- Forests and farmland in early 20th century now desert
- Lake Volta
- Drought since 1974 has dried up headwaters of the River Volta, Lake Volta is decreasing in size

▲ Figure 7.20
Evidence of pluvials in the Sahara

Conclusion: wind versus water — past and present

Satellite photographs have revealed, especially in African deserts, large areas of wind-eroded bedrock (yardangs), longitudinal sand dunes and increased dust storm activity (deflation). Areas where wind appears to be the dominant geomorphological agent are known as **aeolian domains**, although it is still debatable whether such features as dunes are active or form fossil ergs. **Fluvial domains** are those where water processes are dominant or, as evidence increasingly suggests, were dominant in the past. Certainly there appear to be more relict landforms owing their origin to water than to wind.

Yet evidence also suggests that wind and water do interact in arid environments. Landforms produced by each co-exist within the same locality, but the balance between their relative importance has changed significantly over a lengthy period of time. At present it would appear that the rôle of water is declining while that of the wind is increasing, compounded by human mismanagement of areas where the environmental equilibrium is exceptionally fragile.

1 a Describe the circumstances in which moist conditions may be found in hot deserts.

b Discuss the importance of moisture in influencing the rate of rock weathering in hot deserts.

2 Explain why wind is a more important agent of erosion, transportation and deposition in arid areas than in most other environments.

3 Describe and explain how the various processes of aeolian erosion, transportation and deposition have contributed to the development of landforms in hot arid environments.

4 Assess the relative importance of running water in the development of landforms in arid areas.

5 Outline the evidence which has led to the conclusion that hot arid areas have in the past experienced more humid conditions than those which prevail at the present time.

References

Desert Processes: Modern Ideas, Geofile No. 95, Mary Glasgow Publications, 1987.

Desert Processes, Open University, BBC Television.

Modern Concepts in Geomorphology, Patrick McCullagh, Oxford University Press, 1978.

Planet Earth: Arid Lands, Time – Life Books, 1984.

Process and Pattern in Physical Geography, Keith Hilton, UTP, 1979.

The Nature of the Environment, Andrew Goudie, Blackwell, 1984.

CHAPTER 8

Rock types and landforms

'At first sight it may appear that rock type is the dominant influence on most landscapes... As geomorphologists, we are more concerned with the ways in which the characteristics of rocks respond to the processes of erosion and weathering than with the detailed study of rocks themselves.'
The Physical Geography of Landscape, Roy Collard, Unwin Hyman, 1988

Previous chapters have demonstrated how landscapes at both local and global scales have developed from a combination of processes. Plate tectonics, weathering, and the action of moving water, ice and wind both create and destroy landforms. Yet these processes, however important they are at present or have been in the past, are insufficient to explain the many different and dramatic changes of scenery which can occur within short distances, especially in the British Isles.

Each individual rock type is capable of producing its own characteristic scenery.

Landforms are greatly influenced by a rock's vulnerability to weathering, its permeability and its structure.

To show how these three factors affect different rocks and to explain their resultant landforms and potential economic use, four rock types have been selected as exemplars in this chapter. Carboniferous limestone and chalk (both are sedimentary rocks), granite and basalt (both igneous), have been chosen as arguably these produce the most distinctive types of landform and scenery.

> *Lithology* refers to the physical characteristics of a rock

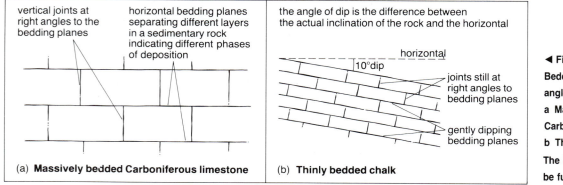

(a) **Massively bedded Carboniferous limestone**
- vertical joints at right angles to the bedding planes
- horizontal bedding planes separating different layers in a sedimentary rock indicating different phases of deposition

(b) **Thinly bedded chalk**
- the angle of dip is the difference between the actual inclination of the rock and the horizontal
- horizontal
- 10°dip
- joints still at right angles to bedding planes
- gently dipping bedding planes

◄ Figure 8.1
Bedding planes with joints and angle of dip
a Massively bedded Carboniferous limestone
b Thinly bedded chalk
The bedding planes in (a) should be further apart than in (b)

Which lithological features affect geomorphology?

Vulnerability to weathering

Mechanical weathering in Britain occurs more readily in rocks which are jointed. Water can penetrate either down **joints** or along **bedding planes** (Figure 8.1) of Carboniferous limestone, or into cracks resulting from pressure release or contraction on cooling within granite and basalt (Figures 1.21 and 1.27). Subsequent freezing and thawing along these lines of weakness causes frost shattering (page 30).

Chemical weathering is a major influence in landforms built of limestone and granite. Limestone, composed mostly of calcium carbonate, is slowly dissolved by the carbonic acid in rainwater, i.e. by the process of carbonation (page 33). Granite consists of quartz, feldspar and mica. It is susceptible to hydration, where water is incorporated into the rock structure causing it to swell and crumble (page 33), and to hydrolysis, when the feldspar is chemically changed into clay (page 33). Quartz, in comparision with other minerals, is one of the least prone to chemical weathering.

Permeability

This is the rate at which water may be stored within a rock or is able to pass through it. Permeability can be divided into two types. **Primary permeability** or **porosity** This depends upon the texture of the rock and the size, shape and arrangement of its mineral particles. The areas between the particles are called **pore spaces**: their size and alignment determine how much water can be absorbed by the rock. Porosity is greatest in rocks which are coarse-grained, e.g. gravels, sands, sandstone, oolitic limestone, and lowest in those which are fine-grained, e.g. clay and granite.

The **infiltration capacity** of sands is estimated to average 200 mm per hour whereas in clay it is only 5 mm per hour. Pore spaces are larger where the grains are rounded rather than angular and compacted (Figure 8.2). Porosity can be given as an index value based upon the percentage of the total volume of the rock which is taken up by pore space, e.g. clay 20 per cent, gravel 50 per cent. When all the pore spaces are filled with water, the rock is said to be **saturated**. The **water table** marks the upper limit of saturation (Figure 8.8). Permeable rocks which store water are called **aquifers**.

Secondary permeability or **perviousness** This occurs in rocks which have joints and fissures along which water can flow. The most pervious rocks are those where the joints have been widened by solution, e.g. Carboniferous limestone, or by cooling e.g. basalt. A rock may be pervious because of its structure, though water may not be able to pass through the rock mass itself.

Where rocks are porous or pervious, water rapidly passes downwards to become groundwater, leaving the surface dry and without evident drainage — chalk and limestone regions have few surface streams. **Impermeable** rocks are those which neither absorb water nor allow it to pass through them, e.g. granite. They therefore have a high drainage density (page 50).

Infiltration capacity is the constant rate at which water percolates into the ground

▶ Figure 8.2

Pore spaces and infiltration capacity

a Large, rounded grains

b Small rounded grains

c Crystals in granite — these fit together more closely than rounded grains, limiting the amount of water held and inhibiting the movement of moisture

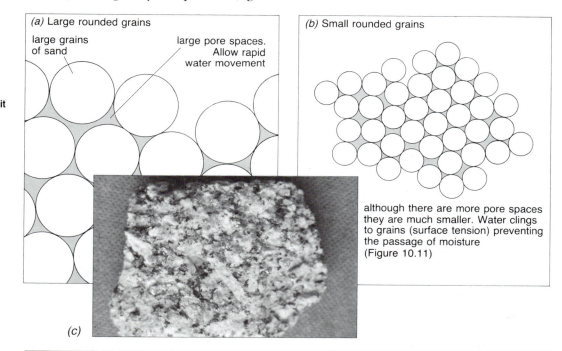

(a) Large rounded grains

large grains of sand

large pore spaces. Allow rapid water movement

(b) Small rounded grains

although there are more pore spaces they are much smaller. Water clings to grains (surface tension) preventing the passage of moisture (Figure 10.11)

(c)

Q Refer to the two storm hydrographs in Figure 3.10. Explain why the first one is likely to be associated with an impervious rock and the second with a permeable rock.

Structure

Resistance to erosion depends upon whether the rock is massive and stratified, folded or faulted. Usually the more massif the rock and the fewer its joints and bedding planes, the more resistant it is to weathering and erosion. Conversely, the softer, more jointed and less compact the rock, the more vulnerable it is to denudation processes. Usually, more resistant rocks remain as upland areas (granite) while those which are less resistant form lowlands (clay).

However, there are exceptions. Chalk, which is relatively soft and may be well-jointed, forms rolling hills because it allows water to pass through it and so fluvial activity is limited. Carboniferous or Mountain limestone, having joints and bedding planes, produces stark plateaux and jagged karst scenery because it is pervious but has very low porosity.

Limestone

Limestone is a rock consisting of at least 80 per cent calcium carbonate. In Britain, most limestone was formed during four geological periods each of which experienced different conditions.

The following list begins with the oldest rocks. Use an atlas to find their location.

Carboniferous limestone is hard, grey, crystalline and well-jointed. It contains many fossils, including corals, crinoids and brachiopods. These indicate that this type of rock was formed on the bed of a warm, clear sea and also provide further evidence that the British Isles once lay in warmer latitudes. Carboniferous limestone has developed its own unique landscape, known as **karst**, and in Britain is seen most clearly in the Peak District and Yorkshire Dales National Parks.

Magnesian limestone, distinctive because it contains a higher proportion of magnesium carbonate, extends in a belt from the mouth of the Tyne to Nottingham. In the Alps it is known as Dolomite.

Jurassic (oolitic) limestone forms a narrow band extending southwards from the North Yorkshire Moors to the Dorset coast. Its scenery is similar to that typical of chalk.

Cretaceous chalk is a pure, soft, well-jointed limestone. Stretching from Flamborough Head in Yorkshire (Figures 6.14 and 6.15), it forms the escarpment of the Lincoln Wolds, the East Anglian Heights and the North and South Downs before ending up as the 'White Cliffs' at Dover, Beachy Head, the Needles and Swanage (Figure 6.17). Cretaceous chalk is assumed to be the remains of small marine organisms which lived in clear, shallow seas.

The most distinctive and extreme of the limestone landforms are found in Carboniferous limestone and chalk.

Carboniferous limestone

This rock develops its own particular type of scenery primarily because of two characteristics. First, it is found in thick beds separated by almost horizontal bedding planes with joints at right angles (Figure 8.1). It is pervious but not porous, meaning that water can pass along the bedding planes and down joints but not through the rock itself. Second, calcium carbonate is soluble. Carbonic acid in rainwater, together with humic acid if the bedrock is overlain by moorland, dissolves the limestone and widens any weaknesses in the rock, i.e. the bedding planes and joints. As there is minimum surface drainage and little breakdown of bedrock to form soil, the vegetation cover tends to be thin or absent. In winter this allows frost shattering to produce scree at the foot of steep cliffs or scars (e.g. Goredale Scar, Figure 2.1).

It is possible to classify landforms into four types.

1 **Surface features caused by solution**
   Limestone pavements are flat areas of exposed limestone. They are flat because they represent the base of a dissolved bedding plane, and exposed because the surface soil may have been removed by glacial activity and never replaced. Where joints reach the surface they may be widened by the acid rainwater (carbonation, page 33) to leave deep gashes called **grikes**. Some grikes at Malham in NW Yorkshire are 0.5 m wide and up to 2 m deep. Between the grikes are flat-topped yet dissected blocks referred to as **clints** (Figure 2.6). In time the grikes widen and the clints are weathered down until a lower bedding plane is exposed and the process of solution—carbonation is repeated.

2 **Drainage features** Rivers which have their source on surrounding impermeable rocks, such as the shales and grits of Northern England, may disappear down **swallow holes** or **sinks** as soon as they flow on to the limestone (Figure 8.3). The streams flow underground finding a pathway down enlarged joints, forming

▼ Figure 8.3
A stream disappearing down a swallow hole near Malham Cove, Yorkshire Dales National Park

▲ Figure 8.4
Resurgence — a river reappears at the foot of Malham Cove.

▶ Figure 8.5 , *above right*
A dry valley above Malham Cove — note the scars and scree of the limestone plateau, and the dry waterfall

▶ Figure 8.6
Stalactites, stalagmites and pillars, Lapa Doce Cave, Chapada, Diamantina National Park, Brazil

potholes, and along bedding planes. Where solution is more active, underground **caves** may form. Corrosion often widens the caverns until parts of the roof collapse, providing the river with angular material ideal for corrasion. Heavy rainfall very quickly infiltrates downwards so caverns and linking passages may become filled within minutes. The resultant turbulent flow can transport large stones and the floodwater may prove fatal to cavers and potholers. Rivers make their way downwards, often leaving caverns abandoned as the water finds a lower level, until they reach an underlying impermeable rock. A **resurgence** is where the river later reappears on the surface. This emergence often occurs at the junction of permeable and impermeable rocks (Figure 8.4)

3 **Surface features resulting from underground drainage** Steep-sided valleys are likely to have been formed as rivers flowed over the surface of the limestone — this was probably during periglacial periods when permafrost acted as an impermeable layer. Once the rivers were able to revert to their subterranean passages, the surface valleys were left dry (Figure 8.5). Should a series of underground caverns become too large to support the rock above them, then the limestone will collapse to form a **gorge**. If the area above an individual cave collapses then a small surface depression called a **doline** is formed. In Yugoslavia, from where the term 'karst' originates, huge depressions called **poljes** may have formed in a similar way to a doline — poljes may be up to 400 km² in area.

4 **Underground depositional features** Groundwater may become saturated with calcium bicarbonate which is formed by the chemical reaction between carbonic acid in the rainwater and the calcium carbonate rock. However, when this 'hard' water reaches a cave, much of the carbon dioxide bubbles out of solution back into the air, i.e. the process of carbonation in reverse. Aided by the loss of some moisture by evaporation, calcium carbonate (calcite) crystals are subsequently precipitated. Water dripping from the ceiling initially forms pendant soda straws which over a very long period of time may grow to produce icicle-shaped **stalactites** (Figure 8.6). Experiments in Yorkshire caves suggest that stalactites grow at about 7.5 mm per

year. As water drips on to the floor, further deposits of calcium carbonate form the more rounded or cone-shaped **stalagmites** which may, in time, join the stalactites to give **pillars**. Elsewhere, water trickles down rocks leaving a layer or curtain of calcite crystals called a **flowstone**.

Economic value of Carboniferous limestone

Human settlement on this type of rock is usually limited and dispersed (page 336) due to the poor natural resources, especially the lack of water and good soil. Villages such as Castleton (Derbyshire) and Malham (Yorkshire) have grown up near to a resurgence.

Limestone is often quarried for the cement and steel industries but the resultant scars have led to considerable controversy. The conflict is between the economic advantages of extracting a valuable raw material which provides local jobs, and the visual eyesore,

noise, dust and extra traffic resulting from the operations, e.g. Hope Valley in N Derbyshire.

Farming is hindered by the dry, thin, poorly developed soils for although most upland limestone areas of Britain receive high rainfall totals, water soon flows underground. The rock does not readily weather into soil-forming particles, such as clay or sand, but is dissolved and the residue is then leached (page 214). On hard limestones a rendzina soil may develop (page 230). These soils are unsuitable for ploughing and their covering of short, coarse, springy grasses favours only sheep grazing. In the absence of hedges and trees, drystone walls are commonly used for field boundaries. The often spectacular rural scenery attracts walkers and school parties while the underground features lure the more adventurous cavers, potholers and **speleologists**.

It is estimated that over 50 per cent of limestone pavements have been damaged by walkers or removed for ornamental stone.

Speleology is the *scientific study of caves*

Figure 8.7 is a model showing typical landforms found in an area of Carboniferous limestone or karst. The features have been labelled *A–Z*.

1 Using the list below, match each term with the appropriate letter.
Abandoned caves; bedding planes; cavern; clints and grikes; dolines; dry valley; gorge; impermeable rock (twice); joints; limestone pavement; limestone plateau; permeable rock; pillar; pothole; resurgence; scar; scree; stalactites; stalagmites; surface river (four times); swallow hole; underground river.

2 Explain the particular location of each feature.

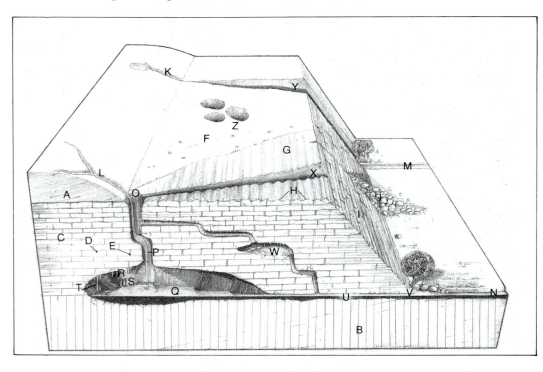

◀ Figure 8.7

Characteristic features of Carboniferous limestone scenery

Chalk

Chalk, in contrast to Carboniferous limestone scenery, consists of gently rolling hills with rounded crest lines. Typically, chalk has steep, rather than gorge-like, dry valleys and is rarely exposed on the surface (Figure 8.8). The most distinctive feature of chalk is probably the **escarpment**, or **cuesta**, e.g. the North and South Downs (Figure 8.9). In these areas the chalk, a pure form of limestone, was gently tilted by the earth movements associated with the collision of the African and Eurasian plates (Chapter 1). Subsequent erosion has left a steep **scarp slope** and a gentle **dip slope**. In SE England, **clay vales** are found at the foot of the escarpment (Figure 3.53).

Like Carboniferous limestone, chalk has very little surface drainage yet its surface is covered in numerous **dry valleys** (Figure 8.10). It has already been explained how chalk, being a primary permeable or porous rock, is able to absorb and to allow rainwater to percolate through it. So how could valleys have been formed without surface rivers?

Goudie lists 16 different hypotheses that have been put forward at various times. He divides these into three categories.

- **Uniformitarian** These theories assume that there have been no major changes in climate or sea level and that 'normal' i.e. fluvial, processes of erosion have operated without interruption. A typical scenario would therefore be that the chalk was once covered by an impermeable rock and so the present rivers were superimposed upon it (page 73).
- **Marine** These hypotheses are related to changes in sea level or base level (page 69). One, which has a measure of support, suggests that when sea levels rose eustatically at the end of the last ice age (page 100), water tables and springs must also have risen higher. As base levels have

fallen since those times, so too have the water table and the spring line causing valleys to become dry.

- **Palaeoclimatic** This group of theories, based on climatic changes during and since the ice age, are the most widely accepted. One hypothesis claims that under periglacial conditions any water in the pore spaces would have been frozen, causing the chalk to behave as an impermeable rock. As temperatures were low, most precipitation would fall in the form of snow and any water from snow melt would have to flow over the surface, cutting characteristic river 'V' shaped valleys (Figure 8.10).

An alternative hypothesis stems from occasions when an area receives excessive amounts of rainfall and streams temporarily reappear in dry valleys. Climatologists have shown that there have been times since the ice age when rainfall was considerably greater than it is today. Figure 8.8 shows the normal water table with its associated spring line. If there is a wetter than average winter, when moisture loss through evaporation is at its minimum, then the level of permanent saturation will rise. Notice that the wet weather water table causes a rise in the spring line and so seasonal rivers, or **bournes**, will flow in the normally dry valleys. Remember also that there will be a considerable lag time between the peak rainfall and the time that the bournes will begin to flow. The springs are the source of obsequent streams (Figure 3.53).

The presence of coombe deposits, resulting from solifluction (page 106), also links chalk landforms with periglacial conditions.

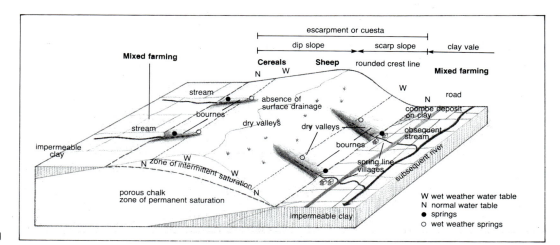

▶ **Figure 8.8**
Scarp and vale scenery — idealised section through a chalk escarpment in SE England

Economic value of chalk

The main commercial use of chalk is in the production of cement but there are objections on environmental grounds to both quarries and the processing works. Settlement tends to be in the form of nucleated villages strung out in lines along the foot of an escarpment, originally being located here to take advantage of the assured water supply from the springs (Figures 8.8 and 8.9). Water-storing chalk aquifers have long been used as a natural, underground reservoir by inhabitants of London although recent increases in demand have exceeded the rate at which this artesian water has been replaced and the result has been a lowering of the water table.

Chalk weathers into a thin, dry, calcareous soil with a high pH reading. Until the twentieth century the springy turf of the downlands was used mostly to graze sheep and for training race horses. Horse racing is still important locally, e.g. at Epsom and Newmarket, but much of the land has been ploughed and is now converted to the growing of wheat and barley. In places, the chalk is covered by a residual deposit of **clay-with-flints** which may have been an insoluble component of the chalk or left from a former overlying rock. This soil is less porous and more acidic than the calcareous soil and several such areas are covered by beech trees — or were before the violent storm of October 1987. Flint has been used as a building material and was the major source for Stone Age tools and weapons.

Granite

Granite was formed when magma was intruded into the earth's crust. Initially, as on Dartmoor and in Northern Arran, the magma created a deep-seated, domed-shaped batholith (page 27), since which time the rock has been exposed by various processes of weathering and erosion. Having been formed at a depth and under pressure, the rate of cooling was slow and this enabled large crystals of quartz, mica and feldspar to form. As the granite cooled it contracted and a series of cracks were created vertically and horizontally, at irregular intervals. These may have been further enlarged, millions of years later, by pressure release (Figure 8.11) as overlying rocks were removed. The coarse-grained crystals render the rock non-porous but while many texts quote granite as an example of an impermeable rock, water may find its way along the many cracks. Certainly, granite often has a high drainage density and as it is usually found in those parts of Britain with heavy rainfall it is often covered by marshy terrain.

◄ Figure 8.9

Scarp slope and clay vale, South Downs near Fulking, Sussex (compare Figure 14.4)

▲ Figure 8.10, *above right*

A dry valley in chalk — Devil's Dyke, South Downs

▼ Figure 8.11

Jointing in granite, Caisteal Abhail, Arran

Although a hard rock, granite is susceptible both to physical and chemical weathering. The joints, which can hold water, are widened by frost shattering while the different rates of expansion and cooling of the various minerals with the rock causes granular disintegration (page 31). The feldspar and, to a lesser extent, mica can be changed chemically by hydrolysis. Hydrolysis is when the H+ ions in water displace mineral cations (cation exchange, page 222). This means that calcium, potassium, sodium, magnesium and, if the pH is less than 5.0, iron and aluminium, are released from the chemical structure. The feldspar is changed into a whitish clay called **kaolin** (or china clay). Quartz, which is not affected by chemical weathering, remains as loose crysals (Figure 2.5).

The most distinctive granite landform is the tor (Figure 8.12). There are two major theories to explain its formation based on physical and chemical weathering respectively. Both, however, suggest the removal of material by solifluction and hence lead to the opinion that tors are relict features.

The first hypothesis suggests that blocks of exposed granite were broken up, sub-aerially, by frost shattering during periglacial times. The weathered material was moved downhill by solifluction leaving the more resistant rock upstanding on hill summits.

The second, proposed by D.L. Linton and more widely accepted, suggests that joints in the granite were widened by sub-surface chemical weathering (Figure 8.13). Linton suggested that deep weathering occurred during the warm Pliocene period (Figure 1.1) when rainwater penetrated the still unexposed granite. As the joints widened, roughly rectangular blocks or **core-stones** were formed. The weathered rock is believed to have been removed by solifluction during periglacial times to leave outcrops of granite tors, separated by shallow depressions. The spacings of the joints is believed to be critical in tor formation — large resistant core-stones have been left where joints were spaced far apart; where they were closely packed and weathering was more active, clay-filled depressions have developed. The rounded nature of the core-stones (Figure 8.13) is caused by spheroidal weathering.

It is thought that a similar process of chemical weathering has produced the much larger **inselbergs**, found in tropical areas, which are similar in structure. Tors have also developed in the millstone grit of the Pennines.

Economic value of granite

As a raw material, granite can be used for building purposes — Aberdeen is the 'granite city' — while kaolin is used in the manufacture of pottery. Peat overlies wide areas of granite bedrock producing a poor, acidic soil which is often saturated with water, forming blanket bogs, and severely gleyed (page 226). The resultant heather-covered moorland is unsuitable for farming but provides ideal terrain for grouse, and army training. With so much surface water and heavy rainfall, granite areas provide ideal sites for reservoirs. Tors, such as Hay Tor on Dartmoor, may become tourist attractions, but granite environments tend to be inhospitable for settlement.

▶ **Figure 8.12**
Hay Tor, Dartmoor

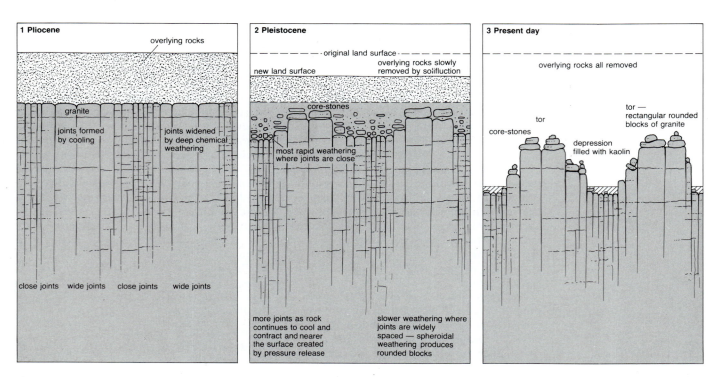

▲ Figure 8.13

The formation of tors

(*after* D.L. Linton)

Basalt

Basalt, unlike granite, was formed on the surface of the earth. As molten rock cooled very rapidly on exposure to the air the resultant crystals were small and fine-grained. As it solidified, the rock contracted, sometimes forming the perfectly shaped hexagonal, columnar jointing as seen at the Giant's Causeway in Northern Ireland (Figure 1.21) and Fingal's Cave on the Isle of Staffa — both of which are important tourist attractions. Elsewhere, basalt has produced extensive lava flows which over a considerable length of time (though faster than in granite) have weathered into a fertile soil, e.g. the Deccan Peninsula of India and the coffee growing regions of SE Brazil.

Q

The following questions are intended to be answered as essays.

1 Discuss the relationship between climate and the process of rock weathering in Carboniferous limestone, chalk, granite and basalt.

2 With reference to Carboniferous limestone, Gault clay and chalk, with granite, discuss the hypothesis that rock type and rock structure are dominant controlling factors in the formation of landforms in these areas.

References

A Dictionary of the Natural Environment, F.J. Monkhouse and J. Small, Edward Arnold, 1978.

Morphology and Landscape, Harry Robinson.

Process and Landform, A. Clowes and P. Comfort, Oliver & Boyd, 1982.

Process and Pattern in Physical Geography, Keith Hilton, UTP, 1979.

The Nature of the Environment, Andrew Goudie, Blackwell, 1984.

The Unquiet Landscape, D. Brunsden and J. Doornkamp (editors), David and Charles, 1977.

Planet Earth: Underground Worlds, Time—Life Books, 1982.

CHAPTER 9

Weather and climate

'When two Englishmen meet, their first talk is of the weather.'
The Idler, Samuel Johnson

'A warmer earth might be a better place to live, but there will be some severe problems of adjustment as some regions warm more than others, and as some areas receive increased rainfall while in others the rainfall declines.'
The Climatic Threat, John Gribbin, Collins Fontana, 1978

The distinction between climate and weather is one of scale. Weather refers to the state of the atmosphere at a local level usually on a time scale of minutes to months. It emphasises aspects of the atmosphere that affect human activity, such as sunshine, cloud, wind, rainfall and temperature. Climate is concerned with the long term behaviour of the atmosphere in a specific area. Climatic characteristics are represented by data on temperature, pressure, wind, precipitation, humidity etc. which are used to calculate daily, monthly and yearly averages (Framework 5, page 208) and to build up global patterns (Chapter 12).

In the study of both climate (**climatology**) and weather, the science of **meteorology**, the study of atmospheres, is used. This has recently been greatly advanced by the use of satellites.

The structure of the atmosphere

The atmosphere is an envelope of transparent, odourless gases held to the earth by gravitational attraction. While the furthest limit of the atmosphere is said by international convention to be at 1000 km, most of the atmosphere, and therefore our climate and weather, is concentrated within 16 km of the earth's surface at the Equator and 8 km at the poles. As shown in Figure 9.1, 50 per cent of atmospheric mass is within 5.5 km of sea level and 99 per cent is within 40 km. Atmospheric pressure also decreases rapidly with height but, as satellite weather stations have recorded, temperature changes are more complicated. It is these changes in temperature which define the four divisions or layers within the atmosphere described below and in Figure 9.1.

Troposphere Temperatures decrease by 6.4°C for every 1000 m in altitude (environmental lapse rate, page 177). This is because the surface is warmed by incoming solar **radiation** which in turn heats the air next to it by conduction. Pressure falls as the effect of gravity decreases but wind speeds usually increase with height. This unstable layer contains most of the atmospheric water vapour, cloud, dust and pollution.

Tropopause This is the upper limit to the earth's climate and weather, marked by an **isothermal** layer where temperatures remain constant despite any increase in height.

Stratosphere This is characterised by a steady increase in temperature (temperature inversion, page 179) caused by a concentration of the gas **ozone** (O_3) which absorbs incoming **ultra-violet (UV) radiation**. Winds, light in the lower parts, increase with height, pressure continues to fall and the air is dry. The stratosphere also acts as a protective shield against meteorites which usually burn themselves out as they enter the earth's gravitational field.

Stratopause This is another isothermal layer where temperatures do not change with an increase in height.

Mesosphere Temperatures fall rapidly as there is no water vapour, cloud, dust or ozone to absorb incoming radiation. This layer has the lowest temperatures (−90°C) and the strongest wind speeds (nearly 3000 km/h) in the atmosphere.

Mesopause This, like the tropopause and stratopause, shows no change in temperature.

Thermosphere Temperatures rise rapidly with height, perhaps reaching 1500°C. This is due to an increasing proportion of atomic oxygen in the atmosphere which, like ozone, absorbs incoming ultra-violet radiation.

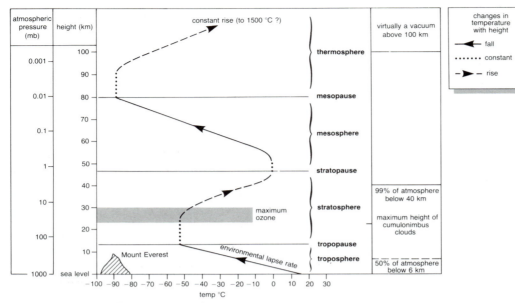

◀ Figure 9.1

The vertical structure of the atmosphere

Composition of the atmosphere

The various gases which combine to form the atmosphere are listed in Figure 9.2. Although nitrogen and oxygen together make up 99 per cent by volume of the atmosphere, changes in the relatively small amounts of ozone and carbon dioxide are causing great concern among scientists (Case Study, page 170).

Energy in the atmosphere

The sun is the earth's prime source of energy. The earth receives energy as incoming **short-wave** solar radiation (also referred to as **insolation**). It is this energy which controls our planet's climate and weather and which, when converted by photosynthesis in green plants, supports all forms of life. The amount of incoming radiation received by the earth is determined by four astronomical factors,

shown in Figure 9.3: the solar constant, the distance from the sun, the altitude of the sun in the sky and the length of night and day. This diagram is theoretical in that it assumes there is no atmosphere around the earth. In reality, much insolation is absorbed, reflected and scattered as it passes through the atmosphere (Figure 9.4).

Absorption of insolation is mainly by ozone, water vapour; carbon dioxide and particles of ice and dust. Clouds and, to a lesser extent, the earth's surface **reflect** considerable amounts of radiation back into space. The ratio between incoming radiation and the amount reflected, expressed as a percentage, is known as the **albedo**. The albedo varies in clouds, from 30 to 40 per cent in thin clouds, to 50 to 70 per cent in thicker stratus and 90 per cent in cumulo-nimbus, when only 10 per cent reaches the atmo-

▼ Figure 9.2

The composition of the atmosphere

Gas		% by volume	Importance for weather and climate	Other functions /source
Permanent gases	nitrogen	78.09		Needed for plant growth
	oxygen	20.95		Produced by photosynthesis — reduced by deforestation
Variable gases	carbon dioxide	0.03	Absorbs long-wave radiation from earth, keeps temperatures steady, 'greenhouse effect'	Used by plants for photosynthesis Increased by burning fossil fuels and by deforestation —
	water vapour	0.20 to 4.0	Source of cloud formation and precipitation, reflects/absorbs incoming radiation	can reach 4% Can be stored as ice/snow
	ozone	0.00006	Absorbs incoming ultra-violet radiation	Reduced/destroyed by chlorofluorocarbons (CFC's)
Inert gases	argon	0.93		
	helium neon krypton	trace		
Non gaseous	dust	trace	Absorbs/reflects incoming radiation Forms condensation nuclei necessary for cloud formation	Volcanic dust Meteoritic dust Soil erosion by wind
Pollutants	sulphur dioxide nitrogen oxide methane	trace	Affects radiation Acid rain	From industry, power stations and car exhausts

(Note: the figures refer to dry air and so the variable amount of water vapour is not usually taken into consideration.)

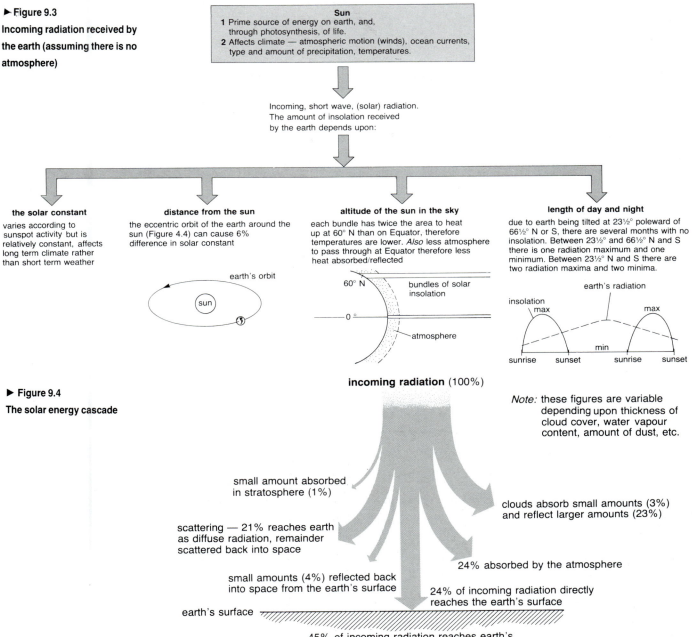

▶ Figure 9.3

Incoming radiation received by the earth (assuming there is no atmosphere)

▶ Figure 9.4

The solar energy cascade

sphere below cloud level. Albedos also vary over different land surfaces, from less than 10 per cent over oceans and dark soil to 15 per cent over coniferous forest and urban areas, 25 per cent over grasslands and deciduous forest, 40 per cent over light-coloured deserts and 85 per cent over reflecting fresh snow. Where deforestation and overgrazing occur, albedo increases. This reduces the possibility of cloud formation and precipitation and increases the risk of **desertification** (page 426).

Scattering occurs when the incoming radiation is diverted by molecules of gas. Scattering takes place in all directions and some will

reach the earth's surface as **diffuse** radiation.

As a result of absorption, reflection and scattering only about 24 per cent of incoming radiation reaches the earth's surface directly, with a further 21 per cent arriving at ground level as diffuse radiation (Figure 9.4).

Incoming radiation is converted into heat energy when it reaches the earth's surface. As the ground warms it radiates energy back into the atmosphere where 94 per cent is absorbed (only 6 per cent is lost to space), mainly by water vapour and carbon dioxide (the greenhouse effect). This outgoing (terrestrial) radiation is **long-wave** or **infra-red** radiation.

169

CASE STUDY

Ozone and carbon dioxide in the atmosphere

Ozone The major concentration of ozone is in the stratosphere 25–30 km above sea level. It acts as a shield, protecting the earth from the damaging effects of ultra-violet radiation from the sun. However, there is serious concern because this shield seems to be breaking down; each spring a hole the size of the USA appears in the ozone layer over Antarctica. In 1989 a similar hole was found to have appeared over the Arctic. It is feared that if more ultra-violet radiation reaches the earth it will increase the incidence of skin cancer (fair skin is at greater risk than dark skin) and inhibit the growing of crops. The damage is believed to be caused by humans releasing into the atmosphere a family of chemicals containing chlorine, which are known as chlorofluorocarbons (CFCs). Chlorine breaks down ozone. Although supersonic aircraft have been blamed, the major culprits include propellants used in aerosols such as hair-spray, deodorants and fly killer, refrigerator coolant and manufacturing processes which produce foam packaging. Scientists claim that a 1 per cent depletion in ozone causes a 5 per cent increase in cases of skin cancer and that this depletion has been 3 per cent since 1970. Following a world conference of 120 nations in 1989, the EC decided to ban all use of CFCs by AD 2000.

In contrast, car exhaust systems generate dangerous quantities of ozone close to the earth's surface causing damage to plant tissues in crops and trees.

Carbon dioxide The burning of fossil fuels and cutting down of the rainforests (which 'absorb' carbon dioxide) have increased the amount of CO_2 (gas) in the atmosphere by 15 per cent in the last century. If this trend continues, CO_2 levels will double by the year AD 2050. Carbon dioxide absorbs outgoing **terrestrial** radiation producing a warming of the lower atmosphere; this is the 'greenhouse effect' (Figure 9.61).

1 Which are the two warm layers in the atmosphere? Why are they warm?

2 Compare characteristics of the troposphere and the stratosphere under the headings:
   a temperature   **b** pressure   **c** water vapour.

3 How do changes in the amount of ozone, water vapour and carbon dioxide in the atmosphere affect the amount of radiation absorbed?

4 Suggest measures which could and should be taken to restrict the potential damage caused by increasing amounts of ozone and carbon dioxide in the atmosphere.

The heat budget

Since the earth is neither warming up nor cooling down, there must be a balance between incoming insolation and outgoing terrestrial radiation. Figure 9.5 shows that:
- there is a net gain in radiation everywhere on the earth's surface except in polar latitudes which have high albedo surfaces,
- there is a net loss in radiation throughout the atmosphere,
- after balancing the incoming and outgoing radiation there is a net surplus between 35°S and 40°N (the difference in latitude is due to the larger land masses of the northern hemisphere) and a net deficit to the poleward side of those latitudes.

There is therefore a **positive heat balance** within the tropics and a **negative heat balance** both at high latitudes (polar regions) and high altitudes, so two major **transfers** of heat take place to prevent the tropical areas from overheating (Figure 9.6).

1 **Horizontal heat transfers** To prevent the Equator from becoming increasingly hot and the poles becoming colder, heat is transferred away from the tropics. Winds (air movements including jet streams, page 190; hurricanes, page 198; and depressions, page 193) are responsible for 80 per cent of this heat transfer, and ocean currents for 20 per cent.

2 **Vertical heat transfers** So that the earth's surface does not get hotter and the atmosphere colder heat must be transferred vertically. This is achieved through **radiation**, **conduction**, **convection** and the transfer of **latent heat** (page 174).

Variations in the radiation balance occur at a number of spatial and temporal scales. Regional differences may be due to the uneven distribution of land and sea and of cloud cover: local variations may be the result of **aspect** or shadow. Seasonal and diurnal (lengths of night and day) variations are related to duration of insolation and are discussed in the following section.

Global factors affecting insolation

Factors which influence the amount of insolation received by any point, and therefore

its radiation balance and heat budget, vary considerably over time and space.

Long term factors
These are relatively constant at a given point.

- **Height above sea level** The atmosphere is not warmed directly by the sun but by heat radiated from the earth's surface and distributed by conduction and convection. As the altitudes of mountains increase they present a decreasing area of land surface from which to heat the surrounding air. In addition, as the density or pressure of the air decreases, so too does its ability to hold heat (Figure 9.1). This is because the molecules in the air which receive and retain heat become fewer more widely spaced as height increases.

- **Altitude of the sun** As the angle of the sun in the sky decreases, the land area to be heated by a given ray and the depth of atmosphere through which that ray has to pass both increase. Consequently the amount of insolation lost through absorption, scattering and reflection is increased. Places in lower latitudes therefore have higher temperatures than those in higher latitudes.

- **Land and sea** Land and sea differ in their ability to absorb, transfer and radiate heat energy. The sea is capable of absorbing heat down to a depth of 10 m as it is more transparent than land, and can then transfer this heat to greater depths through the movement of waves and currents. The sea also has a greater **specific heat capacity** than that of land. Specific heat capacity is the amount of energy required to raise the temperature of one kilogram

▼ **Figure 9.5**
The heat budget

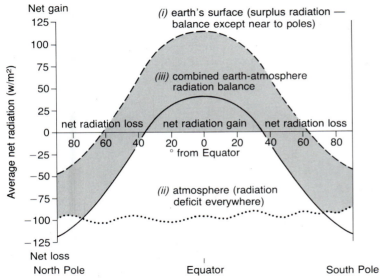

▶ **Figure 9.6**
Heat transfers in the atmosphere

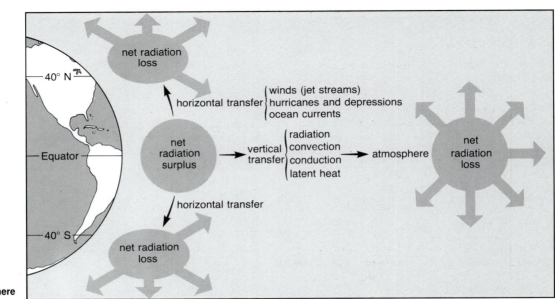

of a substance by 1°C, expressed in kilo-joules per kg per °C. The specific heat capacity of water is 4.2 kJ/kg/°C, soil 2.1 kJ/kg/°C and sand 0.84 kJ/kg/°C. (Expressed in kilo calories, the specific heat capacity of water is 1.0, land 0.5 and sand 0.2.) This means that water requires twice as much energy as soil and five times more than sand to raise an equivalent mass through the same temperature. During summer, therefore, the sea heats up more slowly than the land. On cooling in winter the reverse takes place and land surfaces lose heat energy more rapidly than water. The oceans act as efficient 'thermal reservoirs'. Coastal environments have a lower annual range of temperature than locations at the centre of continents (Figure 9.7).

■ **Prevailing winds** The temperature of the wind is determined by the area where it originates and by characteristics of the surface over which it subsequently blows (Figure 9.8). A wind blowing from the sea tends to be warmer in winter but cooler in summer than the corresponding wind blowing from the land.

■ **Ocean currents** These are a major component in the process of horizontal transfer of heat energy. Warm currents carry water polewards and raise the air temperature of maritime environments to where they flow. Cold currents carry water towards the Equator and so lower the temperature of coastal areas (Figure 9.9).

Temperature **anomalies** show the difference between the mean monthly temperature of a place and the mean monthly temperature for stations with the same latitude. For example, Stornoway (Figure 9.10) has a mean January temperature of 4°C which is 20°C higher than the average for other locations lying at 58°N. Such anomalies result primarily from the uneven heating and cooling rates of land and sea and are intensified by the horizontal transfer of energy by ocean currents and prevailing winds. Temperature anomalies for January are shown in Figure 9.10. Remember that the sun appears overhead in the southern hemisphere at this time of year and isotherms have been reduced to sea level i.e. temperatures are adjusted to eliminate some of the effects of relief, thus emphasising the influence of prevailing winds, ocean currents and continentality.

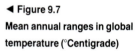

*The **joule** (1 kilojoule = 1000 joules) is a unit of energy or work done and is equivalent to 4.2 calories*

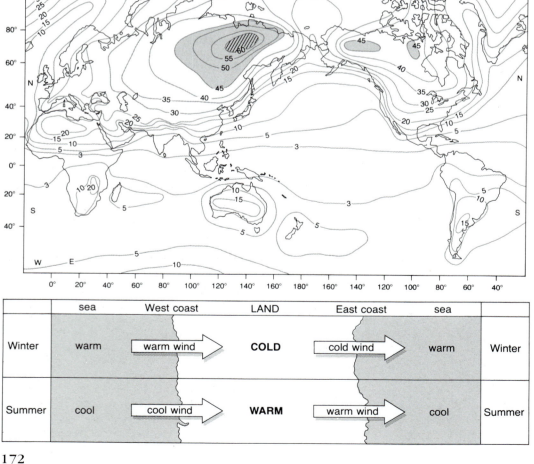

◀ Figure 9.7

Mean annual ranges in global temperature (°Centigrade)

◀ Figure 9.8

Simplified schematic diagram showing the effect of prevailing winds on land and sea temperatures

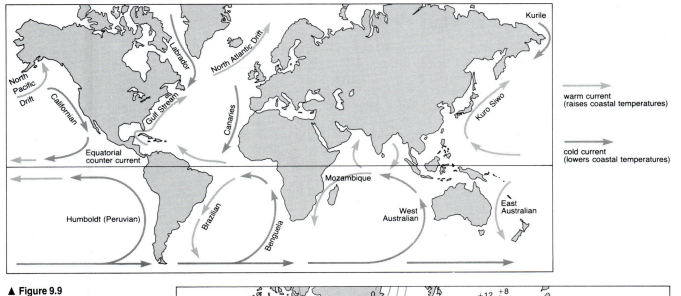

▲ Figure 9.9

Major ocean currents

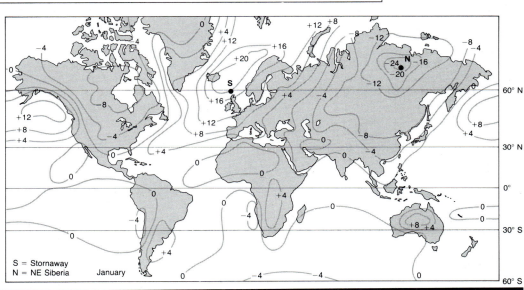

► Figure 9.10

Temperature anomalies
(January) (*after* D.C. Money)

1 With reference to Figure 9.10, how do you account for the following:
   a high negative anomalies of northeast Siberia,
   b high positive anomalies to the north of Scotland,
   c more extreme anomalies which occur in the northern rather than in the southern hemisphere.

2 What difference might you expect to find in temperature anomalies for **a** the British Isles and **b** northeast Siberia in July?

Short term factors

■ **Seasonal changes** At the spring and autumn equinoxes (21 March and 22 September) when the sun is directly over the Equator, insolation is distributed equally between both hemispheres. At the summer and winter solstices (21 June and 22 December) when because of the earth's tilt, the sun is overhead at the tropics, the hemisphere experiencing 'summer' will receive maximum insolation.

■ **Length of day and night** Insolation is only received during daylight hours and reaches its peak at noon. There are no seasonal variations at the Equator, where day and night are of equal length throughout the year. In extreme contrast, polar areas receive no insolation during part of the winter when there is continuous darkness and may receive up to 24 hours of daylight during parts of summer when the sun never sinks below the horizon.

Local influences on insolation

- **Aspect** Hillsides alter the angle at which the sun's rays hit the ground (Figure 9.11). North-facing slopes, being in shadow for most or all of the year, are cooler in the northern hemisphere than those facing south. North and south-facing slopes are referred to, respectively, as the **adret** and **ubac** (the german equivalents for these french terms are sonnenseite and schattenseite). The steeper the south-facing slope, the higher the angle of the sun's rays and the higher the temperature.

- **Cloud cover** The presence of cloud reduces both incoming and outgoing radiation. The thicker the cloud the greater the amount of absorption, reflection and scattering of insolation, and of terrestrial radiation. Clouds may reduce daytime temperatures but they also act as an insulating blanket to retain heat at night. This means that tropical deserts, where skies are clear, are warmer during the day and cooler at night than humid equatorial regions with a greater cloud cover. Tropical deserts therefore have the greatest diurnal range of temperature.

- **Urbanisation** Covering the land with concrete and buildings alters the albedo and creates urban 'heat islands' (page 203).

Atmospheric moisture

Water is a liquid compound which is converted by heat into vapour and by cold into a solid (ice). Its presence serves three essential purposes. First, it maintains life on earth; flora, in the form of natural vegetation (biomes) and crops, and fauna, i.e. all living creatures including humans. Second, water in the atmosphere, mainly as a gas, absorbs, reflects and scatters insolation to keep our planet at a habitable temperature. Third, atmospheric moisture is of vital significance as a means of transferring surplus energy from tropical areas either horizontally to polar latitudes or vertically into the atmosphere to balance the heat budget.

Despite this need for water, its existence in a form readily available to plants and animals is limited. It has been estimated that 97.2 per cent of the world's water is in the oceans and seas; in this form it is only useful to plants tolerant of saline conditions (**halophytes**) and to the populations of a few rich countries who can afford desalinisation plants, e.g. Gulf Oil States. Approximately 2.1 per cent of water in the hydrosphere is held in storage as polar ice and snow. Only 0.7 per cent is fresh water found either in lakes and rivers (0.1 per cent), as soil moisture and groundwater (0.6 per cent) or in the atmosphere (0.001 per cent). At any given time, the atmosphere only holds, on average, sufficient moisture to give every place on the earth 2.5 cm (about 10 days' supply) of rain. There must therefore be a constant recycling of water between the oceans, atmosphere and land. This recycling is achieved through the **hydrological cycle** (Figure 9.12).

Latent heat is the amount of heat energy needed to change the state of a substance without affecting its temperature. When ice changes into water or water into vapour, heat is taken up to help with the processes of melting and evaporation respectively. This absorption of heat results in the cooling of the atmosphere. When the processes work in reverse, i.e. vapour condenses into water or water freezes into ice, then heat energy is released and the atmosphere is warmed.

▼ Figure 9.11

The effect of aspect in an east-west orientated Alpine valley in the northern hemisphere

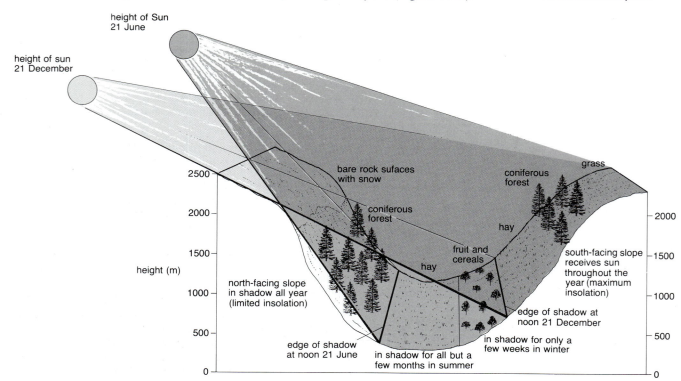

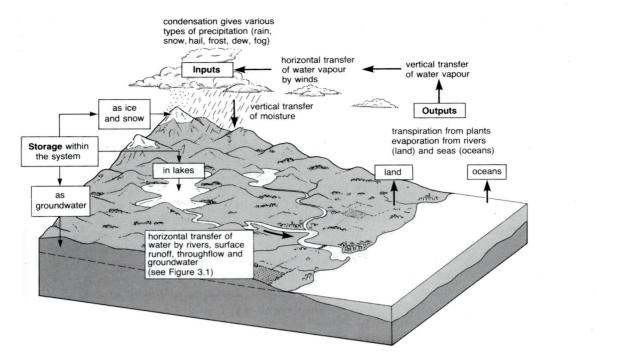

▲ **Figure 9.12**
The hydrological cycle (compare this with Figure 3.1)

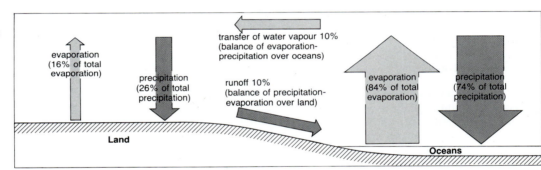

▶ **Figure 9.13**
The world's water balance

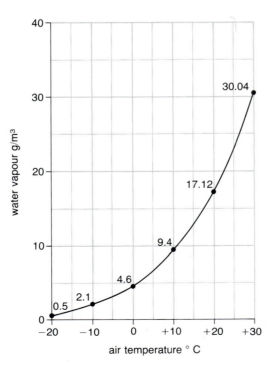

▶ **Figure 9.14**
Air temperatures and absolute humidity

Humidity

Humidity is a measure of the water vapour content in the atmosphere. **Absolute humidity** is the mass of water vapour in a given volume of air measured in grams per cubic metre (g/m^3). **Specific humidity** is similar but is expressed in grams of water per kilogram of air. Humidity depends upon the temperature of the air. At any given temperature there is a limit to the amount of moisture that the air can hold. When this limit is reached the air is said to be **saturated**. Cold air can hold only relatively small quantities of vapour before becoming saturated but this amount increases rapidly as temperatures rise (Figure 9.14). This means that the amount of precipitation obtained from warm air is generally greater than that from cold air.

Relative humidity (RH) is the amount of water vapour in the air at a given temperature as a percentage of the maximum amount of vapour that the air could hold at that temperature. If the RH is 100 per cent then the air is saturated. If it lies between 80 and

99 per cent then the air is said to be 'moist' and the weather is humid or clammy. When the RH drops to 50 per cent the air is 'dry' — figures as low as 10 per cent have been recorded over hot deserts.

If unsaturated air is cooled and atmospheric pressure remains constant, a critical temperature will be reached when the air becomes saturated (i.e. RH = 100 per cent). This is known as **dew point**. Any further cooling will result in the condensation of excess vapour, either into water droplets where condensation nuclei are present (page 181), or into ice crystals if the air temperature is below 0°C. This is shown in the following worked example.

1 The early morning air temperature was 10°C. Although the air could have held 100 units of water at that temperature, at the time of the reading it held only 90. This meant that the RH was 90 per cent.

2 During the day the air temperature rose to 12°C. As the air warmed it became capable of holding more water vapour, up to 120 units. Owing to evaporation, the reading reached a maximum of 108 units which meant that the RH remained at 90 per cent:

$$\left(\frac{108}{120} \times 100\right)$$

3 In the early evening the temperature fell to 10°C at which point, as stated above, it could hold only 100 units. As the air at that time contained 108 units, then as the temperature fell, dew point would be reached and the 8 excess units of water would be lost through condensation.

Condensation

This is the process by which water vapour in the atmosphere is changed into a liquid or, if the temperature is below 0°C, a solid. It usually results from air being cooled until it is saturated. Such cooling may be achieved in four ways.

■ **Radiation** (contact) **cooling** This typically occurs on calm, clear evenings. The ground loses heat rapidly through terrestrial radiation and the air in contact with the ground is then cooled by conduction. If the air is moist, some vapour condenses forming radiation fog, dew, or, if the temperature is below freezing point, hoar frost (page 183).

■ **Advection cooling** This results from warm, moist air moving over a cooler land or sea surface. Advection fogs (page 183) in California and the Atacama desert (page 147) are formed when warm air from the land drifts over adjacent cold coastal ocean currents (Figure 9.9).

As both radiation and advection involve horizontal rather than vertical movements of air the amount of condensation created is limited.

■ **Orographic** and **frontal uplift** Warm, moist air is forced to rise either as it crosses a mountain barrier (orographic ascent) or when it meets a colder, denser mass of air at a front (page 191).

■ **Convective** or **adiabatic cooling** This is when air is warmed during the daytime and rises in pockets as **thermals** (Figure 9.15). As the air expands, it uses energy and so loses heat and the temperature drops. Because air is cooled by the reduction of pressure with height rather than by a loss of heat to the surrounding air, it is said to be adiabatically cooled (see lapse rates below).

As both orographic and adiabatic cooling involve vertical movements of air they are more effective mechanisms of condensation.

Condensation does not occur readily in clean air. Indeed, if air is absolutely pure then it can be cooled below its dew point to become **supersaturated** when it has an RH in excess of 100 per cent. Laboratory tests have shown that clean, saturated air can be cooled to −40°C before condensation, or in this case, **sublimation**. (Sublimation also occurs in the reverse direction from solid to gas.) However, air is rarely pure and usually contains large numbers of condensation nuclei. These microscopic particles are referred to as **hygroscopic nuclei** because they attract water. Hygroscopic nuclei include volcanic dust (heavy rain always accompanies volcanic eruptions); dust from windblown soil; smoke and sulphuric acid originating from urban and industrial areas; and salt from sea spray. Hygroscopic nuclei are most numerous over cities where there may be up to one million per cm^3 and least common over oceans (only 10 per cm^3). Where large concentrations are found,

Sublimation is when vapour condenses directly into ice crystals without passing through the liquid state

▼ Figure 9.15
Convective cooling

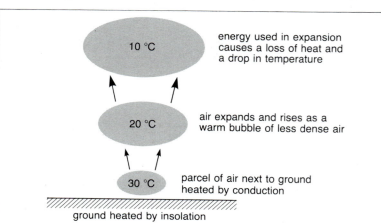

energy used in expansion causes a loss of heat and a drop in temperature

10 °C

air expands and rises as a warm bubble of less dense air

20 °C

parcel of air next to ground heated by conduction

30 °C

ground heated by insolation

condensation can occur with an RH as low as 75 per cent, e.g. as in the smogs of Tokyo and Los Angeles (page 184).

Lapse rates

The **environmental lapse rate** (ELR) is the decrease in temperature usually expected with an increase in height through the troposphere (Figure 9.1). The ELR is approximately 6.5°C/1000 m, but varies according to local air conditions. It may vary with several factors: **height** — ELR is lower nearer ground level, **time** — it is lower in winter or a rainy season, different **surfaces** — it is lower over continental areas, different **air masses** (Figure 9.16a).

The **adiabatic lapse rate** describes what happens when a parcel of air rises and the decrease in pressure is accompanied by an associated increase in volume and a decrease in temperature (Figure 9.15). Conversely, descending air will be subject to an increase in pressure causing a rise in temperature. In either case there is negligible mixing with the surrounding air. There are two adiabatic lapse rates.

1 If the upward movement of air does not lead to condensation then the energy used by expansion will cause the temperature of the parcel of air to fall at the **dry adiabatic lapse rate**, labelled DALR on Figure 9.16b. The DALR, which remains constant, is 9.8°C/1000 m, approximately 1°C/100 m.

2 Where the upwards movement is sufficiently prolonged to enable the air to cool to its dew point and condensation occurs, then the loss in temperature with height is partly compensated by the release of latent heat (Figure 9.16b). Saturated air, which therefore cools at a slower rate than unsaturated air, loses heat at the **saturated adiabatic lapse rate** (SALR). The SALR can vary because the warmer the air the more moisture it can hold and so the greater the amount of latent heat released following condensation. The SALR may be as low as 4°C/1000 m and as high as 9°C/1000 m. It averages about 5°C/1000 m, i.e. 0.5°C/100 m. Should temperatures fall below 0°C, then the air will cool at the freezing adiabatic lapse rate (FALR), which is the same as the DALR as very little moisture is present at low temperatures.

Air stability and instability

Parcels of warm air which rise through the lower atmosphere cool adiabatically. The rate and maintenance of any vertical uplift depends upon the temperature/density balance between the rising parcel and the surrounding air. In a simplified form this balance is the relationship between the environmental lapse rate and the dry and saturated adiabatic lapse rates.

Stability

The state of **stability** is when a rising parcel of unsaturated air cools more rapidly than the air surrounding it. This is shown diagrammatically when the ELR lies to the right of the DALR, as in Figure 9.17. In this example the ELR is 6°C/1000 m and the DALR is 9.8°C/1000 m. By the time the rising air has reached 1000 m it has cooled to 10.2°C which leaves it colder and denser than the surrounding air which has cooled to only 14°C. If there is no mechanism forcing the parcel of air to rise, e.g. mountains or

▼ Figure 9.16

Examples of lapse rates shown in temperature–height diagrams (tephigrams)

a Environmental lapse rates (ELR)

b Adiabatic lapse rates (ALR)

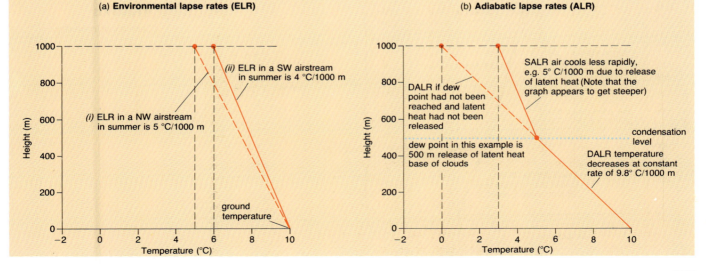

(a) **Environmental lapse rates (ELR)**

(ii) ELR in a SW airstream in summer is 4 °C/1000 m

(i) ELR in a NW airstream in summer is 5 °C/1000 m

ground temperature

(b) **Adiabatic lapse rates (ALR)**

SALR air cools less rapidly, e.g. 5° C/1000 m due to release of latent heat (Note that the graph appears to get steeper)

DALR if dew point had not been reached and latent heat had not been released

dew point in this example is 500 m release of latent heat base of clouds

condensation level

DALR temperature decreases at constant rate of 9.8° C/1000 m

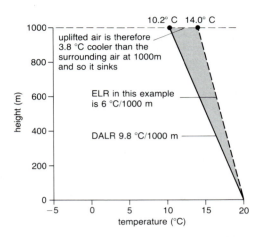

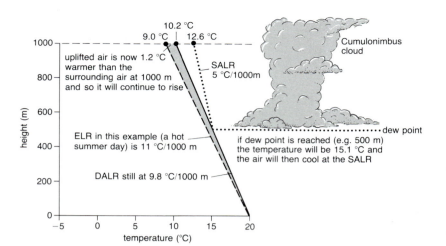

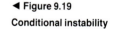

fronts then it sinks back to its starting point. It is described as stable because dew point may not have been reached and the only clouds which might have developed would be shallow, flat-topped cumulus which do not produce precipitation (Figure 9.21). Stability is often linked with anticyclones (page 197) when any convection currents are suppressed by sinking air to give dry, sunny conditions.

Instability

Conditions of **instability** arise in Britain on hot days. Localised heating of the ground warms the adjacent air by conduction, meaning it has a higher lapse rate. The resultant parcel of rising unsaturated air cools less rapidly than the surrounding air. In this case, as shown in Figure 9.18, the ELR lies to the left of the DALR. The rising air remains warmer and lighter than the surrounding air. Should it be sufficiently moist and if dew point is reached, then the upward movement may be accelerated producing towering cumulus or cumulo-nimbus type cloud (Figure 9.21). Thunderstorms are

likely (page 182) and the saturated air will cool at the SALR, following the release of latent heat.

Conditional instability

This type of instability occurs when the ELR is lower than the DALR but higher than the SALR. In Britain it is the most common of the three conditions. The rising air is stable in its lower layers and being cooler than the surrounding air would normally sink back again. However, if the mechanism which initially triggered the uplift remains, then the air will be cooled to its dew point. Beyond this point, cooling takes place at the slower SALR and the parcel may become warmer than the surrounding air (Figure 9.19). It will now continue to rise freely, even if the uplifting mechanism is removed, as it is in an unstable state. Instability is conditional upon the air being forced to rise in the first place, and later becoming saturated so that condensation occurs. The associated weather is usually fine in areas at altitudes below condensation level but cloudy and showery in those above.

◄ Figure 9.17

Stability — changes in lapse rates and air temperature with height

▲ Figure 9.18, *top right*

Instability and the development of clouds

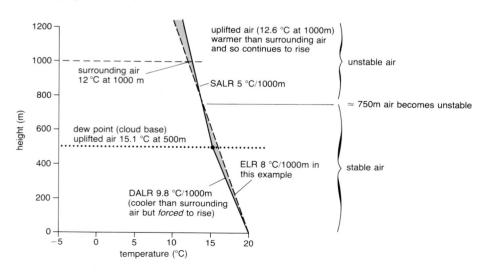

◄ Figure 9.19

Conditional instability

Q

1 Assume that the following conditions occur in three airstreams up to 2000 m above sea level.

Sea level air temperature is always 10°C.
Condensation level is always at 500 m.
DALR is 10°C per 1000 m (1°C/100 m).
SALR is 5°C per 1000 m (0.5°C/100 m).
ELR of airstream 1 is 4°C per 1000 m (0.4°C/100 m).
ELR of airstream 2 is 12°C per 1000 m (1.2°C/100 m).
ELR of airstream 3 is 8°C per 1000 m (0.8°C/100 m).

a On a single sheet of graph paper, plot the data for each of the three air-streams from sea level to 2000 m to show the relationship between the environmental and adiabatic lapse rates.

b Label each graph 'stable', 'unstable', or 'conditionally unstable' as appropriate.

c For the airstream which you have labelled 'conditionally unstable' determine the height above sea level at which the displaced air will become unstable.

2 Study Figure 9.20 which is a temperature/height diagram or **tephigram** showing relationships between adiabatic and environmental lapse rates.

a For the zone below 2000 m, describe the temperature conditions for the environmental situation shown by the curve **C**.

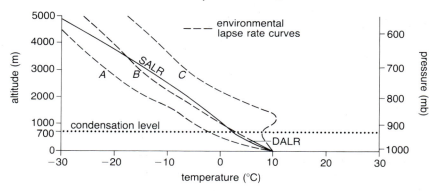

▶ Figure 9.20
Environmental lapse rates (ELR)
and adiabatic lapse rates (ALR)

b State three different mechanisms which are likely to trigger the upward movement of a parcel of air from ground level.

c Given the information in Figure 9.20, for each of the environmental lapse rate curves **A**, **B** and **C**, state which of the three atmospheric conditions of stability, instability and conditional stability is demonstrated.
Justify each of your answers.

d For the curve **B**, assuming that the parcel of air is lifted mechanically to a height of over 1000 m, calculate the lower and upper limits of cloud formation.

e State the weather conditions normally associated with stability and instability.

Temperature inversions

As the lapse rate exercises have shown, the temperature of the air usually decreases with altitude, but there are certain conditions when the reverse occurs. **Temperature inversions**, where warmer air overlies colder air, may occur at three levels in the atmosphere. Figure 9.1 showed that temperatures rise with altitude in both the stratosphere and the thermosphere. Inversions are also found at high and near ground levels in the troposphere. High level inver-sions are found in depressions where warm air overrides cold air at the warm front or is undercut by the colder air at the cold front (page 194). Low level, or ground, inversions usually occur under anticyclonic conditions (page 197) when there is a rapid loss of heat from the ground due to radiation at night, or when warm air is advected over a cold surface (line **C**, Figure 9.20). Under these conditions, fog and frost (page 183) may form in valleys and hollows.

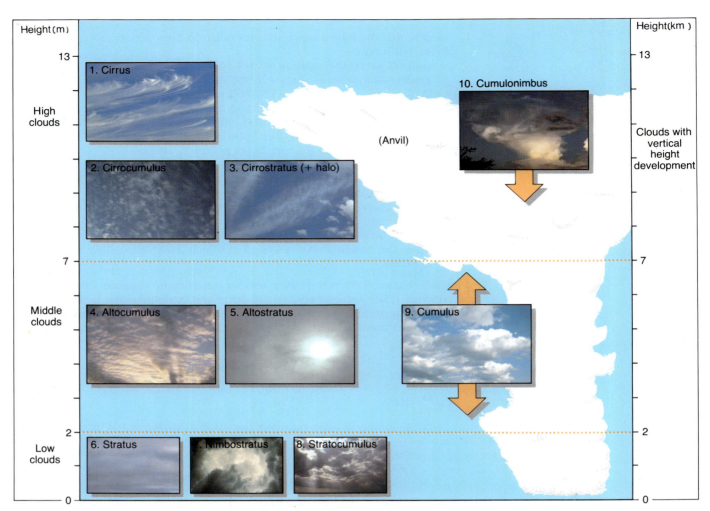

No.	Type	Symbol	Level	Height (km)	Constituent	Description	Precipitation
1	Cirrus	Ci	high	5–13	ice	Detached, wispy, delicate white clouds. May have feathery filaments, known as 'mares tails', indicating strong upper atmosphere winds.	none
2	Cirrocumulus	Cc				Thin layers of small globular masses with a rippled appearance (also known as 'mackerel sky').	
3	Cirrostratus	Cs				A thin, milky layer appearing like a veil. The sun or moon may shine through it with a halo effect.	
4	Altocumulus	Ac	middle	2–7	water and some ice	White-grey cloud usually resembling waves or lumps, separated by patches of blue sky. The sun or moon may be surrounded by a corona.	very occasional and little
5	Altostratus	As				A greyish, uniform sheet of clouds, largely featureless. A 'watery' sun may just be visible.	
6	Stratus	St	low	0–2	water	A persistent, grey uniform sheet of cloud.	drizzle
7	Nimbostratus	Ns				A thick, dark grey-black cloud, usually uniform but may have detached, darker patches beneath it.	continuous rain or snow
8	Stratocumulus	Sc				A grey-white patchy cloud appearing in long rows or in rolls.	occasional showers
9	Cumulus	Cu	high vertical extent	1–12	water	Detached white cloud with a pronounced flat base and sharp outlines — grows vertically and may resemble a cauliflower.	very scattered showers
10	Cumulonimbus	Cb			water and ice	An extreme vertical extension of the cumulus. It may develop an 'anvil' at its head (ice crystal) and become black at its base.	heavy showers, thunderstorms, hail

Clouds

Clouds form when air cools to dew point and vapour condenses into water droplets and/or ice crystals. The general classification of clouds, used in the *International Cloud Atlas*, was proposed by Luke Howard in 1803. His was a descriptive classification, based upon cloud shape and height (Figure 9.21). He used four Latin words: **cirrus** (a lock of curly hair); **cumulus** (a heap or pile); **stratus** (a layer); and **nimbus** (rain bearing). He compiled composite names using these four terms, e.g. cumulonimbus, cirrostratus, and added the prefix alto- for middle level clouds.

Precipitation

How is precipitation formed?

Condensation produces minute water droplets, less than 0.05 mm in radius, and ice crystals if the dew point temperature is below freezing. These droplets are so tiny and weigh so little that they are kept buoyant by the rising air currents which created them. So although condensation forms clouds, clouds do not necessarily produce precipitation. As rising air currents can reach considerable speeds, especially by convection, there has to be a process within the clouds which enables the small water droplets and/or ice crystals to become sufficiently large that they overcome the uplifting mechanism and fall to the ground. One experiment suggested that whereas it would take a droplet of 0.01 mm five days to fall 100 m the equivalent time for one of 3.0 mm would be 1.8 minutes.

There are currently two main theories which attempt to explain the rapid growth of water droplets.

The **ice crystal mechanism** was suggested by Bergeron and later supported by Findeisen and is often referred to as the Bergeron−Findeisen mechanism. It appears that when the temperature of air is between −5°C and −25°C, supercooled water droplets and ice crystals exist together. Supercooling takes place when water remains in the atmosphere when temperatures fall below 0°C — usually due to a lack of condensation neclei. Ice crystals are in a minority because the **freezing nuclei** necessary for their formation are less abundant than condensation nuclei. The relative humidity of air is ten times greater above an ice surface than over water. This means that the water droplets evaporate and the resultant vapour condenses (sublimates) back on to the ice crystals which then grow into hexagonal-shaped snowflakes. The flakes grow in size, as a result of further condensation or by fusion as their numerous edges interlock on collison with other flakes; and increase in number as ice splinters break off and form new nuclei. If the air temperature rises above freezing point as the snow falls to the ground, flakes melt into raindrops. Experiments to produce rainfall artificially by cloud-seeding are based upon this theory.

The Bergeron−Findeisen theory is supported by evidence from temperate latitudes where rainclouds usually extend vertically above the freezing level. Radar and high flying aircraft have reported snow at high altitudes when it is raining at sea level. However, as clouds rarely reach freezing point in the tropics then the formation of ice crystals is impossible in those latitudes.

The **collision and coalescence** process was suggested by Longmuir. 'Warm' clouds, i.e. those containing no ice crystals, found in the tropics, contain numerous water droplets of differing sizes. As a result of friction, droplets of different size are swept upwards at different velocities and collide with other droplets in the process. It is thought that the larger the droplet the greater the chance of collison and subsequent coalescence with smaller droplets. When coalescing droplets reach a radius of 3 mm their motion causes them to disintegrate to form a fresh supply of droplets. The thicker the cloud (cumulonimbus) the greater the time the droplets have in which to grow and the faster they will fall, in the form of more thundery showers.

Latest opinions suggest that these two theories may compliment each other but that a major process of raindrop enlargement has yet to be understood.

Types of precipitation

Although the definition of precipitation includes sleet, hail, dew, hoar frost, fog and rime, only rain and snow provide significant totals in the hydrological cycle.

Rainfall

There are three main types of rainfall, distinguished by the mechanisms which cause the initial uplift of the air. Each mechanism rarely operates in isolation.

1 **Convergent** and **cyclonic** or frontal rainfall results from the meeting of two air streams in areas of low pressure. Within the tropics, the trade winds, blowing towards the Equator, meet at the inter-tropical convergence zone or ITCZ (page 188). The air is forced to rise and in conjunction with convection currents gives rise to the heavy afternoon thunderstorms associated with the equatorial climate. In temperate latitudes, depressions form at the boundary of two air

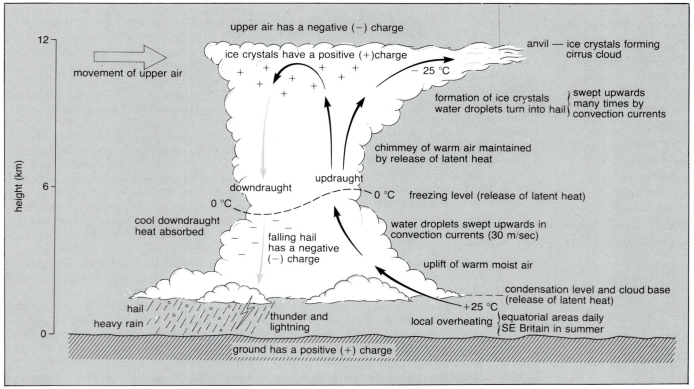

▲ Figure 9.22

Convectional rainfall — the development of a thunderstorm

masses. At the associated fronts, warm, moist, less dense air is forced to rise over colder, denser air to give periods of prolonged and sometimes intense rainfall, often augmented by orographic precipitation.

2 **Orographic** or **relief** rainfall results as near-saturated, warm maritime air is forced to rise when confronted by a coastal mountain barrier. Mountains reduce the water holding capacity of rising air by enforced cooling and increase the amounts of cyclonic rainfall by retarding the speed of depression movement. They also tend to cause air streams to converge and funnel through valleys. Rainfall totals increase where mountains are parallel to the coast, e.g. Canadian Coastal Range, and where winds have crossed warm off-shore water masses, e.g. British Isles. As air descends on the leeward side of a mountain range it becomes compressed, warmed and condensation ceases, creating a **rain shadow** effect where little rain falls.

3 **Convectional** In areas where the ground surface is locally overheated, the adjacent air is heated by conduction, expands and rises. During its ascent the air mass remains warmer than the surrounding environmental air and it is likely to become unstable (page 178) with towering cumulonimbus clouds forming. These unstable conditions, possibly augmented by frontal or orographic

uplift, force air to rise in a 'chimney'. The updraught is maintained by energy released as latent heat at both condensation and freezing levels (Figure 9.22). The cloud summit is characterised by ice crystals in an anvil shape where the movement of upper air flattens the top of the cloud. When the ice crystals and frozen water droplets or hail become large enough, they fall in a downdraught. The air through which they fall remains cool as heat is absorbed by evaporation. The downdraught constitutes a negative feedback to the cloud system (page 44) as it reduces the warm air supply to the 'chimney' and limits the lifespan of the storm. Such storms are usually accompanied by thunder and lightning. One of several theories put forward to account for lightning suggests that the ice crystals in the upper cloud create a positive charge while the earth has a negative charge. **Lightning** is the visible discharge of electricity between clouds or between clouds and the ground.

Convection is one of the processes by which surplus heat and energy from the earth is transferred vertically to the atmosphere in maintaining the heat balance (page 170).

Drizzle is defined as water droplets under 0.5 mm in diameter.

Snow forms under similar conditions to rain (Bergeron–Findeisen process) except that as dew point temperatures are under 0°C,

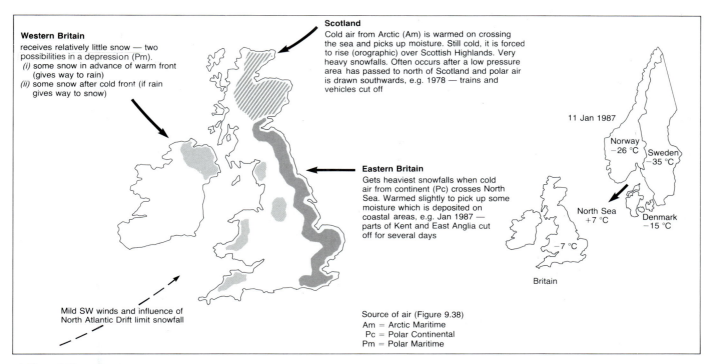

Western Britain
receives relatively little snow — two
possibilities in a depression (Pm).
(i) some snow in advance of warm front
(gives way to rain)
(ii) some snow after cold front (if rain
gives way to snow)

Scotland
Cold air from Arctic (Am) is warmed on crossing
the sea and picks up moisture. Still cold, it is forced
to rise (orographic) over Scottish Highlands. Very
heavy snowfalls. Often occurs after a low pressure
area has passed to north of Scotland and polar air
is drawn southwards, e.g. 1978 — trains and
vehicles cut off

Eastern Britain
Gets heaviest snowfalls when cold
air from continent (Pc) crosses North
Sea. Warmed slightly to pick up some
moisture which is deposited on
coastal areas, e.g. Jan 1987 —
parts of Kent and East Anglia cut
off for several days

Mild SW winds and influence of
North Atlantic Drift limit snowfall

Source of air (Figure 9.38)
Am = Arctic Maritime
Pc = Polar Continental
Pm = Polar Maritime

11 Jan 1987
Norway −26 °C
Sweden −35 °C
North Sea +7 °C
Denmark −15 °C
−7 °C
Britain

▲ Figure 9.23
Causes of uneven snowfall
patterns across Britain

vapour condenses directly into a solid (sublimation). Ice crystals will form if hygroscopic or freezing nuclei are present and these may aggregate to give snowflakes. As warm air holds more moisture than cold, snowfalls are heaviest when the air temperature is just below freezing. As temperatures drop it becomes 'too cold for snow'. Figure 9.23 shows the typical conditions under which snow might fall in Britain.

Sleet is a mixture of ice and snow formed when the upper air temperature is below freezing and allows snowflakes to form and the lower air temperature is around 2° to 4°C which allows partial melting.

Glazed frost is the reverse of sleet and occurs when water droplets form in the upper air but turn to ice on contact with a freezing surface. When glazed frost forms on roads it is known as 'black ice'.

Hail is made up of frozen raindrops which exceed 5 mm in diameter. It usually forms in cumulonimbus clouds, resulting from the uplift of air by convection currents, or at a cold front (page 191). It is more common in areas with warm summers where there is sufficient heat to trigger off the uplift of air, than in colder latitudes. Hail frequently proves a serious climatic hazard in cereal growing areas such as the American Prairies.

Dew, hoar frost and **radiation fog** all form under calm, clear anticyclonic conditions when there is rapid terrestrial radiation at night. Dew point is reached as the air cools by conduction and as moisture in the air, or transpired from plants, condenses. If dew point is above freezing, dew will form; if

it is below freezing then hoar frost develops. Frost may also be frozen dew. Dew and hoar frost rarely occur at more than 1 m above ground level. If the lower air is relatively warm, moist and contains hygroscopic nuclei then if the ground cools rapidly radiation fog may be formed. If visibility is more than 1 km it is mist, if less than 1 km, fog. In order for radiation fog to develop gentle wind is needed to stir the cold air adjacent to the ground so that cooling affects a greater thickness of air. Radiation fogs usually occur in valleys, are densest around sunrise, and consist of droplets which are sufficiently small to remain buoyant in the air. The fog is likely to thicken if temperature inversions take place (Figure 9.24) when cold surface air is trapped by overlying warmer, less dense air. It is under such conditions, in urban and industrial areas, that smoke and other pollutants released into the air are retained as smog (Figure 9.25).

Advection fog results when warm air passes over or meets with cold air to give rapid cooling. In the coastal Atacama Desert (page 147) sufficient droplets fall to the ground as 'fog-drip' to enable some vegetation to grow.

Rime (Figure 9.26) occurs when super-cooled droplets of water, often in the form of fog, come into contact with and freeze upon solid objects, e.g. telegraph poles and trees.

Acid rain is an umbrella term for the presence in rainfall of a series of pollutants which are produced mainly by the burning of fossil fuels. Coal-fired power stations, heavy industry and vehicle exhausts emit sulphur dioxide and nitrogen oxides in

◄ **Figure 9.24**
Temperature inversion in a valley, Iceland

▼ **Figure 9.25**
Formation of radiation fog and smog

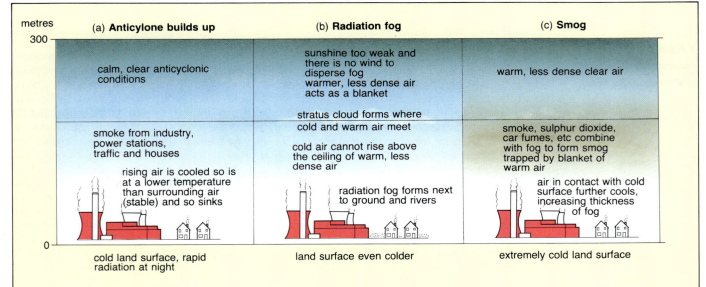

metres 300	(a) **Anticylone builds up**	(b) **Radiation fog**	(c) **Smog**
	calm, clear anticyclonic conditions	sunshine too weak and there is no wind to disperse fog warmer, less dense air acts as a blanket	warm, less dense clear air
		stratus cloud forms where cold and warm air meet	
	smoke from industry, power stations, traffic and houses	cold air cannot rise above the ceiling of warm, less dense air	smoke, sulphur dioxide, car fumes, etc combine with fog to form smog trapped by blanket of warm air
	rising air is cooled so is at a lower temperature than surrounding air (stable) and so sinks	radiation fog forms next to ground and rivers	air in contact with cold surface further cools, increasing thickness of fog
0	cold land surface, rapid radiation at night	land surface even colder	extremely cold land surface

particulate form. These are carried by prevailing winds across seas and national frontiers either to be deposited directly on to the earth's surface as dry deposition or to be converted into acids (sulphur and nitric acid) which fall to the ground in rain as wet deposition. Clean rainwater has a pH value of between 5 and 6 (Figure 10.15). Today rainfall over most of NW Europe is between 4 and 5: the lowest ever recorded is pH 2.4.

The effects of acid rain include increased concentrations of aluminium in lakes, which kills fish — trout and salmon have virtually disappeared from Scandinavia. Forests are being destroyed to the point where forecasts suggest that 90 per cent of West Germany's trees are expected to have died by early next century and consequently wildlife in forest habitats suffers. Water supplies are becoming

◄ **Figure 9.26**
Rime frost, Horseshoe Pass, Llangollen, Wales

184

polluted and acid rain is thought to be related to increasing incidence of Alzheimer's disease (which brings on early senility in humans) bronchitis and lung cancer. Soils are becoming more acidic which reduces future crop yields. Stone buildings are being eroded by chemical action as far afield as St. Paul's Cathedral, the Acropolis and the Taj Mahal and black snow falls annually in the Cairngorms.

Q Describe carefully the factors which are thought to cause acid rain and the effects it has on the environment. How might the problems of acid rain be overcome and what are the difficulties involved in implementing proposed solutions? (You will need to undertake research to answer this question fully.)

World precipitation: distribution and reliability

Although precipitation amounts are unevenly distributed and weather varies greatly, there is an identifiable global climatic pattern (Figure 9.27).

Equatorial areas have high annual totals due to the continuous uplift of air resulting from the convergence of the trade winds and strong convectional currents. There may be a double maximum associated with the passage of the overhead sun and of the ITCZ. Further away from the Equator, rainfall totals

▶ **Figure 9.27**
World precipitation — total amounts and seasonal distribution

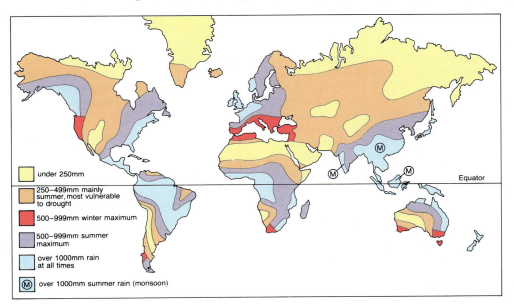

▶ **Figure 9.28**
World precipitation — rainfall reliability

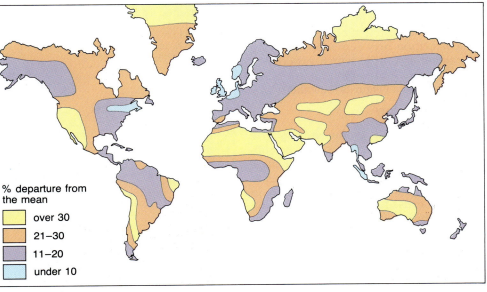

decrease and the length of the dry season increases. These tropical areas experience heavy convectional rainfall in summer when the sun is overhead, followed by a dry winter.

Latitudes bordering the two tropics receive minimal amounts of rainfall as they correspond to high pressure zones of subsiding and therefore warming air. To the poleward side of this arid zone, rainfall quantities begin to increase again and the length of the dry season decreases. Between 30° and 40° north and south (in the west of continents) the Mediterranean climate is characterised by winter rain and summer drought.

Temperate latitudes receive large amounts of heavy precipitation throughout the year due to cyclonic and locally orographic uplift of air. Precipitation totals decrease and rain gives way to snow in polar areas where cold air descends to give stable conditions.

This general pattern is affected locally by the movement of the overhead sun, the presence of mountains or ocean currents, by the monsoon, and by continentality.

However, in many parts of the world, economic development and lifestyles are closely linked to shorter term characteristics of precipitation rather than to annual amounts. Precipitation is more valuable when it falls in the growing season, as in the Canadian Prairies, although it tends to be less effective if it falls when evapotranspiration rates are at their highest, as in the case in savanna areas. In the same way, lengthy episodes of steady rainfall as experienced in Britain provide a more beneficial water supply than storms of a short and intensive duration which occur in the Sahel. This is because moisture is allowed to penetrate the soil more gradually and the risks of soil erosion, flooding and water shortages are reduced.

In Britain, rainfall is reliable with relatively little variation in totals from year to year but unfortunately there appears to be a close correlation between areas receiving limited rainfall and those which experience the greatest annual variability, e.g. the Sahel and northeast Brazil (Figures 9.27 and 9.28).

Atmospheric motion

The movement of air in the atmospheric system may be vertical (i.e. rising or subsiding) or horizontal, in which case it is commonly known as wind. Winds result from differences in air pressure which in turn may be caused by differences in the force exerted by gravity as pressure decreases rapidly with height (Figure 9.1) or temperature. An increase in temperature causes air to heat, expand, become less dense and rise, creating an area of low pressure below.

Conversely, a drop in temperature produces an area of high pressure. Differences in pressure are shown on maps by **isobars** which are lines joining places of equal pressure. To draw isobars, pressure readings are normally reduced to represent pressure at sea level. Pressure is measured in millibars (mb) and it is usual for isobars to be drawn at 4-millibar intervals. Average pressure at sea level is 1013 mb. However, the pattern of the isobars is usually more important in terms of the weather than the actual figures. The closer together the isobars the greater the difference in pressure — the **pressure gradient** — and the stronger the wind. Wind is nature's way of balancing out differences in pressure as well as temperature and humidity.

Figure 9.29 shows the two basic pressure systems which affect the British Isles. In addition to the differences in pressure and wind speeds and direction, the diagrams also show that winds blow neither directly at right angles to the isobars along the pressure gradient, nor parallel to them. This is because the effect of the Coriolis force and friction have to be considered as well as the forces exerted by the pressure gradient and gravity.

The Coriolis force

If the earth did not rotate and was composed entirely of either land or water, then there would be one large convection cell in each hemisphere (Figure 9.30). Surface winds would be parallel to pressure gradients blowing directly from high to low pressure areas. In reality, the earth does rotate and the distribution of land and sea is uneven. Consequently, more than one cell is created (Figure 9.33) as rising air, warmed at the Equator, loses heat to space — there is less cloud cover to retain it — and as it travels further from the source of heat. A further consequence is that moving air appears to

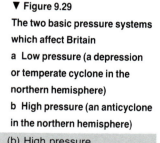

▼ Figure 9.29

The two basic pressure systems which affect Britain

a Low pressure (a depression or temperate cyclone in the northern hemisphere)

b High pressure (an anticyclone in the northern hemisphere)

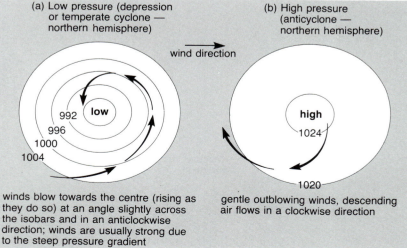

(a) Low pressure (depression or temperate cyclone — northern hemisphere)

wind direction

992
996
1000
1004
low

(b) High pressure (anticyclone — northern hemisphere)

high
1024
1020

winds blow towards the centre (rising as they do so) at an angle slightly across the isobars and in an anticlockwise direction; winds are usually strong due to the steep pressure gradient

gentle outblowing winds, descending air flows in a clockwise direction

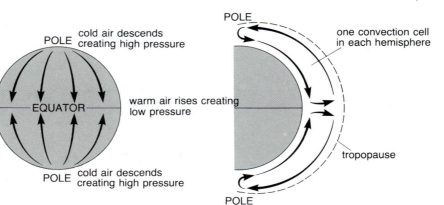

be deflected to the right in the northern hemisphere and to the left in the southern hemisphere as a result of the Coriolis force.

If a person were to stand in the centre of a rotating disc, similar to a giant turntable, and to throw a ball to a second person standing on the edge of that moving disc, the ball would miss. Indeed, to the thrower, the ball would appear to take a curved path away from the receiver as the receiver will have changed position because of the rotation of the disc (Figure 9.31). Similarly, the earth's rotation through 360° every 24 hours means that a wind blowing in a northerly direction in the northern hemisphere appears to have been diverted to the right on a curved trajectory by 15° of longitude for every hour, though to an astronaut in a space shuttle, the path would look straight. This helps to explain why the prevailing winds blowing from the tropical high pressure zone approach Britain from the southwest rather than from the south. In theory, if the Coriolis force acted alone then the resultant wind would blow in a circle.

Winds in the upper troposphere, unaffected by friction with the earth's surface, show that there is a balance between the forces exerted by the pressure gradient and

the Coriolis deflection. The result is the **geostrophic wind** which blows parallel to isobars (Figure 9.32). The existence of the geostrophic wind was recognised in 1857 by a Dutchman, Buys Ballot, whose law states that 'if you stand, in the northern hemisphere, with your back to the wind, low pressure is always to your left and high pressure to your right'.

Friction caused by the earth's surface upsets the balance between the pressure gradient and the Coriolis force by reducing the effect of the latter. As the pressure gradient becomes relatively more important as friction is reduced with altitude, the wind blows across isobars towards the low pressure (Figure 9.29). Deviation from the geostrophic wind is less pronounced over water because its surface is smoother than land.

A hierarchy of atmospheric motion

An appreciation of the movement of air is fundamental to an understanding of the workings of the atmosphere and its effects on our climate and weather. The extent to which atmospheric motion influences local weather and climate depends on winds at a variety of scales and their interaction in a hierarchy of patterns.

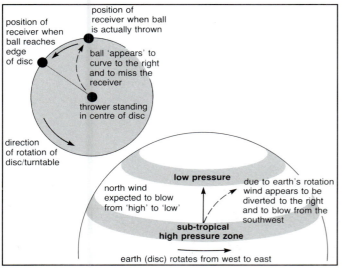

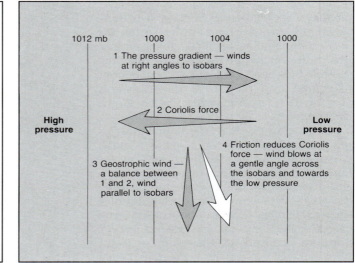

One such hierarchy which is useful in study-ing the influence of atmospheric motion was suggested by B.W. Atkinson in 1988. Al-though defining four levels he stressed that there were important interrelationships be-tween each.

A hierarchy of atmospheric motion systems (Atkinson, 1988)

Scale	Characteristic horizontal size (km)	Systems
1 Planetary	5000 – 10 000	Rossby waves, ITCZ
2 Synoptic (macro)	1000 – 5000	Monsoons, hurricanes, depressions, anticyclones
3 Meso-scale	10 – 1000	Land and sea breezes, mountain and valley winds, föhn, thunderstorms
4 Small (micro)	0.1 – 10	Smoke plumes, urban turbulence

Planetary scale: atmospheric circulation

It has already been shown that there is a surplus of energy at the Equator and a deficit in the outer atmosphere and nearer to the poles (Figure 9.5). Therefore, theoretically, surplus energy is transferred to areas with a deficiency by means of a single convective cell (Figure 9.30). This would be the case for a non-rotating earth, a concept first advanced by Halley (1686) and expanded by Hadley (1735). The discovery of three cells was made by Ferrel (1856) and refined by Rossby (1941). Despite many modern advances using radiosonde readings, satellite imagery and computer modelling, this tri-cellular model still forms the basis of our understandings of the general circulation of the atmosphere.

The tricellular model

The meeting of the trade winds in the equatorial region forms the Intertropical Convergence Zone or ITCZ (page 181). The trade winds, which pick up latent heat as they cross warm, tropical oceans, are forced to rise by violent convection currents. The unstable warm, moist air is rapidly cooled adiabatically to produce the towering cumulonimbus clouds, frequent afternoon thunderstorms and low pressure charac-teristic of the equatorial climate. It is these strong upward currents which form the 'powerhouse of the general global circula-tion' and which turn latent heat first into sensible heat and later into potential energy. At ground level, the ITCZ experiences only very gentle, variable winds known as the **doldrums**.

As rising air cools to the temperature of the surrounding environmental air, uplift ceases and it begins to move away from the Equator. Further cooling, increasing density, and diversion by the Coriolis force causes the air to slow down and subside, forming the descending limb of the **Hadley cell** (Figures 9.33, 9.34). In looking at the northern hemisphere (the southern is the mirror image) it can be seen that the air

subsides at about 30°N of the Equator to create the subtropical high pressure belt with its clear skies and dry, stable conditions (Figure 9.35). On reaching the earth's sur-face the cell is completed as some of the air is returned to the Equator as the northeast trades.

The remaining air diverges polewards as the warm southwesterlies which collect moisture if they cross sea areas. These warm winds meet cold Arctic air at the polar front (about 60°N) and are uplifted to form an area of low pressure and the rising limb of the **Ferrel** and **polar cells** (Figures 9.33,

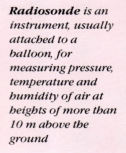

Radiosonde is an instrument, usually attached to a balloon, for measuring pressure, temperature and humidity of air at heights of more than 10 m above the ground

▼ Figure 9.33

Tricellular model showing atmospheric circulation in the northern hemisphere

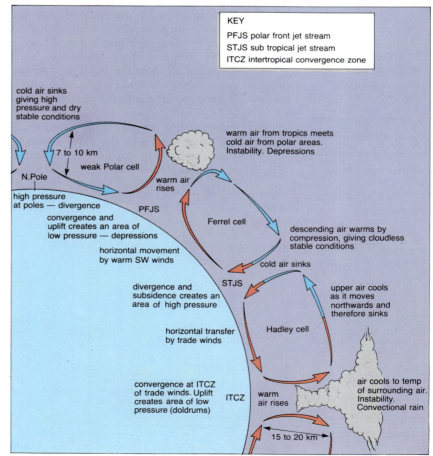

KEY

PFJS polar front jet stream
STJS sub tropical jet stream
ITCZ intertropical convergence zone

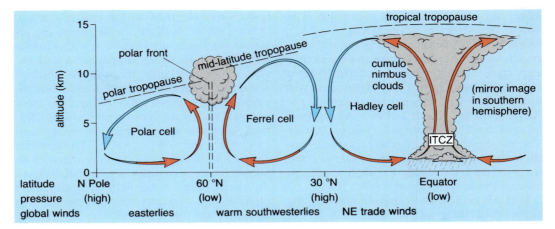

▶ Figure 9.34
Tricellular model to show
atmospheric circulation in the
northern hemisphere and the
altitude of the tropopause

▼ Figure 9.35
Image taken by the Meteosat
Geosynchronous satellite,
25.9.83 — notice the clouds
resulting from uplift at the ITCZ
(not a continuous belt), the clear
skies over the Sahara, the polar
front over the north Atlantic and
a depression over Britain

9.34). The resultant unstable conditions produce the heavy cyclonic rainfall associated with mid-latitude **depressions**. Depressions are another mechanism by which surplus heat is transferred. While some of this rising air eventually returns to the tropics, some travels towards the poles where, having lost all of its heat, it descends to form another stable area of high pressure. Air returning to the polar front does so as the cold easterlies.

This overall pattern is affected by the apparent movement of the overhead sun to the north and south of the Equator. This movement causes the seasonal shift of the heat Equator, the ITCZ, the equatorial low pressure zone and global wind and rainfall belts. Any variation in the characteristics of the ITCZ, i.e. its location or width, can have drastic consequences for the surrounding climates, as seen in the Sahel droughts of the early 1970s and most of the 1980s.

Rossby waves and jet streams

Evidence of strong winds in the upper troposphere first came when World War I Zeppelins were blown off course and several interwar balloons were observed travelling

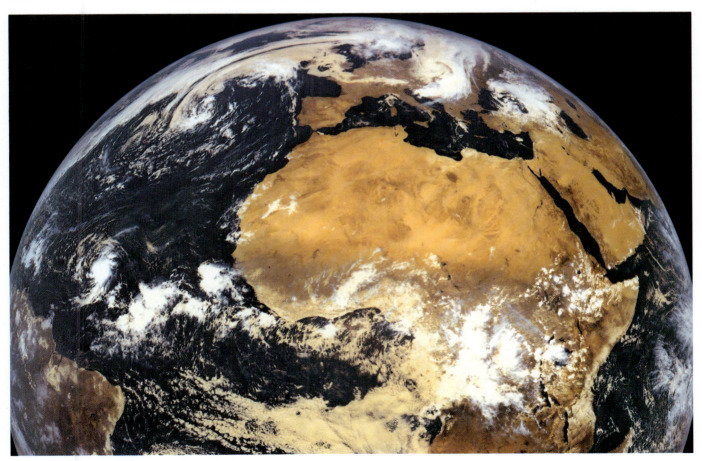

at speeds in excess of 200 km/hr. Pilots in World War II, flying at heights above 8 km, found eastward flights much faster and their return westward journeys much slower than expected, while north–south flights tended to be blown off course. The explanation was found to be a belt of upper air westerlies, the **Rossby waves**, which follow a zig-zag path around both hemispheres (Figure 9.36a). The number of zig-zags, or waves, varies but there are usually four to six in summer and three in winter forming a complete pattern around the globe (Figure 9.36b, c).

Further investigation has shown that the velocity of these upper westerlies is not internally uniform. Within them are narrow bands of extremely fast moving air known as **jet streams**. Jet streams, which help in the rapid tranfer of energy, can exceed speeds of 230 km/hr, which is sufficient to carry a balloon around the earth within a week. Of five recognisable jet streams, two are particularly significant with a third having seasonal importance.

The **polar front jet stream** (PFJS, Figure 9.33) varies between latitudes 40° and 60° in both hemispheres and forms the division between the Ferrel and polar cells, i.e. the boundary of warm tropical and cold polar air. The PFJS varies in extent, location and intensity and is mainly responsible for giving fine or wet weather on the earth's surface. Where, in the northern hemisphere, the jet stream zig-zags south (Figure 9.37a) it brings with it cold air which descends in a clockwise direction to give dry, stable conditions associated with areas of high pressure (**anti-cyclones**, Figure 9.29b). When the now warmed jet stream veers northwards it takes with it warm air which rises in an anti-clockwise direction to give the strong winds and heavy rainfall accompanying areas of low pressure (**depressions**, Figure 9.29a). As the usual path of the PFJS over Britain is oblique, i.e. towards the northeast, this accounts for our frequent wet and windy weather. Occasionally this path may be temporarily altered by a stationary or **blocking anticyclone** (Figure 9.44) which may produce extremes of climate such as the hot, dry summers of 1976 and 1989 or the cold January of 1987 (Figure 9.37b).

The **subtropical jet stream** or STJS occurs at about 25° to 30° from the Equator and forms the boundary between the Hadley and Ferrel cells (Figures 9.33, 9.34). This meanders less than the PFJS and has lower wind velocities but follows a similar west–east path.

The **easterly equatorial jet stream** is more seasonal, being associated with the summer monsoon of the Indian subcontinent.

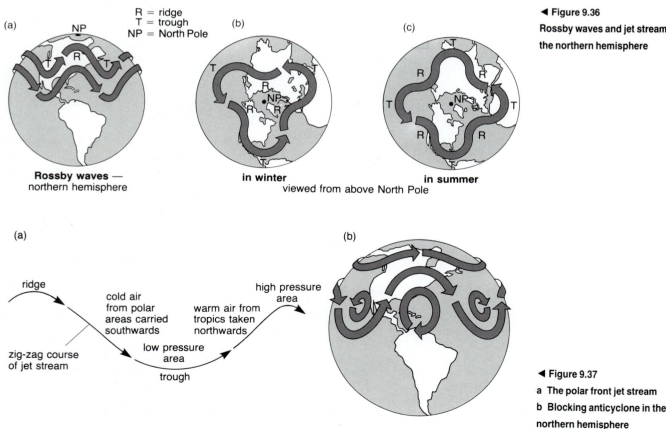

(a)

R = ridge
T = trough
NP = North Pole

NP

(a) Rossby waves —
northern hemisphere

(b) in winter

(c) in summer

viewed from above North Pole

◀ Figure 9.36

Rossby waves and jet streams in the northern hemisphere

(a)

ridge

zig-zag course of jet stream

cold air from polar areas carried southwards

low pressure area

trough

warm air from tropics taken northwards

high pressure area

(b)

◀ Figure 9.37

a The polar front jet stream

b Blocking anticyclone in the northern hemisphere

Q

With reference to the polar front jet stream (PFJS):

1 Explain how the PFJS contributes to the poleward transfer of energy (the heat budget).

2 Using the following labels, complete the flow chart to explain how the jet stream creates areas of high and low pressure in mid-latitudes:

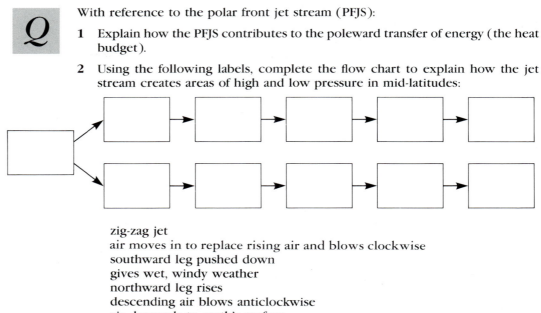

zig-zag jet
air moves in to replace rising air and blows clockwise
southward leg pushed down
gives wet, windy weather
northward leg rises
descending air blows anticlockwise
air descends to earth's surface
suction effect created
rising air creates low pressure, cools and condenses
summer weather hot and dry, winter weather cold and dry
descending air creates high pressure on earth's surface.

Macro-scale: synoptic systems

The concept of air masses is important because these help to categorise world climate types (Chapter 12). In regions where one air mass is dominant all year there is little seasonal variation in weather, e.g. at the tropics and the poles. Areas where air masses constantly interchange, e.g. the British Isles, experience much greater seasonal variation in their weather.

Air masses and fronts — how do they affect the British Isles?

If air remains stationary in an area for several days then it tends to assume the temperature and humidity properties of that region. The major areas of stationary air are the high pressure belts of the subtropics — the Azores and the Sahara — and high latitudes — Siberia and northern Canada. The areas in which homogeneous air masses develop are called **source regions**. Air masses can be classified according to the **latitude** in which they originate and which therefore determines their temperature, i.e. Arctic (*A*), Polar (*P*) or Tropical (*T*), and the nature of the **surface** over which they develop and which affects their moisture content, i.e. maritime (*m*) or continental (*c*).

The five major air masses which affect the British Isles at various times of the year (*Am*,

Pm, Pc, Tm and *Tc*) are derived by combining these characteristics of latitude and humidity (Figure 9.38). When air masses move from their source region they are modified by the surface over which they pass and this alters their temperature, humidity and stability. For example, tropical air moving northwards is cooled and becomes more stable while polar air moving south becomes warmer and increasingly unstable. Each air mass therefore brings its own characteristic weather conditions to the British Isles. The generalised conditions expected in each air mass are described in Figure 9.39. However, it should be remembered that each air mass is unique and dependent on the climatic conditions in the source region at the time of its development, the path which it subsequently follows, the season in which it occurs and, having a three dimensional form, the vertical characteristics of the atmosphere at the time.

When two air masses meet they do not mix readily because they have different temperatures and densities. The junction between them is called a **front**. A **warm front** is when warm air is advancing and being forced to override cold air. A **cold front** occurs when advancing cold air undercuts a body of warm air. In both cases, the rising air cools and usually produces clouds, easily

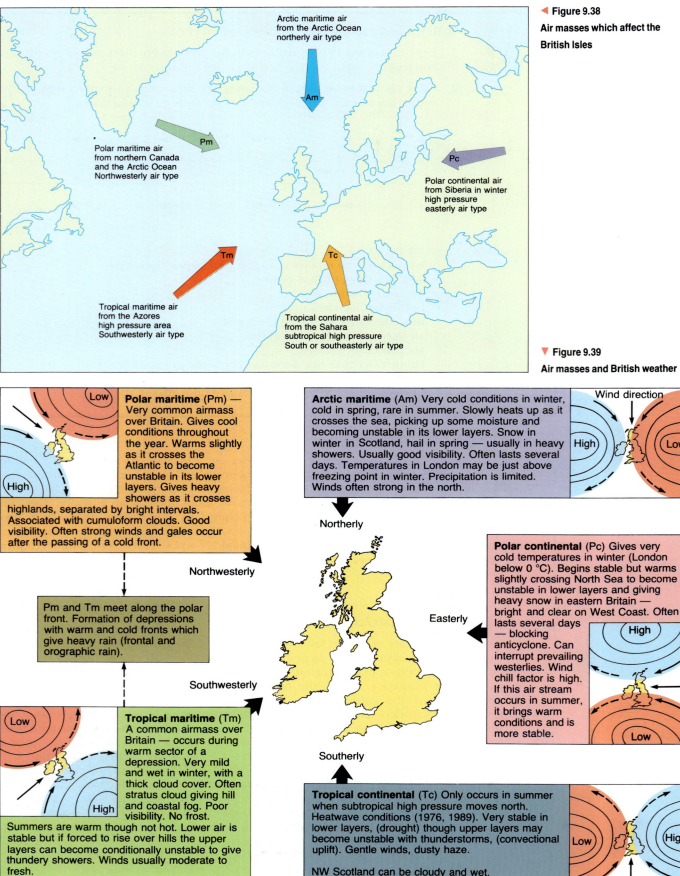

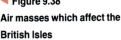

◀ **Figure 9.38**

Air masses which affect the British Isles

Arctic maritime air from the Arctic Ocean northerly air type

Am

Polar maritime air from northern Canada and the Arctic Ocean Northwesterly air type

Pm

Polar continental air from Siberia in winter high pressure easterly air type

Pc

Tropical maritime air from the Azores high pressure area Southwesterly air type

Tm

Tropical continental air from the Sahara subtropical high pressure South or southeasterly air type

Tc

▼ **Figure 9.39**

Air masses and British weather

Polar maritime (Pm) — Very common airmass over Britain. Gives cool conditions throughout the year. Warms slightly as it crosses the Atlantic to become unstable in its lower layers. Gives heavy showers as it crosses highlands, separated by bright intervals. Associated with cumuloform clouds. Good visibility. Often strong winds and gales occur after the passing of a cold front.

Arctic maritime (Am) Very cold conditions in winter, cold in spring, rare in summer. Slowly heats up as it crosses the sea, picking up some moisture and becoming unstable in its lower layers. Snow in winter in Scotland, hail in spring — usually in heavy showers. Usually good visibility. Often lasts several days. Temperatures in London may be just above freezing point in winter. Precipitation is limited. Winds often strong in the north.

Wind direction

Northerly

Northwesterly

Pm and Tm meet along the polar front. Formation of depressions with warm and cold fronts which give heavy rain (frontal and orographic rain).

Southwesterly

Polar continental (Pc) Gives very cold temperatures in winter (London below 0 °C). Begins stable but warms slightly crossing North Sea to become unstable in lower layers and giving heavy snow in eastern Britain — bright and clear on West Coast. Often lasts several days — blocking anticyclone. Can interrupt prevailing westerlies. Wind chill factor is high. If this air stream occurs in summer, it brings warm conditions and is more stable.

Easterly

Tropical maritime (Tm) A common airmass over Britain — occurs during warm sector of a depression. Very mild and wet in winter, with a thick cloud cover. Often stratus cloud giving hill and coastal fog. Poor visibility. No frost. Summers are warm though not hot. Lower air is stable but if forced to rise over hills the upper layers can become conditionally unstable to give thundery showers. Winds usually moderate to fresh.

Southerly

Tropical continental (Tc) Only occurs in summer when subtropical high pressure moves north. Heatwave conditions (1976, 1989). Very stable in lower layers, (drought) though upper layers may become unstable with thunderstorms, (convectional uplift). Gentle winds, dusty haze.

NW Scotland can be cloudy and wet.

seen on satellite weather photographs, which often generate precipitation. Fronts may be several hundred kilometres in width and extend at relatively gentle gradients up into the atmosphere. The most notable type of front, the **polar front**, occurs when warm, moist *Tm* air meets colder, drier *Pm* air. It is at the polar front that depressions or **temperate cyclones** form.

Q

1 What is meant by the terms 'air mass' and 'source regions'?

2 Under what conditions will tropical air become more stable and polar air less stable?

3 Give a reasoned account of the effects of various air masses on the weather of the British Isles in both summer and winter.

How depressions are formed

The Polar Front Theory was forwarded by a group of Norwegian meteorologists in the early 1920s. Although some aspects have been refined recently, since the innovation of radiosonde readings and satellite imagery, the basic model for the formation of frontal depressions is still valid. The following account describes a 'typical' or 'model' depression (Framework 8, page 293). It should be remembered that individual depressions may vary widely from this model.

Depressions follow a life cycle in which three main stages can be identified: formation (embryo), maturity and decay (Figure 9.40).

▼ Figure 9.40

Stages in the formation of a depression a Embryo b Mature c Decay

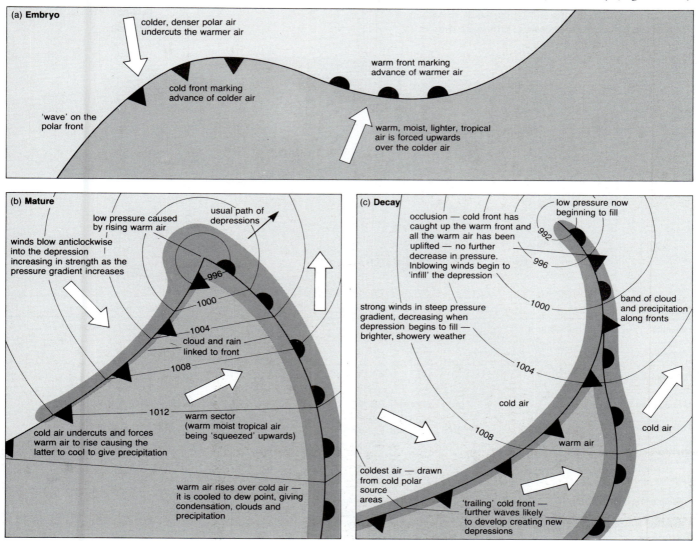

1 The embryo depression begins as a small wave on the polar front. It is here that warm, moist tropical (*Tm*) air meets colder, drier polar (*Pm*) air (Figure 9.40a). Recent studies have shown the boundary between the two air masses to be a zone rather than the simple linear division claimed in early models. The convergence of the two air masses results in the warmer, less dense air being forced to rise in a spiral movement. This upward movement results in 'less' air at the earth's surface creating an area of below average or low pressure. The developing depression, with its warm front (the leading edge of the tropical air) and cold front (the leading edge of the polar air) usually moves in a northeasterly direction under the influence of the upper westerlies, i.e. the polar front jet stream.

2 A mature depression is recognised by the increasing amplitude of the initial wave (Figure 9.40b). Pressure continues to fall as more warm air, in the warm sector, is forced to rise. As pressure falls and the pressure gradient steepens, the inward blowing winds increase in strength. These anticlockwise blowing winds, resulting from the Coriolis force, come from the southwest. As the relatively warm air of the warm sector continues to rise along the warm front it eventually cools to dew point. Some of its vapour will condense to release large amounts of latent heat, and clouds will develop. Continued uplift and cooling will cause precipitation and the clouds will become both thicker and lower (Figure 9.41).

Satellite photographs have shown that there is likely to be a band of 'meso-scale precipitation' extending several hundred kilometres in length and up to 150 km in width along, and just in front of, a warm front. As temperatures rise and the uplift of air decreases within the warm sector then there is less chance of precipitation and the low cloud may break to give some sunshine. The cold front moves faster and has a steeper gradient than the warm front (Figure 9.41).

▼ Figure 9.41
Weather associated with the passing of a typical mid-latitude depression

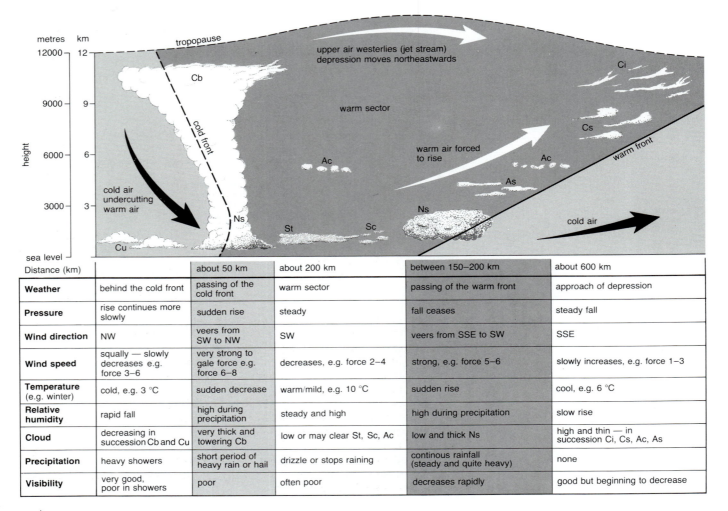

Distance (km)		about 50 km	about 200 km	between 150–200 km	about 600 km
Weather	behind the cold front	passing of the cold front	warm sector	passing of the warm front	approach of depression
Pressure	rise continues more slowly	sudden rise	steady	fall ceases	steady fall
Wind direction	NW	veers from SW to NW	SW	veers from SSE to SW	SSE
Wind speed	squally — slowly decreases e.g. force 3–6	very strong to gale force e.g. force 6–8	decreases, e.g. force 2–4	strong, e.g. force 5–6	slowly increases, e.g. force 1–3
Temperature (e.g. winter)	cold, e.g. 3 °C	sudden decrease	warm/mild, e.g. 10 °C	sudden rise	cool, e.g. 6 °C
Relative humidity	rapid fall	high during precipitation	steady and high	high during precipitation	slow rise
Cloud	decreasing in succession Cb and Cu	very thick and towering Cb	low or may clear St, Sc, Ac	low and thick Ns	high and thin — in succession Ci, Cs, Ac, As
Precipitation	heavy showers	short period of heavy rain or hail	drizzle or stops raining	continous rainfall (steady and quite heavy)	none
Visibility	very good, poor in showers	poor	often poor	decreases rapidly	good but beginning to decrease

Progressive undercutting by cold air at the rear of the warm sector gives a second episode of precipitation although with a greater intensity and a shorter duration than at the warm front. This band of meso-scale precipitation may be only 10–50 km in width. Although the air behind the cold front is colder than that in advance of the warm front, having originated in and travelled through more northerly latitudes, it becomes unstable giving cumulonimbus clouds and heavy showers. Winds often reach their maximum strength at the cold front and change to a more northwesterly direction after its passage (Figure 9.41).

3 The depression begins to decay when the cold front catches up the warm front to form an **occlusion** or occluded front (Figure 9.40c). By this stage the *Tm* air will have been squeezed upwards leaving no warm sector at ground level. As the uplift of air is reduced then so too will be the amount of condensation, the release of latent heat and the amount and pattern of precipitation — there may be only one episode of rain. Cloud cover begins to decrease, pressure rises and wind speeds decrease as the colder air 'infills' the depression by replacing the uplifted air.

Q

1 Reference has already been made to Figure 9.41 which is a model showing the weather associated with the passing of typical mid-latitude depression. Now study Figure 9.42 (**a** to **e**) which plots various observations made at a British weather station during the passage of a non-occluded depression in winter.

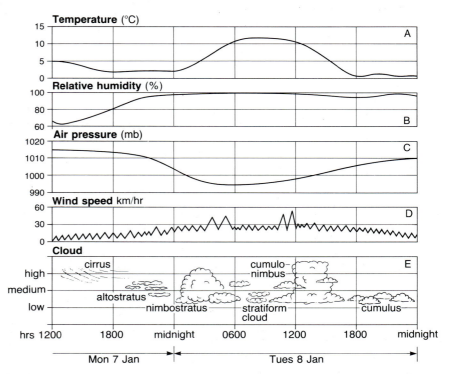

▶ Figure 9.42

The passage over a British weather station of a non-occluded depression in winter

a With reference to Figure 9.42**a**, **b** and **c** only, state the day and time when the warm front and the cold front passed over the weather station. Briefly justify the answers you have given.

b State the relationships between variations in wind speed (Figure 9.42**d**) and the passage of the warm and cold fronts. Account for these relationships.

c Describe the sequence of types and amounts of precipitation you would expect to occur at the weather station during the period from noon on 7 January to midnight 8 January.

d Name one other weather element, not shown in Figure 9.42 which might be used to monitor the passage of the depression.

e Explain how the weather element you have chosen in **d** would indicate the passage of the depression.

2 The actual model of a depression is 3-dimensional. Find a 3-d illustration of the model in a book from the Geography Library and try to explain how it differs from and improves upon the diagram in Figure 9.40.

3 Read the case study on 'The Great Storm in SE England, 16 October 1987'.
   a In what ways was this storm *not* a typical depression?
   b Why is it difficult accurately to forecast an event such as the 'Great Storm'?
   c What could be done to reduce the damage and predict the occurrence of a storm of similar intensity?

4 If every depression behaved as the model in 9.40, then weather forecasting would be relatively simple and highly accurate. Although there have been many recent advances in our knowledge, some explanations are highly complex while others are still incomplete. Those of you with more enquiring, meteorological minds could research:
   a ana and kata fronts
   b warm and cold occlusions
   c the conveyor belt model.

CASE STUDY

The great storm in SE England, 16 October 1987

This storm, the worst to affect SE England since 1703, developed so rapidly that its severity was not predicted in advance weather forecasts.

11 October   High winds and heavy rain were forecast for the end of the week
15 October   1200 hours: depression expected to move along the English Channel with fresh to strong winds
             2130 TV weather forecast: strong winds gusting to 50 km/hour
16 October   0030 Radio weather forecast: warning of severe gales
             0130 Police and fire services warned about extreme winds
             0500 Winds reached 94 km/hour at Heathrow and 100 km/hour on parts of the south coast
             0800 Centre of depression reached the North Sea

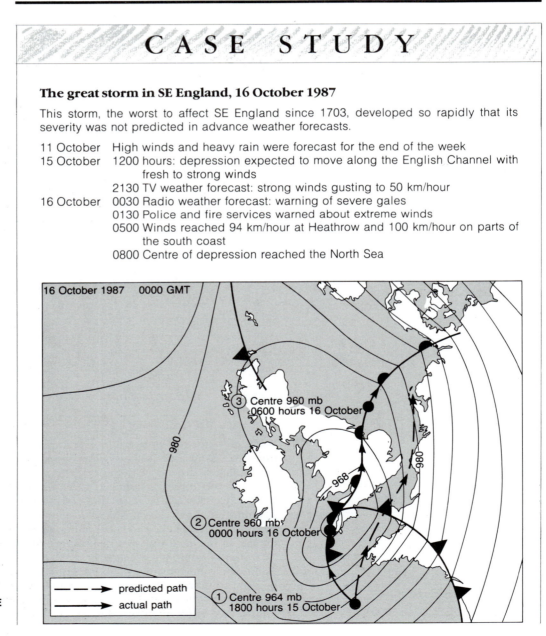

▶ Figure 9.43
The Great Storm over SE England, 16.10.87

The storm began as a small wave on a cold front in the Bay of Biscay, where the few weather ships give only limited information, caused by contact between very warm air from Africa and cold air from the North Atlantic. It appeared to be a 'typical' depression until, at about 1800 hours on 15 October, it unexpectedly deepened giving a central pressure reading of 958 millibars and an exceptionally steep pressure gradient. The exact cause of this is unknown but may have been the result of either (or both) a very strong jeg stream or extreme warming over the sea (hurricanes, page 199). The latter could have caused an excessive release of latent heat energy which North American meteorologists compare to the effect of a 'bomb'. It was this unpredicted deepening, combined with the change of direction from the English Channel towards the Midlands, which caught experts by surprise.

Although the storm passed within a few hours, and luckily during the night when most people were asleep, it left a trail of death and destruction. There were 16 deaths; several houses collapsed and many others lost walls, windows and roofs; thousands of trees were blown over, blocking railways and roads; one-third of the trees in Kew Gardens were destroyed; power lines were cut and, in some remote areas, not restored for several days; few commuters managed to reach London the next day; a ferry was blown ashore at Folkestone; and insurance claims set an all time record.

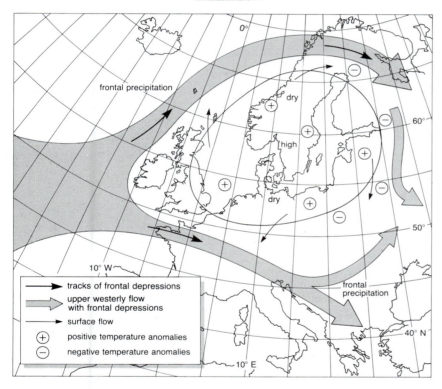

▲ Figure 9.44

A blocking anticyclone over Scandinavia — the upper westerlies divide upwind of the block and flow around it with their associated rainfall: there are positive temperature anomalies within the southerly flow to the west of the block, negative anomalies to the east

Anticyclones

An anticyclone is the name given to a large mass of subsiding air which produces an area of high pressure on the earth's surface. The source of the air is in the upper atmosphere where amounts of water vapour are limited and as it warms at the DALR (page 177) on its descent, dry conditions result. Pressure gradients are gentle resulting in weak winds or calms (Figure 9.29b). Winds which do blow are outwards and clockwise in the northern hemisphere. Anticyclones may be 3000 km in diameter — much larger than depressions — and once established can give several days of settled weather.

Weather conditions due to anticyclones over Britain

Summer Intense insolation gives hot, sunny days without cloud or rain. Rapid radiation at night, under clear skies, can lead to temperature inversions and the formation of dew and mist, although these rapidly clear the following morning. Coastal areas may experience advection fogs and land and sea breezes (page 201) while highlands have mountain and valley winds (page 202). If the air has its source over North Africa, i.e. a *Tc* air mass, then heatwave conditions tend to result. Often, after several days, thermals increase to give thunderstorms.

Winter Although the sinking air again gives cloudless skies there is little incoming radiation during the day because of the low angle of the sun. At night the absence of clouds allows low temperatures causing fog and frost to develop. These may take a long time to disperse the next day in the weak sunshine. Polar continental (*Pc*) air, with its source in central Asia and slow movement over the cold European landmass, is cold, dry and stable until it reaches the North Sea where its lower layers acquire some warmth and moisture, often leading to snowfalls on the east coast (Figure 9.23).

Blocking anticyclones

These occur when cells of high pressure detach themselves from the major high pressure areas of the subtropics or poles (Figure 9.37b). Once created, they last for several days and 'block' eastward-moving depressions (Figure 9.44) to create anomolous conditions, e.g. the extremes of temperature, rainfall and sunshine as in Britain in the summer of 1989 and the winter of 1987.

1 Redraw the outline of NW Europe shown in Figure 9.44 and locate the high pressure over the Irish Sea. Mark on the two possible tracks likely to be followed by any depression. Describe how these tracks will affect Britain's weather.

2 Figures 9.45a and b show two satellite photographs of the British Isles: the first under a depression, the second under an anticyclone. With reference to these photographs and information given in this section draw up a list to show the differences between a depression and an anticyclone using the following headings.

◄ Figure 9.45

Satellite pictures

a, *far left* Depression over the North Atlantic, 30.4.88

b Anticyclone over the British Isles, 19.2.85

cloud type	speed of movement	wind direction
cloud amount	time of life cycle	vertical air movement
precipitation	surface pressure	stability
local winds	wind speed	relative humidity

Name, in each case, the most likely type(s) of air mass.

Tropical cyclones

Tropical cyclones are systems of intensive low pressure known locally as **hurricanes**, **typhoons**, **cyclones** and **willy-willies** (Figure 9.46). As yet there is little certainty as to the process of their formation but they do tend to develop in autumn over warm tropical oceans, when temperatures exceed 26°C, and between 8° and 15° north or south of the Equator. (Nearer to the Equator the Coriolis force is insufficient to enable the feature to 'spin'.) Once formed they move westwards, often on erratic, unpredictable courses, swinging poleward on reaching land (Figure 9.46). They are another mechanism

▼ Figure 9.46

Global location of areas susceptible to tropical cyclones

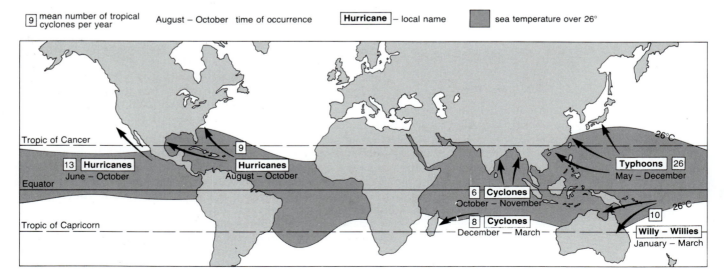

by which surplus energy is transferred away from the tropics (Figure 9.7).

Hurricanes are the tropical cyclone of the Atlantic. They form after the ITCZ has moved to its most northerly extent enabling air to converge at low levels. Unlike depressions, hurricanes occur when temperatures, pressure and humidity are uniform over a wide area for a lengthy period. To enable the hurricane to move there must be a continuous source of heat to maintain the rising air currents. There must also be a large supply of moisture to provide the latent heat, released by condensation, to drive the storm and to provide the heavy rainfall. It is estimated that in a single day a hurricane can release an amount of energy equivalent to that released by 500 000 atomic bombs the size of that dropped on Hiroshima. Only when the storm has reached maturity does the central **eye** develop. This is an area 30—50 km in diameter of subsiding air, light winds, clear skies and anomolous high temperatures (Figure 9.47). The descending air increases instability by warming and exaggerates the storm's intensity.

▼ Figure 9.47

Weather associated with the passage of a hurricane or tropical cyclone

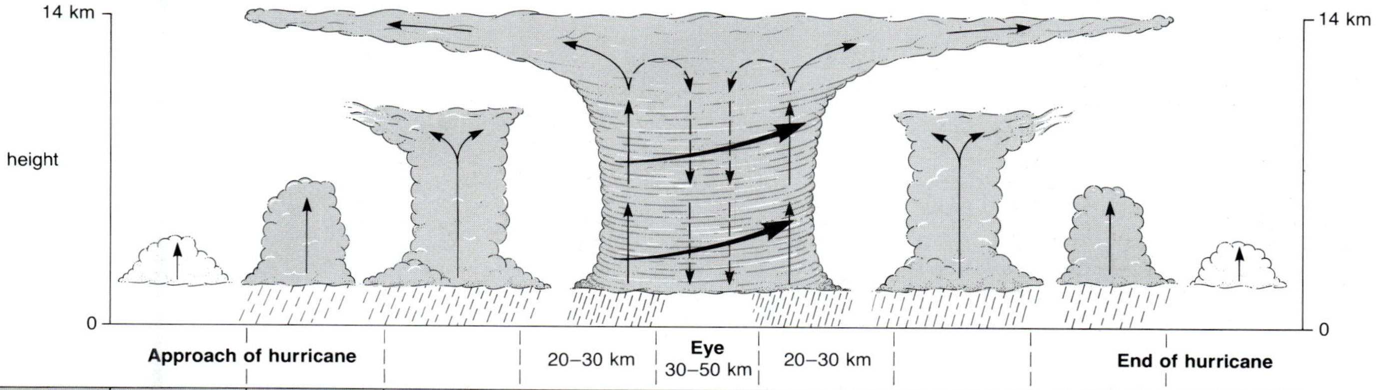

	Approach of hurricane			20–30 km	Eye 30–50 km	20–30 km		End of hurricane	
Vertical movement	updraughts increasing →		spiral uplift		subsiding air	spiral uplift		updraughts decreasing	
Clouds	few Cu	Cu	Cu and some Cb	giant Cb and Ci	none	giant Cb and Ci	Cu and some Cb	Cu	small Cu
Precipitation	none	showers	heavy showers	torrential rain 250 mm/day	none	torrential rain 250 mm/day	heavy showers	showers	none
Wind speed	gentle	fresh gusty	locally very strong	hurricane force 160 km/hour	calm	hurricane force 160 km/hour	locally very strong	fresh gusty	gentle
Wind direction	NNW	NW	WNW	WNW	calm	SSE	SSE	SE	ESE
Temperatures (plus examples)	high 30 °C	still high 30 °C	falling 26 °C	low 24 °C	high 32 °C	low 24 °C	rising 26 °C	high 28 °C	high 30 °C.
Pressure	average 1012 mb	steady 1010 mb	slowly falling 1006 mb	rapid fall	low 960 mb	rapid rise	slowly rising 1004 mb	steady 1010 mb	average 1012 mb

C A S E   S T U D Y

Bangladesh, November 1970

A tropical cyclone moved northwards up the narrowing, shallowing Bay of Bengal. Winds of over 200 km/hour and a storm surge 8 m high hit the densely populated Ganges Delta. Over four million people were affected; 300 000 people died and one million were left homeless; half a million cows and oxen were drowned; two-thirds of the fishing fleet was lost and 80 per cent of the rice crop ruined.

The Caribbean, September 1988

Hurricane Gilbert was the worst tropical storm to hit the Caribbean islands this century. Despite advance warnings winds of 300 km/hr killed 12 people in Jamaica, left one in five homeless, destroyed communications and ruined the banana crop. More deaths were reported as the hurricane crossed the Yucatan Peninsula in Mexico. Yet despite its strength, Gilbert caused fewer fatalities than a cyclone in 1979 which killed 2000 in the Dominican Republic, or another in 1900 which left 6000 dead in Galveston (Texas).

The hurricane rapidly declines once the source of heat is removed, i.e. when it moves over colder water or a landsurface which creates friction and cannot supply sufficient moisture. The average life of a tropical cyclone is seven to 14 days. The characteristic weather conditions associated with the passage of a hurricane are shown diagrammatically in Figure 9.47. Figure 9.48 shows the feature from space.

Tropical cyclones may cause considerable loss of life and damage to property and crops. This may be the direct result of the following features associated with hurricanes.

1 **High winds** which often exceed 160 km/hr and, in extreme cases, over 300 km/hr. Whole villages may be destroyed in less developed countries while even reinforced buildings in the southern USA may be damaged. Countries whose economies rely largely on a single crop may suffer serious economic problems, while electricity and communication links can be severed causing further difficulties.

2 **Ocean storm surges** may inundate coastal areas (page 116), many of which are densely populated.

3 **Flooding** may be caused either by a storm surge or the intensive rainfall. In 1974, 800 000 people died in Honduras as their flimsy homes were washed away.

The monsoon

The word 'monsoon' is derived from the Arabic for a season but the term is more commonly used in meteorology to denote a seasonal reversal of wind direction. The major monsoon occurs in SE Asia and results from three factors.

1 The extreme heating and cooling of large landmasses in relation to smaller heat changes experienced over adjacent sea areas. This in turn affects pressure and winds.

2 The northward movement of the ITCZ during the northern hemisphere summer.

3 The uplift of the Himalayas which, some six million years ago, became sufficiently high to interfere with the general circulation of the atmosphere (Figures 9.33, 9.34).

The southwest or summer monsoon

As the overhead sun appears to move northwards to the Tropic of Cancer in June it draws with it the convergence zone associated with the ITCZ (Figure 9.49a). The increase in insolation over northern India, Pakistan and central Asia means that heated air rises creating a large area of low pressure.

▲ Figure 9.48
Satellite photo of hurricane
Gladys, 12.10.68

Consequently, warm moist *Em* and *Tm* air, from over the Indian Ocean, is drawn first northwards and then, because of the Coriolis force, is diverted northeastwards. The air is humid, unstable and conducive to rainfall. Amounts of precipitation are most substantial on India's west coast, where the air rises over the Western Ghats, and in the Himalayas: Bombay has 2000 mm and Cherrapunji 13 000 mm in four summer months. The advent of monsoon storms may curtail the Himalayan climbing season but allows the planting of rice. Rainfall totals are accentuated as the air rises by both orographic and convectional uplift and the 'wet' monsoon is maintained by the release of substantial amounts of latent heat.

The northeast or winter monsoon

During the northern winter, the overhead sun, the ITCZ and the subtropical jet stream all move southwards (Figure 9.49b). At the same time, central Asia experiences intense cooling which allows a large high pressure system to develop. Airstreams which move outwards from this high pressure area will be dry as their source area is semi-desert. They become even drier as they cross the Himalayas and adiabatically warmer as they descend on to the Indo—Gangetic plain. Bombay receives less than 100 mm of rain in total through eight months of the year.

► Figure 9.49

The monsoon in the Indian Sub-continent

a June–October, the southwest monsoon

b November–May, the northeast monsoon

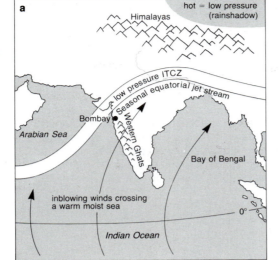

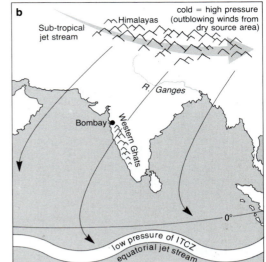

The monsoon, which in reality is much more complex than the model described above, affects the lives of one-quarter of the world's population. Unfortunately, monsoon rainfall, especially in the Indian subcontinent, is unreliable. If the rains fail, then drought and famine ensue: 1987 was the ninth year in a decade where the monsoon failed in NW India. If, conversely, there is excessive rainfall then large areas of land experience extreme flooding, e.g. Bangladesh in 1987 and 1988.

Meso-scale: local winds

Of the three meso-scale circulations described here two, **land and sea breezes** and **mountain and valley winds**, are caused by local temperature differences, while the third, the **föhn**, results from pressure differences either side of a mountain range. All three modes of circulation help to illustrate general principles of atmospheric behaviour.

The land and sea breeze

This is an example, on a diurnal timescale, of a circulation system resulting from differential heating and cooling between land surfaces and adjacent sea areas. The resultant pressure differences, although small and localised, produce gentle breezes which affect coastal areas during calm, clear anticyclonic conditions. The sea breeze builds up in the late morning and dies away in the evening (Figure 9.50a). Although the circulation cell rarely rises above 500 m in height or reaches more than 20 km inland in Britain, the sea breeze is capable of lowering coastal temperatures by 15°C and can produce advection fogs, e.g. the 'sea-fret' of eastern Britain.

At night, when the sea retains heat longer than land, there is a reversal of the pressure gradient and therefore of wind direction (Figure 9.50b). The land breeze, the gentler of the two, begins just after sunset and dies away by sunrise.

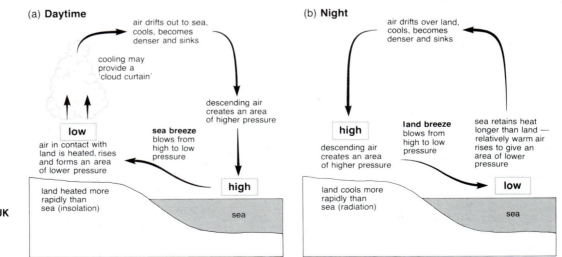

► Figure 9.50

Land and sea breezes in the UK

a Daytime — the sea breeze

b Night — the land breeze

CASE STUDY

Penang, Malaysia

Land and sea breezes occur daily when Penang lies under the influence of the light, variable winds of the doldrums. July has particularly calm and stable conditions. The land heats up rapidly each morning, reaching a midday temperature of around 32°C. By this time a gentle sea breeze has begun to blow. By 1400 hours this breeze has strengthened sufficiently to bring a freshness which is appreciated by windsurfers, para-sailors and European tourists alike, although many of the latter foolishly think this freshness lessens the risk of becoming sunburnt. Waves in the water, almost imperceptible in the morning, increase considerably. Yet by sunset, shortly after 1800 hours, air and sea have both become calm again.

The mountain and valley wind

This wind is likely to blow in mountainous areas during times of calm, clear, settled weather. During the morning, valley sides are heated by the sun especially if they are steep, south-facing (in the northern hemisphere) and lacking vegetation cover. The air in contact with these slopes will heat, expand and rise (Figure 9.51a) creating a pressure gradient. By 1400 hours, the time of maximum heating, a strong uphill or **anabatic** wind is blowing up the valley and the valley-sides — ideal conditions for hang-gliding! The air becomes conditionally unstable (Figure 9.19) often producing cumulus cloud and, under very warm conditions, cumulonimbus with the possibility of thunderstorms on the mountain ridges. A compensatory sinking of air leaves the centre of the valley cloud-free.

During the clear evening, the valley loses heat through radiation. The surrounding air now cools and becomes denser. It begins to drain under gravity down the valley sides and along the valley floor as a mountain or **katabatic** wind (Figure 9.51b). This gives rise to a temperature inversion (Figure 9.24) and if the air is moist enough in winter may create fog or a **frost hollow**. Maximum wind speeds are generated just before dawn which is normally the coldest time of the day. Katabatic winds are usually gentle in Britain but much stronger if they blow over glaciers or in places where the upper slopes are permanently snow-covered. In Antarctica they may reach hurricane force.

The föhn

The **föhn** is a strong, warm and dry wind which blows periodically to the lee of a mountain range. It occurs in the Alps when a depression passes to the north of the mountains and draws in warm, moist air from the Mediterranean. As the air rises

▼ Figure 9.51

The mountain and valley winds

a Daytime — valley wind (anabatic flow)

b Night — mountain wind (katabatic flow)

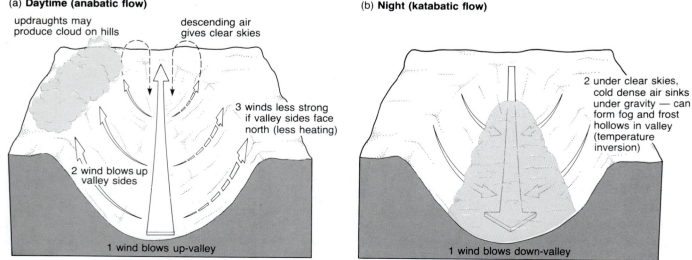

(a) **Daytime (anabatic flow)**

updraughts may produce cloud on hills

descending air gives clear skies

3 winds less strong if valley sides face north (less heating)

2 wind blows up valley sides

1 wind blows up-valley

(b) **Night (katabatic flow)**

2 under clear skies, cold dense air sinks under gravity — can form fog and frost hollows in valley (temperature inversion)

1 wind blows down-valley

(a) **The Föhn**

(b) **Temperature – height graph for Föhn conditions**

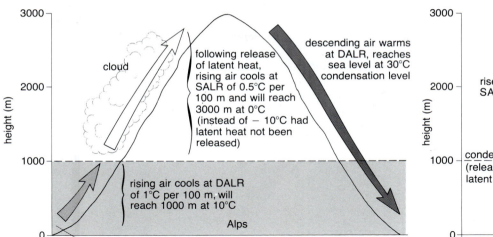

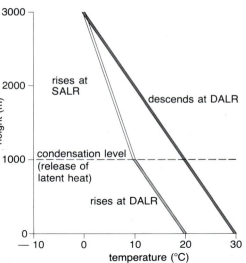

▲ **Figure 9.52**

The Föhn

a **Influences on temperature in the Alps**

b **Temperature–height graph of Föhn conditions**

(Figure 9.52) it cools at the DALR of 1°C/100 m (page 177). If, as in this example, condensation occurs at 1000 m then there will be a release of latent heat and the rising air will cool more slowly at the SALR of 0.5°/100 m. This means that the air reaches 3000 m at a temperature of 0°C instead of the −10°C had latent heat not been released. Having crossed the Alps, the descending air is compressed and warmed at the DALR so that if the land dropped sufficiently, air would reach sea level at 30°C. This is 10°C warmer than when it left the Mediterranean. Temperatures may rise by 20°C within an hour and relative humidity can fall to 10 per cent.

The wind, also known as the **chinook** on the Prairies, has considerable effects on human activity. In spring, when it is most likely to blow, it lives up to its Red Indian name ('chinook' means 'snow-eater') by melting snow and thus enabling wheat to be sown on the Canadian Prairies and Alpine pastures to be cleared in Switzerland. Conversely, its warmth can cause avalanches, forest fires and the premature budding of trees.

Microclimates

Microclimatology is the study of climate over a small area. It includes changes resulting from the construction of large urban centres as well as those existing naturally between different types of land surface, e.g. forests and lakes.

Urban climates

Large cities and conurbations experience climatic conditions different from the surrounding countryside. They generate more dust and condensation nuclei than natural environments, they create heat, alter the chemical composition and the moisture content of the air above them, and affect both the albedo and the flow of air. Urban areas therefore have distinctive climates.

Temperature Although tower block buildings cast more shadow, normal building materials tend to be non-reflective and so absorb heat during the daytime. Dark coloured roofs, concrete or brick walls and tarmac roads all have a high thermal capacity which means that they are capable of storing heat during the day and releasing it slowly during the night. Further heat is obtained from car fumes, factories and power stations, central heating and people themselves. The term **urban heat island** acknowledges that under calm conditions temperatures are warmest in the more built up city centre and decrease towards the suburbs and open countryside (Figure 9.53). Examples of urban areas having higher temperatures than surrounding rural ones are: day temperatures are, on average, 0.6°C warmer; at night, temperatures may be 3° or 4°C warmer as dust and cloud act as a blanket to reduce radiation and the buildings give out heat like a storage radiator; mean urban winter temperature is between 1° and 2°C warmer (rural areas are even colder when snow-covered as this increases their albedo); mean summer temperature may be 5°C warmer; the annual mean is warmer by between 0.6°C in Chicago and 1.3°C in London than the surrounding area.

Note how in Figure 9.53, temperatures not only decrease towards London's boundary but also beside the rivers Thames and Lea. The urban heat island explains why large cities have less snow, fewer frosts, earlier budding and flowering of plants and a greater need, in summer, for air conditioning.

Sunlight Despite having higher mean temperatures, a city area receives less sunshine and more cloud than its rural counterpart. Dust and other particles may absorb and reflect as much as 50 per cent of insolation in winter, when the sun is low in the sky and has to pass through more atmosphere, and 5 per cent in summer. High rise buildings also block out light.

Wind Wind velocity is reduced by buildings which create friction and act as windbreaks. Urban mean annual velocities may be up to 30 per cent lower than in rural areas, and periods of calm may be 10–20 per cent more frequent. In contrast, high rise buildings, including those like New York's skyscrapers, form 'canyons' through which winds may be channelled. These winds may be strong enough to cause tall buildings to sway and pedestrians to be blown over and troubled by dust and litter. The heat island effect may cause local thermals and reduce the wind chill factor. It also tends to generate considerable small-scale turbulence and eddies. In nineteenth-century Britain, industry was traditionally better located to the NE of cities so that smoke was carried away by the prevailing SW winds.

Relative humidity RH is lower in urban areas where the warmer air can hold more moisture and a lack of vegetation and water surface limits evapotranspiration.

Cloud Urban areas appear to receive thicker and up to 10 per cent more frequent cloud cover than rural areas. This may result from convection currents generated by the higher temperatures and the presence of a larger number of condensation nuclei.

Precipitation The mean annual precipitation total and the number of days with less than 5 mm are both about 10 per cent greater in major urban areas. Reasons for this are the same as for cloud formation. Strong thermals increase the likelihood of thunder by 25 per cent and the occurrence of hail by up to 400

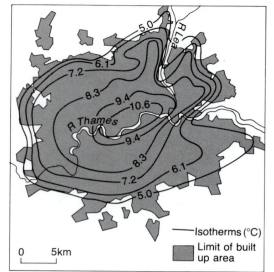

◄ Figure 9.53

Distribution of minimum temperatures over London, 14.5.59 (*after* Chandler)

per cent. The higher urban temperatures may turn the snow of rural areas into sleet and limit by up to 15 per cent the number of days with snow lying on the ground. On the other hand, the frequency, length and intensity of fog, especially under anticyclonic conditions, is much greater: there may be 100 per cent more in winter and 25 per cent in summer because of concentrations of condensation nuclei.

Atmospheric composition There may be three to seven times more dust particles over a city than in rural areas. Large quantities of gaseous and solid impurities are emitted into urban skies by the burning of fossil fuels, industrial processes and from car exhausts. Urban areas may have up to 200 times more sulphur dioxide and ten times more nitrogen oxide (the major components of acid rain) than rural areas as well as ten times more hydrocarbons and twice as much carbon dioxide. These pollutants tend to increase cloud cover and precipitation, cause smog (page 183), give higher temperatures and reduce sunlight.

Forest and lake microclimates

Different land surfaces produce distinctive local climates. The table summarises and compares some of the characteristics of microclimates found in forests and around lakes. As with urban climates, much research is still needed to confirm some of the statements in the table on page 205.

Fog frequencies in London and SE England (hours per year)

	Visibility less than 40 m (very dense fog)	Visibility less than 1000 m (less dense fog)
Kingsway (central London)	19	940
Kew (middle suburbs)	79	633
London Airport (outer suburbs)	46	562
SE England (mean of 7 stations)	20	494

Microclimates of forests and water surfaces

Microclimate feature	Forest — coniferous and deciduous	Water surface — Lake
Incoming radiation and albedo	Much incoming radiation is absorbed and trapped Albedo for coniferous forest is 15%; deciduous 25% in summer, 35% in winter; desert scrub is 40%	Less insolation absorbed and trapped Albedo may be over 60%; i.e. higher than over seas/ocean Higher on calm days
Temperature	Small diurnal range due to blanket effect of canopy Forest floor is protected from direct sunlight Some heat lost by evapotranspiration	Small diurnal range because water has a higher specific heat capacity Cooler summers and milder winters Lakesides have a longer growing season
Relative humidity	Higher during daytime and summer especially in deciduous forest Amount of evapotranspiration depends on length of day, leaf surface area, wind speed, etc.	Very high, especially in summer when evaporation rates are also high
Precipitation	Much evapotranspiration which can give heavy rain, especially in tropical rain forests On average 30–35% of rain is intercepted: more in deciduous woodland in summer	Humid air If forced to rise can be unstable and produce cloud and rain Amounts may not be great because of fewer condensation nuclei Fogs form in calm weather
Wind speed and direction	Trees reduce wind speeds (are often planted as windbreaks), especially at ground level Can produce eddies	May be strong because less friction Large lakes (e.g. Victoria) can create land and sea breezes

Weather maps and forecasting in Britain

A weather map or **synoptic chart** shows the weather for a particular area at one specific time. It is the result of the collection and collation of a considerable amount of data at numerous weather stations, i.e. from a number of sample points (Framework 4, Sampling). This data is then refined, usually as quickly as is possible, and plotted using internationally accepted weather symbols. A selection of these is shown in Figure 9.54. Weather maps are produced for different purposes and at various scales of detail.

1 The daily weather map as seen on television or in a national newspaper aims to give a clear, but highly simplified, impression of the weather.

2 At a higher level, a synoptic map shows selected meteorological characteristics for specific **weather stations**. The **station model** in Figure 9.55 shows six elements: temperature, pressure, cloud cover, present weather (e.g. type of precipitation), wind direction and wind speed.

▼ Figure 9.54

Weather symbols for cloud, precipitation, wind, temperature, pressure and wind direction

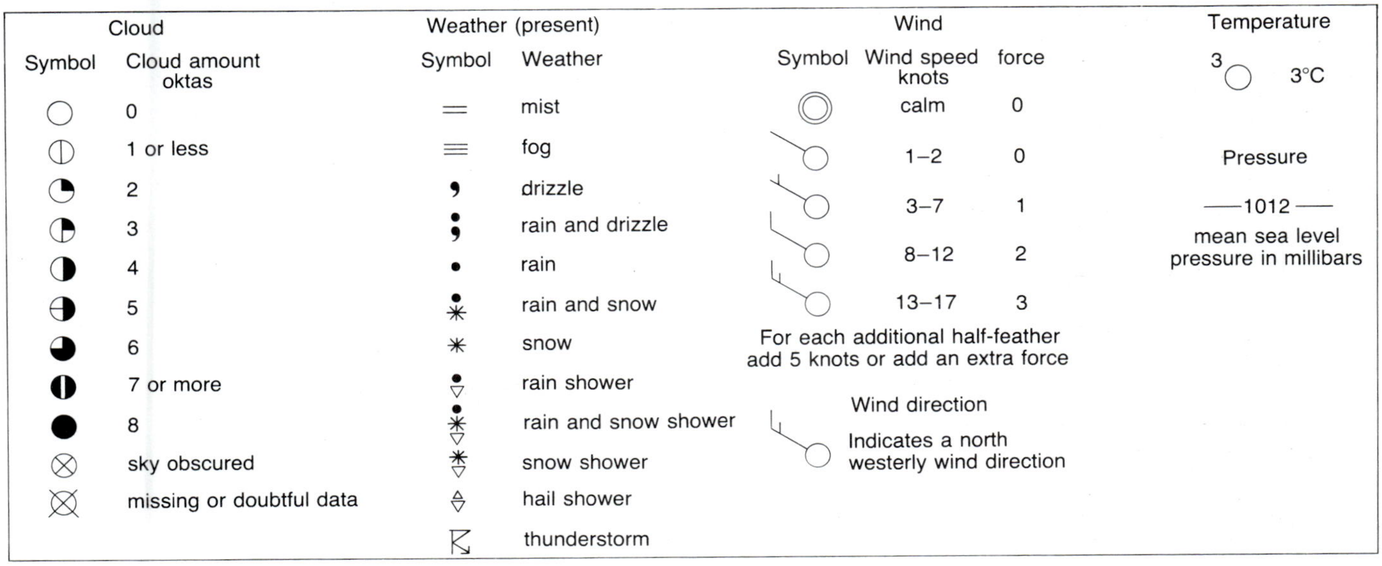

Cloud		Weather (present)		Wind			Temperature
Symbol	Cloud amount oktas	Symbol	Weather	Symbol	Wind speed knots	force	3°C
○	0	═	mist	◎	calm	0	Pressure
◐	1 or less	≡	fog		1–2	0	—1012—
◑	2	'	drizzle		3–7	1	mean sea level pressure in millibars
◑	3	;	rain and drizzle		8–12	2	
◑	4	•	rain		13–17	3	
◑	5	✳	rain and snow	For each additional half-feather add 5 knots or add an extra force			
◑	6	✳	snow				
◑	7 or more	⚏	rain shower	Wind direction			
●	8	✳	rain and snow shower	Indicates a north westerly wind direction			
⊗	sky obscured	✳	snow shower				
⊠	missing or doubtful data	⬧	hail shower				
		⚡	thunderstorm				

205

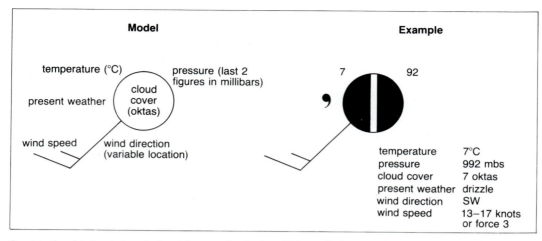

◀ Figure 9.55

A weather station model and an

example of an actual situation

3 At the highest level the Meteorological Office produces maps showing finite detail, e.g. amounts of various types of cloud at low, medium and high levels, dew point temperatures, barometric tendency (trends of pressure change) etc.

The role of the weather forecaster is to try to determine the speed and direction of movement of various air masses and any associated fronts and to predict the type of weather these movements will bring.

Although forecasting is increasingly assisted by information from satellites, radar and high speed computers, which show upper air as well as surface conditions in a 3-dimensional model, the complexity and unpredictability of the atmosphere can still catch the forecaster by surprise (e.g. SE England's storm of 16 October 1987, page 196). Part of this problem is related to the fact that meteorological information is a sample rather than a comprehensive picture of the atmosphere, so there is always a risk of the anomaly becoming a possibility.

◀ Figure 9.56, *bottom left*

Surface weather map at 0000 hrs

19.12.86

Q Study the weather map and infra-red image for 19 December 1986 (Figures 9.56, 9.57).

▼ Figure 9.57

Infra-red image taken at the

same time 19.12.86, showing

the British Isles and Europe

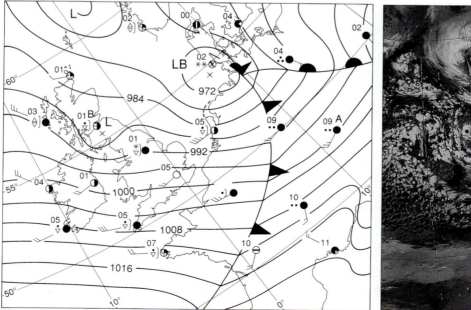

Notice:

□ the position of a warm front, a cold front and an occlusion – these have been superimposed on to the infra-red image,

□ the position of the warm sector (*W*) and two areas of convective cumulus clouds giving showers (*C1*, *C2*),

  the areas of clearer skies associated with the higher pressure over the. Mediterranean Sea.

1 How do you account for the large amount of cloud over Denmark, North Germany and central France?

2 Describe and suggest reasons for the differences in weather between Northern Germany (Station A) and Edinburgh (Station B)?

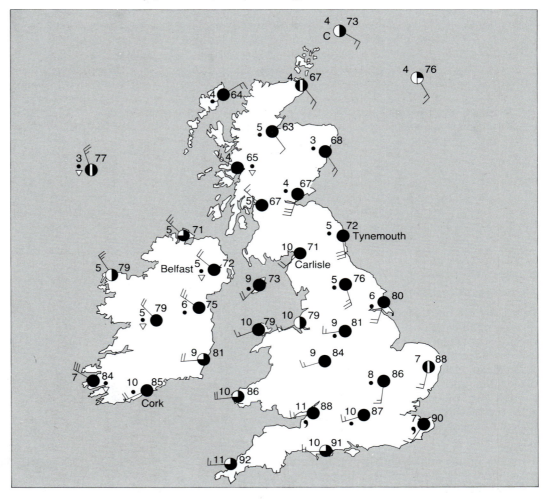

▶ Figure 9.58
Weather map for 0600 hours,
25.2.69

3 **a** On a copy of the map in Figure 9.58 use the information given to insert the probable position of a warm front, a cold front and an occlusion. You should consider differences in temperature, wind direction, wind speed and present weather. You may find Figures 9.40 and 9.41 useful.

  b Plot the likely pattern of isobars at 4 mb intervals on to the map shown in **a**. (If you work backwards from 1012 mb the first isobar you should plot will be for 992 mb.)

4 Using Figure 9.58, describe and account for the differences in the weather between Tynemouth, Carlisle and Belfast.

5 Assuming that the low pressure system in Figure 9.58 is moving northeast, draw two station models to show the weather likely to be experienced by Tynemouth and Carlisle at midday on 25 February 1969.

Measures of dispersion

FRAMEWORK 5

Measures of dispersion

Throughout this chapter on weather and climate, mean climatic figures have been quoted. To build up these pictures of global climate patterns, statistics have been obtained by taking average readings, usually for temperature and precipitation, over a 35 year time-scale. However, these averages themselves are often not as significant as the **range** or the degree to which they vary from, or are dispersed about, the mean.

For example, two tropical weather stations may have equal annual rainfall totals when measured over 35 years. Station A may lie on the Equator and experience reliable rainfall with little variation from one year to the next. Station B may experience a monsoon climate where some years the rains may fail entirely while in others they cause flooding.

The measure of dispersion from the mean can be obtained by using any one of three statistical techniques:
■ the range
■ the interquartile range
■ the standard deviation.

These techniques are included here because meteorological data both requires and benefits from their use but they may be applied to most branches of geography where there is a danger that the mean, taken alone, may be misleading. (The problems of over-generalisation are discussed on page 286.) Again it must be stressed that use of a quantitative technique does not guarantee objective interpretation of data: great care must be taken to ensure an appropriate method of manipulating the data is chosen so as to give a meaningful 'quantity'.

▼ Figure 9.59

Finding the interquartile range
a Temperatures at selected weather stations in the British Isles at 0600 hours, 14.1.79
b The interquartile range for temperatures for the British Isles at 0600 hours, 14.1.79

Range, interquartile range and standard deviation

It has already been seen how it is possible, given a set of data, to calculate the mean and the median (Figure 4.15) but this does not give any idea of the spread, or range, of that data. Mean values, on their own, may be misleading and so the actual dispersion should also be considered. This may be achieved by three methods: the range, inter-quartile range and standard deviation.

Range

This very simple method involves calculating the difference between the highest and lowest value of the sample population, e.g. the annual range in temperature for London is 14°C (July 18°C, January 4°C). The range emphasises the extreme values and ignores the distribution of the remainder.

Interquartile range

This calculation is useful to show how values are grouped around the median. It has the advantages in that it is easy to calculate, is unaffected by extreme values and is a useful means of comparing sets of similar data.

Figure 9.59 shows the temperatures for 19 weather stations in the British Isles at 0600 on 14 January 1979. These temperatures have been ranked in the adjacent table.

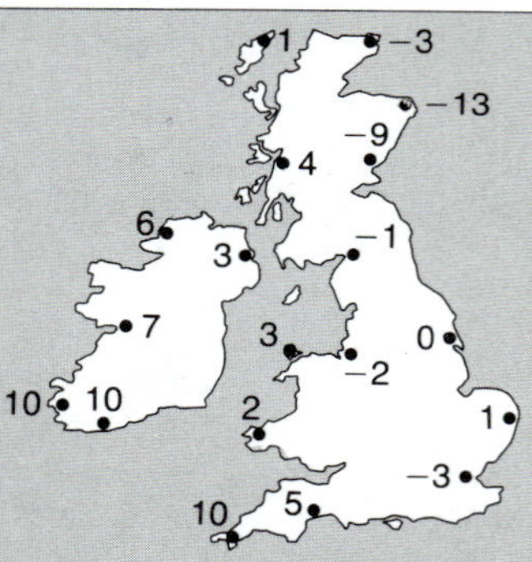

Rank	Temperatures 0°C (ranked)	
	10	
	10	
	10	
	7	
5	6	← upper quartile
	5	
	4	
	3	
	3	
10	2	← median (middle quartile)
	1	
	1	
	0	
	−1	
15	−2	← lower quartile
	−3	
	−3	
	−9	
	−13	

The **upper quartile** is obtained by using the formula:

Upper quartile $(UQ) = \left(\dfrac{n+1}{4}\right)$

i.e. $\left(\dfrac{19+1}{4}\right) = 5$

This means that the UQ is the fifth figure from the top of the ranking order, i.e. 6°C. The **lower quartile** is found by using a slightly different formula:

Lower quartile $(LQ) = \left(\dfrac{n+1}{4}\right) \times 3$

i.e. $\left(\dfrac{19+1}{4}\right) \times 3 = 15$

This shows the LQ to be the 15th figure in the ranking order, i.e. −2°C. You will notice that the middle quartile is the same as the median. The **interquartile range** is the difference between the upper and lower quartiles, i.e. 6°C − −2°C = 8°C.

Another measure of dispersion, the **quartile deviation**, is obtained by dividing the interquartile range by two, i.e.

$\left(\dfrac{8}{2}\right) = 4°C$

The smaller the interquartile range, or quartile deviation, the greater the grouping around the median and the smaller the dispersion or spread.

Standard deviation

This is the most commonly used method of measuring dispersion and although it may involve lengthy calculations it can be used with the arithmetic mean and it removes extreme values. (Computer software is often invaluable in avoiding these complicated calculations.) The formula for standard deviation is:

$$\sigma = \sqrt{\dfrac{\Sigma(x - \bar{x})^2}{n}}$$

where: σ = standard deviation
x = each value in the data set
$\bar{x}$ = mean of all values in the data set
n = number of values in the data set

Let us suppose that the minimum temperatures for 10 weather stations in Britain on a winter's day were, in 0°C, 5, 8, 3, 2, 7, 9, 8, 2, 2, and 4. The procedure is worked out in Figure 9.60.

1 Find the mean.
2 Subtract the mean from each value in the set.
3 Calculate the square of each value in 2. This removes any minus signs.
4 Add together all the values obtained in 3.
5 Divide the sum of the values in 4 by n.
6 Take the square root of 5 to obtain the standard deviation.

The standard deviation of $\sigma = 2.65$ is a low value indicating that the data are closely grouped around the mean.

Minimum temperatures for 10 weather stations in Britain on a winter's day.

The mean will be 5, 8, 3, 2, 7, 9, 8, 2, 2, 4 $\quad \therefore \bar{x} = \dfrac{50}{10} = 5$

Weather station	Temperature at each station (x)	$x - \bar{x}$	$(x - \bar{x})^2$
1	5	5 − 5 = 0	0
2	8	8 − 5 = 3	9
3	3	3 − 5 = −2	4
4	2	2 − 5 = −3	9
5	7	7 − 5 = 2	4
6	9	9 − 5 = 4	16
7	8	8 − 5 = 3	9
8	2	2 − 5 = −3	9
9	2	2 − 5 = −3	9
10	4	4 − 5 = −1	1
			$\Sigma(x - \bar{x})^2 = 70$

$\sigma = \sqrt{\dfrac{70}{10}}$

$\therefore \sigma = \sqrt{\dfrac{70}{10}} = \sqrt{7} = 2.65$

Standard deviation = 2.65

▶ Figure 9.60
Finding the standard deviation
(see Figure 6.46)

Q The table gives the annual precipitation totals for Stornoway (NW Scotland) and Salina Cruz (Mexico) over a period of 15 years.

Year	Annual precipitation totals (mm)	
	Salina Cruz	**Stornoway**
1	1665	877
2	699	1082
3	550	1203
4	1188	963
5	1040	1241
6	886	1194
7	1091	1072
8	578	900
9	762	1146
10	701	1094
11	798	1098
12	1040	1318
13	911	791
14	2356	1035
15	1681	1151

1 For each station calculate:
   a the precipitation range,
   b the upper and lower quartiles of the precipitation data,
   c the interquartile range of the precipitation data,
   d the quartile deviation,
   e the mean annual precipitation total,
   f the standard deviation of the annual precipitation over the 15 year period.

2 Compare carefully the results for the two stations, calculated in **a–e** above. How do you account for these differences?

Climatic change

Climates are constantly changing at all scales and over varying timespans. There have been surges of change over time which meteorologists are trying to clarify and explain.

Evidence of past climatic changes
■ The presence of coal, sandstone, limestone and glacial deposits which must have formed in environments different from the present.

■ Evidence of changes in sea level (Arran, page 133) and lake levels (Sahara, page 156).

■ **Pollen analysis** shows which plants were dominant at a given time. Each plant species has a distinctively shaped pollen grain. If these grains land in an oxygen-free environment, such as a peat bog, they resist decay. It is assumed that grains trapped in peat form a representative sample of the vegetation which was growing in the surrounding area at a given time. The extent and type of vegetation is assumed to be a response to the temperature and humidity of that time and is used as an indication of the prevailing climate. As peat accumulated it has preserved the pollen as a vegetation-climatic timescale (Figure 11.13).

■ **Dendrochronology** or tree-ring dating is the technique of obtaining a core from a trunk from which it is possible to determine the age of a tree. Tree growth is rapid in spring, slower by the autumn and, in temperate latitudes, stops in winter. Each year's growth is shown by a single ring but if the year is warm and wet then the ring will be larger, because the tree grows more quickly, than if the year is cold and wet. Tree-rings therefore reflect climatic changes. Samples from existing trees can be cross-dated with timbers in old houses, if the date of their construction is known. The oldest reliable dating in Britain goes back about 1500 years. Bristlecone pines, still living after 5000 years, provide a very accurate measure in California.

■ **Isotope analysis** An isotope is one of two or more forms of an element which differ from each other in atomic mass. For example, two isotopes in oxygen are O-16 and O-18. O-16, which is slightly lighter, vaporises more readily whereas O-18, being heavier, condenses more easily. During warm, dry periods the evaporation of

O-16 will leave water enriched with O-18 which, if it freezes into polar ice will be preserved as a later record. Colder, wetter periods will be indicated by ice with a higher level of O-16.

The most accurate form of dating is based on C-14, a radioactive isotope of carbon. Carbon is taken in by plants during the carbon cycle (Figure 11.19). Carbon-14 decays radioactively at a known rate and if compared with C-12, which does not decay, in a dead plant then scientists can determine the date of death to a standard error of ± 5 per cent. This method can accurately date organic matter to some 50 000 years ago.

- ■ **Historical records** Human evidence of climatic change includes:
  1 Cave paintings of elephants in the central Sahara (page 156)
  2 Vines growing successfully in southern England between AD 1000 and 1300
  3 Graves in which bodies were buried in Greenland were dug to a depth of 2 m in the thirteenth century, 1 m in the fourteenth and not at all, because of the extension of permafrost, in the fifteenth century
  4 Fairs were held on the frozen River Thames in Tudor times
  5 Measurement of recent advances and retreats of Alpine glaciers.

Causes of climatic change

A number of theories covering varying time-scales have been put forward to explain climatic changes. It is possible that a combination of several ideas, or one not yet propounded, may be responsible for climatic changes. At present there is no clear concensus of opinion. Some of these hypotheses are summarised below.

1 **Variations in solar energy** Although it was initially believed that solar energy output did not vary over time (hence the term 'solar constant' in Figure 9.3), increasing evidence suggests that sunspot activity, which occurs in cycles, may significantly affect our climate.
2 **Astronomical relationships between the sun and the earth** There is increasing evidence supporting Milankovitch's cycles of change in the earth's orbit, tilt and wobble (Figure 4.4).
3 **Composition of the atmosphere** It has long been accepted that volcanic activity has influenced climate in the past and continues to do so. World temperatures have been lowered after any large eruption (e.g. Krakatoa), or after a period of several eruptions, when the extra dust particles absorb and scatter incoming radiation. At present, increasing concern is being expressed at the build up of CO_2 gas in the atmosphere and the resultant greenhouse effect (see below), and the use of aerosols (page 170) with the potential consequences for world climates.
4 Plate tectonics have had a long term effect on climate, especially where landmasses have 'drifted' into different latitudes (e.g. British Isles, page 19) or where the sea bed has been pushed upwards to form fold mountains (e.g. the Himalayas, page 18).

Climatic change in Britain
The major changes in Britain's climate since the ice age are described in Figure 11.13. In more recent times, following the 'little ice age' which lasted from about AD 1540 to 1700, temperatures have generally increased and reached a peak in about 1940. Since then there has been a tendency for summers to become cooler and wetter, springs to be later, autumns milder and winters more unpredictable. However, the 1980s have seen a rise in mean annual temperatures.

Future trends
There are two major trends causing considerable environmental concern worldwide.

Slight changes in the movement and positioning of the ITCZ alter the structure and location of the Hadley Cell (Figures 9.33, 9.34). It is believed, though not universally accepted, that two contributary causes of the several prolonged droughts in the Sahel countries of Africa since 1970 have been the failure of the ITCZ to migrate as far north as usual and its replacement by the easterly jet stream, which has moved further south. The replacement of rising air with subsiding air has reduced rainfall totals.

The **greenhouse effect** is the result of an increasing build up in the amount of carbon dioxide in the air. It is caused by increased consumption of fossil fuels, burning of rain-forests, car exhaust emissions, the release of chlorofluorocarbons (CFCs) and methane gas resulting from animal rearing and rice cultivation (Figure 9.61). Although figures vary it is claimed that the amount of CO_2 in the atmosphere increased by 20 per cent between 1880 and 1980. It is estimated that this will increase by a further 10 per cent before AD 2000, and by an additional 10 per cent by AD 2010. Incoming, short-wave radiation from the sun is able to pass through the greenhouse gases, but terrestrial long-wave infra-red radiation is either absorbed by gases or re-radiated back to earth (page 169).

Based on computer calculations, this may mean a rise in global temperature of 1°C by the year AD2000 (Figure 9.62) and by as much as 2°C to 5°C by the end of the next century — the rise in temperature since the last ice age has been only between 4°C and 5°C. Although this might mean that Britain is as warm as the present day Costa Brava the effect on the world as a whole is likely to be less beneficial. The higher temperatures would melt much, if not all, of the polar ice caps thus causing a significant eustatic rise in sea level and the flooding of many coastal areas — is this already happening in Bangladesh? Figure 9.62 shows that the polar regions are likely to experience the greatest warming and reduced areas of snow-cover would alter the albedo which might further raise temperatures. Many areas of the world which at present have adequate water supplies may find themselves short of water and the rate of desertification is predicted to increase. Computer predictions suggest that areas around 40°N will become drier (Figure 9.62) and as these latitudes contain many

important cereal growing regions there could be a world food shortage. There could also be a northward migration of climate belts in the northern hemisphere so that while the Sahel countries would receive more rainfall, the Mediterranean and the USSR virgin lands (page 385, Figure 16.7) might turn into deserts.

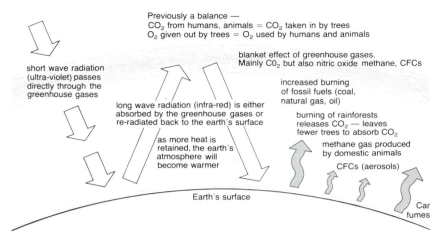

▲ Figure 9.61

The greenhouse effect

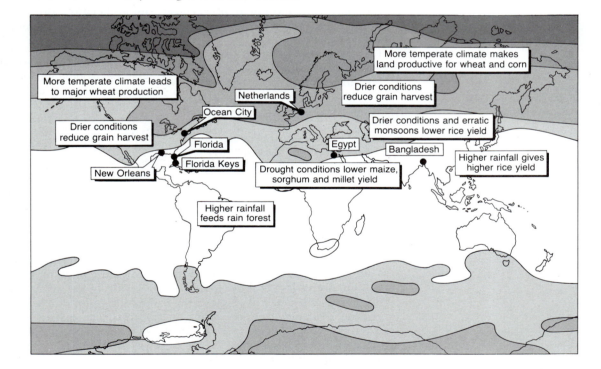

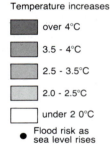

▲ Figure 9.62

Weather forecast for the year 2000 — the world's 'winners' and 'losers'

References

Atmosphere, Weather and Climate, R.G. Barry and R.J. Chorley, Methuen, 1968.

Climate, Soils and Vegetation, D.C. Money, U.T.P., 1965.

Modern Meterology and Climatology, T.J. Chandler, Thomas Nelson & Sons, 1981.

Planet Earth: Storms, Time–Life Books, 1982.

Planet Earth: The Atmosphere, Time–Life Books, 1983.

Process and Pattern in Physical Geography, Keith Hilton, U.T.P., 1979

Statistics in Geography for 'A' level Students, John Wilson, Schofield & Sims, 1984.

The Atmospheric System, Greg O'Hare and John Sweeney, Oliver & Boyd, 1986.

The Nature of the Environment, Andrew Goudie, Blackwell, 1984.

The South-east Asia Monsoon, Geofile No. 116, Mary Glasgow Publications, September 1988.

Weather Systems, Leslie Musk, Cambridge U.P., 1988.

CHAPTER 10

Soils

'To many people who do not live on the land, soil appears to be an inert, uniform, dark-brown coloured, uninteresting material in which plants happen to grow. In fact little could be further from the truth.'

Soil Processes, Brian Knapp, 1979

What is soil?

Soil forms the thin surface layer to the earth's crust. It provides the foundation for plant and, consequently, animal life on land. The most widely accepted scientific definition is that by J. Joffe (1949) who stated that:

> 'the soil is a natural body of animal, mineral and organic constituents differentiated into horizons [Figure 10.4] of variable depth which differ from the material below in morphology, physical make up, chemical properties and composition, and biological characteristics'.

A simpler definition is: 'soil results from the interrelationship between, and interaction of, several physical, chemical and biological processes all of which vary according to different natural environments'.

The study of soil, their origin, characteristics and utilisation (**pedology**) is a science in itself.

How is soil formed?

The first stage in the formation of soil is the weathering of parent rock to give a layer of loose, broken material known as **regolith**.

Regolith may also be derived from the deposition of alluvium, drift, loess and volcanic material. The second stage, the formation of **true soil** or **topsoil**, results from the addition of water, gases (air), living organisms (biota) and decayed organic matter (humus). Pedologists have identified five main factors involved in soil formation. As all of these are closely interconnected and interdependent, their relationship may be summarised as follows:

$$\text{soil} = f(\text{parent material} + \text{climate} + \text{topography} + \text{organisms} + \text{time})$$

where: f = function of

Parent material When a soil develops from an underlying rock, its supply of minerals is largely dependent on the parent rock. These minerals are susceptible to different rates and processes of weathering, e.g. granite, Figure 10.1. Parent rock contributes to control of the depth, texture, drainage (permeability) and quality (nutrient content) of a soil and also influences its colour (Figure

▼ Figure 10.1

The influence of parent rock (granite) on soil formation (see Figure 10.4 for explanation of terms)

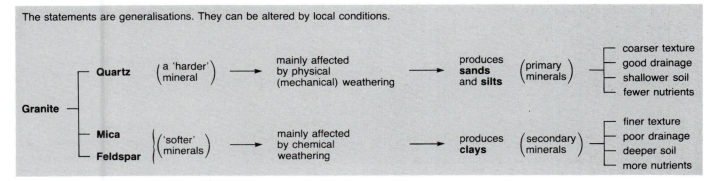

The statements are generalisations. They can be altered by local conditions.

Granite
- **Quartz** — (a 'harder' mineral) → mainly affected by physical (mechanical) weathering → produces **sands** and **silts** (primary minerals) —
  - coarser texture
  - good drainage
  - shallower soil
  - fewer nutrients
- **Mica** / **Feldspar** — ('softer' minerals) → mainly affected by chemical weathering → produces **clays** (secondary minerals) —
  - finer texture
  - poor drainage
  - deeper soil
  - more nutrients

10.1). In parts of Britain, parent rock may be the major factor in determining the soil type (e.g. limestone or granite).

Climate Climate determines the type of soil at a global scale. The distribution and location of world soils corresponds closely to patterns of climate and vegetation. Climate affects the rate of parent rock weathering with the most rapid breakdown being in hot, humid environments.

Precipitation affects the type of vegetation which grows in an area which in turn provides humus; more humus is found in tropical rainforests than in the tundra. Rainfall totals and intensity are also important. Where rainfall is heavy the downward movement of water through the soil transports mineral salts with it, a process known as **leaching**. Where rainfall is light or where evapotranspiration exceeds precipitation, water and mineral salts are drawn upwards towards the surface, by the process of **capillary action**. Leaching tends to produce acidic soils; capillary action results in alkaline soils (page 223).

Temperatures determine the length of the growing season and affect the supply of humus. The speed of vegetation decay is fastest in hot, wet climates as temperature further influences the activity and number of soil organisms and the rate of evaporation, i.e. whether leaching or capillary action is dominant.

Topography or relief As the height of the land increases, so do amounts of precipitation, cloud cover and wind while the temperature and length of the growing season both decrease. Aspect is an important local factor (page 174) with south-facing slopes in the northern hemisphere being warmer and drier than those facing north. The angle of slope affects drainage and soil depth. The steeper the slope, the faster the rate of throughflow and surface runoff of water which may accelerate mass movement and increase the risk of soil erosion. Soils on steep slopes are likely to be thin, poorly developed and relatively dry.

The more gentle the slope, the slower the rate of movement of water through the soil and the greater the likelihood of waterlogging and the formation of peat. There is little risk of erosion but the increased rate of weathering, because of the extra water, and the receipt of material moved downslope tends to produce deep soils. A **catena** is where soils are related to the topography of a hillside and is a sequence of soil types down a slope. The catena (Figure 10.2) is described in more detail on page 231.

Organisms (biota) Plants, bacteria, fungi and animals all interact in the **nutrient cycle** (page 265). Plants take up mineral nutrients from the soil and return them to it after they die. This recycling of plant nutrients (Figure 11.19) is achieved by the activity of microorganisms, such as bacteria and fungi, which assist in the decomposition and decay of dead vegetation. At the same time, macroorganisms, which include worms and termites, mix and aerate the soil. Human activity increasingly affects soil development through the addition of fertiliser, the breaking up of horizons by ploughing, draining or irrigating land, and by unwittingly accelerating or deliberately controlling soil erosion.

Time It takes up to 400 years for 10 mm of soil to form and between 3000 and 12 000 years to produce sufficient depth of mature soil for farming. Young soils tend to retain many characteristics of the parent material from which they are derived. With time they acquire new characteristics resulting from

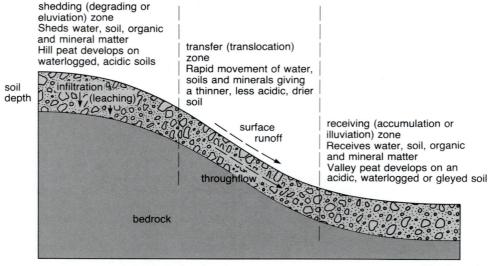

soil depth

shedding (degrading or eluviation) zone
Sheds water, soil, organic and mineral matter
Hill peat develops on waterlogged, acidic soils

infiltration (leaching)

transfer (translocation) zone
Rapid movement of water, soils and minerals giving a thinner, less acidic, drier soil

surface runoff

receiving (accumulation or illuviation) zone
Receives water, soil, organic and mineral matter
Valley peat develops on an acidic, waterlogged or gleyed soil

throughflow

bedrock

- - - - -> movement of water

(Not drawn to scale)

◀ Figure 10.2
A catena: the relationship between soils and topography (not drawn to scale)

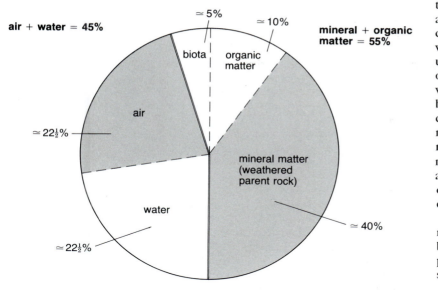

air + water = 45%

≈5%

≈10%

mineral + organic
matter = 55%

biota

organic
matter

air

≈22½%

mineral matter
(weathered
parent rock)

water

≈22½%

≈40%

the addition of organic matter and the activity of organisms. **Horizons** or layers develop as soils reach a stage of equilibrium with their environment. In northern Britain, upland soils must be less than 10 000 years old, as that was the time of the last glaciation when any existing soil cover was removed by ice. The time taken for a mature soil to develop depends primarily upon parent material and climate. Soils develop more rapidly on rocks which weather into sandy material than on those producing clays and they develop more rapidly in hot, wet climates compared with colder and/or drier environments.

A mature soil consists of four components: mineral matter, organic matter including biota (page 221), water and air. The relative proportions of these components in a 'normal' soil, by volume, is given in Figure 10.3.

▲ Figure 10.3

Relative proportions, by volume,

of components in a 'normal' soil

(*after* Courtney and Trudgill)

1 Although air and water together account for some 45 per cent of the total volume, under what conditions may water account for **a** 45 per cent, **b** zero?

2 Why is the proportion of mineral matter much greater in colder, drier climates than in warmer, wetter areas?

3 Why does the amount of organic matter, including biota, increase as rainfall and temperatures increase?

4 How would you expect the proportions of the four components in a peat soil to differ from those given in Figure 10.3?

The soil profile

The **soil profile** is a vertical section through the soil showing its different horizons (Figure 10.4). It is a product of the balance between inputs and outputs into the soil system (Figure 10.5) and the redistribution of, and chemical changes in, the various constituents within the soil. Different soil profiles are described in Chapter 12, but an idealised profile is given here to aid familiarisation with several new terms.

▶ Figure 10.4

An idealised soil profile in

Britain

t o p s o i l	surface horizons	O	L F H		leaf litter / fermentation layer / humus layer
					Depth of soil measured from this point
	A horizon or zone of eluviation (outwashing)	A		dark coloured, stained by down-washed humus	mixed mineral/organic layer
		E		lighter colour due to removal of clay and sesquioxides	some organic material, as well as clay and calcium, removed by water, a process known as translocation
s u b s o i l	B horizon or zone of illuviation (inwashing)	B		fe / brighter colouring due to deposition of sesquioxides — iron and aluminium	possible iron accumulation / organic enriched mineral layer / (accumulation of mechanically and chemically downwashed material)
regolith	weathered parent material (*in situ*)	C			mineral layer
	BEDROCK (unaltered)	R or D			

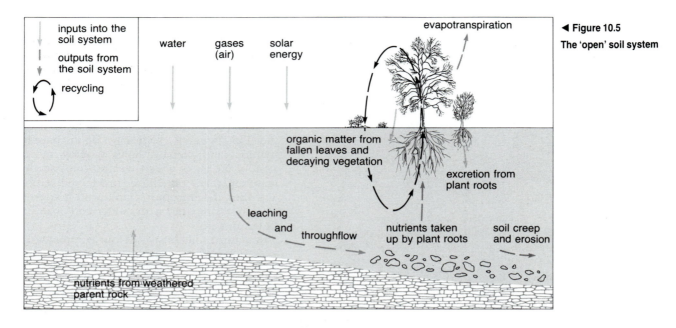

The various soil horizons are referred to by specific letters to indicate their genetic origin. The surface, or **A horizon**, is where biological activity and humus content are at their maximum. It is also the zone which is most affected by the leaching of soluble materials and by the downward movement, or **eluviation**, of clay particles.

Beneath the upper layer the **B horizon** is the zone of accumulation, or **illuviation**, where clays and other materials removed from the A horizon are re-deposited. The A and B horizons together make up the true soil.

The **C horizon** consists mainly of recently weathered material (regolith) resting on the bedrock.

While this threefold division is useful and convenient it is, as will be seen later, over-simplified. Several examples show this.

Humus may be mixed throughout the depth of the soil or it may form a distinct layer. Where humus is mixed with the soil to give a crumbly, black, nutrient-rich layer it is known as **mull**. Where humus is slow to decompose, e.g. in cold, wet upland areas, it produces a fibrous, acidic and nutrient-deficient soil known as **mor** (peat moorlands).

The junctions of horizons may not always be clear.

All horizons need not always be present.

The depth of soil and of each horizon varies at different sites. Local conditions produce soils with characteristic horizons differing from the basic A, B, C pattern, e.g. a water-logged soil, suffering from a shortage of oxygen, develops a gleyed (G) horizon (page 226).

The soil system

Figure 10.5 is a model showing the soil as an open system where materials and energy are gained and lost at its boundaries. The system comprises inputs, stores and recycling or feedback loops (Framework 1, page 44).
Inputs into the system include:
- water from the atmosphere or through flow from higher up the slope,
- gases from the atmosphere and the respiration of soil animals,
- mineral nutrients from weathered parent materials, needed as plant food,
- organic matter and nutrients from decaying plants and animals,
- solar energy and heat.

Outputs from the system include:
- water lost to the atmosphere through evapotranspiration,
- nutrients taken up by plants as food,
- nutrients lost through leaching and throughflow,
- loss of soil particles through soil creep and erosion.

Recycling
Some of the nutrients taken up and stored by plants may later be returned to the soil via leaf litter the following autumn or when plants die and decompose.

Soil properties

The four major components of soil — water, air, mineral and organic matter — (Figure 10.3) are all closely interlinked. The resultant interrelationships produce a series of 'properties', ten of which are listed and described below. It is necessary to understand the workings of these properties to appreciate how a particular soil can best be managed.

Ten properties of soil:

1 Mineral (inorganic) matter
2 Texture
3 Structure
4 Organic matter (humus)
5 Moisture
6 Air
7 Organisms (biota)
8 Nutrients
9 Acidity (pH value)
10 Temperature

1 Mineral (inorganic) matter

As shown in Figure 10.1, soil minerals are obtained mainly by the weathering of parent rock. Weathering is the major process by which nutrients, essential for plant growth, are released. **Primary minerals** are those resistant to chemical weathering but vulnerable to physical weathering. Minerals such as quartz (sands), resulting from temperature extremes as found in deserts, or frost action, as in upland Britain, remain unchanged from the original parent material. **Secondary minerals** are those which have been broken down and altered by various processes of chemical weathering: oxidation, carbonation, hydrolysis and hydration (Chapter 2). Chemical weathering forms clays, of which several varieties exist, and **sesquioxides** (Figure 10.4) which are the oxides of two primary minerals, iron and aluminium.

2 Soil texture

The term 'texture' refers to the degree of coarseness or fineness of the mineral matter in the soil. It is determined by the proportion of **sand**, **silt** and **clay** particles. Particles larger than sand are grouped together and described as stones. In the field it is possible to decide whether a soil sample is mainly sand, silt or clay by its 'feel'. As shown in Figure 10.6, a sandy soil feels gritty and lacks cohesion; a silty soil has a smoother, soap-like feel as well as having some cohesion; and a clay soil when wet is sticky and plastic and may, being very cohesive, be rolled into various shapes.

This method gives a quick guide to the texture but it lacks the precision needed to determine the proportion of particles in a given soil with any accuracy. This precision may be obtained from either of two laboratory measurements, both of which are dependent upon particle size (Figure 10.6). The Soil Survey of Great Britain accepts the International scale in preference to the American scale which gives the following diameter sizes:

coarse sand   between 2.0 and 0.2 mm
fine sand   between 0.2 and 0.02 mm
silt   between 0.02 and 0.002 mm
clay   less than 0.002 mm

One method of measuring texture involves the use of sieves with differing meshes (Figure 10.6). A mesh of 0.2 mm allows fine sand, silt and clay particles to pass through it while trapping the coarse sand. The sample must be dry and needs to be well-shaken. The weight of particles remaining in each sieve is expressed as a percentage of the total sample.

The second method, sedimentation, is when a weighed sample is placed into a beaker of water, thoroughly shaken and then allowed to settle. According to **Stoke's Law**, 'the settling rate of a particle is proportional to the diameter of that particle'. Consequently the larger, coarser sand grains settle quickly at the bottom of the beaker and the finer clay particles settle last, closer to the surface (compare Figure 3.28). The Soil Survey tends to use both methods because sieving is

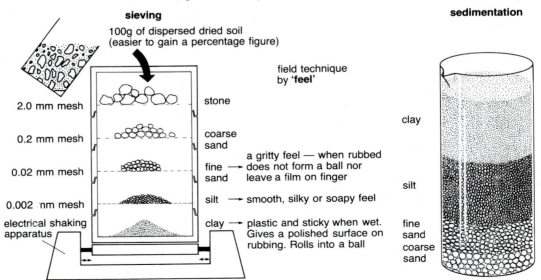

▶ Figure 10.6

Measuring soil texture (*after* Courtney and Trudgill)

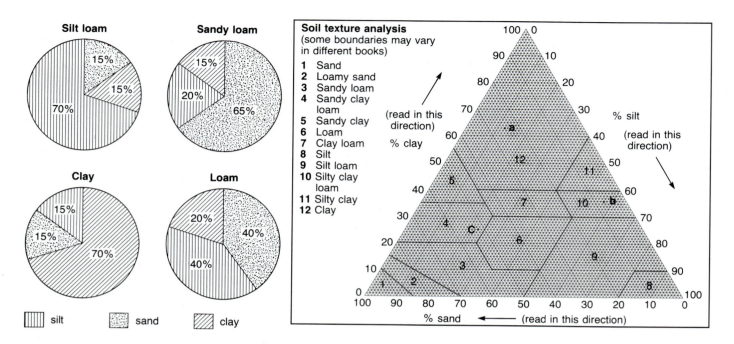

◄ Figure 10.7, *above left*
The texture of different soil types

▲ Figure 10.8
Soil texture analysis — the use of a triangular graph

less accurate in measuring the finer material and sedimentation is less accurate with coarser particles.

The results of sieving and sedimention are usually plotted either as a pie chart (Figure 10.7) or a triangular graph (Figure 10.8). As the proportions of sand, silt and clay vary considerably it is traditional to have 12 texture categories (Figure 10.8).

Q

1 Using Figure 10.8 give the percentage constituents of soils **a**, **b** and **c**.

2 On a sheet of triangular graph paper, plot and match correct names to the following samples:

Sample	% clay	% silt	% sand
a	61	26	13
b	33	7	60
c	8	79	13
d	5	5	90
e	34	36	30

3 What type of weathering is likely to produce **a** sand, **b** clay? What is the connection between weathering and soil texture?

4 **a** What is meant by primary minerals and secondary minerals?
  b Name one type of parent rock likely to give soil *1* and one type likely to give soil *12* on the graph. State your reasons.

The importance of texture

As texture controls the size and spacing of soil pores then it directly affects the soil water content, water flow and extent of aeration. Clay soils tend to hold more water and are less well drained and aerated than sandy soils (page 220).

Texture also controls the availability and retention of nutrients within the soil. Nutrients stick to, i.e. are **adsorbed** by, clay particles and are less easily leached by infiltration or throughflow than in sandy soils.

Plant roots can penetrate coarser soils more easily than finer soils, and 'lighter' sandy soils are easier to plough for arable farming than 'heavier' clays.

Texture greatly influences soil structure (see following section).

How does texture affect farming?

The following comments are generalised as it must be remembered that soils vary enormously.

Sandy soils, being well-drained and aerated, are easy to cultivate and permit crop roots (e.g. carrots) to penetrate. However, they are vulnerable to drought because they lack humus. They also need considerable amounts of fertiliser as nutrients and organic matter are often leached out and not replaced.

Silty soils also tend to lack mineral and organic nutrients. The smaller pore size means that more moisture is retained than in sands but heavy rain tends to 'seal' or cement the surface increasing the risk of sheetwash and erosion.

Clay soils are rich in nutrients and organic matter but they are difficult to plough and after heavy rain are prone to waterlogging and may become gleyed (page 226). Plant roots find difficulty in penetration. Clays expand when wet, shrink when dry and take longer to warm up.

The ideal soil for agriculture is a **loam** (Figure 10.8). This has sufficient clay (20 per cent) to hold moisture and retain nutrients; sufficient sand (40 per cent) to prevent waterlogging, to be well-aerated and to be light enough to work; and silt (40 per cent) to act as an adhesive holding the sand and clay together. A loam is likely to be less susceptible to erosion.

3 Soil structure

It is the aggregation of individual particles which give the soil its structure. In undis-turbed soils these aggregates form different shapes known as **peds**. It is the shape and alignment of these peds which (combined with particle size/texture) determine the size and number of the pore spaces through which water, air, roots and soil organisms can pass. The size, shape, location and suggested agricultural value of each ped type is given in Figure 10.9. While six different types of structure are listed here it should be noted that some authorities only give five — some combine the crumb and granular structures, others the columnar and prismatic.

There is some uncertainty as to exactly how peds form but it is accepted that soils with a good crumb structure give the highest agricultural yield, are more resistant to erosion and develop best under grasses: this is one reason why fallow should be included in a crop rotation. Sandy soils have the weakest structures as they lack the chemical cement from calcium carbonate (or the secretions of organisms, page 222), clays and humus needed to cause the individual particles to aggregate. The crumb structure is the ideal as it provides the optimum balance between air, water and nutrients.

▼ Figure 10.9

Soil structures

Type of structure (ped)	Size of structure	Description of peds	Shape of peds	Location (horizon–texture)	Agricultural value
crumb	1 to 5 mm	Small individual particles similar to breadcrumbs Porous		*A* horizon Loam soil	The most productive. Well aerated and drained — good for roots
granular	1 to 5 mm	Small individual particles Usually non-porous	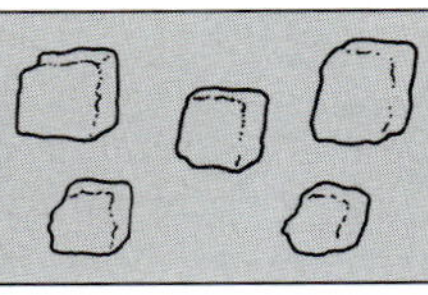	*A* horizon Clay soil	Fairly productive. Problems with drainage and aeration
platy	1 to 10 mm	Vertical axis much shorter than horizontal, like overlapping plates preventing flow of water	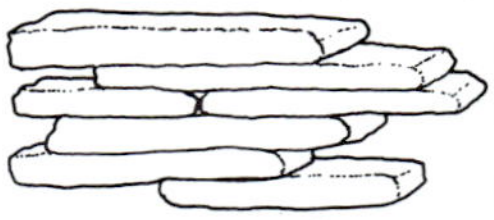	*B* horizon Silts and clays or when compacted by farm tractors	The least productive. Hinders water and air movement. Restricts roots
blocky	10 to 75 mm	Irregular shape Horizontal and vertical axes about equal May be rounded or angular but closely fitting	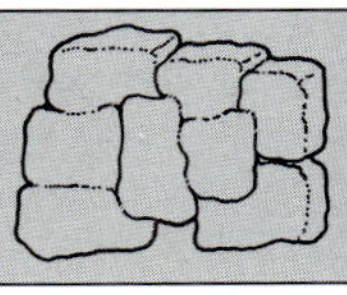	*B* horizon Clay-loam soils	Productive. Usually well drained and aerated
prismatic	20 to 100 mm	Vertical axis much larger than horizontal Angular caps and sides to columns	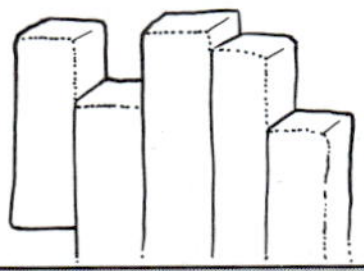	*B* and *C* horizons Often limestones or clays	Usually quite productive. Formed by wetting and drying. Adequate water movement and root development
columnar	20 to 100 mm	Vertical axis much larger than horizontal Rounded caps and sides to column	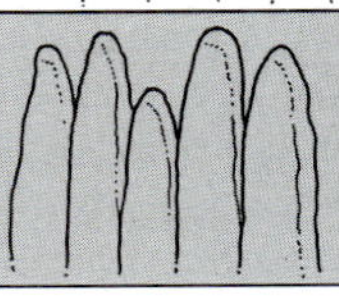	*B* and *C* horizons Alkaline and desert soils	Quite productive (if water available)

Note 1 Some soils may be structureless, e.g. sands
2 Some soils may have more than one structure (ped) in a horizon
3 Each horizon is likely to have its own distinctive ped

Q

1 Distinguish between the terms soil 'texture' and soil 'structure'.

2 How does soil texture influence soil structure?

3 Which types of soil structure (peds) are most beneficial to farmers?

4 Match the six structures named in Figure 10.9 to the letters used in Figure 10.10.

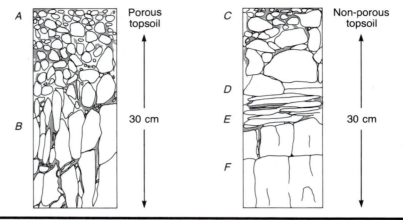

◄ Figure 10.10

4 Organic matter

Organic matter, or humus, is derived mainly from decaying plants and animals or from the secretions of living organisms. Fallen leaves and decaying grasses and roots are the main sources of humus. As bacteria and fungi break down the organic matter, three distinct layers can be seen in the soil profile (Figure 10.4).

1 *L* or **leaf litter** layer is where plant remains are still visible.
2 *F* or **fermentation** layer is where decay is most rapid although some plant remains are still visible.
3 *H* or **humus** layer is where the process of decay has been completed and no plant remains are visible.

Humus gives the soil a black or dark brown colour. The highest amounts of humus are found in areas of temperate grassland forming the **chernozems**, or black earths, of the North American Prairies, the Russian Steppes (page 275) and the Argentinian Pampas. In tropical rainforests, heavy rainfall soon leaches the humus remaining after the considerable uptake and storage by plants from the soil. In drier climates there may be insufficient vegetation to give an adequate supply of humus.

Humus is a major source of nutrients and it combines with clays to form the **clay–humus complex** (soil nutrients, below). The clay–humus complex is essential for a fertile soil as it provides it with a high water- and nutrient-holding capacity. Humus acts as a cement, binding the soil particles together and thus reduces the risk of erosion by improving cohesion.

5 Soil moisture

Soil moisture is important because it affects the upward and downward movement of water. It helps the development of horizons; it supplies water for living plants and organisms; it provides a solvent for plant nutrients; and it controls soil temperature and determines the incidence of erosion. The amount of water in a soil at a given time can be expressed as:

$$W \propto R - (E + T + D)$$
$$(\text{input}) - (\text{outputs})$$

where: W = water in the soil
R = rainfall/precipitation
T = transpiration
$\propto$ = proportional to
E = evaporation
D = drainage

Drainage depends upon the balance between the **water retention capacity** or storage of a soil and the infiltration rate. This may be controlled by the soil texture and structure. It has already been shown how texture and structure affect the size and distribution of pore spaces. Clays have numerous small pores (**micropores**) which may retain water for lengthy periods but which also restrict infiltration rates (page 45). Sands have fewer but much larger **macropores** which permit water to pass through more quickly (a rapid infiltration rate), but have a low retention capacity. A loam provides a more balanced supply of water in the micropores and air in the macropores.

The presence of moisture in the soil does not necessarily mean that it is available for plant use. Plants growing in clays may still

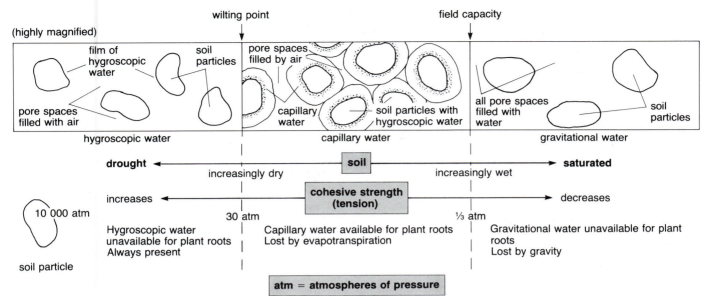

atm = atmospheres of pressure

▲ Figure 10.11

Availability of soil moisture for plant use

suffer from water stress even though clay has a high water-holding capacity and is usually wet when 'squeezed'. Soil water can be classified according to the tension at which it is held and it is measured in atmospheres of pressure (ATM). Following a heavy storm or lengthy episode of rain or snowmelt, all the pore spaces may be filled, with the result that the soil becomes saturated. When infiltration ceases, water with a low cohesive strength (low surface tension) drains away rapidly under gravity. This is called **gravitational**, or **free**, water and is not available for use by plants. Once this excess water has drained away the remaining moisture that the soil holds is said to be its **field capacity** (Figure 10.11).

Moisture at field capacity is held either as **hygroscopic** water or as **capillary** water. Hygroscopic water is always present, regardless of how dry a soil becomes but it is unavailable for plant use. It is found as a thin film around the soil particles to which it sticks due to the strength of its surface tension. Capillary water is attracted to, and forms a film around, the hygroscopic water but has a lower cohesive strength. It is capillary water which is freely available to plant roots. However, this water can be lost to the soil by evapotranspiration. When a plant loses more water through transpiration than it can take up through its roots it is said to suffer **water stress** and it begins to wilt. At

wilting point, photosynthesis (page 252) is reduced but, provided water can be obtained relatively soon or if the plant is adapted to drought conditions (page 267), this need not be fatal. Figure 10.11 shows the different water-holding characteristics of soil.

6 Air

Air fills the pore spaces left unoccupied by soil moisture. It is essential for plant growth and living organisms. Compared with atmospheric air, air in the soil contains more carbon dioxide, released by plants and soil biota, and more water vapour, but less oxygen as this is consumed by bacteria. Biota need oxygen and give off carbon dioxide — these gases are exchanged through the process of diffusion.

7 Soil organisms (biota)

Soil organisms include bacteria, fungi, and earthworms. They are more active and plentiful in warmer, well-drained and aerated soils (e.g. mull) than they are in colder, more acidic and less well-drained and aerated soils (e.g. mor).

Organisms are responsible for three important soil processes.

Decomposition Detritivores, such as earthworms, mites, woodlice and slugs, begin this process by burying leaf litter (detritus), which hastens its decay, and eating some of it. Their faeces (wormcasts) increase the sur-

1 What effect are large amounts of carbon dioxide in the soil likely to have upon parent rock? (See weathering processes page 33.)

2 What effect might a lack of oxygen and low soil porosity have upon the soil, soil biota activity, and plant roots?

3 Why are sandy soils better aerated than clay soils?

face area of detritus upon which fungi and bacteria can act. Fungi and bacteria secrete enzymes which break down the organic compounds in the detritus. This releases nutrient ions essential for plant growth (Soil nutrients, below), into the soil while some organic compounds remain as humus.

Fixation By this process bacteria can transform nitrogen in the air into nitrate which is an essential nutrient for plant growth.

Development of structure Fungi help to bind individual soil particles together to give a crumb structure, while burrowing animals create passageways which help the circulation of air and water and facilitate root penetration.

8 Soil nutrients

Nutrient is the term given to chemical elements found in the soil which are essential for plant growth and to maintain the fertility of a soil. There are four **primary elements** (carbon: C, hydrogen: H, nitrogen: N and oxygen: O) which are needed in considerable quantities; five **secondary elements** (calcium: Ca, magnesium: Mg, potassium: K sulphur: S, and Phosphorus: P) in smaller amounts and several trace elements (e.g. molybdenum: Mo), required in minute amounts. Ca helps the growth of roots and new shoots; Mg is a component of chlorophyll; K and Na help in the formation of starches and oils, and Mo activates enzymes. Of these N, K and Ca are most likely to become deficient in areas of cultivation.

Nutrients may be obtained from four sources.

1 Nutrients in solution may originate in rainwater. Although these are readily available to plants they may be leached out of the soil in gravitational water. Secondary elements are especially vulnerable.

2 Fertiliser may be added artificially to the soil.

3 Minerals may be released from the parent rock by weathering or from decaying organic matter by soil organisms. These are soluble and dissolve into the soil solution, producing positively (+) charged ions called **cations**.

4 Nutrients attached to the clay–humus complex provide the major reserves. The particles of clay and humus develop negatively (−) charged ions known **anions**. The negatively charged clay and humus particles attract the positively charged minerals and the cations are adsorbed or stuck to the particles (Figure 10.12). The resultant double layer of negative and positive ions is called the **Gouy Layer** (Figure 10.13).

Cation exchange This is the important process by which nutrients, adsorbed to particles of clay and humus, become detached so that they can be absorbed by plant roots (Figure 10.14). Cations of Ca^{++}, Mg^{++}, K^+, and Na^+ can be released from the clay–humus particles and replace an equal number of H^+ cations which were initially attached to plant roots. As well as providing nutrients for plant roots the cation exchange releases hydrogen which in turn increases the acidity of the soil (see next section). This acidity accelerates the rate of weathering of the parent rock, releasing more minerals to replace those used by the plants or lost through leaching by gravitational water.

Different soils have different **cation exchange capacities** (CEC), i.e. different abilities to retain cations for plant use. As the table shows, sands have a much lower ability to retain plant nutrients than does humus.

Cation exchange capacity (CEC) (measured in mille-equivalents per 100 g of soil)	
Sandy soils	1 to 5
Clays	3 to 50
Humus	150 to 400

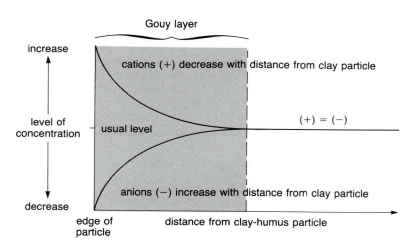

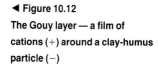

◄ **Figure 10.12**
The Gouy layer — a film of cations (+) around a clay-humus particle (−)

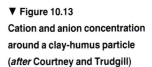

▼ **Figure 10.13**
Cation and anion concentration around a clay-humus particle (*after* Courtney and Trudgill)

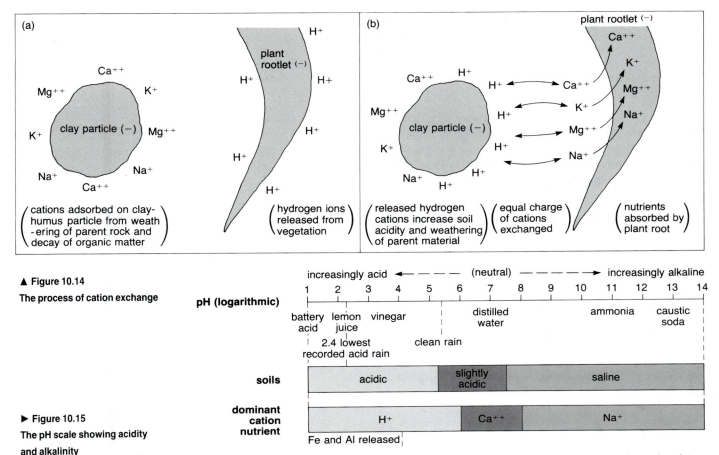

▲ Figure 10.14

The process of cation exchange

▶ Figure 10.15

The pH scale showing acidity and alkalinity

9 Acidity (pH)

As mentioned in the previous section, soil contains positively charged hydrogen cations. **Acidity** or **alkalinity** is a measure of the degree of concentration of these cations and is measured on the pH scale (Figure 10.15). The scale, ranging from 1 to 14, is logarithmic (compare the Richter scale, Figure 1.4). This means that a reading of 6 is ten times more acidic than a reading of 7 (which is neutral), and one hundred times more acid than one of 8. Most British soils are slightly acidic though in upland Britain, acidity increases as the heavier rainfall of those areas leaches out the alkaline calcium. Acid soils therefore tend to need constant liming if they are to be farmed successfully.

A slightly acid soil is the optimum for farming in Britain as this helps to release the secondary elements. However, if a soil becomes too acidic it releases iron and aluminium which, in excess, may become toxic and poisonous to plants and organisms. Increased acidity makes organic matter more soluble and therefore vulnerable to leaching and it discourages living organisms, thus reducing the rate of breakdown of plant litter and causing the formation of peat.

In areas where there is a balance between precipitation and evapotranspiration soils are often neutral (American Prairies, Figure 12.32, 12.33), while in areas with a water deficiency (deserts, Figure 3.50), soils are alkaline.

10 Soil temperature

Incoming radiation can be absorbed, reflected or emitted by the earth's surface (Figure 9.4). The topsoil, especially if vegetation cover is limited, heats up more rapidly than the subsoil during the daytime and loses heat more rapidly at night. A 'warm' soil will have greater biota activity giving a more rapid breakdown of organic matter. It is more likely to contain nutrients because chemical weathering of parent material may be faster, and seeds will germinate more readily than in a 'cold' soil.

Q With reference to the ten properties above describe what makes the optimum soil for farming in two contrasting environments of your choice.

CASE STUDY

How are soils studied?

Studying soils in the field: the soil pit

Begin by reading a book which describes in detail how to dig a soil pit and describe and explain the resultant profile.

First make sure you obtain permission to dig a pit. The site must be carefully chosen. You will need to find an undisturbed soil so avoid digging near to hedges, trees, footpaths or on recently ploughed land. Ideally make the surface of the pit 0.7 m² and 1 m deep (unless you hit bedrock first). Carefully lay the turf and soil on plastic sheets. Clear one face of the pit to get a 'clean' profile so that you can complete your recording sheet — the one in Figure 10.16 is very detailed. Sometimes not all the readings may be taken due to problems of clarity, lack of time and equipment, or irrelevence to a particular enquiry.

Make a detailed field sketch before replacing the soil and turf. You may have to complete several tasks in the laboratory before writing up your description. Remember it is unlikely that your answer will exactly fit a model profile. It may show the characteristics of a podsol (Figure 12.47) the further north you live in Britain (cooler, wetter uplands) or of a brown forest earth (Figure 12.41) if you live in the south, but do not force your profile to fit a model.

▼ Figure 10.16

A soil recording sheet for

a soil site

b soil profile

a The soil site

Recorded by	Date		Locality		Six-figure grid reference
Parent rock (geological map)	Altitude (estimate from Ordnance Survey map)	Angle of slope (Abney level)	Aspect (bearing or compass point)		Relief (uniform, concave, convex, terrace)
Exposure (exposed, sheltered)	Drainage (shedding or receiving site, floodplain, terrace, boggy)	Natural vegetation or type of farming (tree species, ground vegetation, crops, animals)	Previous few days' weather (warm, cold, wet, dry)		Other local details (remember your labelled field sketch)

b The soil profile

Horizon	Depth of horizon (cm)	Lower boundary of horizon	Colour	Texture	Stoniness	Structure (peds)	Consistency	pH	Moisture content	Porosity	% organic water	% roots	% carbonates	Soil biota and/or animals
How to read	measure from base of humus layer	sharp abrupt clear indistinct gradual irregular smooth broken	Use Munsell colour chart	% clay, silt or sand 'feel' sieves sedimentation	(i) size of stones (ii) number of stones (iii) shape of stones	structureless crumb etc	loose friable firm hard plastic sticky soft	litmus paper or soil testing kit	weigh sample evaporate water reweigh sample moisture meter	time taken for a beakerful of water to infiltrate	type and amount	weigh burn sample (and roots) reweigh sample	add dilute hydrochloric acid — if it effervesces it is over 1% carbonate	number type
A														
B														
C														

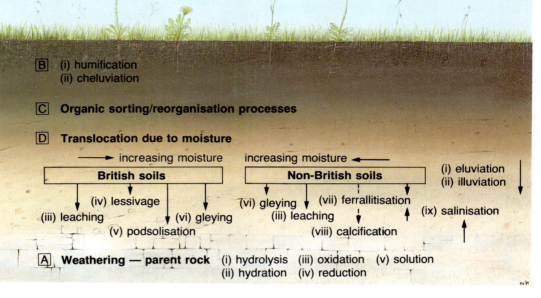

▶ Figure 10.17
Soil forming processes in
British and non-British soils

Eluviation *means the washing out of material, i.e. the removal of organic and mineral matter from the A horizon (Figure 10.4).* **Illuviation** *is the process of inwashing, i.e. the redeposition of organic and mineral matter in the B horizon.*

Processes of soil formation

Numerous processes are involved in the formation of soil and the creation of the profiles, structures and other features described above. Soil forming processes depend on all five of the factors described on page 213. Figure 10.17 shows some of the more important processes.

A Weathering

As described on page 217, weathering produces primary and secondary minerals as well as determining the rates of release of nutrients and the soil depth, texture and drainage. In systems terms this means that minerals are released as inputs into the soil system from the bedrock store and transferred into the soil store (Figure 10.5).

B Humification and cheluviation

Humification is the process by which organic matter is decomposed to form humus (see 4 Organic matter, above), a task performed by soil organisms. Humification is most active in the *H* horizon of the soil profile (Figure 10.4) where it can result in

mull with pH 5.5 to 6.5, mor, pH 3.5 to 4.5 (page 216), or the intermediate moder, pH 4.5 to 5.5.

As organic matter decomposes it releases nutrients and organic acids. These acids, known as **chelating agents**, then attack clays, releasing iron and aluminium. The dissolved minerals are transported downwards under the influence of the chelating agents — this is the process of **cheluviation**. (Note the similarity with leaching.)

C Organic sorting

Several processes operate within the soil to reorganise mineral and organic matter into horizons and contribute to the aggregation of particles and the formation of peds. Earthworm activity is a significant factor in sorting material into different particle size.

D Translocation of soil materials

Translocation is the movement, in any form (e.g. solution, suspension) or direction (e.g. downward, upward) of soil components and usually takes place in association with soil moisture.

1a List the five processes of chemical weathering shown in Figure 10.17.
b Refer back to Chapter 2 and describe each process, explaining how it helps in the formation of soil.

2 Which process of chemical weathering is not listed in Figure 10.17? Suggest a reason for its omission.

Note: Among other functions, hydrolysis is responsible for the formation of the ($^+$) cations of Ca, Na, K and Mg and several ($^-$) anions referred to under cation exchange.

British soils

In Britain there is usually a soil moisture budget surplus in the ground resulting from an excess of precipitation over evapotranspiration (water balance, Figure 3.3) or locally from poor drainage. This results in the three translocation processes of leaching, podsolisation and gleying, to which some pedologists add a fourth: lessivage.

Leaching is the removal of soluble material in solution. Where precipitation exceeds evapotranspiration and soil drainage is good, rainwater — containing oxygen and carbonic acid plus organic acids collected as it passes through the surface vegetation — causes chemical weathering, the breakdown of clays and dissolving of soluble salts (bases). Ca and Mg are eluviated from the A horizon, making it increasingly acid as these are replaced by hydrogen ions, and are subsequently illuviated in the B horizon below (Figure 10.18a).

Podsolisation operates as an intense form of leaching. It is most common in cool climates where precipitation is greatly in excess of evapotranspiration and where soils are well-drained or sandy. Podsolisation is also defined as the removal of sesquioxides (and chelates) under extreme leaching. As the surface vegetation is often coniferous forest, heathland or moors, rain percolating through it becomes progressively more acidic and may reach a pH of below 4.5. This in turn (Figure 10.15) dissolves an increasing amount and number of bases (Ca, Mg, Na and K), silica, and, ultimately, the sesquioxides of iron and aluminium (Figure 10.18b). The resultant **podsol** soil (Figure 12.11) therefore has two distinct horizons: the bleached A horizon, drained of coloured minerals by leaching, and the reddish-brown B horizon where the sesquioxides have been illuviated. Often the iron deposits form a **hard pan** which is a characteristic of a podsol.

Gleying occurs when the output of water from the soil system is restricted giving **anaerobic** or **waterlogged** conditions. This is most likely to occur on gentle slopes, in depressions where the underlying rock is impermeable, or following periods of heavy rain. The pore spaces fill with stagnant water which becomes de-oxygenised. The reddish coloured ferric (iron) compounds are chemically reduced to give a grey-blue ferrous compound. Occasionally, pockets of air re-oxygenise the ferrous compound to give scatterings of red mottles. Although many British soils show some evidence of gleying, the conditions develop most extensively on moorland plateaux.

Lessivage is a particular type of leaching which applies to clay particles carried downwards in suspension. This process can lead to the breakdown of peds.

Courtney and Trudgill have summarised the relationship of leaching, podsolisation and gleying with precipitation and drainage (Figure 10.19).

Non-British soils

Leaching occurs in most areas, such as the tropical rainforests, where precipitation exceeds evapotranspiration. It is the rapid translocation of bases that makes these soils infertile once the protective forest cover has been removed.

Gleying occurs worldwide where environmental conditions are similar to those described for its operation in Britain.

▼ Figure 10.18
The process of leaching, *left*, and podsolisation, *right* (see also Figure 12.47)

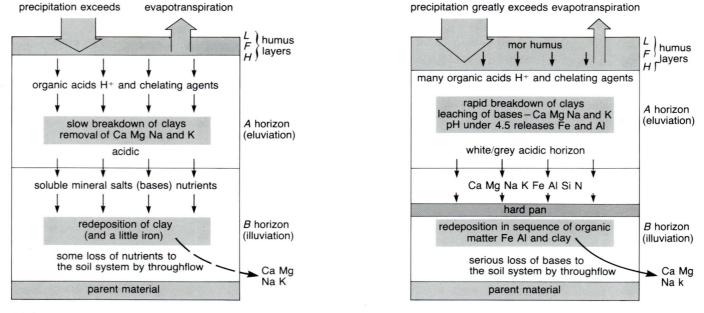

▶ Figure 10.19

The water balance (Figure 3.3) and soil forming processes

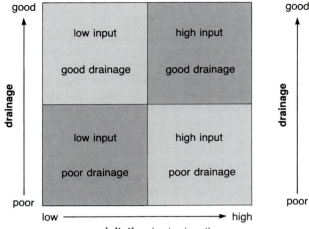

Ferrallitisation is the process by which parent rock is changed into a soil consisting of clays (kaolinite) and sesquioxides (hydrated oxides of iron and aluminium). In humid tropical areas, with constantly high temperatures and rainfall for all or most of the year, there is rapid chemical weathering. This produces clay. The clay then breaks down to form silica, which is removed by leaching, and the sesquioxides of iron and aluminium, which remain to give the characteristic red colour of many tropical soils. This is the reverse of podsolisation where the silica remains and the iron and aluminium are removed. In tropical rainforests, with rain throughout the year, **ferrallitic** soils develop (page 265). In savanna areas, with an alternating dry and wet season, **ferruginous** soils form (page 268).

Calcification is a process typical of low rainfall areas where precipitation is either equal to, or slightly higher than, evapotranspiration. Although there may be some leaching it is insufficient totally to remove calcium which then accumulates, in relatively small amounts, in the *B* horizon (Figure 10.20b, chernozems).

Salinisation occurs when potential evapotranspiration is greater than precipitation where the water table is near to the surface. It is therefore found locally in dry climates and is not a characteristic of desert soils. As moisture is evaporated from the surface, salts are drawn upwards in solution by capillary action. Further evaporation results in the deposition of salt as a hard crust (Figure 10.20c). Salinisation has become a critical problem in some irrigated areas, e.g. California, page 413.

▼ Figure 10.20

Soil forming processes (see also Figures 12.19, 10.22, 16.43)

a Ferrallitisation

b Calcification

c Salinisation

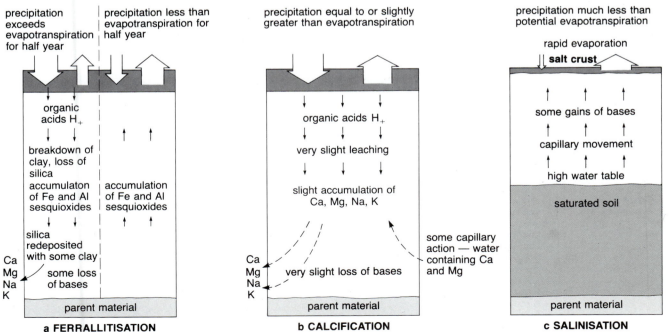

Soil classification

The need for geographers to classify material has already been discussed (page 130). However, there appear to be more numerous bases for the classification of soils than, say, coasts, climates and settlements: indeed, new ones seem to be proposed annually. As every soil is unique, attempts have been made to introduce terms (e.g. podsol, gleyed) which summarise the main characteristics of each particular soil type. A classification of soils means that predictive statements can be made to enable farmers, among others, to appreciate the capabilities, limitations and management requirements of a particular soil. As with other classification systems, soils must be grouped so that there is minimum diversity within each group and maximum diversity between groups. What constitutes a working basis is complicated by the fact that on a global scale different soil forming processes operate from those on a **regional** (Britain) or **local** (within a valley) scale.

The following reference to only three of the numerous classifications shows a degree of simplification and convenience on the one hand and bias on the other. These three have been selected because each is more appropriate at a particular scale (global, regional and local).

- The earliest classification, in the 1880s, was based upon the major soil types found across present day USSR. Still widely used in accounting for the distribution of soils on a **global** scale, it takes climate as being the most significant single factor in soil development. The resultant types, which include the Russian terms 'podsol' and 'chernozem', are recognised by their distinctive 'zones' or profiles. This classification tends to ignore local factors such as parent rock and changes in relief.

- In 1940 the Soil Survey of Great Britain adopted a scheme based on six main soil groups found (**regionally**) in Britain. These groups (brown earths, podsols, gleyed, calcareous, organic and undifferentiated alluvium) form a classification most appropriate to the British Isles but which cannot be applied effectively on a global scale and does not have the detail to be useful at a more local level. This was revised in the 1970s.

- On a **local** scale, factors other than climate become increasingly important, e.g. parent rock, drainage and relief. The United States Department of Agriculture uses its '7th Approximation' classification (so-called because it is the seventh revision), based upon a description of the soil as it is today rather than by its genesis. It is very technical, full of new terminology and contains 10 major types and 42 sub-types.

Q With reference to any soil classification which you have studied, describe the basis for this classification and briefly list its advantages and disadvantages. (See references at the end of this chapter for soil textbooks.)

Zonal, intrazonal and azonal soil classification

This 'compromise' classification has been selected because it links world soil types with the global climate and vegetation belts. It acknowledges that some soils develop mainly due to climate (**zonal**) while others result from local factors (**intrazonal**) and that while some soil types are mature, others are still young (**azonal**) and relatively undeveloped.

Criteria for classification	Zonal (climate and good drainage)	Intrazonal (parent rock, extremes of drainage, relief)		Azonal (immature)
Soil type	Tundra Podsol Brown earths Chernozems Chestnut and prairie Mediterranean Ferrallitic Ferruginous Desert soils	Rendzina Terra rossa } calcimorphic Gleyed Peat (or bog) } hydromorphic Saline } halomorphic		Alluvium Scree Till Sand dunes Salt marsh Volcanic

Zonal soils

These result from the maximum effects of climate and living matter upon parent rock in areas where there are no extremes of weathering, relief or drainage and where the landscape and climate have been stable for a long time. Consequently these soils have had time to develop distinctive profiles with, usually, clear horizons. The relationship between the world's major climate, soil and vegetation zones is described in Chapter 12 and summarised in Figure 12.3.

Intrazonal soils

These soils reflect the dominance of a single local factor and exist in different parts of the world rather than in zones. They are commonly grouped into three types.

- **Calcimorphic** or **calcareous** soils develop upon a limestone parent rock, e.g. rendzina and terra rossa.
- **Hydromorphic** soils are those having a continuous presence of water, e.g. gleyed soils and peat.

▼ **Figure 10.21**

Calcimorphic soils — terra rossa and rendzina

▶ **Figure 10.22**, *below right*

A rendzina, Grude Imotsk Polje, Yugoslavia

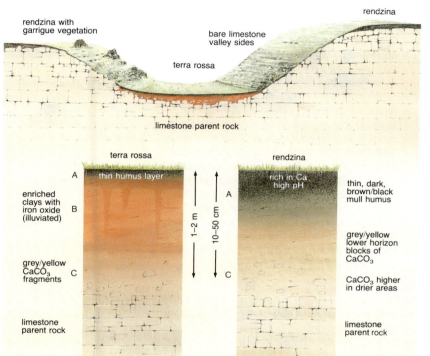

▶ **Figure 10.23**

Hydromorphic soils

a Gleying

b Peat (Sutherland Flows)

■ **Halomorphic** soils have a continued high level of soluble salts which renders them saline.

Reference has been made to some of these soils already but they are re-considered here to present a complete picture.

Calcimorphic

1 **Rendzina** (Figure 10.2, page 229) The rendzina develops in areas where limestones or chalk form the parent material and where grasses form the surface vegetation (e.g. the English Downs). The grasses produce a leaf litter rich in bases. The leaf litter encourages considerable activity by organisms which help with the rapid recycling of nutrients. The *A* horizon therefore consists of a black or dark brown mull humus. Due to the high concentration of calcium, the soils are alkaline with a pH of between 7.0 and 8.0. The calcium-saturated clays, which have a crumb or blocky structure, tend to limit the movement of water and so there is little leaching. Consequently there is no *B* horizon. The underlying limestones, affected by chemical weathering, leave very little insoluble residue and this, together with the permeable nature of the bedrock, results in a thin soil with limited moisture reserves.

2 **Terra rossa** (Figure 10.22) As its name suggests, this is a red-coloured soil (it has been called a 'red rendzina'). It is found in areas of heavy, even if seasonal, rainfall where the calcium carbonate parent rock is chemically weathered (carbonation) and silicates are leached out of the soil to leave a residual deposit rich in iron hydroxides. It usually occurs in depressions within the limestone and where the vegetation is garrigue (Figure 12.28).

Hydromorphic

1 **Gley** soils (Figure 10.23) Gleying occurs in saturated soils when the pore spaces are filled with water to the exclusion of air. The lack of oxygen leads to anaerobic conditions (page 32) and the reduction (chemical weathering) of iron compounds from a ferric to a ferrous form. The resultant soil has a grey/blue colour with scatterings of red mottles (page 226). Because gleying is a result of poor drainage, it can occur in zonal soils anywhere in the world almost independent of climate. Pedologists often differentiate between **surface gleys**, caused by slow infiltration rates through the topsoil, and **groundwater gleys**, resulting from a seasonal rise in the water table or the presence of an impermeable parent rock.

2 **Peat** (Figure 10.23b) Where a soil is waterlogged and the climate is too cold and/or wet for organisms to break down vegetation fully, layers of peat accumulate. These conditions mean that litter input (supply) is greater than the rate of decomposition by organisms whose activity rates are slowed down at low temperatures and under anaerobic conditions. Peat is a soil in its own right when the layer of poorly decomposed material exceeds 38 cm. Peat can be divided according to its location and acidity. **Blanket peat** is very acidic, covers large areas of wet upland plateaux in Britain and is believed to have formed 5000 to 8000 years ago during the Atlantic climatic phase (Figure 11.17). **Raised bogs**, also composed of acidic peat, occur in lowlands with a heavy rainfall. Here the peat accumulates until it builds up above the surrounding countryside. **Valley**, or **basin**, peat may be almost neutral or only slightly acidic if water has drained off surrounding calcareous uplands (e.g. Somerset Levels and the Fens) otherwise it too (e.g. Rannoch Moor in Scotland) will be acid. Fen peat is a high quality agricultural soil.

Halomorphic

Halomorphic soils contain high levels of soluble salts and have developed through the process of salinisation (page 227, Figure 10.20c). They are most likely to occur in dry climates where, in the absence of leaching, the mineral salts are brought to the surface by capillary action and where the parent rock or groundwater contains sodium chloride (common salt). Such alkaline soils are also known by their Russian name, *solonchak*. If, at a later stage, some of these salts are leached downwards through the profile (e.g. under irrigation) the resultant soil is called by its Russian name, *solonetz*.

Azonal soils

These are immature soils which show the characteristics of their origin (parent rock or agent of deposition) and where soil forming processes have not had time to operate fully. More traditionally, in Britain these have been listed as **scree** (weathering); **alluvium** (fluvial); **till** (glacial); **sands** and **gravels** (fluvio-glacial); **sand dune** (aeolian and marine); **salt marsh** (marine) and **volcanic** (tectonic) soils. The American '7th Approximation' classification (page 228), divides azonal soils into **lithosols**, developed

▶ Figure 10.24
Readings taken on a catena, Arran

Site	Interval (metres)	Altitude (metres)	Slope angle	Depth of soil (cm)	pH	Soil (moisture meter)	Vegetation	
							dom	sub dom
A	100	320	2°	170	4.4	6.0	N Cv	Sph Et
B	100	310	10°	110	3.8	5.2	N Cv Et	Pe
C	120	280	16°	45	4.4	1.5	N Cv	Et Pe
D	100	240	19°	42	4.7	2.0	N	Cv Et j l
E	60	195	28°	24	5.0	2.2	N Cv b Et	Vm
F	60	165	27°	18	5.6	2.0	N b	Cv
G	60	125	30°	28	5.9	2.5	b	f Ag N Pe
H	60	95	33°	20	5.7	1.5	N b	Vm Et Pe
I	60	80	20°	21	5.8	1.5	b	Ag Fe Ev Vm
J	60	55	17°	70	4.5	3.5	Ag	Cv F Sph s
k	40 to river	40	11°	70	4.8	3.8	Ag	Fe j Sph Et

N = *Nardus*   Cv = *Calluna vulgaris*   Et = *Erica tetralix*   b = bracken   ag = *Agrostis*
Sph = *Sphagnum*   pe = *Potentilla erecta*   f = fern   j = juncus   l = lichen   s = sedge
Vm = *Vaccillium myrtilus*   Fe = *Fescue* (bracken — *Pteridium aquilinum*)

at high altitude by weathering (scree and rendzina); **regosols**, formed on unconsolidated material (sand dunes and volcanic ash); and **alluvial**, where successive flooding builds up layers (a riverine floodplain or a salt marsh).

The soil catena

A catena (Latin for 'chain') describes a series of soils where each facet on a slope is different from, but linked to, its adjacent facets. Catenas therefore illustrate the way in which soils can change down a slope where there are no marked changes in climate or parent rock. It is an example of a small scale open system involving inputs, processes and outputs (Figure 10.2). The slope itself is in a delicate state of dynamic equilibrium (Figure 2.11).

1 With reference to Figure 10.2 and Chapter 10, draw a soil catena and on it locate the following.
  a The shedding, transfer and receiving slopes.
  b The area with:
     the highest pH   lowest pH
     most leaching   maximum throughflow
     basin peat   brown earths
     blanket peat   gleying
     most organic material most soil organisms

  In each case give a reason for your answer.
  c Your answers to **b** were based on the bedrock being impermeable. What difference would it make if the bedrock was limestone (calcareous)?

2 A sixth form field course to the Isle of Arran included taking various measurements down a valley side in Glen Rosa. The results are given in Figure 10.24. Two further tasks have to be completed.
   The first is to choose the most appropriate and effective graphical techniques to illustrate the data. The second is the interpretation of the results. Use the data in Figure 10.24 to complete the following tasks.

231

a Draw a true scale transect and by methods of your choice show on it all the information given in the table.

b Account for the differences in acidity and moisture at different points on the slope.

c From the information given identify the plant associations on this slope and describe the conditions favouring their growth.

d Draw a simple soil profile to show the likely conditions at sites *A*, *F* and *K*.

e If drainage were to be improved on the flood plain, what long term effects might this have on the character of the soil at site *K*?

The techniques used by one student to illustrate her answer to **2a** is shown in Figure 10.25.

3 Figure 10.26 shows the relief and surface geology of an upland edge and adjacent lowland in eastern England. The upland receives about 730 mm of precipitation per annum and mean monthly temperatures vary from 2.7°C in January to 15.5°C in July and August. Six sites, lettered *A* to *F* on Figure 10.26, have their soil profile characteristics shown on the accompanying graphs.

a Describe and explain the characteristics of the soil profile at site *F*.

b Suggest why from the evidence in Figure 10.36, the soil at site *E* is less prone to waterlogging than that at site *F*.

c Describe and explain the changes in soil profile characteristics that occur downhill from site *B*, through site *C*, to site *D*.

d Comment upon the advantages and limitations for arable farming of the area above 120 m containing site *A*.

▶ **Figure 10.26**

Relief and surface geology of an upland edge and adjacent lowland in eastern England

▼ **Figure 10.25**

Graphical representation of a catena — one possible method of mapping the soil and slope characteristics suggested by a sixth form student, Joanne Morton

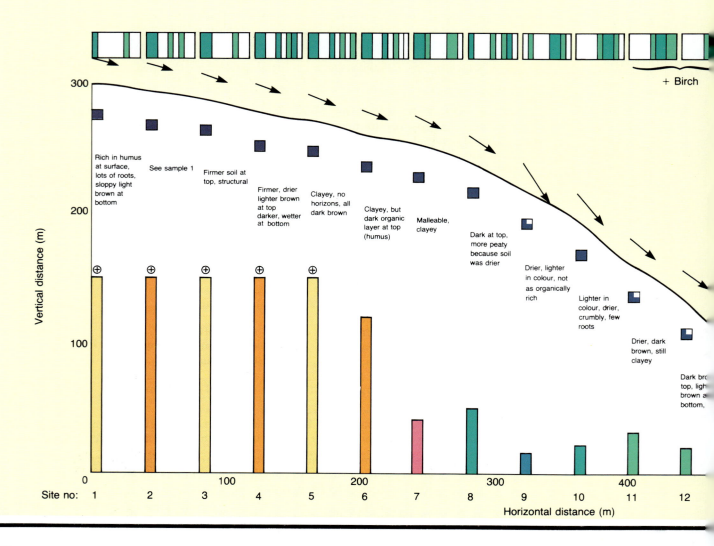

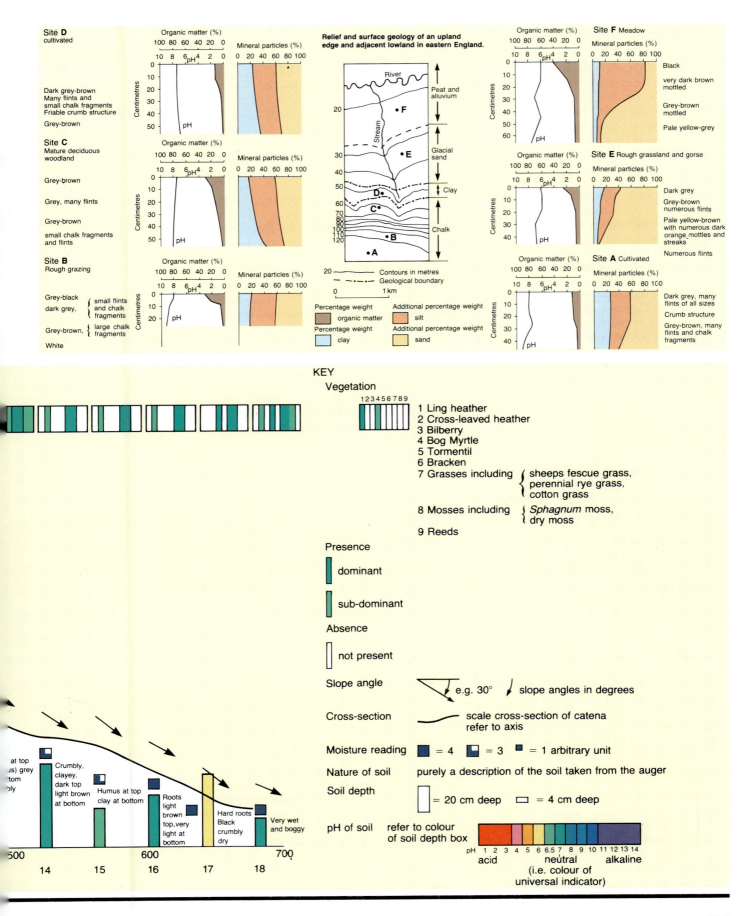

Relief and surface geology of an upland edge and adjacent lowland in eastern England.

Site D — cultivated
Dark grey-brown
Many flints and small chalk fragments
Friable crumb structure
Grey-brown

Site C — Mature deciduous woodland
Grey-brown
Grey, many flints
Grey-brown
small chalk fragments and flints

Site B — Rough grazing
Grey-black
dark grey, small flints and chalk fragments
Grey-brown, large chalk fragments
White

Site F — Meadow
Black
very dark brown mottled
Grey-brown mottled
Pale yellow-grey

Site E — Rough grassland and gorse
Dark grey
Grey-brown numerous flints
Pale yellow-brown with numerous dark orange mottles and streaks
Numerous flints

Site A — Cultivated
Dark grey, many flints of all sizes
Crumb structure
Grey-brown, many flints and chalk fragments

Contours in metres
Geological boundary
0   1 km

Percentage weight
■ organic matter
Percentage weight
■ clay

Additional percentage weight
■ silt
Additional percentage weight
■ sand

KEY

Vegetation

1 Ling heather
2 Cross-leaved heather
3 Bilberry
4 Bog Myrtle
5 Tormentil
6 Bracken
7 Grasses including { sheeps fescue grass, perennial rye grass, cotton grass }
8 Mosses including { *Sphagnum* moss, dry moss }
9 Reeds

Presence
■ dominant
■ sub-dominant

Absence
□ not present

Slope angle   e.g. 30° ↙ slope angles in degrees

Cross-section   ⌐ scale cross-section of catena refer to axis

Moisture reading   ■ = 4   ◪ = 3   ▪ = 1 arbitrary unit

Nature of soil   purely a description of the soil taken from the auger

Soil depth   ▯ = 20 cm deep   ▭ = 4 cm deep

pH of soil   refer to colour of soil depth box

pH 1 2 3 4 5 6 6.5 7 8 9 10 11 12 13 14
acid   neutral (i.e. colour of universal indicator)   alkaline

at top
(us) grey
tom
ply

Crumbly, clayey, dark top light brown at bottom

Humus at top clay at bottom

Roots light brown top, very light at bottom

Hard roots Black crumbly dry

Very wet and boggy

500   600   700
14   15   16   17   18

Soil erosion and conservation

As we have seen, soil usually takes thousands of years to become sufficiently deep and mature for economic use. During that time there is always some natural loss through leaching, mass movement and erosion. Normally there is an equilibrium, however fragile, between the rate at which soil forms and that at which it is eroded or degraded. That balance is being disturbed by human mismanagement with increasing frequency and to serious detrimental effect.

Recently increasing concern about the environment has led to the introduction of soil improvement and conservation schemes. Nevertheless, it is estimated that 7 per cent of the world's topsoil is lost each year. The World Resources Institute claims that Burkina Faso loses 35 tonnes of soil per hectare per year. Other comparable figures are: Ethiopia, 42; Nepal, 70; Deccan Plateau (India), 100; El Salvador, 192; the Loess Plateau of North China, 251 (Case Study). Soil removed during a single rain or dust storm may never be replaced.

Soil degradation

The major cause of soil erosion is the removal of natural vegetation cover to leave the ground exposed to the elements. The most serious of these removals is deforestation. In countries such as Ethiopia (page 425) the loss of trees, used for fuel, means that the heavy rains, when they do occur, are no longer intercepted by the vegetation. Rainsplash (the direct impact of raindrops, page 35) loosens the topsoil and prepares it for removal by sheetwash (overland flow). Water flowing over the surface has little time to infiltrate into the soil, or recharge the soil moisture store (page 221). Where this water evaporates, a hard crust may form, making the surface less porous and increasing the amount of runoff. More topsoil tends to be carried away where there is little vegetation as there are neither plant roots nor organic matter to bind it together. Small channels or rills may be formed which in time may develop into large gulleys making the land useless for agriculture (Figure 10.27).

Even where the soil is not actually washed away, heavy rain may accelerate leaching and remove nutrients and organic matter at a rate faster than that at which they can be replaced by the weathering of bedrock or vegetation breakdown (e.g. Amazon Basin, page 266 and Figure 12.9). The loss of trees

also reduces the rate of transpiration and therefore the amount of moisture in the air. It has already been seen that rainfall in Panama has decreased with the forest clearances (page 426). There are fears that large scale deforestation will turn areas at present under rainforest into deserts. Although the North American Prairies and the African savannas were grassland when the European settlers arrived, it is now believed that these areas too were once forested and were cleared, mainly by firing, by the local Indians and Africans. The burning of vegetation initially provides nutrients for the soil but once these have been leached by the rain or utilised by crops there is little replacement of organic material. Where the grasslands have been ploughed up for cereal cropping the breakdown of soil structure has often led to their drying out and becoming easy prey to wind erosion. Large quantities of topsoil were blown away to create the American Dust Bowl in the 1930s, while a similar fate has more recently been experienced by the virgin lands of southern Siberia (Figure 16.43). In Britain, the removal of hedges to create larger fields which are more manageable for modern machinery, has led to accelerated soil erosion by wind (page 422).

Ploughing land for crops can have adverse effects on soils. Deep ploughing destroys the soil structure by breaking up peds and burying organic material too deep for plant use. It also loosens the topsoil for future wind and water erosion. The weight of farm machinery can compact the soil surface or produce **platy** peds, both of which reduce infiltration capacity and inhibit aeration of the soil. Ploughing up- and downhill creates furrows which increase the rate of runoff.

Overgrazing, especially on the African savannas, also accelerates soil erosion (e.g. the Rendille of Kenya, page 400). Not only has the reduction of the natural fauna — carnivores — through 'big game' hunting led to an increase in herbivores, but many African tribes have long measured their wealth in terms of the numbers, but not the quality, of their animal herds. As the human population of these areas continues to expand rapidly, so too do the numbers of herbivorous animals needed to support them. This almost inevitably leads to overgrazing, the reduction of grass cover and an increase in exposed surface. When new shoots appear after the rains they are eaten immediately by cattle, goats and camels. The arrival of the rains causes erosion; the failure of the rains causes animal deaths.

CASE STUDY

Loess plateau of North China

This region experiences the most rapid soil loss in the world. During and following the ice age, Arctic winds transported large amounts of loess and deposited this fine, yellow material to a depth of 200 m in the Huang He basin. Following the removal of the subsequent vegetation cover of trees and grasses to allow farming, the unconsolidated material has been washed away by the heavy summer monsoon rains at the rate of 1 cm per year. It is estimated that 1.6 billion tonnes of soil reach the Huang He River during each annual summer flood. This material is the most carried by any river in the world and has given the Huang He its name which means the 'Yellow River'. A further problem is that 6 cm of silt settles annually on the river's bed so that it now flows 10 m above its flood plain. When the large flood banks are breached, the river can drown many people and ruin all crops: it is also known as 'China's Sorrow'.

Burkina Faso

As the size of cattle and goat herds have grown the already scant dry scrub savanna vegetation on the southern fringes of the Sahara has been totally removed over increasing areas. As the Sahara 'advances' the herders are forced to move southwards into moister environments where they compete for land with sedentary farmers who are already struggling to produce sufficient food for their own increasing numbers. This disruption of equilibrium further reduces the land carrying capacity (page 319), i.e. the number of people that the soil and climate of an area can permanently support when the land is planted with staple crops. These farmers have long been aware that three years' cropping had to be followed by at least eight fallow years in order for grass and trees to re-establish themselves and to replenish organic matter. The arrival of the herders has brought a land shortage resulting in crops being grown on the same plots every year; the nutrient-deficient soil, typical of most of tropical Africa, is rapidly becoming even less productive. This over-cropping, increasingly a problem in many global subsistence areas, uses up humus and other nutrients, weakens soil structures and leaves the soil exposed and so susceptible to accelerated erosion.

Monoculture, cultivating the same crop each year on the same piece of land repeatedly uses up the same nutrients. The replacement of forest by plantation crops has increased the build up of laterite soils in the tropics and thus reduced their agricultural productivity.

In many parts of the world where animals are kept and firewood is at a premium, dung has to be used as a fuel instead of being applied to the land. In parts of Ethiopia, dung is mixed with straw, dried into 'cakes' and sold, as the sole income of rural dwellers, to the towns. If this dung were to be applied to the fields, harvests could be increased by over 20 per cent. However, for most of these farmers, their concern is survival today rather than planning for tomorrow.

Water is essential for a productive soil. The first known civilisations, which grew up in river valleys (Figure 14.1), relied upon irrigation, as do many areas of the modern world. Unfortunately, the overuse of water in a hot, dry climate tends to lead to salinisation and a reduction of soil quality (Figures 10.20c and 16.43). Salinisation is also a problem in coastal areas, such as Bangladesh, which are subject to marine flooding.

Wells sunk in dry climates use up reserves of groundwater which may have taken many centuries to accumulate and which cannot quickly be replaced (fossilwater stores, Figure 7.16). This results in a fall in the water table making it harder for plant roots to obtain moisture. The sinking of wells in East Africa, following the drought of the early 1980s, has also created difficulties. The presence of an assured water supply has attracted numerous migrants and their animals which has accelerated the destruction of the remaining trees and exacerbated the problems of overgrazing (Figure 10.28). Even well-intentioned aid projects may therefore be environmentally damaging.

Fertiliser and pesticide are not always beneficial to a soil if applied repeatedly over lengthy periods. Chemical fertiliser does not add humus and so fails to improve or maintain soil structure and there are fears that nitrogen added to the soil may find its way into groundwater, some of which may later be used as drinking water. Nitrate and

Figure 10.27

Gulleys in loess, northern China

phosphate runoff causes **eutrophication** (Figure 16.52). The excessive use of pesticide can cause a build up of DDT, which is not yet banned in all developing countries. It kills soil organisms and slows down the decomposition of organic matter and the release of nutrients. Chemical pesticides are blamed for a 60 to 80 per cent reduction in the 800 species of fauna found in the Paris Basin as well as for the decline in Britain's bee population. Acid rain increases the pH of the soil which means extra quantities of iron and aluminium are released. These may eventually reach toxic levels.

Soil conservation

Fertility refers to the ability of a soil to provide an unconstrained or optimum growth of plants. The capacity to produce high or low yields depends upon the nutrient content, structure, texture, drainage, acidity and organic content of a particular soil as well as the relief, climate and farming techniques. The **law of minimum** states that 'there is an optimum minimum level for all nutrients'. After water, nitrogen is the most significant limiting factor and this must be carefully controlled if soil fertility is to be conserved. In order to grow, plants must have access to nine primary and secondary elements and several trace elements (page 222).

Under normal recycling, these nutrients will be returned to the soil as the vegetation dies and decomposes. When a crop is harvested there is less organic material left to be recycled. As nutrients are taken out of the soil system and not replaced, there will be an increasing shortage of macro-nutrients, particularly nitrogen, calcium, phosphorus and potassium. Where this occurs, and when other nutrients are dissolved and leached from the soil, fertiliser is essential to maintain yields.

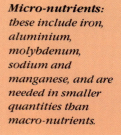

Micro-nutrients: *these include iron, aluminium, molybdenum, sodium and manganese, and are needed in smaller quantities than macro-nutrients.*

Macro-nutrients: *nitrogen, oxygen, hydrogen, and carbon, are the more important group, also phosphorus, sulphur, potassium, calcium and magnesium.*

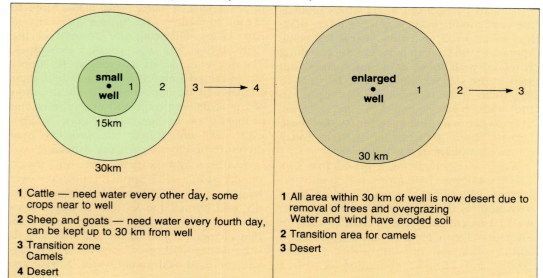

1 Cattle — need water every other day, some crops near to well

2 Sheep and goats — need water every fourth day, can be kept up to 30 km from well

3 Transition zone Camels

4 Desert

1 All area within 30 km of well is now desert due to removal of trees and overgrazing Water and wind have eroded soil

2 Transition area for camels

3 Desert

Figure 10.28

Some consequences of providing a larger and more reliable water supply in northern Kenya

▲ Figure 10.29

Planting marram grass to 'fix' (stabilise) sand dunes, near North Bay airport, Barra, Western Isles, Scotland

Conserving and upgrading soils

If the most serious cause of erosion is the removal of vegetation cover, then the best way to protect the soil is likely to be the addition of vegetation. Afforestation provides a long term solution since once the trees have grown their leaves intercept rainfall and their roots help to bind the soil together and reduce surface runoff (Case Study, page 238).

The growing of cover crops reduces rainsplash and surface runoff, and can protect newly ploughed land from exposure to climatic extremes. Marram grass anchors sand (Figure 10.29) and gulleys can be seeded and planted with brushwood. Certain crops and plants, especially leguminous species, e.g. peas, beans, clover and gorse are capable of fixing atmospheric nitrogen in the soil, thus improving its quality.

Soil can also be conserved by improving farming methods. Most arable areas benefit from a rotation of crops, including grasses, which improve soil structures and reduce the likelihood of soil-borne diseases which may develop under monoculture. Many tropical soils need a recovery period of five to 15 years in shrub or forest for each three to six years in crops. In areas where slopes reach up to 12°, ploughing should be done following the contours to prevent excessive erosion. On even steeper slopes terracing helps to slow down runoff, giving water more time to infiltrate and reducing its erosive ability (Figure 16.28). Strip cropping involves the planting of two or more crops in the same field (Figure 10.30). The crops may differ in height, times of harvest and use of nutrients. In areas where lateritic soils have developed, deep ploughing may be able to break up the hard pan.

Where evapotranspiration exceeds precipitation, dry farming can be adopted. This entails covering the soil with a mulch of straw and/or weeds to reduce moisture loss and limit erosion. In the Sahel countries the drastic depopulation of cattle because of drought gives the herders a chance to restock with smaller herds (reducing overgrazing) of better quality (giving more meat and milk) so that incomes do not fall and the soils are allowed to recover.

▶ Figure 10.30

Strip farming along contours in southern USA

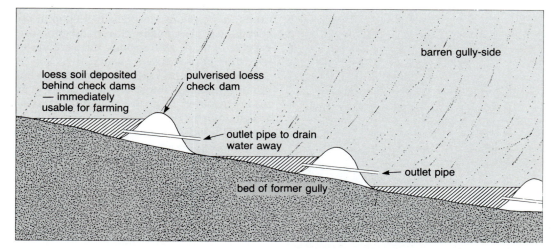

Many soils suffer from either a shortage or a surfeit of water. In East Africa there is a need to construct more wells not just to provide a reliable supply but to prevent overcrowding at the few existing ones (Figure 10.28). Several Sahel countries have built small dams to trap water long enough for infiltration to restore the groundwater store and eventually to replenish wells (Case Study). In North China, 'check dams', 6 m high and built from loess, have been constructed by communal effort. Silt washed from the hills is trapped behind the dams, the surplus water is drained away through pipes, and new land areas are created (Figure 10.31).

CASE STUDY

Afforestation in Nepal

Deforestation in this Himalayan country has resulted in serious soil erosion within its borders and extreme flooding in Bangladesh to the south east.

In the 1970s, a district forestry officer met a group of village leaders from one small valley. The villagers agreed to stall-feed their cattle and buffalo, when they were not needed for work. This meant that the animals ate what they were given rather than anything and everything, as they did in the forest where they had previously been allowed to graze. Their manure could now be collected and spread on the land instead of being deposited on the forest floor. Tree seedlings were grown in nurseries using dung as a fertiliser, and were then transplanted. Hardy pines were planted but a deciduous undergrowth was encouraged. As well as reducing erosion the trees provided a new source of income and began to re-establish an economic equilibrium as well as a more stable environmental balance.

Stone lines in Burkina Faso

This project, begun by Oxfam in 1979, aimed to introduce water harvesting techniques for tree planting. It met with resistance from local people who were reluctant to divert land and labour from food production, or to risk wasting dry season water needed for drinking.

So attention was diverted to improving food production by using the traditional local technique of placing lines of stones across slopes to reduce runoff (Figure 16.50). Where these lines were aligned with the contours they dammed rainfall giving it time to infiltrate. Unfortunately most slopes were so gentle, under 2°, that local farmers could not determine the contours. An innovation costing less than £3 solved the problem.

A calibrated transparent hose, 15 m long, is fixed at each end to stakes and filled with water. When the water level is equal at both ends of the hose then the bottom of the stakes must be on the same contour. These lines can be made during the dry season when labour is not needed for farming. They take up only 1 or 2 per cent of cropland, yet may increase yields by over 50 per cent. They help to replenish falling water tables and can regenerate the barren, crusted earth because soil, organic matter and seeds collect uphill of the stone lines and plants begin to grow again.

In irrigated areas, water must be continually flushed through the system to prevent salinisation. In areas of heavy rainfall, dams may be built to control flooding. Where soils are waterlogged then the addition of field drains can improve drainage.

The erection of shelter-belts (or windbreaks) and the addition of humus to loose soil, can reduce wind erosion. Soil structure and texture may be improved (theoretically) by adding lime to acid soils, which improves their drainage and therefore makes them warmer; by adding humus, clay or peat to sands, to give body and to improve their water holding capacity; and by adding sand to heavy clays, so improving drainage, aeration and making them lighter to work. In practice such methods are rarely used because of the expense involved.

Chemical (inorganic) fertilisers help to replenish deficient macro-nutrients, especially nitrogen, potassium and phosphorus. However, their use is expensive, especially to farmers in Third World countries, and can cause environmental damage. Many farmers in the poorer countries rely on organic fertiliser because of the cost of the alternative (as do an increasing number in Britain but on environmental grounds). Animal dung and straw left after the cereal harvest are mixed together and spread over the ground. This improves soil structure and, as it decays, returns nutrients to the soil. Where crop rotations are practised, grasses add organic matter and legume crops provide nitrogen.

Q

1 Describe the environmental and demographic factors which encourage desertification. How has economic development accelerated the process of desertification?

2 With reference to specific examples, explain how economic development has led to the degradation and erosion of soils in some areas and the conservation and even the upgrading of soils elsewhere.

References

Climate, Soils and Vegetation, D.C. Money, U.T.P., 1965.
Earth, the Living Planet, Michael Bradshaw, Hodder & Stoughton, 1977.
Only One Earth, Lloyd Timberlake, BBC/Earthscan, 1987.
Soils, D. Briggs, Butterworth, 1977.
Soil Processes, Brian Knapp, Allen & Unwin, 1979.
Soils — Profiles and Processes, Geography Today, A Sense of Order, BBC Television, 1982.
Soils, Vegetation and Ecosystems, Greg O'Hare, Oliver & Boyd, 1987.
The Living Planet, BBC Television/Earthscan, 1987.
The Nature of the Environment, Andrew Goudie, Blackwell, 1984.
The Soil, F.M. Courtney and S.T. Trudgill, Arnold, 1976.
The Study of Soil in the Field, G.R. Clarke, Oxford University Press, 1971.
World Resources 1987, World Resources Institute, Basic Books, 1988.
World Soils, E.M. Bridges, Cambridge University Press, 1970.

CHAPTER 11

Biogeography

'The Earth's green cover is a prerequisite for the rest of life. Plants alone, through the alchemy of photosynthesis, can use sunlight energy, and convert it to the chemical energy animals need for survival.'
***The Gaia Atlas of Planet Management*, James Lovelock, Pan 1985**

Biogeography may be defined as the study of the distribution of plants and animals over the earth's surface. The geographer is interested in describing meaningful patterns of plant and animal distributions in a given area either at a particular time or through the passage of time and in trying to account for how those distributions occurred or those changes evolved. The concept of the **ecosystem** or ecological system (page 250) is fundamental to biogeography.

Seres

A **sere** is a stage in a sequence of events by which the vegetation of an area develops over a period of time. A **prisere** is the complete chain of successive seres beginning with a **pioneer community** and ending with a **climatic climax vegetation** (Figure 11.1). The first plants to develop in and to colonise an area are called the pioneer community (or species). The climatic climax is when the ultimate vegetation has become in harmony or equilibrium with the local environment, i.e. when the natural vegetation has reached a delicate but stable balance with the climate and soils of an area. Each successive seral community shows an increase in the number of species and the height of the plants.

However, there are very few parts of the world today with a climatic climax. This is partly due to local changes in the physical environment (e.g. drought) but is mainly a result of human interference (e.g. deforestation, farming, acid rain).

Where human activity has permanently arrested and altered the natural vegetation, the resultant community is said to be a **plagioclimax**.

Each successive sere is referred to by one or more of the larger species within that community — the so-called **dominant species**. The dominant species may be the **largest** plant or tree in the community which exerts the maximum influence on the local environment or habitat, or the **most numerous** species in the community. In parts of the world where the climatic climax is forest, i.e. areas with higher rainfall, the plant community tends to be structured into layers (Figure 11.2).

The climatic climax may develop from one of two starting points — as a **primary succession** or as a **secondary succession**.

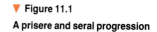
▼ Figure 11.1

A prisere and seral progression

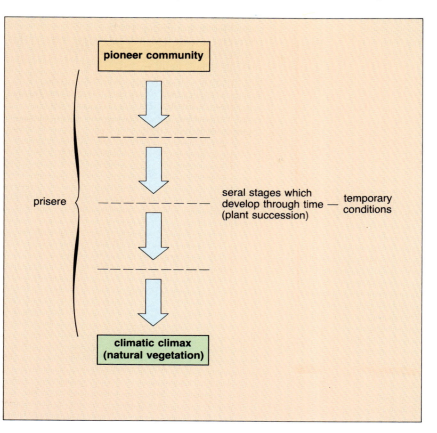

▲ Figure 11.2

Layers of vegetation typical of forested areas

a, *above left* Tropical rainforest (Amazonia)

b, *above right* Temperate deciduous woodland (Strid Woods, Yorkshire)

Primary succession

This occurs on a new or previously sterile land surface or in water. Figure 11.3 shows how the four more commonly accepted non-vegetated environments in Britain develop until they all reach the same climatic climax, the oak—ash woodland.

Secondary succession

Here the natural vegetation of an area has been replaced or altered by a second type of vegetation, either naturally or by human activity. Examples include the deforestation of lowland Britain for farming, the replacement of natural grasses by cereal crops, the results of a forest or heathland fire, a mudflow in a river valley and shifting cultivation and abandonment. Although the early stages in a secondary succession are distinctive to the particular local environment, as the seres develop they increasingly resemble the latter stages of the primary succession.

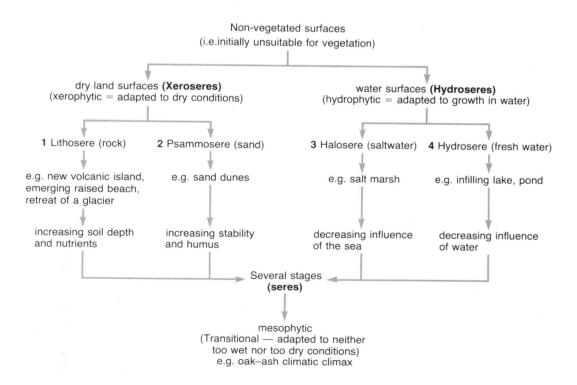

► Figure 11.3

The development of a climatic climax vegetation

Four basic seres forming a primary succession

1 Lithosere

Areas of bare rock will initially be colonised by blue-green bacteria and single-celled photosynthesisers which have no root systems and can survive where there are few mineral nutrients. Blue-green bacteria are autotrophs (page 252), photosynthesising and producing their own organic food source. Lichens and mosses also make up the pioneer community (Figure 11.4). These plants are capable of living in areas lacking soil, devoid of a permanent supply of water and experiencing extremes of temperature. Lichen and various forms of weathering help to break up the rock to form a veneer of soil in which more advanced plant life can then grow. As these plants die they will be converted by bacteria into humus which develops a richer soil. Seeds, mainly of grasses, then colonise the area. As these plants are taller than the pioneer species they will replace the lichen and mosses as the dominants although the lichens and moss will still continue to grow in the community. As the plant succession evolves over a period of time the grasses will give way as dominants to fast-growing shrubs, which in turn will be replaced by relatively fast-growing trees (e.g. rowan). These will eventually face competition from slower-growing trees in which the oak and ash will form the climax vegetation. It should be noted that while each stage of the succession is marked by a new dominant, the earlier species remain growing.

Figure 11.4 shows an idealised primary succession across a newly emerging rocky coastline. It excludes the increasing number of species found at each stage of the seral succession. The species are determined by local differences in rainfall, temperature and sunlight, bedrock and type of soil, aspect and relief. Lithoseres can develop on bare rock exposed by a retreating glacier, on ash or lava following a volcanic eruption on land (e.g. Krakatoa), or forming a new island (e.g. Surtsey), or, as in Figure 11.4, on land emerging from the sea as a result of isostatic uplift following the melting of an ice cap (page 132).

Over time, the area shown to have the pioneer community passes through several stages until the climatic climax is reached — this assumes that the land continues to rise, that there is no significant change in the local climate, and that there is no human interference. Figure 11.5a and b shows two stages

◄ **Figure 11.4**
Fieldsketch of a lithosere on a newly emerging rocky coastline (raised beach), Arran

▼ **Figure 11.5**
Primary succession on a lithosere, Arran
a, *below left* Lichen, mosses and grasses on a rocky coastline
b Bracken, rowan and deciduous trees (climatic climax) behind a rocky beach

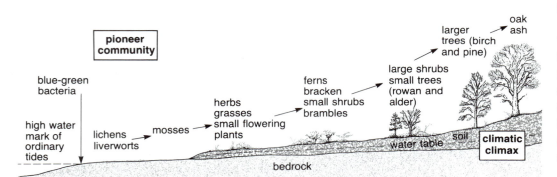

in the succession taken on a raised beach on the east coast of Arran. Photograph *a* shows lichen, favouring a south-facing aspect on gently dipping rocks, and mosses growing in darker north facing hollows. Beyond, where soil has begun to form and where the water table is high, grasses and bog myrtle have entered the succession. Photograph *b* is taken where the soil depth and amount of humus have increased and the water table is lower, as indicated by the presence of bracken. The reeds to the right are found in a hollow where the water table is nearer to the surface. In the middle distance a small rowan can be seen, while in the further distance taller deciduous trees indicate an approaching climax vegetation.

CASE STUDY

Krakatoa

In August 1883, a series of volcanic eruptions reduced the island of Krakatoa to one-third of its previous size and left a layer of ash over 50 m deep. No vegetation or animal life was left on the island or in the surrounding sea. Yet within three years (Figure 11.6), 26 species had reappeared and in 1933, 271 plant and 720 insect species were recorded. The first colonisers arrived in three ways. Most were seeds blown by the wind from surrounding islands, while others drifted in from the sea or were carried by birds. However, in this example, the concept of plant succession, later put forward by F.E. Clements in 1916, seemed open to dispute, as many of the plants which colonised Krakatoa after 1883 arrived there by chance, i.e. a piece of driftwood with a particular seed type just happened to be washed ashore whereas it could just as easily have missed the island altogether.

▼ Figure 11.6

Primary succession, Krakatoa

a, *top* Vegetation, 1983

b Vegetation distribution according to height above sea level, 1983

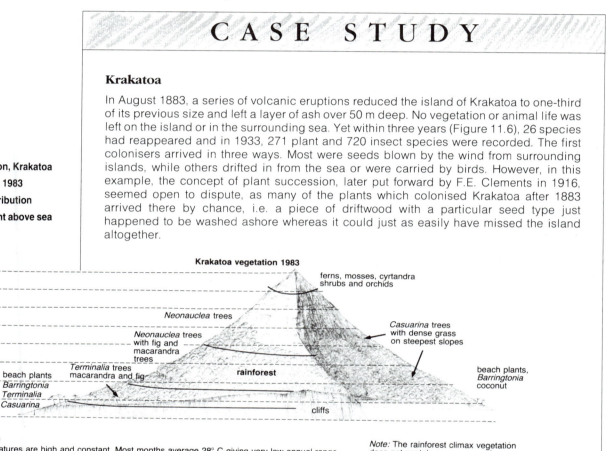

Krakatoa vegetation 1983

Climate
Temperatures are high and constant. Most months average 28° C giving very low annual range. Rain is heavy, falling in convectional storms most afternoons throughout the year.

Note: The rainforest climax vegetation does not contain as many species as the rainforests on surrounding islands.

VEGETATION DISTRIBUTION ACCORDING TO HEIGHT

Height (m)	1883	1886	1908	1918	1933	1983
600–800	(no vegetation)	(no vegetation)	(no vegetation)	ferns, *Cyrtandra* shrubs, mosses and orchids	mosses and orchids, *Cyrtandra* shrubs, woodland in ravines	cyrtandra shrubs, ferns, small trees
500–700						neonauclea trees
400–600			coarse grassland,	savanna grassland, grass 3 m high	mixed woodland	
200–450			*Neonauclea* trees, ferns, shrubs, dense grass, some macarandra and figs	macarandra trees with figs increasing in number some *Neonauclea*	*Neonauclea* trees taking over from macarandra and figs	rainforest climax *Neonauclea* with fig and macarandra
120–200		ferns growing on blue–green bacteria				*Terminalia* and macarandra
0–120		beach plants *Barringtonia*	beach plants and coconuts *Barringtonia* tussock grass	beach plants *Barringtonia* tussock grass, coconut	coastal woodland climax (types as 1918)	*Barringtonia*, beach plants, *Casuarina*
Year	1883	1886	1908	1918	1933	1983
Number of Species	0	26	115	132	271	?

Q

1 With reference to Chapter 1 explain why a volcanic eruption occurred on Krakatoa.

2 Using Figure 11.6, describe carefully and give reasons for the primary succession on Krakatoa following the eruption in 1883.

3 With reference to Figure 11.6, say to what extent you consider that a climax vegetation has again been reached on the island. Give reasons for your answer.

2 Psammoseres

A psammosere succession develops on sand and is best illustrated by taking a transect across coastal dunes (Figure 6.27). The first plants to colonise, indeed to initiate dune formation, are usually lyme grass, sea twitch and marram grass. Sea twitch grows on berms around the high water mark of tides and is often responsible for the formation of embryo dunes. On the yellow fore-dunes, which are arid, being above the highest of tides and experiencing rapid percolation by rainwater, marram grass becomes equally important.

The main dune ridge, which is extremely arid and exposed to wind, is likely to be vegetated exclusively by marram grass. Marram has adapted to these harsh conditions by having leaves which can fold to reduce surface area; these are shiny and can be aligned to the wind direction, all of which helps to limit evapotranspiration. Marram also has long roots to tap underground water supplies and is able to grow as fast as sand deposition can cover it. Grey dunes behind the main ridge have lost their supply of sand and are sheltered from the prevailing wind. Their greater humus content from the decomposition of earlier marram grass enables the soil to hold more moisture. Although marram is still present here it faces increasing competition from small flowering plants and herbs such as red fescue, sea spurge (with succulent leaves to store water) and heather.

The older ridges, further from the water, have both more and taller species and the presence of creeping willow, yellow iris, reeds, rushes and shrubs indicates a deeper and wetter soil. On the landward side of the dunes, perhaps 400 m from the beach, are small trees including ash, hawthorn and especially coniferous species which prefer sandy soils. Furthest inland comes the oak–ash climax. Figure 11.7 is a psammosere based on sand dunes at Morfa Harlech (North Wales) and in southeast Lancashire. Figures 11.8 and 6.28 show marram and lyme grass forming the yellow fore-dunes, with gorse and heather on the greyer dunes behind (these are grey because of their humus content). Figures 11.9 and 6.29 show vegetation on the inland ridges.

3 Hydroseres

Lakes and ponds originate as clear water which contains few plant nutrients. Any sediment carried into the lake will enrich the water with nutrients and begin to infill the lake. The earliest colonisers will probably be algae and mosses whose spores have been blown on to the water surface by the wind. These grow to form vegetation rafts which provide a habitat for bacteria and insects. Next will be water-loving plants which may either grow on the surface, e.g. water lilies and pondweed, or be totally submerged (Figure 11.10). Bacteria and unicells recycle the nutrients from the pioneer community and marsh plants such as bullrushes, sedges and reeds begin to encroach into the lake. As these marsh plants grow outwards into the

▼ Figure 11.7

Transect across sand dunes to show a psammosere, Morfa Harlech, North Wales

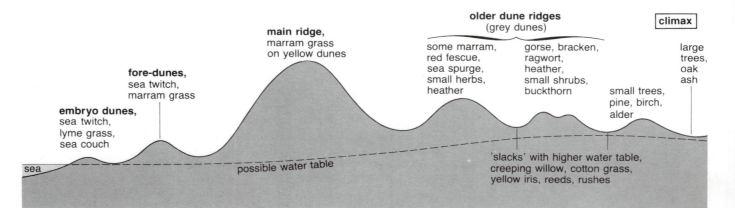

older dune ridges (grey dunes) climax

main ridge, marram grass on yellow dunes

fore-dunes, sea twitch, marram grass

embryo dunes, sea twitch, lyme grass, sea couch

some marram, red fescue, sea spurge, small herbs, heather

gorse, bracken, ragwort, heather, small shrubs, buckthorn

small trees, pine, birch, alder

large trees, oak ash

sea

possible water table

'slacks' with higher water table, creeping willow, cotton grass, yellow iris, reeds, rushes

▲ Figure 11.8
Primary succession on a psammosere — colonisation on embryo and foredunes, Arran (compare with Figures 6.28 and 6.29)

lake and further sediment builds upwards at the expense of the water, small trees may take root to form a marshy thicket.

In time, the lake is likely to contract in size, to become deoxygenised by the decaying vegetation and eventually to disappear, to be replaced by an oak–ash climax. This primary succession is shown in Figure 11.10. Figure 11.11 shows land plants encroaching at the head of a reservoir and Figure 11.12 illustrates the water, marsh and land plant succession around a small pond.

4 Haloseres

In river estuaries, large amounts of silt and mud are deposited by the ebbing tide and inflowing rivers. The earliest plant colonisers are green algae and eel grass which can tolerate submergence by the tide for most of the 12 hour cycle and which trap mud, causing it to accumulate. Two other colonisers are *Salicornia* and *Spartina townsendii* which are **halophytes** meaning that they can tolerate saline conditions. They grow on the slob zone (page 127), with a maximum of four hours exposure to the air in every 12. *Spartina* was brought to Britain in the 1870s. Its long roots enable it to trap more mud than the initial colonisers, algae and *Salicornia*, and in most *Spartina* has now replaced these indigenous plants as the dominant vegetation. The slob zone receives new sediment each day, is water-logged to the exclusion of oxygen, and has a high pH value.

The sward zone, in contrast, is inhabited by plants which can only tolerate a maximum of four hours submergence in every 12. Here the dominant species are sea lavender, sea aster and grasses, including the 'bowling green turf' of the Solway Firth. However, although the vegetation here tends to form a thick mat, it is not continuous. Hollows may remain where the sea water has been trapped and has then evaporated, leaving saltpans where the salinity is too great for plants. As the tide ebbs, water draining off the land may be concentrated into creeks. The upper sward zone is only covered by spring tides and here juncus and other rushes grow. Further inland, non-halophytic grasses and shrubs enter the succession, to be followed by small trees and ultimately by the climax of oak and ash. Figure 11.13 is a

▶ Figure 11.9
Primary succession on a psammosere — vegetation succession on dune ridges, Braunton Burrows, Devon, England (notice the dune slack)

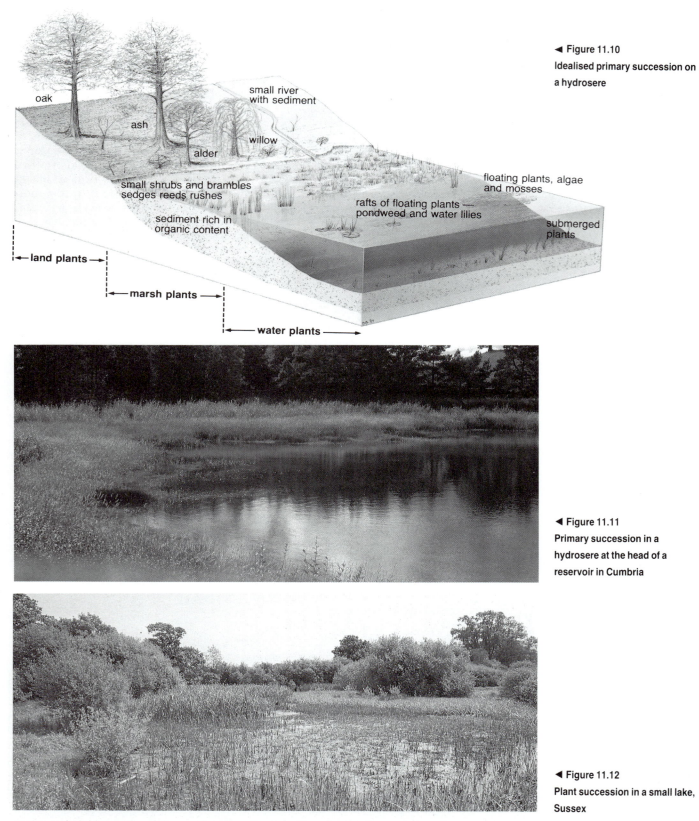

◄ Figure 11.10

Idealised primary succession on a hydrosere

◄ Figure 11.11

Primary succession in a hydrosere at the head of a reservoir in Cumbria

◄ Figure 11.12

Plant succession in a small lake, Sussex

transect based on the salt marshes on the north coast of the Gower Peninsula in South Wales. The photographs in Figure 11.14 show three different stages in the seral succession.

You do not have to be an expert botanist to recognise the plants mentioned in the primary successions described here: you just need access to a good plant recognition book!

246

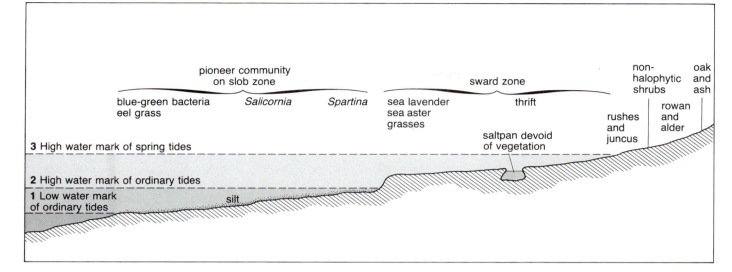

pioneer community
on slob zone

sward zone

non-
halophytic
shrubs

oak
and
ash

blue-green bacteria
eel grass

Salicornia

Spartina

sea lavender
sea aster
grasses

thrift

rowan
and
alder

rushes
and
juncus

saltpan devoid
of vegetation

3 High water mark of spring tides

2 High water mark of ordinary tides

1 Low water mark
of ordinary tides

silt

▲ Figure 11.13
Fieldsketch of primary
succession in a halosere,
Llanhridian Marsh, Gower
Penninsular, South Wales
(compare Figures 6.30 and 6.31)

► Figure 11.14
Primary succession in a
halosere
a Spartina, a pioneer species
b A saltpan on the Suffolk
coast, covered only by the
highest of tides
c The whole succession is
shown here in the gradiation
from Spartina through to
established trees

Q   1 Study Figure 11.15 which shows the vegetation succession in a lake (hydrosere) in lowland England and the graph which plots the number of species in the various communities over time. The lake covers an area of 6 km² and has a maximum depth of 5 m. Describe and suggest reasons for the changes in vegetation shown in the graph and transects.

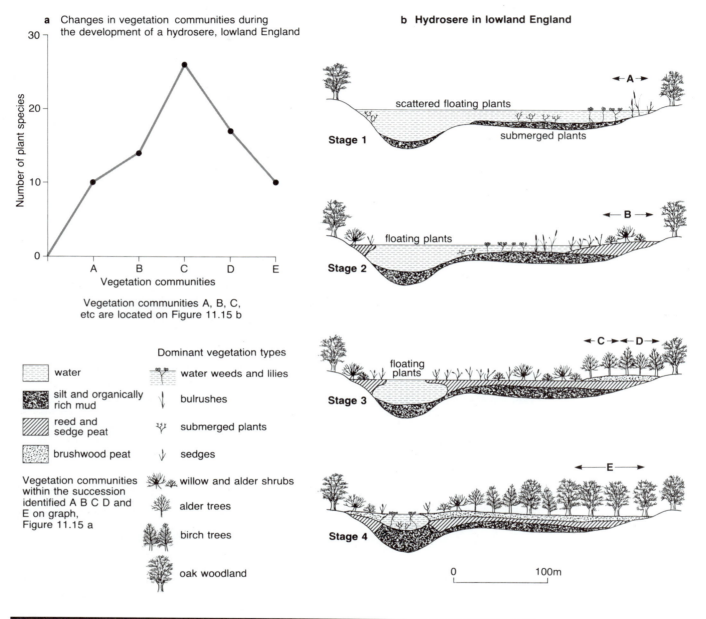

Secondary succession

A climatic climax occurs when there is stability in transfers of material and energy in the ecosystem between the plant cover and the physical environment. However there are several factors which can arrest the plant succession before it has achieved this dynamic equilibrium, or which may alter the climax after it has been reached. These include:

- a mudflow or landslide
- deforestation or afforestation
- overgrazing by animals or the ploughing up of grasslands
- burning grasslands, moorlands, forests and heaths
- draining wetlands
- disease (e.g. Dutch elm) and pests (e.g. locusts)
- changes in climate.

▲ Figure 11.15

Vegetation succession in a lake (hydrosere) in lowland England

CASE STUDY

Mudflow on Arran

The mudflow shown in Figure 2.15 occurred in October 1981 and completely covered all of the existing vegetation. Twelve months later it was estimated than 20 per cent of the flow had been recolonised, a figure which had grown to 40 per cent in 1984 and 70 per cent in 1988. Had this been a primary succession, lichens and mosses would have formed the pioneer community and would probably have covered only a small area; the pioneer plants would probably have been randomly distributed and after seven years the species would have been few in number and small in height.

Instead, by 1988 much of the flow was already recolonised; most plants were found to the edges of the flow and were not randomly distributed; there were already several species including grasses, heather, bog myrtle and mosses, some of which exceeded half a metre in height.

This suggests a secondary succession with plants from the surrounding climax area having **invaded** the flow, mainly due to the dispersal of their seeds by the wind.

---

*The **Holocene** is the most recent of the geological periods (Figures 1.1 and 11.17).*

The effect of fire

The severity of a fire and its effect on the ecosystem depends largely upon the climate at the time. The fire is likely to be hottest in dry weather and, in the northern hemisphere, on sunny south-facing slopes where the vegetation is driest. The spread of a fire is fastest when the wind is strong and blowing uphill. The extent of disruption also depends upon the type and the state of the vegetation. The following is a list of examples in rank order of severity.

1 Areas with Mediterranean climate where the **chaparral** of California and the **maquis-garrigue** of southern Europe is tinder dry and densest in late summer after the seasonal drought. Examples in the 1980s include the fires in southeast Australia, the south of France and around Los Angeles.
2 Coniferous forests, because dry conifers burn readily.
3 Deciduous woodlands, because they may have a thick litter layer.
4 Ungrazed grasslands which have a low biomass (page 254) but a thick litter layer.
5 Grazed grasslands which have a lower biomass and a limited litter layer.

Following a fire — the blackened soil has a lower albedo and absorbs heat more readily, the soil is more vulnerable to erosion without the protection of vegetation. Ash considerably increases the quantity of inorganic nutrients in the soil and bacterial activity is accelerated. Any seedlings left in the soil will grow rapidly as there is now plenty of light, no smothering layer of leaf litter, plenty of nutrients, a warmer soil and, at first, less competition from other species. Heaths and moors which have been fired are conspicuous by their greener, more vigorous growth. A fire climax community contains plants with seeds which have a thick protective coat and which may germinate because of the heat of the fire. The community may have a high proportion of species which can sprout quickly after the fire, which are protected by thick, insulating barks (e.g. cork oak in the chaparral and baobab in the savannas), or which have underground tubers or rhizomes insulated by the soil. It is suggested that the grasslands of the American Prairies and the African savannas are not climatic climax vegetations but the result of firing by the Indians and Africans.

Vegetation changes in the Holocene

The last glacial advance in Britain ended about 18 000 BP. Although the extreme south of England was covered with hardy tundra plants, most of the northern parts of our islands were left as bare rock or glacial till. Had the climate gradually and constantly ameliorated then it could have been assumed that a primary succession would have taken place from south to north, as previously described for a lithosere. However, we know that there have been several major changes in climate in those last 10 000 years (page 251), which caused significant fluctuations in the climax vegetation.

There are several techniques for determining vegetation change, including pollen analysis, dendrochronology, radio-carbon dating and historic evidence (page 210). Each plant has a characteristic pollen grain. Where pollen is blown by the wind on to peat bogs, such as at Tregarron in west Wales, the grains are trapped by the soil. As, over the years, more peat accumulates, the pollen of successively later times indicates which were the dominant and subdominant plants of the period (Figures 11.16 and 11.17). As each plant grows best under certain defined temperature and precipitation limits, it is possible to determine when the climate either improved or deteriorated.

Dendrochronology is dating by means of the annual growth rings of trees. Apart from the bristle-combe pine which can date back some 5000 years in California, few British trees can go back even as far as written historic records. Radio-carbon dating is a relatively new method based on changing amounts of radioactivity in both the atmosphere and in plants. Notice in Figure 11.17, which links climatic and vegetation changes, how forests increase as the climate ameliorates, and how heathland and peat moors take over when the climate deteriorates.

Q Figure 11.16 shows how the surface of Britain has changed in the last 12 000 years. Describe the conditions in

a 12 000 BP   **d** 3000 BP
b 9000 BP   **e** the present.
c 6000 BP

In each case give reasons for your answer.

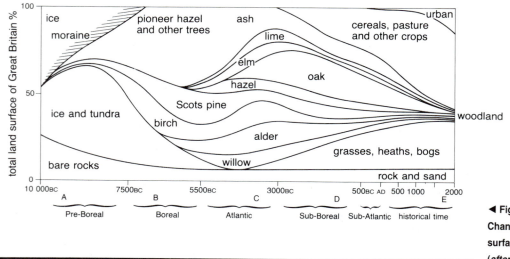

◄ Figure 11.16
Changes in Britain's land surface, 10 000 BC to the present (*after* G. Wilkinson)

Ecology and ecosystems

The term **ecology**, which comes from the Greek word *oikos* meaning 'home', refers to the study of where organisms live. An organism's home or **habitat** lies in the biosphere, which is the surface zone of the earth and its adjacent atmosphere, in which all organic life exists. The scale of these homes varies from small **microhabitats**, e.g. under a stone or a leaf, to **biomes**, e.g. tropical rainforests or deserts (Figure 11.18). Fundamental to the four ecological units listed in Figure 11.18 is the concept of the environment. The **environment** is a collective term to include all the conditions in which an organism lives and it can be divided into two. The **physical, non-living** or **abiotic environment** includes properties of temperature, water, light, humidity, wind, carbon dioxide, oxygen, pH, rocks and nutrients in the soil. The **living** or **biotic environment** refers to all organisms — plants, animals, bacteria and fungi.

The ecosystem

An ecosystem is a natural unit in which the life cycles of plants, animals and other organisms are linked to each other and to the non-living constituents of the environment to form a natural system (Framework 1). The **community** is made up of all the different species in a habitat or ecosystem. The **population** is all the individuals of a particular species in a habitat. An ecosystem may vary in scale from a freshwater pond or an oak tree to a climax community extending across several continents — the concept of an ecosystem can be applied to

Date	Phase – period	Climate	Vegetation	Cultures
Pre–15 000	final glaciation	glacial	none	none
15 000–12 000	periglacial	cold, 6° C summer	tundra	none
12 000–10 000	Allerød	warming slowly to 12° C summer	tundra with hardy trees, e.g. willow and birch	none
10 000–8000	pre-Boreal	glacial advance, colder, 4° C summer	Arctic/Alpine plants, tundra	none
8000–6000	Boreal	continental, winters colder and drier, summers warmer than today	forests — juniper first then pine and birch and finally oak, elm and lime	none
6000–3000	Atlantic	maritime — warm summers 20° C. mild winters 5° C, wet	our 'optimum' climate and vegetation oak-ash-elm and lime (too cold for lime today) peat on moors	beginning of Neolithic
3000–500	sub-Boreal	continental — warmer and drier	elm and lime declined as birch flourished and fine peat dried out	Neolithic times first deforestations, beginning of Bronze Age
500–0	sub-Atlantic	maritime — cooler stormy and wet	peat bogs re-formed, decline in forests due to climate and farming	settled agriculture
0–1000	historical times	improvement — warmer and drier	clearances for farming	early part — Roman occupation
1000–1550		decline — much cooler and wetter	further clearances — little climax vegetation left, medieval farming	
1550–1700		'little ice age' colder than today		
Post-1700		gradual improvement	recently some afforestation — coniferous trees	Agrarian and Industrial Revolution

BC / AD

▲ Figure 11.17
Climatic and vegetation changes in Britain during the Holocene

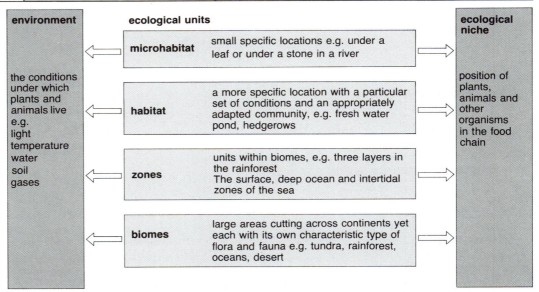

▶ Figure 11.18
A hierarchical structure of ecological units

micro-organisms in a drop of water, or the entire earth as one system (the *Gaia* concept).

An ecosystem depends on two basic processes: energy flow and material cycling. The flow of energy is in only one direction and because it crosses the system boundaries, this aspect of the ecosystem behaves as an open system. Nutrients are circulated in a series of closed systems as they are constantly recycled for future use.

Energy flows

The sun is the primary source of energy for all living things on earth. As energy is retained only briefly in the biosphere before being returned to space, ecosystems have to rely upon a continual supply. The sun provides *heat* energy which cannot be captured by plants or animals but which does warm up the communities and their non-living surroundings. The sun is also a source of *light* energy which can be captured by green plants and transformed into chemical energy through the process of photosynthesis. Without photosynthesis, there would be no life on earth. Light, chlorophyll, warmth, water and carbon dioxide are required for this process to operate. Carbon dioxide which is absorbed through stomata in the leaves reacts, indirectly, with water taken up by the roots when temperatures are suitably high, to form carbohydrate. The energy for this comes from sunlight which is 'trapped' by chlorophyll. Oxygen is a by-product. The carbohydrate is then available as food for the plant.

Food chains and trophic levels

A food chain arises when energy, trapped in the carbon compounds initially produced by plants through photosynthesis, is transferred through an ecosystem. Each link in the chain feeds on and obtains energy from the one preceding it and in turn is consumed by and provides energy for the following one (Figure 11.19). There are usually, but not always, four links in the chain. Each link or stage is known as a **trophic** or **energy level** (Figure 11.20). In order for the first link in the chain to develop, the non-living environment has to receive both energy from the sun and other factors (water, CO_2 etc) needed for photosynthesis.

The **first trophic level** is occupied by the **producers** or **autotrophs** ('self-feeders') which include green plants capable of producing their own food by photosynthesis. All other levels are occupied by **consumers** or **heterotrophs** ('other-feeders'). These include animals which obtain their energy either by eating green plants directly or by

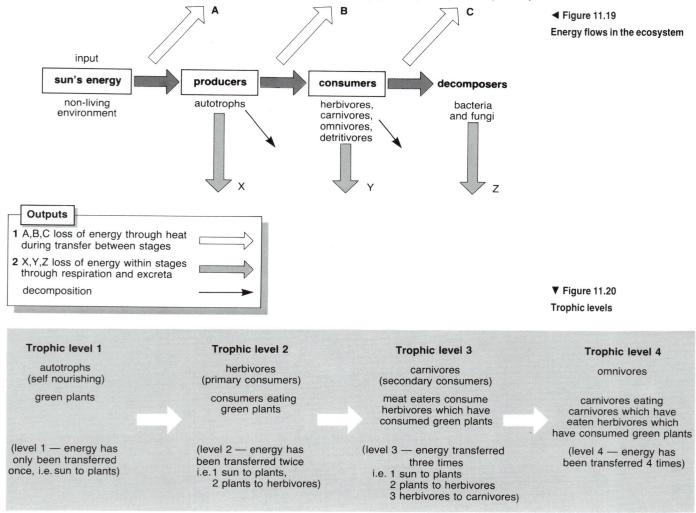

◀ Figure 11.19
Energy flows in the ecosystem

▼ Figure 11.20
Trophic levels

Trophic level 1	Trophic level 2	Trophic level 3	Trophic level 4
autotrophs (self nourishing)	herbivores (primary consumers)	carnivores (secondary consumers)	omnivores
green plants	consumers eating green plants	meat eaters consume herbivores which have consumed green plants	carnivores eating carnivores which have eaten herbivores which have consumed green plants
(level 1 — energy has only been transferred once, i.e. sun to plants)	(level 2 — energy has been transferred twice i.e. 1 sun to plants, 2 plants to herbivores)	(level 3 — energy transferred three times i.e. 1 sun to plants 2 plants to herbivores 3 herbivores to carnivores)	(level 4 — energy has been transferred 4 times)

eating animals which have previously eaten green plants. The **second trophic level** is where **herbivores**, the primary consumers, eat the producers. The **third trophic level** is where smaller **carnivores** (meat eaters) act as secondary consumers feeding upon the herbivores. The **fourth trophic level** is occupied by the larger carnivores, the tertiary consumers. Also known as **omnivores**, these eat both plants and animals and so have two sources of food (e.g. humans). Figure 11.20 shows the main trophic or feeding levels in a food chain.

Three examples of food chains through four trophic levels:

Level 1	Level 2	Level 3	Level 4
grass leaf phytoplankton	worm caterpillar zooplankton	blackbird shrew fish	hawk badger human

Detritivores, such as bacteria and fungi, are consumers which operate at all trophic levels.

However, no transfer of energy is 100 per cent efficient and, as Figure 11.19 shows, energy is lost through respiration or in excreta within each unit of the food chain, and as heat given off when energy is passed from one level to the next. Consequently, at each higher level, fewer organisms can be supported than at the previous level, even though their individual size generally increases. Simple food chains are rare; there is usually a variety of plants and animals at each level forming a more complicated **food web**. This range of species is necessary as a sole species occupying a particular trophic level in a simple food chain could be

'consumed' and this would adversely affect the organisms in the succeeding stages.

Trophic levels and energy pyramids

The progressive loss of energy through the food chain imposes a natural limit on the total mass of living matter (the **biomass**) and on the number of organisms that can exist at each level. It is convenient to show these changes in the form of a pyramid (Figure 11.21). A pyramid of numbers is of limited value for comparing ecosystems for two reasons. First, it is difficult to count the numbers of grasses or algae per unit area. Second, it does not take into account the relative sizes of organisms — a bacterium would count the same as a whale! A pyramid of biomass takes into account the difference in size between organisms but cannot be used to compare masses at different trophic levels in the same ecosystem or at similar trophic levels in different ecosystems. This is because biomass will have accumulated over different periods of time.

Humans are found at the end of a food chain and human population is dependent upon the length of the chain (and therefore the amount of energy lost). In other words, in a shorter food chain, less energy will have been lost by the time it reaches humans and so the land can support a higher density of population, e.g. in parts of SE Asia where the inhabitants live on rice rather than animal products. In a longer food chain more energy will have been lost by the time the food is consumed by humans which means that the carrying capacity (page 319) is lower and fewer people can be supported by a given area of land, e.g. in western Europe where most of the population rely on animal products as well as crops.

▶ Figure 11.21

The trophic pyramid

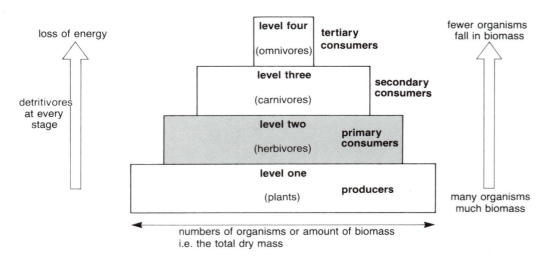

Material cycling

Chemicals needed to produce organic material are circulated around the ecosystem and recycled continually. Various chemicals can be absorbed by plants either as gases from the air or as soluble salts from the soil (page 222). Each cycle consists at its simplest of plants taking up chemical nutrients which once they have been used are passed on to the herbivores and then the carnivores which feed upon them. As organisms at each of these trophic levels die, they decompose and nutrients are returned to the system. Two of these cycles, the carbon and nitrogen, are illustrated in Figures 11.22 and 11.23. In each case the most basic cycle has been given (a) followed by a more detailed example (b), though still not in its total complexity.

Model of the mineral nutrient cycle

This model, proposed by P.F. Gersmehl in 1976, attempts to show the differences between ecosystems in terms of nutrients stored in three compartments (Figure 11.24). Gersmehl identified these storage compartments as: **litter** — the surface layer of vegetation which may eventually become humus — **biomass** — the total mass of living organisms, mainly plant tissue, per unit area — and **soil**.

Figure 11.25 shows the mineral nutrient cycles for three biomes: the coniferous forest (taiga), the temperate grassland (prairies and steppe) and the tropical rainforest (selvas).

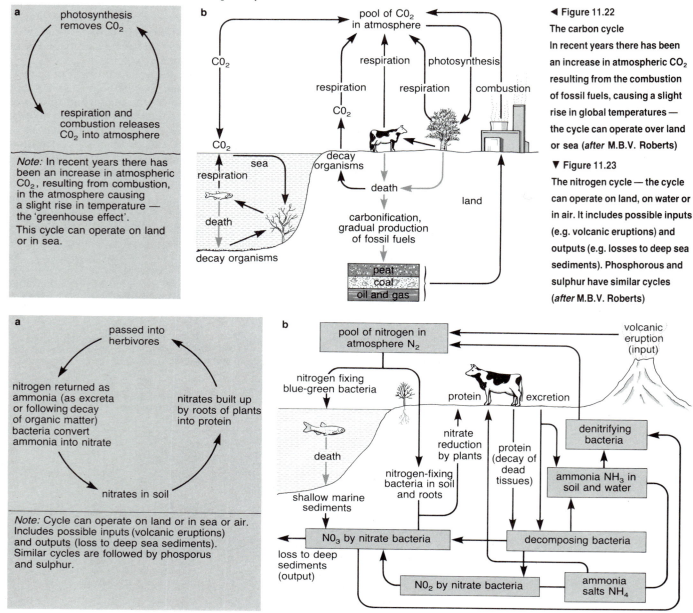

◀ Figure 11.22

The carbon cycle

In recent years there has been an increase in atmospheric CO_2 resulting from the combustion of fossil fuels, causing a slight rise in global temperatures — the cycle can operate over land or sea (*after* M.B.V. Roberts)

▼ Figure 11.23

The nitrogen cycle — the cycle can operate on land, on water or in air. It includes possible inputs (e.g. volcanic eruptions) and outputs (e.g. losses to deep sea sediments). Phosphorous and sulphur have similar cycles (*after* M.B.V. Roberts)

▶ Figure 11.24

A model of the mineral nutrient cycle (*after* Gersmehl)

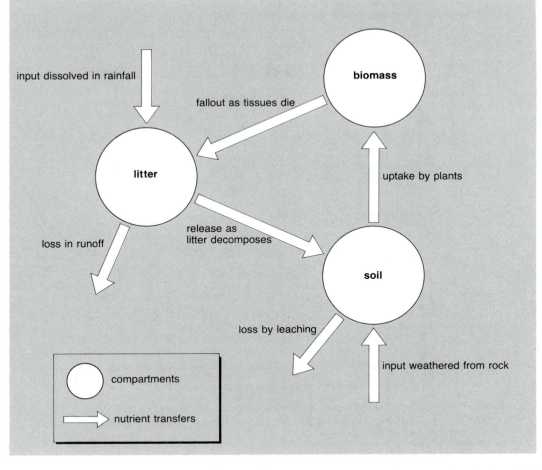

1 Litter is the largest store of mineral nutrients in the taiga. The area is forest but the biomass is relatively low as the coniferous trees form one layer with little undergrowth, contain little variety of species and have needle-like leaves. The soil contains few nutrients because following their loss through leaching replacement is slow in the low temperatures. In comparison, the layer of needles is often thick but the low temperatures and the thick cuticles of the needles discourage the action of the decomposers (page 221) and the breakdown of litter into humus is very slow.

2 Soil is the largest store of mineral nutrients in the grassland. The biomass store is small due to the climate which provides insufficient moisture to support trees and temperatures are cold enough to reduce the growing season to approximately six months. Indeed, much of the biomass is found beneath the surface as rhizomes and roots. The grass dies back in winter and nutrients are returned rapidly to the soil. The soil retains most of these nutrients because the rainfall is insufficient for effective leaching and the

climate is conducive to both chemical and physical weathering which release further nutrients from the bedrock. The presence of bacteria also speeds up the return of nutrients from the litter to the soil.

3 The biomass is the largest store of mineral nutrients in the tropical rainforests. High annual temperatures, the heavy, evenly distributed rainfall and the year long growing season all contribute to the tall, dense and rapid growth of vegetation. The biomass is composed of several layers of plants as well as countless different species. The many plant roots take up vast amounts of nutrients. In comparison the litter is limited as the hot, wet climate provides the ideal environment for bacterial action and the decomposition of dead vegetation.

In areas where the forest is cleared, heavy rain soon removes the nutrients from the soil by leaching or surface runoff. The litter layer is rapidly reduced and so the soil is left to rely upon nutrient replacement from the bedrock. Initially nutrients such as phosphorus may increase if the vegetation was burnt.

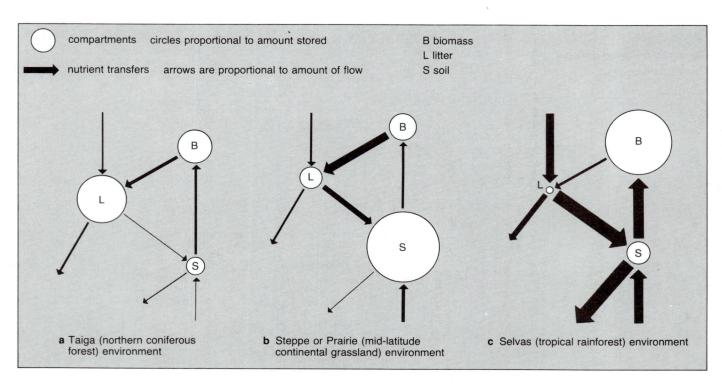

a Taiga (northern coniferous forest) environment

b Steppe or Prairie (mid-latitude continental grassland) environment

c Selvas (tropical rainforest) environment

Q

Use Figure 11.25 to answer the following questions.

1 Explain the differences between tropical rainforest (selvas) and coniferous forest and taiga ecosystem in terms of the major nutrient stores.

2 Explain the differences between the tropical rainforest (selvas) and temperate grassland (prairie) ecosystems in terms of the three major nutrient stores.

3 Why are the transfer flows in the taiga ecosystem so small?

4 Why are the transfer flows in the tropical rainforest ecosystem so high?

5 For each of the three ecosystems explain how human activities can interrupt or alter the mineral nutrient cycle.

6 The following figures refer to a tropical grassland ecosystem:
   nutrients stored in biomass 978;
      stored in litter 300,
      stored in soil 502.
   Annual nutrient transfer between soil and biomass 319;
      between biomass and litter 312
      between litter and soil 266.
   Draw an accurate diagram to show the amounts of nutrients stored in each compartment and the quantities of nutrient flow within the tropical grassland ecosystem.

▲ Figure 11.25

Mineral nutrient cycle in three different environments (*after* Gersmehl)

a Taiga (coniferous forest)

b Temperate grassland (steppe or prairie)

c Tropical rainforest (selvas)

Biomes

The biome is the largest ecosystem unit. Each one obtains its name from the dominant type of vegetation found within it (e.g. grassland, coniferous forest, etc) but also refers in its definition to soil types and animal communities. Usually climate has been the major controlling factor in their location and distribution but economic development has transformed many of these natural systems. A biome may cover a larger area of a continent and its characteristics may be found in several continents (e.g. deserts and tropical rainforests). Although some authorities suggest that it is 'old fashioned' to link together climate, vegetation and soils in a 'natural region' the concept is still useful and convenient as a framework of study and as a valid hypothesis for investigation.

There are four main factors which interrelate to produce each biome.
- **Climatic**, e.g. water supply, temperatures, light and wind,
- **Topographic**, e.g. altitude, angle of slope and aspect,
- **Edaphic**, which are related to soils,
- **Biotic**, including the living environment.

Climatic controls on the biome

1 Precipitation This largely determines the vegetation type, e.g. forest, grassland or desert. The annual amount of precipitation is usually less important than its effectiveness for plant growth. How long is any dry season? Does the area receive steady, beneficial, or short, heavy and destructive periods of rain? Is rainfall concentrated in summer when evapotranspiration rates are higher? Does most rain fall during the growing season? Is there sufficient moisture for photosynthesis?

Heavy rainfall throughout the year enables forests to grow. These may be tropical rainforests, where the plants need a constant and heavy supply of water, or coniferous forests where trees can grow due to the lower rates of evapotranspiration.

Rain in winter only may be effective because evapotranspiration is low, but areas such as the Mediterranean lands often suffer from drought in summer and so plants need to be **xerophytic** (drought resistant).

Rain in summer only is less effective as this is when most moisture is lost through evapotranspiration. Effective precipitation is insufficient for trees and so savanna grassland grows in tropical latitudes and prairie grasslands in temperate areas.

Little rain throughout the year produces either desert biomes, where **ephemerals** (plants with very short life cycles, Figure 12.22) dominate the vegetation, or tundra,

where precipitation falling as snow, which is unavailable to plants, combines with the low temperatures to discourage plant growth.

2 Temperature This has a great influence on the flora, e.g. whether it is tropical or coniferous forest, prairie or savanna grassland.

Where mean monthly temperatures are above 15°C and there is a continuous growing and rainy season, broad-leaved evergreen trees tend to dominate, i.e. tropical rainforests.

Where there is a resting period in tree growth, either in hot climates with a dry season or cooler climates with between one and five months below 6°C, broad-leaved deciduous trees grow.

Where there are more than six months with temperatures below 6°C, giving a short growing season, coniferous trees are likely to dominate.

Grasses, including most cereals, can only grow when the mean monthly temperature exceeds 6°C.

Most plants prefer temperatures between 10°C, which is the minimum for effective photosynthesis, and 35°C.

The higher the temperature, the greater the need for water to combat losses through evapotranspiration and the sooner the wilting point is reached.

The lower the temperature, the fewer worms and bacteria there are in the soil, meaning a slower break down of humus and recycling of nutrients.

3 Light Intensity of light affects the process of photosynthesis. In locations nearer to the Equator and receiving most incoming radiation, the tropical ecosystems have a much higher energy input than do polar ecosystems. As light decreases there are few plants capable of living on the floor of the tropical rainforest in the shade of the larger trees or in seas more than 200 m deep.

Quality of light also affects plant growth: the increase of ultraviolet light on mountains reduces the number of species found there.

Duration of light (hours of daylight) varies between seasons with distance from the Equator.

4 Winds Wind can increase the rate of evapotranspiration and the wind chill factor, as well as 'bending' trees in areas of exposed, strong, prevailing winds (Figure 11.26).

Topographic

1 **Altitude** As altitude increases there are fewer species; they grow less tall and they form less dense cover. Relief may provide protection against heavy rain (rainshadow areas) and wind.

▼ Figure 11.26

Wind-distorted trees at Norfolk Island, Virginia, showing the effect of strong prevailing winds

2 Slope angle This influences soil depth, acidity (pH) and drainage. Steeper slopes usually have thinner soil, are less water-logged and less acidic than gentler slopes (catena, page 231).

3 Aspect Orientation of the slope alters sunlight and temperatures, with south-facing slopes in the northern hemisphere being more favourable to plant growth than those facing north.

Edaphic factors: soils

In Britain there is considerable local variation in vegetation due to differences in soil and underlying parent rock, e.g. grass on chalk, conifers on sand, and deciduous trees on clay. Plant growth is affected by soil texture, structure, acidity, organic content, depth, water and oxygen content, and nutrients (Chapter 10).

Biotic

Biotic factors include the element of competition: between plants for light, root space and water, and between animals. Competition increases with density of vegetation. Natural selection is an important biotic factor (page 245, where *Spartina* becomes a dominant in the slob zone of the salt marsh). The composition of seral communities and the degree of reliance upon other plants and animals for food (parasites) or energy (heterotrophs feeding on autotrophs) are also biotic factors.

Modification of biomes by humans

There are very few areas of climax vegetation or biomes left in the world today as most have been altered by human activity. In many parts the natural biome has been entirely replaced by a 'created environment'.

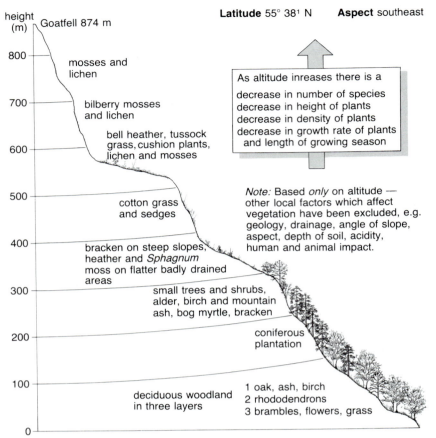

Latitude 55° 38¹ N   Aspect southeast

As altitude inreases there is a
decrease in number of species
decrease in height of plants
decrease in density of plants
decrease in growth rate of plants
and length of growing season

Note: Based *only* on altitude — other local factors which affect vegetation have been excluded, e.g. geology, drainage, angle of slope, aspect, depth of soil, acidity, human and animal impact.

mosses and lichen

bilberry mosses and lichen

bell heather, tussock grass, cushion plants, lichen and mosses

cotton grass and sedges

bracken on steep slopes, heather and *Sphagnum* moss on flatter badly drained areas

small trees and shrubs, alder, birch and mountain ash, bog myrtle, bracken

coniferous plantation

deciduous woodland in three layers

1 oak, ash, birch
2 rhododendrons
3 brambles, flowers, grass

▲ Figure 11.28

Mapping the effect of altitude on vegetation, Goatfell, Arran

1 Atmosphere The equilibrium of biomes has been altered by increasing levels of carbon dioxide, causing temperatures to rise through the greenhouse effect, as well as by increasing sulphur dioxide and nitrogen oxide emissions from power stations and factories (acid rain).

2 Landscape Erosion of the land, subsidence from mining, urbanisation, construction of reservoirs, exhaustion of soils, deforestation or afforestation, fires and clearing land for farming all 'create' a non-natural environment.

3 Ecological The use of fertilisers and pesticides, and the grazing of natural vegetation by animals mean that the balance between and nature of inputs and outputs to and from the system are changed.

The spatial pattern of world biomes

Figure 11.29 shows the distribution of the world's major biomes. When looking at maps of biomes in an atlas (they usually come under the heading 'Vegetation'), you should remember that all vegetation maps are very generalised (problems of generalisation, page 286) and do not show local variations, transition zones or, except in extreme cases, the influence of relief. There is no universal consensus as to the precise number of

◄ Figure 11.27, *far left*

The effect of altitude on vegetation, Cumbria— notice deciduous trees in the valley, grassland on the upper slopes and moorland on the summit

FRAMEWORK 6

Scientific enquiry: hypothesis testing

Since the 1960s geographers have felt an increasing need to adopt a more scientific approach to their studies. This stemmed from a number of changes which were taking place in attitudes to the study of geography and to science in a broader sense:

■ The increasing scale and complexity of the subject's material and the data available which impeded accurate observations and prevented the recording of the totality.

■ The rapid development of theory, often using computer modelling, from which order and patterns could be found and from which predictions could be made.

■ The realisation that, despite great care, all human observers have their own, subjective, opinions which influence an assessment or conclusion (i.e. scientific objectivity could not be guaranteed.)

The scientific approach to geography involved a series of logical steps, already practised in the physical sciences, which enabled conclusions to be drawn from precise and unbiased data (Framework 5). This approach may be summarised in a flow diagram.

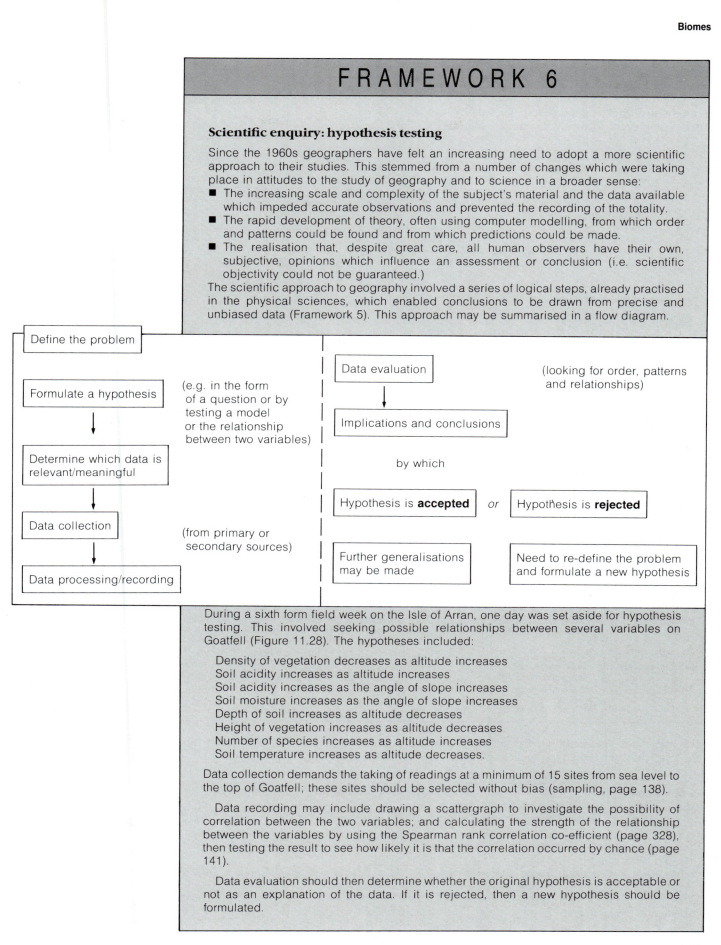

Define the problem

Formulate a hypothesis

(e.g. in the form of a question or by testing a model or the relationship between two variables)

Determine which data is relevant/meaningful

Data collection

(from primary or secondary sources)

Data processing/recording

Data evaluation

(looking for order, patterns and relationships)

Implications and conclusions

by which

Hypothesis is **accepted**   *or*   Hypothesis is **rejected**

Further generalisations may be made

Need to re-define the problem and formulate a new hypothesis

During a sixth form field week on the Isle of Arran, one day was set aside for hypothesis testing. This involved seeking possible relationships between several variables on Goatfell (Figure 11.28). The hypotheses included:

Density of vegetation decreases as altitude increases
Soil acidity increases as altitude increases
Soil acidity increases as the angle of slope increases
Soil moisture increases as the angle of slope increases
Depth of soil increases as altitude decreases
Height of vegetation increases as altitude decreases
Number of species increases as altitude increases
Soil temperature increases as altitude decreases.

Data collection demands the taking of readings at a minimum of 15 sites from sea level to the top of Goatfell; these sites should be selected without bias (sampling, page 138).

Data recording may include drawing a scattergraph to investigate the possibility of correlation between the two variables; and calculating the strength of the relationship between the variables by using the Spearman rank correlation co-efficient (page 328), then testing the result to see how likely it is that the correlation occurred by chance (page 141).

Data evaluation should then determine whether the original hypothesis is acceptable or not as an explanation of the data. If it is rejected, then a new hypothesis should be formulated.

biomes. Bradshaw has suggested 16 land biomes and five marine; Simmons describes 11 land biomes together with islands and seas; while more generalised geography texts (and examination syllabuses) usually accept eight land biomes.

It is possible to identify these eight biomes by at least two different methods, described in the table.

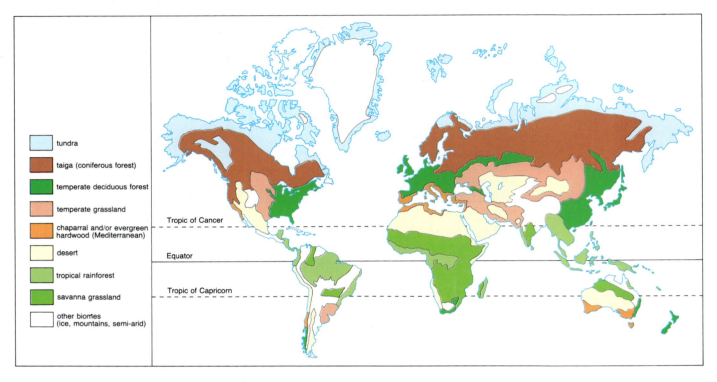

Legend:
- tundra
- taiga (coniferous forest)
- temperate deciduous forest
- temperate grassland
- chaparral and/or evergreen hardwood (Mediterranean)
- desert
- tropical rainforest
- savanna grassland
- other biomes (ice, mountains, semi-arid)

Tropic of Cancer

Equator

Tropic of Capricorn

▲ Figure 11.29

World biomes (*after* Strahler)

Method one	Method two
Traditional geographical method based on latitude	Modern approach based on differences in the production of organic matter (net primary production or NPP). NPP, given in brackets is equivalent to grams of dry organic matter per square metre per year ($g/m^2/yr$)

Method one			Method two			
Tropical	1	Rainforests	**High energy**	1	Rainforests	(2200)
	2	Tropical grasslands		2	Deciduous forest	(1200)
	3	Deserts	**Average energy**	3	Tropical grassland	(900)
Warm temperate	4	Mediterranean		4	Coniferous forest	(800)
Cool temperate	5	Deciduous forest		5	Mediterranean	(700)
	6	Temperate grasslands		6	Temperate grassland	(600)
Cold	7	Coniferous forest	**Low energy**	7	Tundra	(140)
	8	Tundra		8	Deserts	(90)
					(*after* I. Simmonds)	

The type and distribution of vegetation is extremely closely linked to climate and soils. The interrelationships of the three are described and explained in Chapter 12.

References

Annual Abstract of Statistics 1988, HMSO, 1988.

Biogeographical Processes, Ian Simmons, Allen & Unwin, 1982.

Climate, Soils and Vegetation, D.C. Money, U.T.P., 1965.

Coastal Dunes: Geography Today, BBC Television, 1982.

Earth, the Living Planet, Michael Bradshaw, Hodder & Stoughton, 1977.

Ecological Succession and Climax, Geofile No. 129, Mary Glasgow Publications, April 1989.

Nature at Work: Introducing Ecology, British Natural History Museum Publications, 1978.

Soils, Vegetation and Ecosystems, Greg O'Hare, Oliver & Boyd, 1987.

The Nature of the Environment, Andrew Goudie, Blackwell, 1984.

World Vegetation, D. Riley and A. Young, Cambridge University Press, 1966.

CHAPTER 12

World Climate, Soils and Vegetation

'There was ... an instant in the distant past when the living things, the rocks, the air and the oceans merged to form the new entity, Gaia.'
The Ages of Gaia, James Lovelock, O.U.P. 1989

Although it is possible to study climatic phenomena in isolation (Chapter 9), an understanding of the development of soils (Chapter 10) and vegetation (Chapter 11) necessitates an appreciation of the inter-relationships between all three. This chapter attempts to show how the integration and interaction of climate, soils and vegetation gives the world its major zones, or environments, and how these have often been modified by human activity.

The purpose of grouping together different soil types, the resultant difficulties in trying to achieve this, and a suggested classification, were described on page 228. The major vegetation and fauna zones (biomes) were classified on page 260 and their global location shown in Figure 11.29. In a similar way, geographers seek to classify climates (Framework 3, page 130).

Why classify climates?

By studying the weather, i.e. the atmospheric conditions prevailing at a given time in a specific place or area, it is possible to make generalisations about the area's climate, i.e. the average conditions over a period of time. Any area may experience temporary departures from its 'normal' climate but at the same time it may have long term similarities with regions in other parts of the world.

In seeking a sense of order the geographer tries to group together those parts of the world which have similar measurable climatic characteristics (temperature, rainfall distribution, winds, etc) and to identify and to explain similarities and differences in spatial and temporal distributions and patterns. The geographer may then compare areas on a global scale by using the most convenient methods available which are easy to use and practicable to apply in order to help map and explain distributions of soil, vegetation and crops.

Bases for classifications

The early Greeks divided the world into three zones based upon a simple temperature description: torrid (tropical), temperate and frigid (polar); they ignored precipitation.

In 1918 **Koppen** advanced the first modern classification. To support his claim that natural vegetation boundaries were determined by climate he selected as his basis appropriate temperature and seasonal precipitation values. This remains today a useful classification even though it makes no attempt to explain the causes of the climate. Since then a modification by **Trewartha** has become widely accepted despite the apparent drawback of its having 23 regions!

Q

1 Explain why geographers find it useful and necessary to classify climates.

2 Outline the distinguishing characteristics of any one climatic classification with which you are familiar and explain why it is useful.

3 Indicate what you consider to be the main strengths and limitations of the classification which you have chosen in answer to **2**.

Thornthwaite, in the 1930s and 1940s, suggested and modified a classification with a more quantitative basis. He introduced the term 'effectiveness of precipitation' (his P/E index) which was obtained by dividing the mean monthly precipitation by the mean monthly evapotranspiration and taking the sum of the 12 months. Here the difficulty was, and still is, in obtaining accurate evapotranspiration figures. The classification results in a large number (32) of regions.

In Britain in the 1930s, **Miller** proposed a relatively simple classification based upon five latitudinal temperature zones subdivided longitudinally by seasonal precipitation distributions (Figure 12.1). Its advantages include the following: by using only three temperature figures — 21°C (the limit for growth of coconut palms), 10°C (the minimum for tree growth) and 6°C (the minimum for grasses and cereals) — it is easy and convenient to apply; it is related to vegetation zones; and as these are a response on a global scale to climate and vegetation, it is related also to soils.

The weaknesses of Miller's classification are common to many. It does not show transition zones between climates and often division lines are purely arbitrary; some areas lack sufficient climatic data to categorise them; it does not allow for mesoscale (the Lake District and London do *not* have exactly the same climate) or microscale variations; it is criticised for being simplistic and having several anomalies; it ignores human influence and climatic change; and, being based upon temperature and precipitation figures, it neglects recent studies in heat and water budgets, air mass movement and the transfer of energy.

Note: Miller labelled his hot and cold deserts **F**, whereas in Figure 12.1 they have been placed in a latitudinal slot. He also introduced a mountain climate which he referred to as **G**.

If this classification were applied to a theoretical, large, isotrophic continent in the northern hemisphere then the climates would be arranged as indicated in Figure 12.2.

A Tropical (hot) climates: no mean monthly temperature below 21° C
  1 Equatorial: rain throughout the year with a double maxima
  2 Continental savanna: summer rain, winter drought
  3 Hot deserts: constant drought
  4 a Tropical eastern margins: rain all year, later summer maximum
    b Monsoon: heavy summer rain, winter drought

B Warm temperate climates (sub-tropical): no mean monthly temperature below 6° C
  1 Western margins (Mediterranean) winter rain, summer drought
  2 Eastern margins: rainfall all year, summer maximum (China type monsoon)

C Cool temperate climates: 1 to 5 months below 6° C
  1 NW European: rain all year, winter maximum
  2 Continental: summer maximum
  3 Eastern margins: rain all year
  4 Cold deserts: constant drought

D Cold climates: 6 months or more under 6° C (seasonal precipitation not significant as temperatures too low for plant growth)

E Arctic climates: no month above 10° C

◀ Figure 12.1
A classification of climate (*after* Miller)

▼ Figure 12.2
Model showing the generalised pattern of world climates (based on Miller's classification)

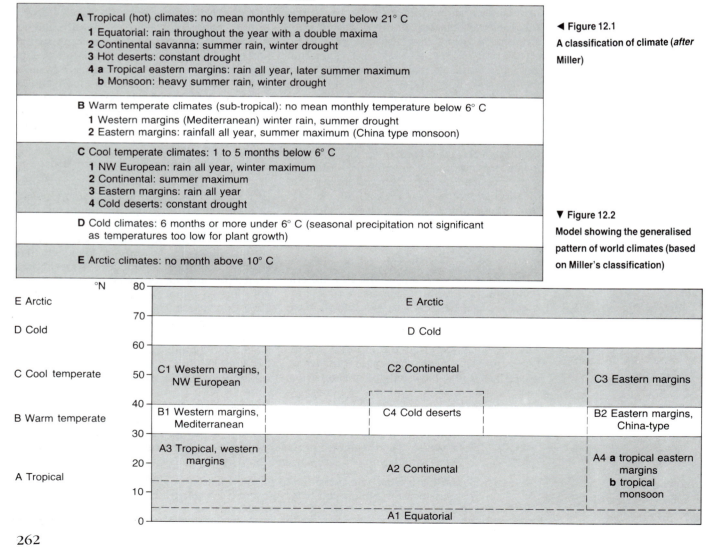

▶ Figure 12.3

Relationships between climate, soil and vegetation at the global scale

Climate		Text reference number	Climate type	Biome (based on NPP)	Soil (zonal type)	
Very cold all year		8	Arctic	tundra	tundra	
Cold all year		7	cold	coniferous forest	podsol	
Cool temperate,	rain all year, winter maximum	6	cool temperate western margin,	deciduous forest	brown earths	
	summer maximum	5	cool temperate continental	temperate grassland	chernozems	prairie
		4A	eastern margins (monsoon)			chestnut
Warm temperate	winter rain	4	western margins Mediterranean	Mediterranean	Mediterranean	
	all year, summer maximum	see 4A	eastern margins (monsoon)			
Tropical	little rain	3	desert	desert (xerophytic)	red yellow desert	
	summer rain	2	continental	tropical grassland (savanna)	ferruginous	
		1B	monsoon	jungle		
	rain all year	1A	east coasts	rainforests	ferrallitic	
		1	Equatorial			

The world's major climatic zones

One major advantage of Miller's classification is the similarity between world maps showing climate, soils and vegetation. Figure 12.3 illustrates how the eight major environmental zones are derived by linking these three components. The following section describes each of these eight zones in turn. How useful are these descriptions? Think how helpful they are in developing our understanding of the processes which operate in each environmental system.

1 Equatorial

Areas with equatorial climate are located within 5° either side of the Equator. They include the Amazon and Zaire Basins and the coastal lands of Ecuador and West Africa.

Climate

Temperatures are high and constant throughout the year because the sun is always high in the sky. The annual temperature range is under 3°C inland, e.g. Manaus, and 1°C on the coast, e.g. Belem, (Figure 12.4). Mean monthly temperatures, ranging from 26°C to 28°C, reflect the lack of seasonal change. Slightly higher temperatures may occur during any 'drier' season. Insolation is evenly distributed throughout the year with each day having approximately 12 hours of daylight and 12 hours of darkness. The diurnal range is also small, about 10°C. Evening temperatures rarely fall below 22°C and, due to the presence of afternoon cloud, daytime temperatures rarely rise above 32°C. It is the

high humidity, with its sticky, unhealthy heat, which is least appreciated by Europeans.

Annual rainfall totals usually exceed 2000 mm (Belem, 2732 mm) and most afternoons have a heavy shower (Belem has 243 rainy days per year). This results from the convergence of the trade winds at the ITCZ and the subsequent enforced ascent of warm, moist, unstable air in strong convection currents. Evapotranspiration is rapid from the many rivers, swamps and trees. Most storms are violent with the heavy rain, accompanied by thunder and lightning, coming from cumulonimbus clouds. Some areas may have a drier season when the ITCZ moves a few degrees away from the Equator at the winter and summer solstice and a double maxima when the sun is directly overhead at the spring and autumn equinox. The high daytime humidity needs only a little night time radiation to give condensation in the form of dew. The winds at ground level at the ITCZ are light and variable (doldrums) allowing land and sea breezes to develop in coastal areas (Figure 9.50).

Tropical rainforests

It is estimated that the rainforests provide 40 per cent of the net primary production of terrestrial energy (NPP, page 260). This is a result of the high solar radiation, heavy rainfall, a constant moisture budget surplus, the rapid decay of leaf litter and the recycling of nutrients.

Vegetation consists of trees of many dif-

▼ Figure 12.4

Climate graph for equatorial regions

Belem (Brazil) 1° S
altitude 24m
annual range of temperature 1° C
annual precipitation 2732 mm

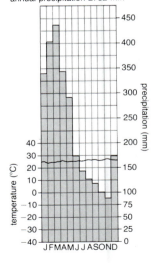

263

ferent species. In Amazonia there may be over 300 species in 1 sq km, including rosewood, mahogony, ebony, greenheart, palm and rubber. The trees, which are mainly hardwoods, have an evergreen appearance, for although they are deciduous they shed their leaves at any time during the continuous growing season. The very tallest trees, **emergents**, may reach up to 50 m in height and form the habitat for numerous birds and insects. Below the emergents are three layers all competing for sunlight.

The top layer, or **canopy**, forms an almost continuous cover which absorbs over 70 per cent of the light and intercepts 80 per cent of the rainfall (Figure 12.5). The crowns of these trees merge some 30 m above ground level and shade the underlying species, they protect the soil from erosion, and they are inhabited by most of the birds, animals and insects of the rainforest.

The second layer, or **undercanopy**, consists of trees growing up to 20 m, similar in size to deciduous trees in Britain. The lowest, or **shrub layer**, consists of shrubs and small trees which are adapted to living in the shade of their taller neighbours.

The trees grow tall to try to reach the sunlight and the climate is at the optimum for photosynthesis. Taller trees have buttress roots which emerge over 3 m above ground level to give support (Figure 12.6). The trunks are usually slender and branchless. Some, like the cacao, have flowers growing on them, and the bark is thin as there is no

need for protection against adverse climatic conditions. Tree trunks also provide support for lianas, vine-like plants, which can grow to 200 m in length and up the tree. Lianas climb along branches before plunging back down to the forest floor. Leaves are dark green, smooth and often have **drip tips** to shed excess water.

Epiphytes are plants which do not have their roots in the soil. They grow on trunks, branches and even the leaves of trees and shrubs. Epiphytes simply 'hang on' to the tree: they derive no nourishment from the host and are *not* parasites. On the darker forest floor are large numbers of **saprophytes** which can exist without photosynthesis by relying upon certain fungi for their nutrients. Less than 5 per cent of insolation reaches the forest floor with the result that undergrowth is thin apart from in areas where trees may have been felled by shifting cultivators or where a giant emergent has fallen, dragging with it several of the top canopy trees.

Vegetation is also dense along the many river banks, again because sunlight can penetrate the canopy. Alongside the Amazon, many trees spend several months of the year growing in water (Figure 12.7) as the river and its tributaries rise over 15 m in the rainy season. In such conditions live huge water lilies which have leaves of over 2 m in width, capable of supporting the weight of a young person (Figure 12.8). Mangrove swamps are found in coastal areas.

Although ground animals are relatively

◄ Figure 12.5, *far left*
Emergents rising above the rainforest canopy, Malaysia

◄ Figure 12.6
Buttress roots and lianas at Bombacaceae in the Amazon rainforest

▲ Figure 12.7
Rainforest growing in water in the Peruvian Amazon — the vegetation has adapted to the wet environment

▶ Figure 12.8, *above right*
Large water lilies, *Vitoria Regia*, native to the Amazon Basin

▼ Figure 12.9
The nutrient cycle

▶ Figure 12.10, *below right*
The broken nutrient cycle

few in number, the rainforests of Brazil alone are said to be the habitat for 2000 species of bird, 600 species of insects and mosquitoes, and 1500 species of fish.

Yet the productivity of this biome, one which on its own replaces much of the world's used oxygen, largely depends upon the rapid and unbroken recycling of nutrients. Figure 12.9 shows the natural nutrient cycle whereas Figure 12.10 constructs the consequences of human clearances of the forest which break the system. In areas where the forest has been cleared, the secondary succession differs from that of the original climax vegetation (Krakatoa, Figure 11.6). The dominants are less tall in these areas. The trees are less stratified, there are many fewer species, and most are intolerant of shade although there is more light at ground level which encourages a dense undergrowth.

Ferrallitic soils (latosols)

These soils result from the high annual temperature and rainfall which cause rapid chemical weathering of bedrock and create the optimum conditions for organic activity to decompose the luxuriant vegetation.

Continuous leaf fall within the forest gives a thick litter layer but the underlying humus layer is thin due to the rapid decomposition of organic matter by intensive biota activity. There is a rapid recycling of nutrients within the humus cycle (Figure 12.9) but the release of bases prevents the soil from

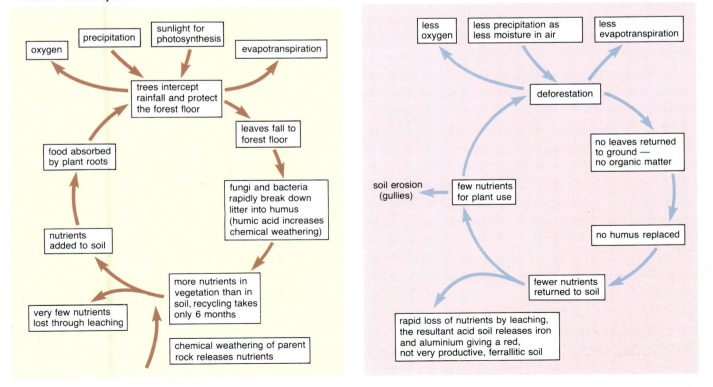

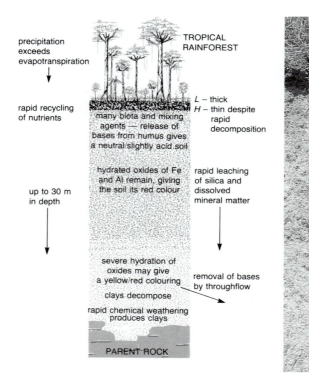

precipitation exceeds evapotranspiration

TROPICAL RAINFOREST

rapid recycling of nutrients

L – thick
H – thin despite rapid decomposition

many biota and mixing agents — release of bases from humus gives a neutral/slightly acid soil

hydrated oxides of Fe and Al remain, giving the soil its red colour

rapid leaching of silica and dissolved mineral matter

up to 30 m in depth

severe hydration of oxides may give a yellow/red colouring

removal of bases by throughflow

clays decompose

rapid chemical weathering produces clays

PARENT ROCK

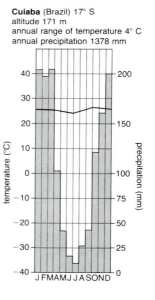

◀ Figure 12.11, *far left*

Ferrallitic soils (latosols) characteristic of a tropical rainforest

◀ Figure 12.12

Ferrallitic soils on clay, Hawaii

becoming acidic. The soils have a relatively low nutrient status which is maintained only through the rapid and continuous replacement from the lush vegetation. Once this supply has been removed, by deforestation, these soils soon lose their fertility. Despite some protection from the tree canopy the heavy rainfall causes severe leaching. Silica is more soluble than the sesquioxides and is removed by ferrallitisation leaving iron and aluminum which give the soil its deep red colour (Figure 12.12).

The continual leaching and abundance of mixing agents inhibits the formation of horizons. The lower parts of the profile may have a more yellowish-red tint due to the extreme hydration of iron oxides. The clay-rich soils are also very deep, often up to 30 m, mainly because of the rapid breakdown of parent rock under the ideal conditions for chemical weathering (Figure 2.9). Ferrallitic soils have a loose structure and if exposed to heavy rainfall are easily gullied and eroded (Figure 10.27).

Despite their depth the soils of the rainforest are not agriculturally productive. Once the source of nutrients (the trees) has been removed, the soil rapidly loses its fertility and local farmers, often shifting cultivators, have to move to clear new plots (page 402).

1A Tropical eastern margins

Located within the tropics, the eastern coasts of Central America, Brazil, Madagascar and Queensland Australia, like the equatorial climate zones, receive rain throughout the year. This rain is brought by the trade winds, blowing across warm ocean currents (Figure 9.9). Temperatures are generally very high although there is a slightly cooler season when the overhead sun appears to have migrated into the opposite hemisphere. The resultant vegetation and soil types are, therefore, similar to those found in the equatorial belt, i.e. rainforest and ferrallitic.

2 Tropical continental

These are mainly located between latitudes 5° and 15° north and south of the Equator and within central parts of continents, i.e. the Llanos (Venezuela), the Campos (Brazilian Highlands), most of Central Africa surrounding the Zaire Basin and parts of Mexico and northern Australia (Figure 11.29).

Climate

Although temperatures are high throughout the year there is a short, slightly cooler season (in comparison with the equatorial) when the sun is overhead at the tropic in the opposite hemisphere (Figure 12.13). The annual range is also slightly greater (Cuiaba 5°C) owing to the sun's being at a slightly lower angle in the sky for part of the year, the greater distance from the sea, and less cloud and vegetation cover. Temperatures may drop slightly at the onset of the rainy season. For most of the year, cloud is almost absent and diurnal temperatures are high, up to 25°C.

▼ Figure 12.13

Climate graph for a tropical continental (savanna) location

Cuiaba (Brazil) 17° S
altitude 171 m
annual range of temperature 4° C
annual precipitation 1378 mm

► Figure 12.14

Water balance for a tropical continental climate in the southern hemisphere (refer also to Figure 3.3)

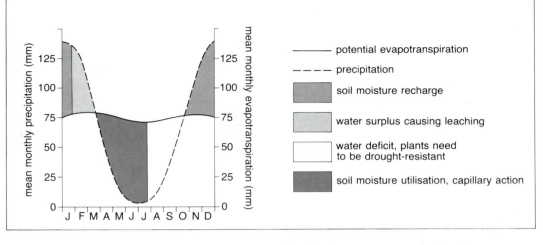

▼ Figure 12.15

Causes of seasonal rainfall in the tropical continental (savanna) climate

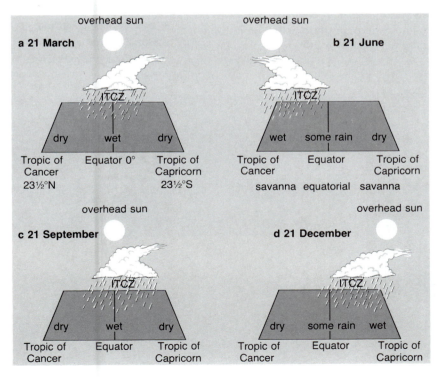

The main characteristic of this climate is the alternate wet and dry season. The wet season occurs when the sun moves overhead bringing with it the heat Equator, the ITCZ, and the equatorial low pressure belt (Figure 12.15). Heavy convectional storms can give 80 per cent of the annual rainfall total in four or five months. The dry season corresponds to the moving away of the ITCZ leaving the area with the strong and steady trade winds. These are dry because they warm up as they blow towards the Equator and will have shed any moisture on distant coastal areas. Areas nearer to the desert margins tend to experience dry, stable conditions (the subtropical high pressure) caused by the migration of the descending limb of the Hadley cell (page 188). Humidity is also low during this season.

Tropical or savanna grasslands

The tropical grasslands are estimated to have a mean NPP of 900 g/m²/yr (page 260). This is considerably less than the rainforest partly because of the smaller number of trees, species and layers and partly because although the grasslands have the potential to return organic matter back to the soil, the rate of decomposition is reduced during the winter drought so much is left stored in the litter.

As shown in Figure 12.16, the savanna includes a series of transitions between the rainforests and the desert. At one extreme, the 'closed' savanna is mainly trees with areas of grasses; at the other, the 'open' savanna is vegetated only by scattered tufts of grass. The trees are deciduous and, like those in Britain, lose their leaves and reduce transpiration, but, unlike in Britain, this is due to the winter drought rather than the cold. Trees are **xerophytic** or drought resistant. Even when leaves do appear they are small, waxy and sometimes thorn-like. Roots are long and extend to tap any underground water. Trunks are gnarled and the bark is usually thick to reduce moisture loss.

The baobab tree (also known as the 'upside-down tree') has a trunk of up to 10 m in diameter which holds a supply of water, while its branches which are root-like in appearance have only the minimum number of tiny leaves (Figure 12.17). Some baobabs are estimated to be several thousand years old, and like other savanna trees are **pyrophytic**, meaning that their trunks are resistant to the many local fires. Acacias, with their flattened crowns (Figure 12.18) which result from the steady savanna winds, provide welcome though limited shade — as do the eucalyptus in Australia. Savanna trees reach 6–12 m in height. Many trees have 'Y' shaped, branching trunks, ideal for the lion to rest in after its meal! The number of trees increases near to rivers and waterholes.

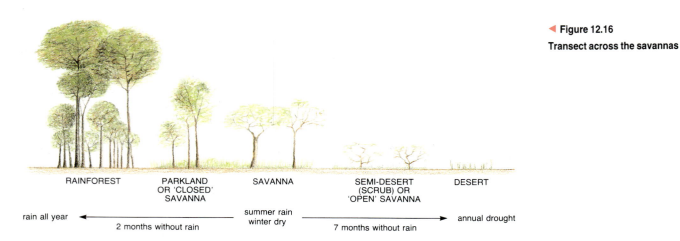

Grasses grow in tufts and tend to have inward-curving blades and silvery spikes. After the onset of the summer rains, grasses grow very quickly to over 3 m in height: elephant grass reaches 5 m. However, the yellowish colour becomes more pronounced as the sun dries up the vegetation. By early winter the straw-like grass has died down, leaving seeds dormant on the surface until the following season's rain. By the end of winter only the roots remain and the surface is exposed to wind and rain.

Over 40 different species of large herbivore graze on the grasslands, e.g. wildebeeste, zebra and antelope, and it is the home of several carnivores, both predators, e.g. lions, and scavengers e.g. hyenas. Termites and microbes are the major decomposers. As has been mentioned earlier (page 249) fire is possibly the major determinant of the savanna biome — caused either deliberately by farmers or by lightning associated with the frequent summer electrical storms. It is the fringes of the savannas, those bordering the deserts, which are now experiencing extremes of desertification. As more trees are being removed for fuel and as animals over-graze and reduce the productivity of grasslands, the heavy rain forms gulleys and wind blows away the surface soil. Those parts of the savanna which are not farmed have more trees which suggests that grass may not be the natural climatic climax vegetation (page 240).

Ferruginous soils

As savanna grasses die back during the dry season they provide organic matter which is readily broken down to give a thin, dark-brown layer of humus. At the same time, capillary action brings bases in solution towards the surface (Figure 12.14). During the following wet season rapid leaching and the relatively high pH removes silica from the upper profile, leaving behind the red-coloured oxides of iron and aluminium. This alternating leading capillary action often produces a cemented layer, or **laterite**, just below the surface, which impedes drainage, plant root penetration and ploughing. The parent material weathers into a clay which tends to become sticky and plastic during the wet season but which can become

◀ Figure 12.17, *below left*
The Baobab tree, Western Australia

▼ Figure 12.18
Savanna grassland in Masai Mara, Kenya, during the rainy season when the grass is long and green

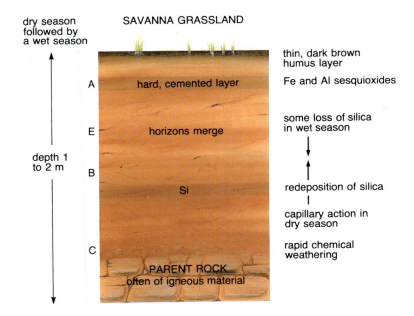

SAVANNA GRASSLAND

dry season followed by a wet season

depth 1 to 2 m

A — hard, cemented layer — thin, dark brown humus layer — Fe and Al sesquioxides

E — horizons merge — some loss of silica in wet season

B — Si — redeposition of silica — capillary action in dry season

C — PARENT ROCK often of igneous material — rapid chemical weathering

▲ Figure 12.19
Ferruginous soils (see also Figure 10.20a)

▼ Figure 12.20
Climate graph for a hot desert

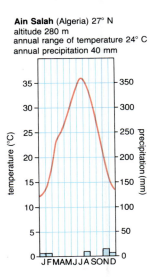

Ain Salah (Algeria) 27° N
altitude 280 m
annual range of temperature 24° C
annual precipitation 40 mm

'brick hard' during the drought. The lower part of the profile may develop a prismatic structure with a layer of redeposited silica.

Ferruginous soils contain few nutrients and so are not particularly suitable for agriculture. Indeed, grasslands are better suited to animals than to arable. When a lateritic crust forms near to the surface it often causes the soil above to dry out and become highly vulnerable to erosion by wind and, when the rains return, water. The word 'laterite' is derived from the Latin for 'brick' and this deposit is indeed still used as a building material.

3   Hot deserts

Miller defined hot deserts as areas having no month with a mean temperature below 6°C. Several arid areas therefore extend beyond the limits of the tropics. Major deserts are usually found on the west coast of continents, between 15° and 30° north or south of the Equator and in the trade wind belt (Figure 7.2 and 11.29). The exception is the extensive Sahara-Arabian-Thar desert which owes its existence to the size of the Afro-Asian continent. All have temperatures far in excess of Miller's critical limit.

Climate

Desert temperatures are characterised by their extremes. The annual range is often 20–30°C and the diurnal range over 50°C (Figure 12.20). Daytimes, especially in summer, receive intense insolation from the overhead sun, intensified by the lack of cloud cover and the bare rock or sand ground surface. In contrast, nights may be extremely

cold with temperatures likely to fall below 0°C. Coastal areas, however, have much colder monthly temperatures (Arica in the Atacama has a warmest month of only 22°C) due to the presence of offshore cold ocean currents (Figure 9.9).

Although all deserts suffer an acute water shortage, none are truly dry. Aridity and extreme aridity have been defined by using Thornthwaite's P/E index (Figure 7.1). The four main causes of deserts are discussed on page 146. Amounts of moisture are usually small and precipitation is extremely unreliable. Death Valley, California, averages 40 mm a year, yet rain may fall only once every two or three years; whereas mean annual totals vary by less than 20 per cent a year in NW Europe, the equivalent figure for the Sahel is 80–150 per cent. Rain, when it does fall, is heavy. The resultant rapid surface runoff (page 45), together with low infiltration and high evaporation rates, minimises the effectiveness of the rain for vegetation. The Atacama, an almost rainless desert, has some vegetation because moisture is available in the form of advection fog (page 183). The subsiding air, forming the descending limb of the Hadley cell, gives high pressure and produces the trade winds which are strong, persistant and likely to cause localised dust storms.

Vegetation

Hot deserts have the lowest organic productivity levels of any biome. The average NPP is 90 g/m²/yr, most of which occurs underground away from the direct heat of the sun. Vegetation has to have a high tolerance to the moisture budget deficit, intense heat and, often, salinity. Although desert plants are few in species, have simple structures, no stratification by height and provide a low density cover, few areas are totally devoid of vegetation. However, plants are xerophytic (page 267) as the lack of water hinders the ability of roots to absorb nutrients and of the green parts of plants to photosynthesise.

Many plants are **succulents**, i.e. they can store water in their tissues. Many succulents have fleshy stems and some have swollen leaves. Cacti (Figure 12.21) absorb large amounts of water during the infrequent but heavy storms. Their stems swell up, only to contract later as moisture is slowly lost through transpiration. Transpiration takes place from the stems and is further reduced because the stomata close during the day and open nocturnally. The stems also have a thick, waxy cuticle. Australian eucalyptids have thick, protective bark for the same purpose.

Most plants, e.g. cactus and thornbush have

269

small spiky or waxy leaves to reduce transpiration and deter animals. Roots are very long to tap groundwater supplies — those of the acacia exceed 15 m — or they spread out over wide areas near to the surface to take the maximum advantage of any rain or dew e.g. creosote bush. This means that bushes tend to be widely spaced to avoid competition for water. Some plants have bulbous roots to store water. The seeds, which usually have a thick case protecting a pulpy centre, can lie dormant for months or several years until the next rainfall.

Following a storm, the desert blooms (Figure 12.22). Many plants are **ephemerals** and can complete their life cycles in two or three weeks. Others are **halophytic** and can survive in salty depressions, e.g. saltbush, or where the water table is near enough to the surface to form oases, e.g. date-palm.

Grass is absent from the desert and, with the general lack of green plants, this means that there are only limited food chains, i.e. desert biomes have a low capacity to sustain life. There is insufficient plant food to support an abundance of animal life. Food chains (page 252) are simple, often just a single linear sequence (in contrast to the interlocking webs characteristic of, e.g. forests). This is why the ecosystem is 'fragile' — organisms do not have all the alternative sources of food that are available in more complex ecosystems. Many animals are small and nocturnal (the camel is an exception) and burrow into the sand during the heat of the day. Reptiles are more adaptable

but bird life is limited. The desert fringes form a delicately balanced ecosystem which is being disturbed by human development and population growth. Removal of woody plants for fuel and overgrazing by herds of animals is causing the desert areas of several parts of the world to expand, i.e. the process of desertification (e.g. the Sahel).

Grey desert soils

In desert areas the climate is too dry and the vegetation too sparse for any significant chemical weathering or accumulation of organic material. Any moisture in the soil is likely to be drawn upwards by capillary action causing salts and bases to be deposited near to the surface. Soils are therefore alkaline, with a high concentration of magnesium, sodium and calcium. The grey colour results from the lack of moisture which with the high pH value restricts hydrolysis and the release of red-coloured iron. Soils are usually thin, lacking horizons and structure.

Desert soils are unproductive mainly because of the lack of moisture and humus but potentially they are not particularly infertile. Areas under irrigation are capable of producing high quality crops although this farming technique is being threatened by salinisation (Figure 10.20c).

4 Mediterranean (warm temperate western margins)

This type of biome is found on the west coasts of continents between 30° and 40° north and

◄ Figure 12.21, *far left*
Saguaros cacti in the Arizona desert

◄ Figure 12.22
Ephemerals in flower following a desert rainstorm, Saudi Arabia

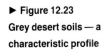

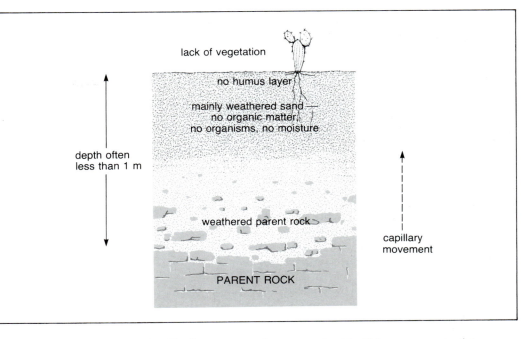

south of the Equator, i.e. in Mediterranean Europe, (which is the only area where the climate penetrates far inland) California, central Chile, Cape Province (Republic of South Africa), and parts of southern Australia.

Climate

The climate is noted for hot, dry summers and warm, wet winters. Summers in southern Europe are hot: the sun is high in the sky, though never directly overhead, and there is little cloud. Winters are mild, partly because the sun's angle is still quite high but mainly due to the moderating influence of the sea. Other 'Mediterranean' areas are less warm in summer and have a smaller annual range

because of cold offshore currents. (Compare San Francisco, 8°C in January and 15°C in July, with Athens, Figures 12.25, 12.26.) Diurnal temperature ranges are often high as many days, even in winter, are cloudless.

As the ITCZ moves northwards in the northern summer the subtropical high pressure area migrates with it to affect these latitudes. The trade winds bring arid conditions with the length of the dry season increasing towards the desert margins. In winter the ITCZ and subsequently the subtropical jet stream (page 190) move southwards allowing the westerlies, which blow from the sea, to bring moisture. Most areas are backed by coastal mountains and so the combined effects of orographic and frontal

▶ Figure 12.24

Climate graph for a
Mediterranean climate (Athens,
Greece)

▶ Figure 12.25, *far right*

Soil moisture balance for a
Mediterranean climate (San
Francisco, USA)

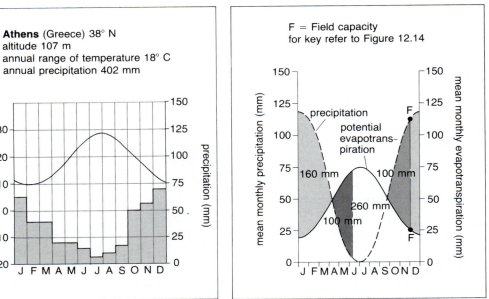

Athens (Greece) 38° N
altitude 107 m
annual range of temperature 18° C
annual precipitation 402 mm

F = Field capacity
for key refer to Figure 12.14

precipitation give high seasonal totals. Areas with adjacent cold offshore currents experience advection fogs, e.g. California, Atacama. The Mediterranean Sea region is noted for its local winds (Figure 12.26). The **sirocco** and **khamsin** are two of the hot, dry winds which blow from the Sahara and can raise temperatures to over 40°C. The **mistral** is a cold, katabatic wind which originates over the Alps and is funnelled at considerable speed down the Rhône valley.

Vegetation

The xerophytic or drought resistant vegetation of those regions of the world experiencing a Mediterranean climate is locally called **chaparral** in California, **maquis** or **garrigue** in Europe, and **mallee** in Australia. The NPP of these ecosystems is about 700 g/m²/yr; it is limited by the summer dry season, and has probably been reduced over the centuries by human activity. Indeed this human activity, together with frequent fires, has left very little of the original climatic climax vegetation.

The climax vegetation is believed to have been open woodland comprising a mixture of broad-leaved evergreen trees (e.g. cork oak and holm oak) and conifers (e.g. aleppo pines, cypresses and cedars). The present vegetation is described as **sclerophyllous** scrub. Many trees are evergreen, maximising the potential for photosynthesis. Trees such as the cork oak have thick and often gnarled barks to help reduce transpiration. Others, such as olive and eucalyptus, have long tap roots to reach groundwater supplies and in some cases the roots may be bulbous to store water. High temperatures during the dry summer limit the amount and quality of grass. Citrus fruits are adapted to the climate by having thick skins to preserve moisture. Most trees only grow to 3 to 4 m in height. They provide little shade as they grow widely spaced and are **pyrophytic** (fire resistant, page 267).

In Mediterranean Europe there are two main variations in the present vegetation ecosystem. First, maquis, which grows in areas of impermeable rock (e.g. granite), consists of shrubs growing to 3 m high, heathers and broom (Figure 12.27). Second, garrigue, found on drier more permeable rocks (e.g. limestone), is much lower and less dense (Figure 12.28). Apart from gorse, with its prickles, some of the more common plants are aromatic shrubs such as thyme, lavender and rosemary.

The limited leaf litter tends to decompose slowly during the dry summer, even though temperatures are high enough for year-round bacterial activity. Wildlife and climax vegetation have retreated as human activity advanced, so, arguably, Mediterranean regions of Europe and California form the biome most altered by human activity in today's world.

Soils

Mediterranean soils are transitional between brown earths on the wetter margins and grey desert soils at the drier fringes. Initially

Sclerophyll *means an evergreen tree or shrub with small, hard, leathery, waxy or thorn-like leaves effective at reducing transpiration during the dry summer season.*

▶ Figure 12.28
Garrigue vegetation in
Provence, France

▶ Figure 12.29
Mediterranean soils

▶ Figure 12.30 *far right*
Mediterranean soils—a
characteristic profile (refer also
to Figure 10.22)

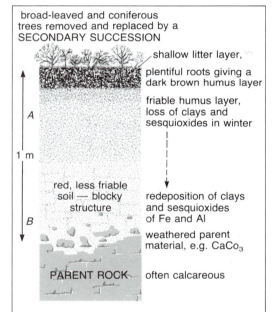

formed under broad leaved and coniferous woodland, the soil is partly a relict feature from a previously forested landscape.

There are often sufficient roots and decaying plant material to provide a significant humus layer. Winter rains cause the leaching of bases, sesquioxides of iron and aluminium and the translocation of clays (lessivage, page 226). The *B* horizon is therefore clay-enriched and may be coloured a bright red by the redeposition of iron and aluminium. The soils, which are often thin, are less acid than the brown earths as there is less leaching in the dry season and calcium is often released, especially in limestone areas.

In many Mediterranean biomes parent rock is locally a more important factor in soil formation than climate. This results in the development of intrazonal soils such as rendzina and terra rossa (Figure 10.22).

4A Eastern margin climates in Asia

Southeast and eastern Asia is dominated by the monsoon (page 200) Temperature figures and rainfall distributions are similar to those of tropical continental areas although annual amounts of rain are appreciably higher (Figure 9.49). The natural vegetation is jungle and is similar in many ways to tropical rainforests.

273

5   Cool temperate continental

This biome lies in the centre of continents between approximately 40° and 60° north of the Equator and according to Miller has at least one month with a mean temperature below 6°C. The two main areas are the North American Prairies and the Russian Steppes (Figure 11.29).

Climate

The annual range of temperature is high as there is no moderating influence from the sea (38°C at Saskatoon, Figure 12.31). The land warms up rapidly in summer as insolation is not intercepted by cloud, to give maximum mean monthly readings of around 20°C. However, the rapid radiation of heat from midcontinental areas in winter means there are several months when the temperature remains below freezing point. The clear skies result in a large diurnal temperature range.

Precipitation decreases rapidly towards the east in Russia as distance from the sea increases, whereas in North America totals are lowest to the west in the rainshadow of the Rockies. Annual amounts in both areas average only 500 mm and there is a threat of drought, e.g. North America, 1988. Although, fortuitously, 75 per cent of precipitation falls during the summer growing season, it may be in the form of harmful heavy thunderstorms and hailshowers. The ground is snow-covered for several months between October and April. Overall there is a close balance between precipitation and evapotranspiration (Figure 12.32). Cool temperate continental areas in the USA and USSR are open to cold blasts of arctic air, though the chinook may bring warmer spells to the Prairies (Figure 9.52).

Temperate grasslands

This type of vegetation lies to the south of the coniferous forest belt in the dry interiors of North America and the USSR. Temperate grasslands are also found in parts of the southern hemisphere where their locations are more sporadic. They are usually between 30° and 40° south, e.g. the Pampas, South America, and Canterbury Plains, New Zealand are on eastern coasts; the Murray–Darling Basin, Australia, and the South African Veldt are more continental. NPP of 600 g/m²/yr is considerably less than the tropical grasslands because vegetation grows neither as rapidly nor as tall.

Whatever the original climax vegetation of the biome may have been — the ecosystem has been significantly altered by fire and human exploitation (Figure 11.25) — today the dominants are grasses, e.g. grama and buffalo grass. There are two main types of grass. Feather grasses grow to 50 cm and form a relatively even coverage whereas tufted (tussock) grasses, reaching up to 2 m, are found in more compact clumps (Figure 12.33). The grass forms a tightly knit sod which was believed to have restricted tree growth and made early ploughing difficult.

The deep roots, which are often over 2 m long to reach down to the water table, help to bind the soil together and so reduce erosion. Most of the organic material is in the grass roots: indeed, the roots and rhizomes provide the largest store of nutrients (Figure 11.25b). During the autumn the grass dies down to form a turf mat in which seeds lie dormant until the snowmelt, rains and warmer temperatures of the following spring. Growth in early summer is rapid and the grasses produce narrow, inward-curving blades to limit transpiration. By the end of summer their blue-green colour may have turned more parched. Herbaceous plants are found and some trees, e.g. willow, grow along water courses. In response to the windy climate many prairie and steppe farms are protected by trees planted as windbreaks.

The decay of the grasses in summer causes a rapid accumulation of humus in the soil, making the area ideal for cereals or, in drier areas, for cattle ranching (Figure 12.35). The

◀ Figure 12.31, *below left*
Climate graph for temperate grasslands (Saskatoon, USA)

▼ Figure 12.32, *below*
The soil moisture balance for temperate grasslands

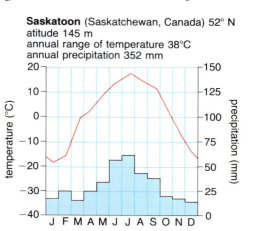

Saskatoon (Saskatchewan, Canada) 52° N
altitude 145 m
annual range of temperature 38°C
annual precipitation 352 mm

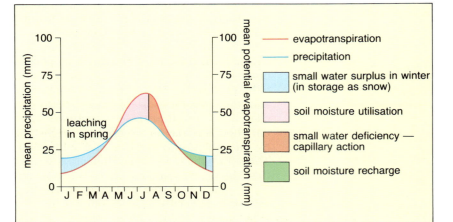

▶ Figure 12.33
Tufted grasses on the North American Prairies, Wyoming — the Oregon Trail

temperate grasslands form a resilient ecosystem. The grasses provide food for burrowing animals such as rabbits and gophers, and large herbivores such as antelopes, bison and kangaroos. These in turn, may be consumed by carnivores (e.g. wolves and coyotes) or by predatory birds (e.g. hawks and eagles).

Chernozems or black earths

The thick grass cover supplies plentiful mull humus which forms a black, crumbly topsoil. While the abundance of biota, especially earthworms, causes the rapid decay of organic matter during the warm summer, decomposition is arrested in any drier spells and during the long cold winter. Humus is therefore retained near the surface. There is effective recycling as the grasses take up and return many minerals from the soil. The late spring snowmelt and early summer storms cause some leaching (Figure 12.35), but bases such as potassium and nitrogen are moved downwards only slowly. In later summer this is compensated by the upward movement of capillary water bringing bases nearer the surface and maintaining a neutral or slightly alkaline soil (pH 7 to 7.5). The grasses have an extensive root system which gives a deep (up to 1 m) dark brown to black A horizon.

The alternating dry and wet seasons immobilise iron and aluminium sesquioxides and clay within aggregates (peds) in the upper horizon and this, together with the large number of mixing agents, limits the formation of a recognisable B horizon. The subsoil, often of loess origin (page 107), is usually porous and this, with the capillary moisture movement in summer, means that it usually remains dry. This upward movement of moisture causes calcium carbonate to be deposited, often in the form of nodules, in the upper C horizon.

▼ Figure 12.34
Land uses of the temperate grassland, a changed biome

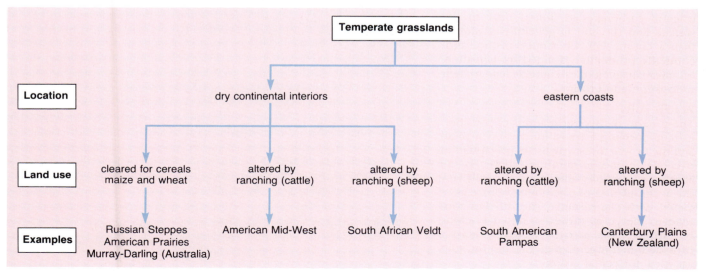

		Temperate grasslands			
Location		dry continental interiors		eastern coasts	
Land use	cleared for cereals maize and wheat	altered by ranching (cattle)	altered by ranching (sheep)	altered by ranching (cattle)	altered by ranching (sheep)
Examples	Russian Steppes American Prairies Murray-Darling (Australia)	American Mid-West	South African Veldt	South American Pampas	Canterbury Plains (New Zealand)

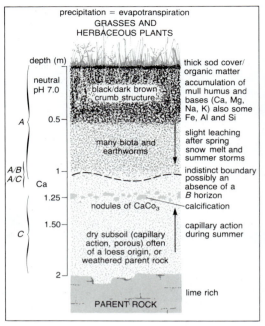

precipitation = evapotranspiration
GRASSES AND
HERBACEOUS PLANTS

depth (m)

thick sod cover/
organic matter

neutral
pH 7.0

black/dark brown
crumb structure

accumulation of
mull humus and
bases (Ca, Mg,
Na, K) also some
Fe, Al and Si

A   0.5

many biota and
earthworms

slight leaching
after spring
snow melt and
summer storms

A/B
A/C   1
Ca

indistinct boundary
possibly an
absence of a
B horizon

1.25

nodules of CaCo₃

calcification

C   1.50

dry subsoil (capillary
action, porous) often
of a loess origin, or
weathered parent rock

capillary action
during summer

2

lime rich

PARENT ROCK

◄ Figure 12.35, *far left*

Characteristic profile of a
chernozem (black earth)

◄ Figure 12.36
A chernozem soil

Chernozems are regarded as the optimum soil for agriculture as they are deep, rich in organic matter, retain moisture, and have an ideal crumb structure and well-formed peds. After intensive ploughing chernozems may require the addition of potassium and nitrates.

Chestnut soils

These are found in juxtaposition with the chernozems but where the climate is drier so evapotranspiration slightly exceeds precipitation and the resultant vegetation is sparser and more xerophytic. As the root system is less dense both the amount and depth of organic matter decrease and the colour becomes a lighter brown than chernozems. Chestnut soils are more alkaline because of increased capillary action and suffer from more frequent summer droughts (e.g. North America, 1988). Deposits of calcium carbonate are found near to the surface and the soil is generally shallower than a chernozem. Chestnut soils are agriculturally productive if aided by irrigation, but mismanagement can quickly lead to their exhaustion and erosion.

Prairie soils

These lie to the wetter margins of the chernozems forming a transition between them and the brown forest earths. As precipitation exceeds evapotranspiration there is an absence of capillary action and the soils lack the accumulation of calcium carbonate associated with chernozems. The A/B horizons tend to merge as there is limited leaching but strong biota activity. Decaying grasses provide much organic material and the soils are ideal for cereal crops.

6   Cool temperate western margins

Sometimes referred to as the northwest European, this climate is experienced on west coasts between approximately 40° and 60° north and south of the Equator. Other areas with similar climatic characteristics are the NW USA and British Columbia, southern Chile, Tasmania and New Zealand's South Island (Figure 11.29).

◄ Figure 12.37
A chestnut soil near Volgograd,
USSR

Climate

Summers are cool (Figure 12.38) with the warmest month between 15°C and 17°C. This is a result of the low angle of the sun in the sky combined with frequent cloud cover and the cooling influence of the sea. Winters,

▶ Figure 12.38

Climate graph for cool temperate western margins (Shannon, Eire)

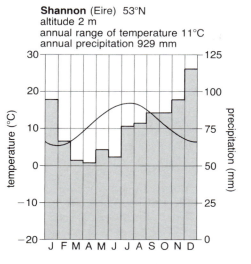

Shannon (Eire) 53°N
altitude 2 m
annual range of temperature 11°C
annual precipitation 929 mm

▼ Figure 12.39

Broad-leaved, deciduous woodland in Surrey, England

in comparison, are mild. Mean monthly temperatures remain a few degrees above freezing due to the warming effect of the sea, the presence of warm offshore ocean currents and the insulating cloud cover. Diurnal ranges are extremely low, autumns are warmer than springs, and seasonal temperature variations depend on prevailing air masses (Fig. 9.39).

This climatic zone lies at the confluence of the Ferrel and Polar cells (Figures 9.33, 9.34) where tropical and polar air converge at the Polar Front. Warmer tropical air is forced to rise, creating an area of low pressure and forming depressions with their associated fronts. The prevailing south-westerlies, laden with vapour after crossing warm offshore currents, give heavy orographic and frontal rain. Precipitation, often exceeding 2000 mm annually, falls throughout the year but with a winter maximum when depressions are more frequent and intense. Although snow is common in the mountains, it rarely lies for long at sea level. Fog, most common in autumn, forms under anticyclonic conditions (page 183).

Deciduous forests

Although having the second highest NPP of all biomes (1200 g/m^2/yr) deciduous forest falls well short of the figure for tropical rainforests, mainly because of the dormant winter season when the deciduous trees shed their leaves. Leaf fall has the effect of reducing transpiration when colder weather reduces the effectiveness of photosynthesis and when roots find it harder to take up water and nutrients.

The dominant tree in Britain's natural ecosystem, the oak, grows to heights of 30 to 40 m, while others such as the elm (common before its population was diminished by Dutch elm disease), ash, beech and chestnut grow a little less high. They all produce large crowns and have broad but thin leaves. Unlike the rainforests there are relatively few species. The maximum number per km^2 in southern Britain is eight and on occasions some woodlands may only have a single dominant. The trees have between six to eight months in which to bud, leaf, flower and fruit, and may only grow by about 50 cm a year. Most woodlands show some stratification (Figure 11.2) with a lower shrub layer varying between 5 m (holly, hazel and hawthorne) and 20 m (ash and birch). This layer can be quite dense as the open mosaic of branches of the taller trees allows more light to penetrate than in the rainforests. The forest floor is often covered in a thick undergrowth of brambles, grass, bracken and ferns. Flowering plants, such as bluebells, bloom early in the year before the taller trees have developed their full foliage. Epiphytes, which include mosses, lichen and algae, often grow on tree trunks.

The forest floor has a thick leaf litter which is readily broken down by the numerous mixing agents living in the relatively warm soil. There is a rapid recycling of nutrients although some are lost through leaching. The leaching of humus and nutrients and the mixing by biota produces a brown-coloured soil. Geology contributes to determine the dominant tree with oaks and elms preferring loams, beech the drier chalk, ash the drier gravels and limestone, and willows and alder wetter soils. There is a well-developed food chain in these forests with many autotrophs, herbivores (e.g. rabbits, deer, mice and hedgehogs) and carnivores (e.g. foxes).

Most of Britain's natural deciduous woodland has been cleared for farming, for use as fuel and for building and for urban development. Northwards, deciduous trees give way to coniferous towards polar latitudes and with increasing altitude and steepness of slope.

Brown forest earths

The considerable leaf litter which accumulates in autumn decomposes relatively quickly in the presence of organisms and the less acidic mull humus. Organic matter is incorporated into the A horizon, by the action of earthworms, giving it a dark brown colour. Precipitation exceeds evapotranspiration sufficiently to cause leaching but not enough for podsolisation (Figure 10.18). Bases, especially calcium and magnesium, are absent in the upper horizons and there is some loss of clay and sesquioxides.

The horizons merge more gradually than in the podsol, assisted by increased biota activity. The colour becomes increasingly reddish-brown with depth due to the re-deposition of iron and aluminium. There is no hard pan so the brown earths tend to be free draining. There is considerable recycling as the deciduous trees take up many nutrients from the soil only to return them later through fallen leaves. The soil is deeper and more productive than the podsol and tree roots may penetrate and break up the bedrock. Due to the relatively high clay content throughout the profile most areas of brown earth are potentially fertile though they may benefit from liming.

7 Cold climates

Areas of cold climates, having over six months with a mean temperature of below 6°C, are found in the sub-arctic regions, poleward of 60°N, of Eurasia and North America. They also occur at higher altitudes in more temperate latitudes and in southern Chile.

Climate

Winters in this biome are long and cold. Minimum mean monthly temperatures may be as low as −30°C (−28°C at Dawson, Figure 12.42). There is little moderating influence from the sea and insolation is nil because north of the Arctic Circle there is a period each year when the sun never rises. The wind-chill factor is high with strong winds evaporating moisture, freezing the skin and causing frostbite. Summers are short

◄ Figure 12.40, *below left*
A brown earth in the UK

▼ Figure 12.41
Characteristic profile of a brown earth

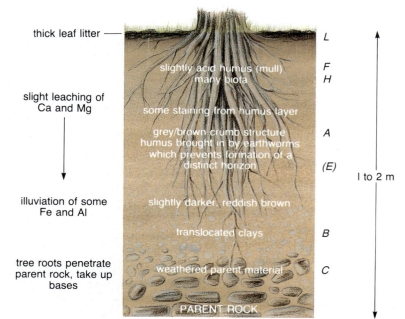

DECIDUOUS TREE WITH UNDERGROWTH

thick leaf litter —

slight leaching of Ca and Mg

illuviation of some Fe and Al

tree roots penetrate parent rock, take up bases

slightly acid humus (mull) many biota

some staining from humus layer

grey/brown crumb structure humus brought in by earthworms which prevents formation of a distinct horizon

slightly darker, reddish brown

translocated clays

weathered parent material

PARENT ROCK

L

F
H

A

(E)

B

C

l to 2 m

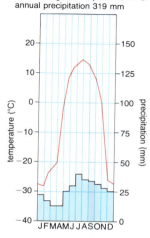

Dawson (Yukon Territory, Canada) 64°N
altitude 324 m
annual range of temperature 44°C
annual precipitation 319 mm

▲ **Figure 12.42**
Graph for a cold climate (Dawson, Yukon Territory, Canada)

▼ **Figure 12.43**
Coniferous forest from the air — notice the clear boundary of the tree line (Mt Adams, Gifford Pinchot National Park, Washington State, USA)

► **Figure 12.44**, *below right*
Forest floor in a coniferous forest, Cumbria, England

but the long hours of daylight and clear skies mean they are relatively warm. Precipitation is light as the cold air can hold only limited moisture and the small amount of winter snowfall is frequently blown about in blizzards. Isolated convectional rainstorms give these areas a summer precipitation maximum.

Coniferous forest or taiga

Coniferous trees have developed distinctive adaptation which enables them to tolerate long, cold winters, cool summers with a short growing season, limited precipitation and podsolic soils. The size of the dominant trees and the fact that they are evergreen, which gives them the potential for year-round photosynthesis, results in the relatively high NPP of 800 $g/m^2/yr$. The trees, which are softwoods, rarely number more than two or three species per km^2. Often there may be extensive stands of a single species, e.g. spruce, fir or pine. In colder areas, such as Siberia, the larch tends to dominate. Although it is cone-bearing, larch is deciduous and sheds its leaves in winter. All trees in the taiga, some of which attain a height of 40 m, are adapted to living in a harsh environment (Figure 12.43).

Most tree species are evergreen and so can photosynthesise as soon as conditions become suitable, without having to wait for new growth. Conditions become favourable in spring as incoming radiation increases and water becomes available through snowmelt — days in winter are long and dark and soil moisture is frozen. The needle-like leaves are small and the thick cuticles help to reduce transpiration during times of strong winds and during the winter when moisture is in a form unavailable for absorption by tree roots. Cones protect the seeds and thick resinous

bark protects the trunk from the extreme cold of winter. The conical shape of the tree and its downward sloping, springy branches allow the winter snows to slide off without breaking the branches. The conical shape also gives some stability against the strong winds, since the tree roots are usually shallow.

There is usually only one layer of vegetation in the coniferous forest. The amount of ground cover is limited partly by the lack of sunlight reaching the forest floor and partly by the deep, acidic layer of non-decomposed needles. Plants which can survive on the forest floor include mosses, lichens, and wood sorrel (Figure 12.44).

The cold climate and soils discourage earthworms and bacteria. Needles therefore decompose very slowly — less rapidly than they accumulate — to give an acid mor humus (page 216). Most of the nutrients are held in the litter (Figure 11.25). Although precipitation is limited, evapotranspiration rates are also low with the result that leaching occurs and the few nutrients which are returned to the podsol soil are soon lost. Conifers require few nutrients, taking only 225 kg of plant nutrient annually from each hectare compared with 430 kg by deciduous trees. The limited food supply means that animal life is not abundant. The dark woods are not favoured by bird life, although deer, wolves, brown bears and beavers are found in certain areas.

In North America and Eurasia the coniferous forest merges into the tundra on its northern fringes (Figure 12.45). The **tree line**, the point beyond which trees are unable to grow, corresponds closely with locations where the mean monthly temperature fails to reach 10°C for at least one

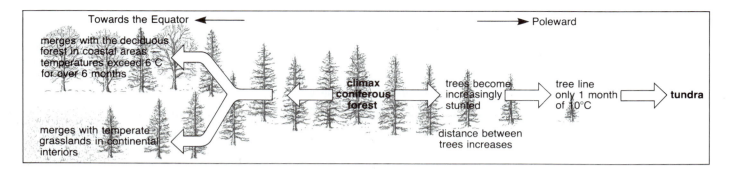

▲ Figure 12.45

The coniferous forest and its

transition zones

month in the year. (Inability to support tree growth can also be due to thin soils in mountainous areas.) South of the taiga lie either the deciduous forest or the temperate grassland ecosystems (Figure 11.29), depending on whether the location is coastal or inland.

Podsols

Podsols develop in areas where precipitation exceeds evapotranspiration (Figure 3.4a), where coniferous forest or heathland provide the vegetation cover and where soils are sandy. Podsols usually, but not exclusively, form in cool climates. Podsolisation is a process which also operates within tropical ferruginous soils (page 268).

Pine needles, with their thick cuticles, provide only a thin leaf litter and inhibit the formation of humus. The humus formed is very acid (mor) and provides chelating agents and humic acid which help to make the iron, aluminium and silica minerals more soluble. The cold climate discourages organisms and the soil is too acidic for earthworms. Consequently the decomposition of leaf litter is slow and definite horizons develop as there are few mixing agents. The downward percolation of water through the soil, especially following snowmelt, causes the leaching of bases, the translocation of clays and organic matter and the eluviation of the sesquioxides of clay and aluminium. This leaves a narrow ash-grey bleached A horizon (podsol is Russian for 'ash-like') composed mainly of quartz sand and silica (Figure 12.47). There is some dispute among pedologists as to whether the translocated materials are moved by physical, chemical or biological processes or a combination of all three.

The dark coloured organic matter is redeposited at the top of the B horizon. Beneath this the sesquioxides of, first, iron and then aluminium are deposited as a rust-coloured hard pan. Where this hard pan, which often has a convoluted shape, becomes marked, it acts as an impermeable layer restricting the downward movement of moisture and the penetration of plant roots. This can cause some waterlogging in the E horizon to give a 'gleyed podsol'. The lower B horizon, composed mainly of redeposited clays, has an orange-brown colour and overlies weathered parent material. Any throughflow from this horizon is likely to contain bases in solution. Although they are not naturally fertile soils, podsols can be improved by the addition of lime and fertiliser.

Q Several readings were made in a soil pit within the coniferous Delamere Forest in Cheshire. Some of these results have been given in Figure 12.48.

Explain how these readings suggest that the profile is that of a podsol.

---

8 Arctic climates

Areas with continuous permafrost (Chapter 5) include the extreme north of Alaska, Canada and the USSR as well as all of Greenland and Antarctica. Summers may have continuous periods of daylight but monthly temperatures struggle to rise above freezing point (Figure 12.46) while nearer the poles the climate is perpetual frost. Although winters are severe and the sea freezes, the water still has a moderating effect on temperatures which do not fall as low as more inland areas with a cold climate. Precipitation is light: Barrow (Figure 12.46) could classify as a desert if temperatures were warm enough for plant growth.

▼ Figure 12.46

Climate graph for an Arctic

climate (Barrow, Alaska, USA)

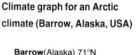

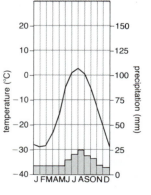

Barrow (Alaska) 71°N
altitude 7 m
annual range of temperature 32°C
annual precipitation 110 mm

▶ Figure 12.47

A podsol (compare Figure 10.18)

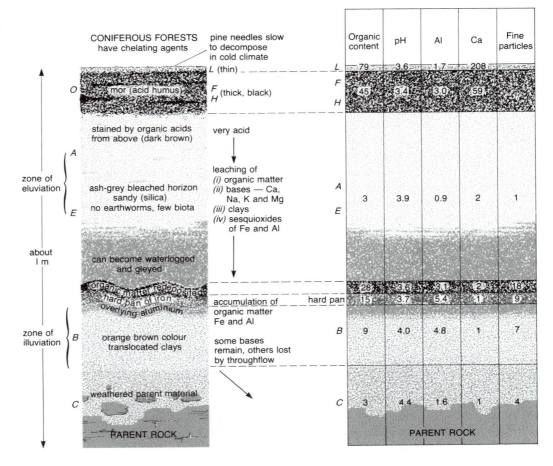

▶ Figure 12.48, *far right*

Readings taken in a soil pit, Delamere Forest, Cheshire England

Tundra

The tundra ecosystem is one with very low organic productivity. The NPP of only 140 g/m²/yr is the second lowest of the major land biomes. Lying poleward of the taiga the tundra has long, severe winters, a short growing season and limited precipitation which falls mainly as snow. The ground, apart from the top 50 cm in summer, is permanently frozen.

In Finnish, tundra means a 'barren or treeless land', which accurately describes its winter appearance, and in Russian it means a 'marshy plain', which it is in summer. Any vegetation must have a high degree of tolerance to extreme cold and to moisture-deficient conditions — the latter because water is unavailable for most of the year when it is stored as ice or snow. There are fewer species of plants in the tundra than in any other biome. All are very low-growing, compact and rounded to gain protection against the wind (plants as well as humans are affected by wind-chill), and most complete their life cycles within 50 to 60 days. There is no stratification by height.

The five main dominants, each with its specialised local habitat are lichen, mosses, grasses, cushion plants and low shrubs (Figure 12.49). Most have small leaves to limit transpiration and short roots to avoid the permafrost. Lichen are pioneer plants in areas where the ice is retreating and they can help date the chronology of an area following deglaciation (page 242). Much of the tundra is waterlogged in summer (Figure 12.51) because the impermeable permafrost prevents infiltration. Relief is gentle and the evaporation rates are low. In such areas mosses, cotton grass and sedges thrive. On south-facing slopes and in better drained soils, cushion plants provide a mass of colour in summer (Figure 12.50). These 'bloom mats' include arctic poppies, anemones, pink saxifrages and gentians — none of which grows higher than 10 cm. Where decaying vegetation accumulates (there is little bacterial action to decompose dead plants) the resultant peat is likely to be covered in heather whereas on drier gravels berried plants (e.g. bilberry and crowberry) are the dominants. Adjacent to the seasonal snow-melt rivers, dwarf willows and stunted birch grow but only to a maximum of around 30 cm and even so their crowns are often

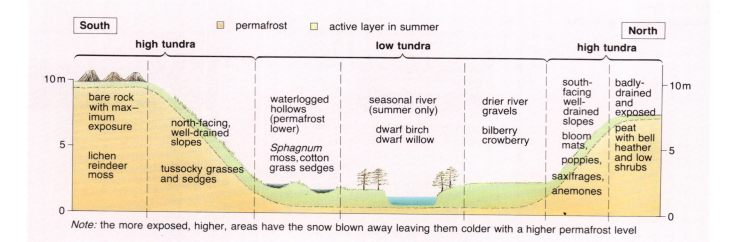

| South | permafrost | active layer in summer | | North |

high tundra — **low tundra** — **high tundra**

- bare rock with max-imum exposure
- lichen reindeer moss
- north-facing, well-drained slopes
- tussocky grasses and sedges
- waterlogged hollows (permafrost lower)
- *Sphagnum* moss, cotton grass sedges
- seasonal river (summer only)
- dwarf birch dwarf willow
- drier river gravels
- bilberry crowberry
- south-facing well-drained slopes
- bloom mats, poppies, saxifrages, anemones
- badly-drained and exposed
- peat with bell heather and low shrubs

Note: the more exposed, higher, areas have the snow blown away leaving them colder with a higher permafrost level

distorted and mishapen by the wind. In winter the whole biome is covered in snow although this acts as an insulator for the plants.

The lack of nitrogen-fixing plants limits fertility and the cold, wet conditions inhibit the breakdown of plant material. Photo-synthesis is hindered by the lack of sunlight and water for most of the year, though the presence of autotrophs, such as lichen and mosses, does provide the basis for a food-chain longer than might be expected. Herbivores such as reindeer, caribou and musk-ox survive because plants like reindeer moss have a high sugar content. However, these animals have to migrate in winter to find pasture not covered by snow. The major carnivores are wolves, arctic fox and owls.

The tundra is an extremely fragile eco-system in a delicate balance (page 250): once disturbed by human activity, such as oil exploration and extraction or tourism, it may take many years before it becomes re-established.

▲ Figure 12.49
The importance of site factors on vegetation in the tundra

◄ Figure 12.50
'Bloom mats' in the tundra of NW Canada

◄ Figure 12.51
Waterlogged tundra in the summer season

▶ Figure 12.52

Profile typical of tundra soils

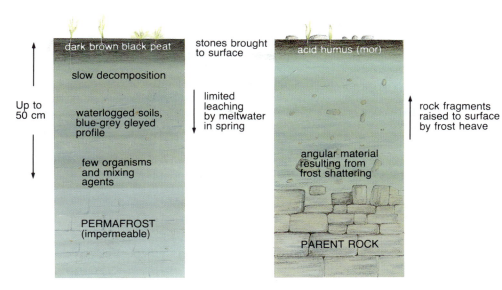

Tundra soils

The limited plant growth of this biome only produces a small amount of litter and as there are few soil biota in the cold soil, organic matter decomposes very slowly to give a thin layer of peat — a very acidic humus or mor. When water percolates downwards, usually as meltwater in late spring, the humic acid within it (pH less than 4.5) releases iron. Underlying the soil, often at a depth of less than 50 cm, is permafrost. This, acting as an impermeable layer, severely restricts moisture percolation and causes extreme waterlogging and gleying. Few mixing agents can survive in the cold, wet tundra soils which are thin and have no developed horizons. Where bedrock is near to the surface the parent material is physically weathered during times of freeze–thaw action. The shattered angular fragments are raised to the surface by frost heave, preventing the formation of horizons and creating a range of periglacial landforms (Figure 5.13).

Q

1 **a** Describe the composition and structure of one of the following:
        boreal forest   tropical rainforest   temperate deciduous forest.
   b Show how the characteristics of the vegetation of your chosen area have enabled it to become dominant.

2 For any one biome describe and explain the relationship between climate, soils and vegetation.

3 Draw a fully labelled diagram to show the vertical structure and composition of **a** a mature deciduous woodland in lowland Britain and **b** a tropical rainforest.

4 Describe and account for the differences in net primary production (NPP) between the major biomes.

5 **a** Discuss the nature of the vegetation in the tundra, hot desert, Mediterranean, equatorial, and tropical continental environments in relation to their soil-moisture budgets and any other climatic factors.
   b Describe the distribution of vegetation types within an area of Mediterranean climate and show how this distribution is related to variations in relief, climate, soils and human activities.

6 Describe the characteristics of the vegetation of the tropical grasslands and the Mediterranean lands. Discuss the extent to which these characteristics are a response to seasonal variations in climate.

7 Account for the relationships between the vegetation communities of the tundra and both soil and other site factors.

8   a   Explain what is meant by the term 'climatic climax vegetation'.
    b   Why are there relatively few areas in the world today with a natural climatic climax vegetation?

9   a   How has human development been responsible for the degradation of climatic climax vegetation communities and soils in the tropical rainforests?
    b   Discuss the view that many tropical grasslands and temperate grasslands are not composed of climatic climax communities.
    c   Explain why 'secondary vegetation' commonly results in areas which have been seriously affected by human activity.

10 Explain what is meant by the following terms:

ephemerals	epiphytes
xerophytic	dominants
halophytes	food chain
net primary production (NPP)	photosynthesis
stratification	sclerophyllous
biomes	

11 How does an understanding of soil-moisture budgets help to explain the characteristics of the world's major zonal soils?

References

A Dictionary of the Natural Environment, F.J. Monkhouse and J. Small, E.J. Arnold, 1978.

Biogeographical Processes, Ian Simmons, Allen & Unwin, 1982.

Biology: a Functional Approach, Michael Roberts, Thomas Nelson, 1986 (Fourth edition).

Climate, Soils and Vegetation, D.C. Money, U.T.P., 1965.

Earth, the Living Planet, Michael Bradshaw, Hodder & Stoughton, 1977.

Ecology, T.J. King, Thomas Nelson, 1989 (Second edition).

Science in Geography, Books 1–4, Oxford University Press, 1974.

Soils, Vegetation and Ecosystems, Greg O'Hare, Oliver & Boyd, 1987.

Statistics in Geography for 'A' level Students, John Wilson, Schofield & Sims, 1984.

The Nature of the Environment, Andrew Goudie, Blackwell, 1984.

The Unquiet Landscape, D. Brunsden and J. Doornkamp, David & Charles, 1977.

World Vegetation, D. Riley and A. Young, Cambridge University Press, 1966.

CHAPTER 13

Population

'There is a real danger that in the year 2000 a large part of the world's population will still be living in poverty. The world may become overpopulated and will certainly be overcrowded.'

North—South: a Programme for Survival, Willy Brandt, Pan 1980

Distribution describes the way in which people are spread out across the earth's surface.

Density describes the number of people living in a given area.

In the study of human population it is important to remember that the situation is dynamic, not static. The number of people constantly changes in time, in space and at micro-meso-and macro-scales in the population system.

Distribution and density

The distribution of population over the world's surface is uneven and there are considerable variations in density.

One of the best means of illustrating distributions is using a dot map, where each point or symbol represents a given number of people. However, this method creates the problem of having to select the value and size of the dot.

In Figure 13.1, this method effectively shows the concentration of people in the Nile Valley in Egypt, where 99 per cent of the country's population live on 4 per cent of the total land area, but it does suggest that other areas are uninhabited because they have insufficient numbers to warrant a symbol.

Densities are usually shown by a choropleth map. Although such maps are easy to read, once a class interval has been chosen they tend to hide concentrations. Figure 13.2 gives the impression that the number of people living in each square kilometre is equally distributed across the country. It also suggests that there is an abrupt change in the population density on crossing national boundaries, because the total population of Egypt has been divided by its total area.

▼ Figure 13.1
World distribution of population

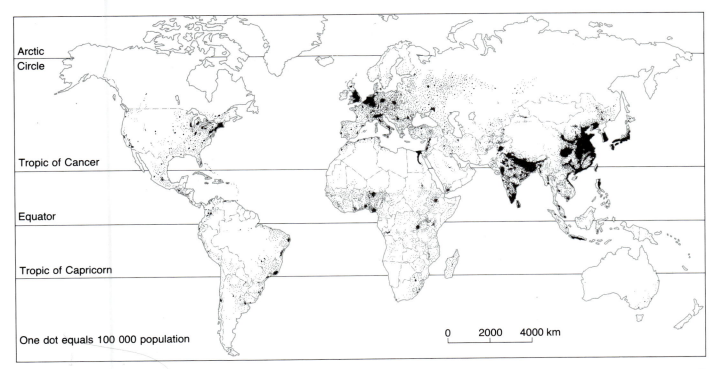

Arctic Circle

Tropic of Cancer

Equator

Tropic of Capricorn

One dot equals 100 000 population

0   2000   4000 km

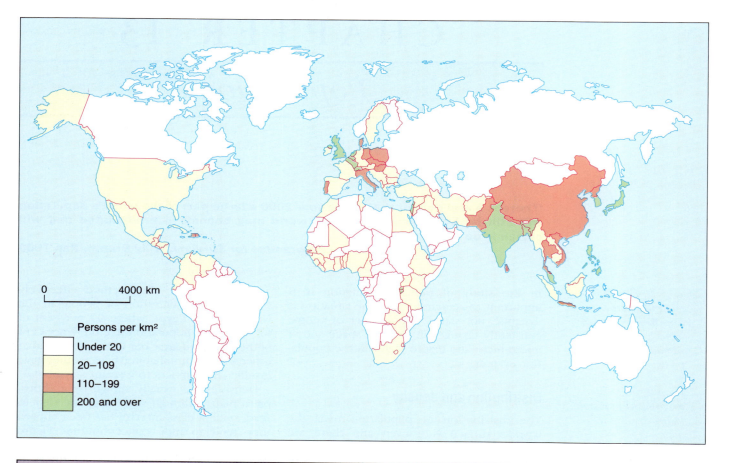

Persons per km²

- ⬜ Under 20
- ⬜ 20–109
- 🟧 110–199
- 🟩 200 and over

▲ Figure 13.2

World density of population

FRAMEWORK 7

Scale and generalisations

The study of any environment, whether natural or altered by human activity, involves numerous different and interacting processes. The relative importance of each process may vary according to the scale of the area under study: global or **macroscale**; intermediate or **mesoscale**; and local or **microscale**.

For example in a study of soils (Chapters 10 and 12) it is climate which tends to impose the greatest influence upon the formation and distribution of the major global types (e.g. the podsol and chernozem). At a smaller, regional level (e.g. Mediterranean), rock type may be the major influencing factor (e.g. terra rossa and rendzina). Within a small area, such as a river valley where climate and rock type are homogeneous, relief may be dominant, as seen in the catena (pages 214, 231).

A common problem with scale, as with models (Framework 8), is that a chosen level of detail may become either too large and generalised or too small and complex to be meaningful in addressing the problem. Population distributions and densities may be studied at a variety of scales. At a world scale (Figures 13.1 and 13.2) the pattern shown is so general and deterministic that it may lead the student into an over-simplified understanding of the processes which produced the apparent distribution and density. Indeed some of the generalised patterns do not stand up to careful scrutiny when studied at a more local scale or over a period of time.

Although it may be easier to identify and account for distributions, densities and anomalies at the national level, it is more difficult in the case of a country the size of Brazil (Figure 13.4) than it is for a smaller country such as Uruguay. Yet it is often only by looking at a smaller region (Figure 13.5) or an urban area (Figure 13.6) that the complexities of the different and multiple processes become most apparent.

▶ Figure 13.3

The uninhabitable Earth — how valuable are the world's soils for food production?

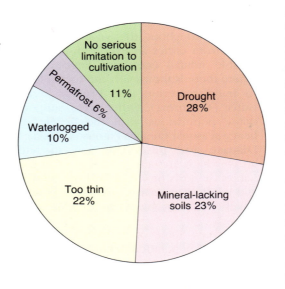

Figures 13.1 and 13.2 both show that there are parts of the world which are sparsely populated and others which are densely populated. One useful generalisation that may be made, remembering the pitfalls of generalisations (page 286), is that on a global scale this distribution is affected mainly by physical opportunities and constraints, whereas on a local or regional scale it is more likely to be influenced by economic, political and social factors.

Land accounts for about 30 per cent of the earth's surface (70.9 per cent is water). Of this, about 11 per cent presents no serious limitations to settlement and agriculture (Figure 13.3). The remainder is desert, snow and ice, high or steep-sided mountains, and tropical rainforests. Usually there is a combination of reasons why an area should be sparsely or densely populated.

Q

1 Read the section on sparsely populated areas. Certain types of environment have traditionally supported relatively low overall population densities. Is this still true today or have constraining factors been modified or overcome?

2 Read the section on densely populated areas and Framework 7.
   a Comment critically on the accuracy and value of the listed factors which, individually or collectively, have been offered as explanations why certain areas have become densely populated.
   b Why do geographers make generalisations like that in **a** above?

Sparsely populated areas

The following is a list of how various factors may lead to an area being sparsely populated. Compare these factors with the patterns evident in Figures 13.1 and 13.2.

- **Physical** Rugged mountains with active volcanoes and a decrease in temperatures and pressure with altitude usually hinder development (Andes): so do high plateaux (Tibet) and worn down shield lands (page 16, Canadian Shield).
- **Climate** Areas where climate limits population include those which receive little rainfall during the year (Sahara Desert); experience a long seasonal drought or unreliable, irregular amounts of rainfall (the Sahel countries); suffer from high humidity (Amazon Basin); or have very low temperatures throughout the year with an associated short growing season (northern Canada).
- **Vegetation** Both the coniferous forests of northern Eurasia and the rainforests of the tropics have relatively few permanant inhabitants.
- **Soils** An increasingly large area of the world has soils which are unsuitable for cultivation. In addition to the frozen soils of the Arctic (permafrost in Siberia), the thin soils of mountains (Nepal), and the leached soils of the tropical rainforests (Amazon Basin), other areas are experiencing severe soil erosion resulting from deforestation and overgrazing (the Sahel).
- **Water supplies** Many parts of the world lack a permanant supply of clean fresh water mainly because of insufficient, irregular rainfall or a lack of money and technology to build reservoirs and wells or to lay pipelines (Ethiopia).
- **Disease and pests** These may limit the areas in which people can live or may seriously curtail the lives of those who populate such areas (malaria in central Africa).
- **Resources** Areas devoid of minerals and easily obtainable sources of energy tend to attract neither people nor industry (Paraguay).
- **Communications** Areas where it is difficult to construct and to maintain transport systems tend to be sparsely populated, e.g. mountains (Bolivia), deserts (Sahara) and forests (Amazon).

■ **Economic** Areas with less developed, subsistence economies usually need large areas of land to support few people, though this is not applicable in SE Asia. Such areas tend to fall into three belts: tundra (the Lapps), desert fringes (the Rendille, page 400) and tropical rain-forests (shifting cultivators, page 401).

■ **Political** Areas where the state fails to invest money or to encourage develop-ment, economically or socially (interior of Brazil) are likely to support sparse populations.

Densely populated areas

This list shows how different attributes of the same environmental and human factors can affect population with opposite effects to those described in the list above.

■ **Physical** Flat, lowland plains are attrac-tive to settlement (Netherlands) as indeed are some areas surrounding volcanoes (Etna).

■ **Climate** Population is attracted to areas with evenly distributed rainfall through-out the year, with no temperature ex-tremes and a lengthy growing season (NW Europe); to those with plenty of sun (Costa del Sol) or snow (Alps) which attracts tourists; and to monsoon lands (SE Asia).

■ **Vegetation** Areas of grassland tend to have higher densities than dense forests or deserts.

■ **Soils** Deep, humus-filled soils (Paris Basin) and especially river-deposited silt (Ganges Delta) as both favour farming.

■ **Water supply** Population is more likely to expand in areas with a reliable water supply throughout the year. This may result from either an evenly distributed, reliable rainfall (northern England) or where there is the wealth and technology to build reservoirs and provide clean water (California). Other areas support dense populations with the benefit of heavy seasonal rainfalls (monsoon Asia).

■ **Disease and pests** This may include areas which initially had few of these natural hazards or which had the money and medical expertise to eradicate those which were a problem (Pontine Marshes, Rome).

■ **Resources** Some of the major concen-trations of population occur where there are, or were, large mineral deposits or energy supplies (the Ruhr) which in turn gave rise to large scale industry (Pitts-burgh region, USA).

■ **Communications** Areas where it is easier to construct canals, railways, roads and airports have attracted settlements (North European Plain), as have large natural ports out of which trade with other countries can take place (Singapore).

■ **Economic** Regions with intensive farm-ing or industry can support large numbers of people on a small area of land (Nether-lands).

■ **Political** Decisions may affect popula-tion distribution by creating new towns (Brasilia) or opening up new 'pioneer' lands for development (USSR).

C A S E   S T U D Y

National level: Brazil

Even a quick look at the population density map of Brazil (Figure 13.4) shows a relatively simple general pattern. Over 90 per cent of Brazilians live in a discontinuous 500 km strip adjacent to the east coast and widest in the south: this accounts for less than 25 per cent of the country's total area. Towards the north-west, the density declines very rapidly with several remote areas almost entirely lacking in permanant settlement.

The area marked *1A* on Figure 13.4 is the dry northeast or Sertao. Here the long and frequent water balance deficits (droughts), high temperatures and poor soils combine to make the area unsuitable for growing high yield crops or rearing good quality animals. The Sertao also lacks known mineral or energy reserves, communications are poor

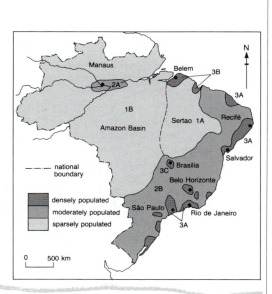

◀ Figure 13.4

Population density in Brazil — the national scale

and the basic services of health, education, clean water and electricity are lacking. Although birth rates are exceptionally high (many mothers have more than ten babies), there is a rapid outward migration to the urban areas (page 307), a high infant mortality rate and a short life expectancy (page 295).

Area *1B* is the tropical rainforest drained by the Amazon and its tributaries. Here the climate is hot, wet and humid, rivers flood annually, and there is a high incidence of disease. In the past the forest has proved difficult to clear and once the protective trees have gone, soils are rapidly leached and become infertile. Land communications are difficult to construct and maintain. Areas *1A* and *1B* have both suffered from a lack of federal investment and are able to support only subsistence economies.

An anomaly in this interior area is a zone along the Amazon centred on Manaus (*2A* on Figure 13.4). Originally a Portuguese trading post, Manaus has had two growth periods. The first was associated with the rubber boom at the turn of this century, and the second was in the 1980s with the development of tourism and its new status as a free.port (page 465). The more easterly parts of the Brazilian Plateau are moderately populated (area *2B*). The climate is cooler and considered healthier than on the coast and in the rainforests; the soil in parts is a rich terra rossa (page 229) which here is a weathered volcanic soil. Several precious minerals have been found here. However rainfall is irregular with a long winter drought, communications are still limited, and federal investment has been inadequate to stimulate much population growth.

Except where the highland reaches the sea, the eastern parts of the plateau around São Paulo and Belo Horizonte, and the east coast have the highest population densities. Although the coastal area is often hot and humid, the water supply is good. There are several natural harbours ideal for ports and encouraging to trade and the growth of industry. Salvador grew as the first capital at the centre of the slave trade. Rio became the second capital, developing as an economic, cultural and administrative centre. More recently it has received increasing numbers of tourists from overseas and migrants from the north of Brazil.

The fastest growing city in the world, other than Mexico City, is São Paulo. The cooler climate and terra rossa soils initially led to the growth of commercial farming based on coffee. Access to minerals such as iron ore and to energy supplies later made it a major industrial centre. All of this region has had high levels of federal investment, leading to the development of a good communications network and the provision of modern services.

Area *3B* is a recent growth pole (page 446) based on the discovery and exploitation of vast deposits of iron ore and bauxite, the construction of hydro-electric power stations and the advantages of access along the coastal strip and Amazon corridor. *3C* is the new federal capital, Brasilia, built to try to redress the imbalance in population density and wealth between the east of the country and the interior.

▼ **Figure 13.5**
Population density in the 'North' economic planning region of England — the regional scale

▶ **Figure 13.6,** *below right*
Population density in Greater London — the local (urban) scale

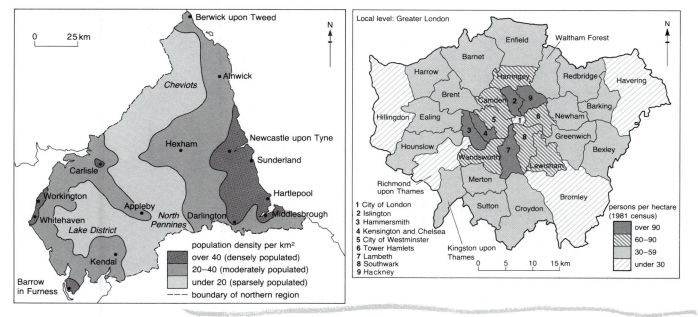

▶ Figure 13.8
World population growth

Q

1 With reference to Figure 13.5, describe and account for variations in population density in the Northern Economic Region of Britain. (You may find it helpful to consult physical, climatic and economic maps in an atlas in order to answer this question.)

2 With reference to Figure 13.6, describe and account for the differences in population density evident in Greater London.

Changes in time: population growth

Referring again to page 9 is a reminder that, of Mother Earth's 46 years, it was only 'in the middle of last week, in Africa, that some man-like apes turned into ape-like men' and the world's human population slowly began to grow. In the absence of any census, this population is estimated to have been about 500 million by 1650 (Figure 13.8). It was only after the Industrial Revolution in western Europe that numbers began to multiply prodigiously, with the most rapid increase having taken place over the last 50 years.

Birth rates, death rates and natural increase

The total population of an area is a balance between two forces of change. The **natural change** is the difference between birth rate and death rate. The **birth rate** is the number of live births per 1000 people per year and the **death rate** is the number of deaths per 1000 people per year. Throughout history, until the last few years in a handful of countries, birth rates have nearly always exceeded death rates. Exceptions are major outbreaks of disease, e.g. the Black Death, or wars. The resultant natural increase in population is usually expressed as a percentage rate. This relationship between birth and death rates is affected by a second force: **migration**. Although migration does not affect world population figures, it has considerable influence in areas where immigrants are more numerous than emigrants (population increase) or where emigrants are more numerous than immigrants (population decrease).

Population change can be shown as an example of a dynamic open system (Figure 13.7).

Q

1 a According to Figure 13.8, when did the world's population reach 1000 million?

b How long did it take this population to double to 2000 million and how long to double again to 4000 million?

c What, according to the United Nations, was the total estimated world population on 11 July 1987?

2 Figure 13.8 also shows the uneven distribution in population growth. What do you notice about the three continents whose populations are growing by more than 1.7 per cent per year, and the three continents which are growing by less than 1.7 per cent per year?

inputs	processes	outputs
births →	natural change	→ deaths
	total population	
immigrants →	migration	→ emigrants

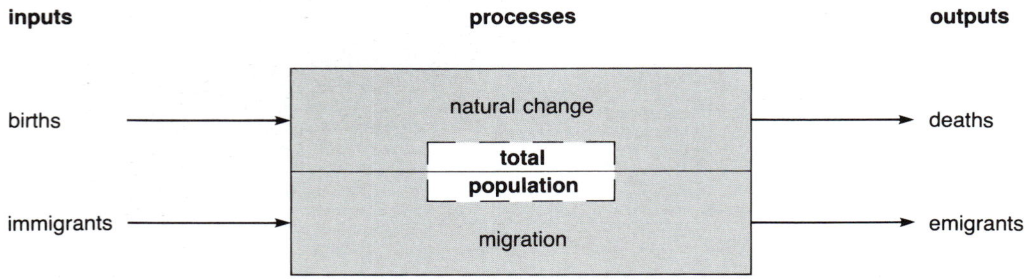

◀ Figure 13.7
Population change as an open system

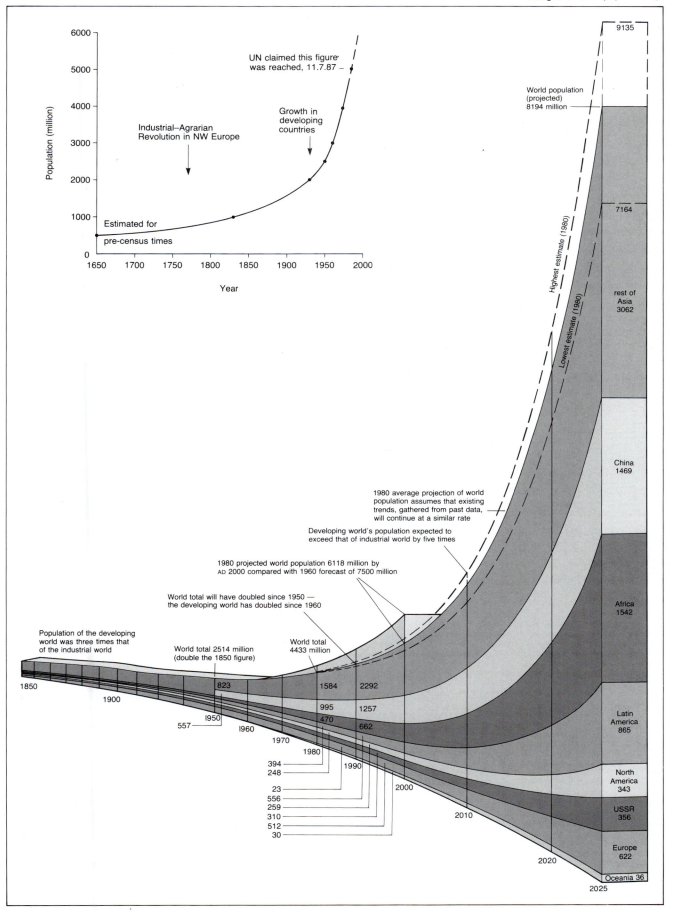

Population (million)

6000
5000
4000
3000
2000
1000
0

UN claimed this figure
was reached, 11.7.87 –

Industrial–Agrarian
Revolution in NW Europe

Growth in
developing
countries

Estimated for
pre-census times

1650 1700 1750 1800 1850 1900 1950 2000

Year

9135

World population
(projected)
8194 million

Highest estimate (1980)

Lowest estimate (1980)

7164

rest of
Asia
3062

China
1469

1980 average projection of world
population assumes that existing
trends, gathered from past data,
will continue at a similar rate

Developing world's population expected to
exceed that of industrial world by five times

1980 projected world population 6118 million by
AD 2000 compared with 1960 forecast of 7500 million

World total will have doubled since 1950 —
the developing world has doubled since 1960

Africa
1542

Population of the developing
world was three times that
of the industrial world

World total 2514 million
(double the 1850 figure)

World total
4433 million

Latin
America
865

1850

1900

823

1584

2292

North
America
343

557

1950

995

1257

USSR
356

1960

470

662

1970

Europe
622

1980

394
248

1990

23
556
259
310
512
30

2000

2010

Oceania 36

2020

2025

The demographic transition model

From the study of birth and death rates for several industrialised countries in western Europe and North America, a model has been developed suggesting that *all* countries pass through similar demographic transition stages or **population cycles**. Figure 13.9 shows the model and illustrates countries which fit into each stage for their inclusion as well as the reasons in that part of the cycle.

However, the model fails to take into consideration three factors:

1 Birth rates in several countries have recently fallen below death rates to give a population decline for the first time (e.g. West Germany). Perhaps this indicates a fifth stage in the model?

2 It was assumed that in time all countries, like those in western Europe and North America, would experience a fall in the death rate during stage two as this was believed to be a consequence of industrialisation. However, it now seems increasingly unlikely that many of the less developed African countries will ever become industrialised (Figure 17.29). Might this cause a modification of the model?

3 The timescale of the model, especially stages two and three, is being squashed where it applies to those countries which are currently considered to be the least developed, i.e. the IDCs (page 325).

▼ Figure 13.9
The demographic transition model

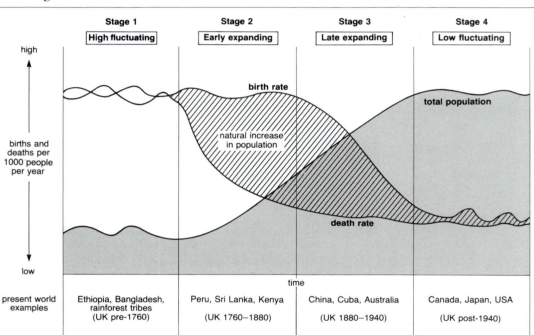

Stage 1 Here both birth rates and death rates fluctuate at a high level (about 35 per 1000) giving a small population growth.

Birth rates are high because:
• No birth control or family planning.
• So many children die in infancy that parents tend to produce more in the hope that several will live.
• Many children are needed to work on the land.
• Children are regarded as a sign of virility.
• Religous beliefs (e.g. Roman Catholics, Muslims and Hindus) encourage large families.

High death rates, especially among children, are due to:
• Disease and plague (bubonic, cholera, kwashiorkor).
• Famine, uncertain food supplies, poor diet.
• Poor hygiene —no piped, clean water and no sewage disposal.
• Little medical science —few doctors, hospitals. drugs.

Stage 2 Birth rates remain high, but death rates fall rapidly to about 20 per 1000 people giving a rapid population growth.

The fall in death rates results from:
• Improved medical care—vaccinations, hospitals, doctors, new drugs and scientific inventions.
• Improved sanitation and water supply.
• Improvements in food production, both quality and quantity.
• Improved transport to move food, doctors, etc.
• A decrease in child mortality.

Stage 3 Birth rates now fall rapidly, to perhaps 20 per 1000 people, while death rates continue to fall slightly (15 per 1000 people,) to give a slowly increasing population.

The fall in birth rate may be due to:
• Family planning — contraceptives, sterilisation, abortion and government incentives.
• A lower infant mortality rate meaning less pressure to have so many children.
• Increased industrialisation and mechanisation meaning fewer labourers are needed.
• Increased desire for material possessions (cars, holidays, bigger homes) and less for large families.
• An increased incentive for smaller families.
• Emancipation of women, enabling them to follow their own careers rather than being solely child bearers.

Stage 4 Both birth rates (16 per 1000) and death rates (12 per 1000) remain low, fluctuating slightly to give a steady population.

(Will there ever be a **Stage 5** where birth rates fall below death rates to give a declining population? Some evidence suggests that this might be occurring in several western European countries.)

FRAMEWORK 8

Models

Although models form an integral and accepted part of present day geographical thinking and teaching, claims asserting their value, made by Chorley and Haggett in the mid-1960s, initially provoked an outburst of opposition from traditional geographers.

Nature is highly complex. In an attempt to seek order in this complexity, geographers try to produce models which are a simplified representation of reality.

Chorley and Haggett described a model as:

'a simplified structuring of reality which presents supposedly significant features or relationships in a generalised form ... as such they are valuable in obscuring incidental detail and in allowing fundamental aspects of reality to appear.'

They stated that a model:

'can be a theory or a law, an hypothesis or structured idea, a rôle, a relationship, or equation, a synthesis of data, a word, a graph, or some other type of hardware arranged for experimental purposes.'

A good model will stand up to being tested in the real world and should fall between two extremes:

very simple and easy to work but too generalised to be of real value ← **model** → very difficult to use being almost as complex as reality

To achieve this balance (several though sometimes only one) critical criteria or variables are selected as a basis for the model. J.H. von Thünen (page 387) chose distance from a market as his critical variable and then tried to show the relationship between this variable and the intensity of land use. If necessary, other variables may be added which, as in the case of von Thünen's navigable river and a rival market, may add both a greater reality and complexity. Models can be used in all fields of geography. Some applications are shown in the table.

Physical (landforms)	Climate, soils and vegetation	Human and economic

beach profile
slope form
corrie development
geomorphological
  systems
  (Framework 1)
  e.g. drainage basin
glacier budget
	atmospheric circulation
	heat budget
	seres
	soil profiles (e.g. podsol)
	catena
	depression
	biome
		urban models (e.g. Burgess)
		von Thünen
		Weber
		Christaller
		gravity models
		Malthus
		demographic transition
		Rostow

Throughout this book models and theories are presented, their advantages and limitations are examined and their application to a real world situation is demonstrated, together with their usefulness in explaining that situation.

Use Figures 13.9 and 13.10 to answer the following questions.

1 Explain why Britain had a high birth rate and a high death rate between 1700 and 1760 (stage *A*).

2 Why did Britain have a rapidly falling death rate between 1760 and 1880 (stage *B*)? (Historians should be able to give dates of social legislation and health improvements.)

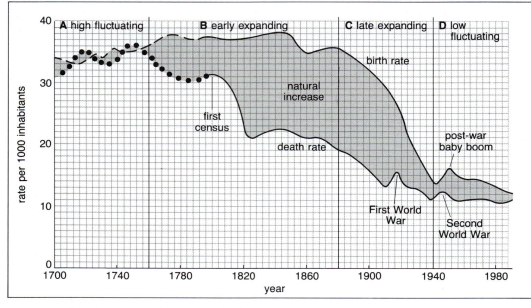

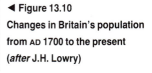

◄ Figure 13.10
Changes in Britain's population from AD 1700 to the present (*after* J.H. Lowry)

3 'Between 1880 and 1940 (stage *C*) Britain's birth rate declined rapidly while the death rate fell relatively slowly.' Suggest an explanation for this trend.

4 Explain why there has been a slightly fluctuating population growth with both birth and death rates remaining low since 1940.

5 Attempt to predict Britain's natural increase for the year 2000.

6 How closely do you consider that Britain's population growth can be compared with the demographic transition model?

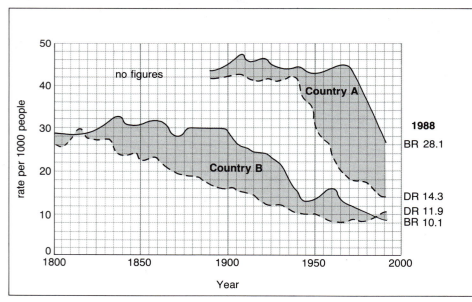

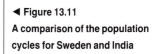

◄ Figure 13.11
A comparison of the population cycles for Sweden and India

Figure 13.11 shows the population cycle for Sweden and India.

1 Is country *A* Sweden or India?

2 Describe and give reasons for the differences between the two graphs.

3 For each country give the approximate year in which the natural increase was at its greatest.

4 What is likely to happen to the total population of each country in the next few years?

Population structure

The rate of natural increase or decrease resulting from the difference between the birth and death rates of a country represents only one aspect of the study of population. A second method is to look at the population structure of that country. This is important because the make-up of the population by its age and sex, together with its **life expectancy**, has implications for the future growth, economic development and social policy of that country. Differences in language, race, religion, family size, etc. will also affect a country's socio-economic welfare.

Population pyramids

The population structure of a country is best illustrated by a population or **age-sex pyramid**. The technique normally divides the population into five-year age groups, e.g. 5–9; 10–14, on the vertical scale, and into males and females on the horizontal scale. The number in each age group is given as a percentage of the total population and is shown by horizontal bars with males located to the left and females to the right. As well as showing past changes the pyramid can predict both short and long term future changes in population.

Whereas the demographic transition model shows only the natural increase or decrease resulting from the balance between births and deaths, the population pyramid shows the results of migration, the age and sex of migrants (page 310), and the effects of large scale wars and major epidemics of disease. Figure 13.12 is a partly completed pyramid for the UK. If you complete it, you should notice the following: a narrow pyramid indicating approximately equal numbers in each age group; a low birth rate (meaning fewer school places will be needed) and a low death rate (suggesting a need for more old people's homes) indicating a steady growth, or even a static population. There are more males among the younger age groups though the females live longer.

A model has also been produced to try to show the characteristics of four basic types of pyramid (Figure 13.13). As with most models, however, many countries show a transitional shape which does not fit precisely into any pattern. Figure 13.14 shows the pyramids for *selected* countries, chosen because they do conform closely to the model!

Stage 1 Mexico's pyramid has a concave shape showing that the birth rate is very high. There are many inhabitants (45 per cent) under 15 years old; there is a rapid fall upwards in each age group showing a high death rate; and a low life expectancy as only 3 per cent expect to live beyond 65.

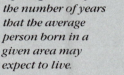

Life expectancy is the number of years that the average person born in a given area may expect to live.

Net migration is the balance between immigrants and emigrants.

▶ Figure 13.12

Constructing a population pyramid

▼ Figure 13.13

Model showing population pyramids related to different stages of the demographic transition model

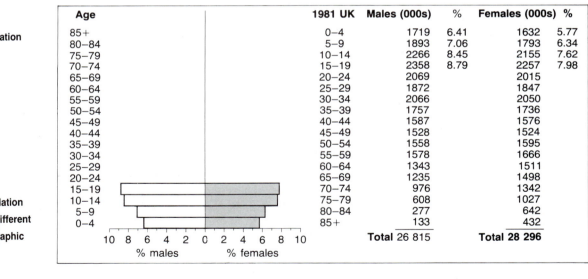

Age	1981 UK	Males (000s)	%	Females (000s)	%
85+	0–4	1719	6.41	1632	5.77
80–84	5–9	1893	7.06	1793	6.34
75–79	10–14	2266	8.45	2155	7.62
70–74	15–19	2358	8.79	2257	7.98
65–69	20–24	2069		2015	
60–64	25–29	1872		1847	
55–59	30–34	2066		2050	
50–54	35–39	1757		1736	
45–49	40–44	1587		1576	
40–44	45–49	1528		1524	
35–39	50–54	1558		1595	
30–34	55–59	1578		1666	
25–29	60–64	1343		1511	
20–24	65–69	1235		1498	
15–19	70–74	976		1342	
10–14	75–79	608		1027	
5–9	80–84	277		642	
0–4	85+	133		432	
		Total 26 815		**Total 28 296**	

10 8 6 4 2 0 2 4 6 8 10
% males   % females

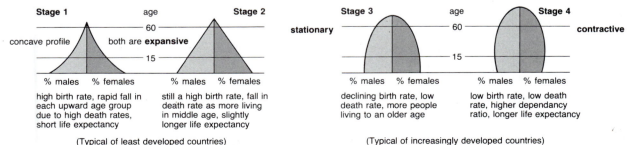

Stage 1	Stage 2	Stage 3	Stage 4
concave profile	both are **expansive**	stationary	contractive
high birth rate, rapid fall in each upward age group due to high death rates, short life expectancy	still a high birth rate, fall in death rate as more living in middle age, slightly longer life expectancy	declining birth rate, low death rate, more people living to an older age	low birth rate, low death rate, higher dependancy ratio, longer life expectancy
(Typical of least developed countries)		(Typical of increasingly developed countries)	

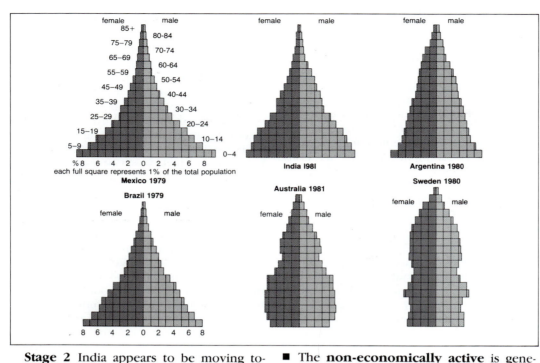

each full square represents 1% of the total population

Mexico 1979 · India 1981 · Argentina 1980 · Brazil 1979 · Australia 1981 · Sweden 1980

◀ Figure 13.14
Population pyramids for
selected countries

Stage 2 India appears to be moving towards stage 2 (shown by a less concave slope than that of Mexico) and Argentina and Brazil appear to have reached it (shown by the more uniform sides). All pyramids in this stage have a broad base indicating a high birth rate, but as the death rate declines then more people are living into middle age and life expectancy is slightly longer. The result is that although the actual numbers of children may be the same they are fewer as a percentage of the total population. This is shown by the narrower base. The large youthful population will soon enter the reproductive period and become economically active. India has 39 per cent under 15 and 3 per cent over 65. The corresponding figures for Brazil are 38 and 5; and Argentina 30 and 8.

Stage 3 Australia has probably reached this stage as its birth rate is declining (shown by the almost equal groups in the lower age groups), the death rate is much lower, allowing more people to live to an older age, and the actual growth rate has become stable: 26 per cent under 15 and 9 per cent over 65.

Stage 4 Sweden has a smaller proportion of its population in the pre-reproductive age groups (19 per cent under 15) and a larger proportion in the post-reproductive groups (16 per cent over 65) indicating a low birth and a low death rate. As fewer people enter the reproductive age groups there will be a fall in the total population.

Dependency ratios
The population of a country can be divided into two categories according to the contribution to economic productivity.

■ The **non-economically active** is generally said to include those under 15 and those over 65 years of age. (Perhaps in Britain it should be those under 16, as that is the school-leaving age, whereas in developing countries the cut-off point is much lower because many children try to earn money from a very young age.)
■ The **economically active** or working population aged 15 to 64 years.

The dependency ratio can be expressed as:

$$\frac{\text{children } (0-14) \text{ and elderly } (65 \text{ and over})}{\text{those of working age}} \times 100$$

e.g. UK 1971 (figures in millions)

$$\frac{13.387 + 10.512}{31.616} \times 100 = 75.59$$

So, for every 100 people of working age there were 75.59 people dependent on them.

By 1981 the dependency ratio had changed to:

$$\frac{11.455 + 11.023}{32.635} \times 100 = 68.87$$

So although the number of elderly people had increased this was more than offset by the larger drop in the number of children. (The dependency ratio does not take account of those who are unemployed.) The dependency ratio for most developed countries is between 50 and 70, whereas for Third World countries it is often over 100.

Future trends

The United Nations Fund for Population Activities (UNFPA) designated 11 July, 1987 as the date of the arrival of the five billionth human being on earth. Of course that 'celebration' was fictitious as nobody knows exactly how many people are living on the earth at a given moment: in many areas census figures are either inaccurate or non-existent. However, although that figure is approximate, it is certain that the world's population is still growing by 150 people a minute.

Recent evidence has shown that fertility in the Third World countries has begun to fall. The 1985 UNFPA estimate claimed that the annual growth rate of the world's population (2.1 per cent in 1965) had fallen to 1.6 per cent, mainly due to China's one child per family policy. It may continue to fall to under 1.5 per cent by the year AD 2000. This would mean that by the end of the century the world's population is only 6100 million instead of the 7600 million it would have reached had the growth rate of 1950–1980 continued.

What these figures fail to show is the marked variations between different areas in the world, especially between the developed and Third World continents. At present the growth rate for the developed countries averages 0.64 per cent per year compared with 2.02 per cent in those described as developing (Figures 13.15 and 13.16). To achieve population stability the average family of today would, worldwide, have to consist of 2.3 children. This is currently under two in western Europe and North America but more than four in Asia (excluding China and the USSR) and Latin America, and over six in Africa. These differences pose very different problems to countries in the developed and developing worlds.

Population problems

1 **The world's ageing population** Due to improvements in medical facilities, hygiene and vaccines, life expectancy has increased considerably. Whereas in 1970 there were 291 million people over 60 years of age (7.8 per cent of the world's total) and 26 million over 80 years of age (0.78 per cent of the total), these figures are expected to rise to 600 million (9.8 per cent) and 58 million (0.9 per cent) respectively by AD 2000. In the developed countries this is likely to mean a greater demand for services, which will have to be provided by a smaller percentage of people of working age. In Third World countries it means, initially, a rapid increase in population with an associated strain on their stretched resources.

2 **Zero growth rate** Throughout history the replacement rate has always been exceeded — hence the continuous expansion of world population. Recently, several European countries have been producing insufficient numbers of children to maintain numbers and have begun to show a population decline (Figure 13.15). The West Germans and Swiss are concerned that their population may decrease by several million by AD 2035 which could reduce competitive advantage in science and technology, unskilled labour and the security of future pensions. (Perhaps the problems of skilled labour shortages will in the short term be alleviated by a migration, e.g. the influx of East Germans in 1989.) On a continental basis it is estimated that zero growth will be the norm in Europe by 2010, North America by 2030, China by 2070, South East Asia and Latin America by 2090 and Africa by 2100.

3 **High birth rates** It appears that those countries which have the fewest resources (food, minerals, education) have the highest birth rates. A survey of younger mothers in Brazil indicates that those with secondary education have, on average 2.5 children, and those without, 6.5. However, in 1986 UNFPA claimed that 'a high birth rate was the consequence, not the cause, of poverty', and 'our theory that underdevelopment is caused by population growth is a catastrophe'. It seems that the resulting practice of sending shiploads of contraceptives to the Third World with no real commitment from the developed countries to discover and understand the living conditions, interests, and cultural framework of the recipients has turned out to be a failure. The paradox is that the more children who survive infancy, the smaller the number who need to be born for the population to maintain its growth rate, so birth rates are likely to fall.

4 **Government legislation and family planning** Several governments, especially in SE Asian countries, have attempted in recent years to encourage couples to have fewer children. In Sri Lanka, males who are sterilised after two or three children receive a bonus child allowance. In South Korea, couples with two children or less are given housing priority, while in the Philippines, paid maternity leave is stopped after four confinements. The case studies (page 300) summarise the degrees of success achieved by the Chinese and Singapore governments in trying to reach zero birth rate.

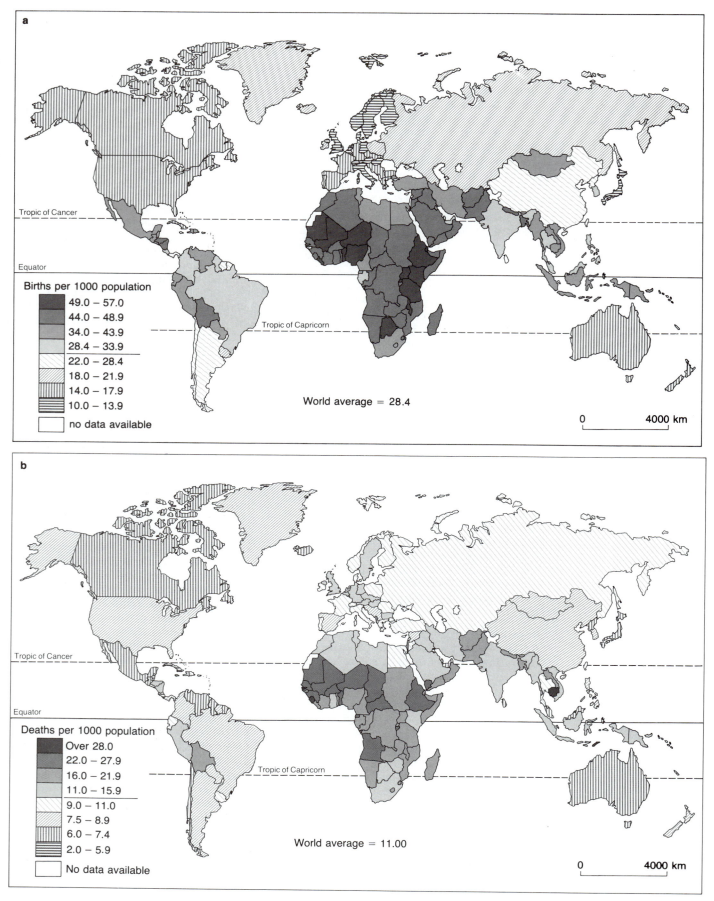

a

Tropic of Cancer

Equator

Births per 1000 population

- 49.0 – 57.0
- 44.0 – 48.9
- 34.0 – 43.9
- 28.4 – 33.9
- 22.0 – 28.4
- 18.0 – 21.9
- 14.0 – 17.9
- 10.0 – 13.9
- no data available

Tropic of Capricorn

World average = 28.4

0   4000 km

b

Tropic of Cancer

Equator

Deaths per 1000 population

- Over 28.0
- 22.0 – 27.9
- 16.0 – 21.9
- 11.0 – 15.9
- 9.0 – 11.0
- 7.5 – 8.9
- 6.0 – 7.4
- 2.0 – 5.9
- No data available

Tropic of Capricorn

World average = 11.00

0   4000 km

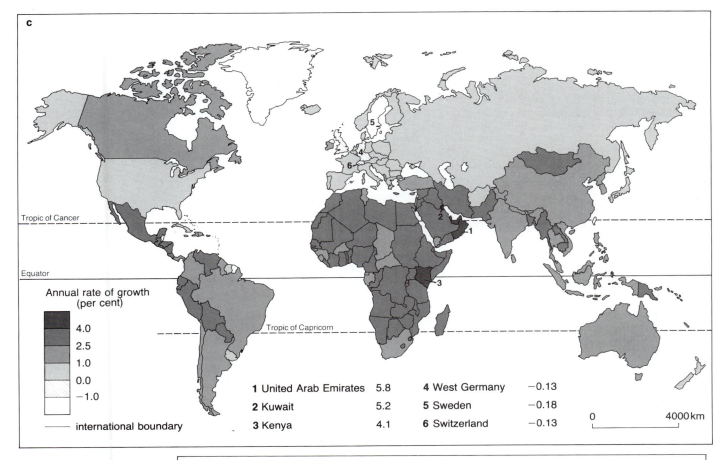

c

Annual rate of growth
(per cent)

- 4.0
- 2.5
- 1.0
- 0.0
- −1.0

—— international boundary

1 United Arab Emirates	5.8		**4** West Germany	−0.13	
2 Kuwait	5.2		**5** Sweden	−0.18	
3 Kenya	4.1		**6** Switzerland	−0.13	

0   4000 km

◄◄▲ **Figure 13.15**

a, *top left* World birth rates, 1980

b, *left* World death rates, 1980

c, *above* Population growth, 1980–1985

Growth rates	1985
Developed continents	0.64
North America	1.04
Europe and USSR	0.34
Oceania	1.57
Developing continents	2.07
Africa	3.08
Latin America	2.38
Asia	1.73
(China	1.17)
(Rest of SE Asia	2.30)

Life expectancy	1960	1970	1980
Developed continents	62	70	76
Third World continents	42	57	66

▶ **Figure 13.16**

a Table of growth rates

b Map of life expectancy in developed and Third World countries

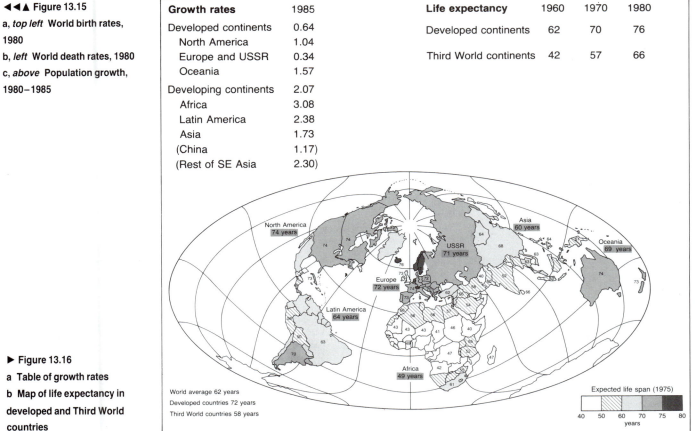

World average 62 years
Developed countries 72 years
Third World countries 58 years

Expected life span (1975)

40 50 60 70 75 80
years

CASE STUDY

China

Although by 1975 the average family size in China had fallen to three children, this was still regarded as being too many. Indeed, even had the family size been reduced to two, it would still have meant that China's 1000 million population would have doubled within 50 years. As a result, the government tried to encourage a 'one child only' policy. Inducements to have only one child included free education, priority housing, pension and family benefits — inducements lost after a second child was born. In addition the marriageable age for men was set at 22 and for women at 20, with couples having to apply to the state to be married and again to have a child.

The birth rate fell from 37 in 1960 to 17 in 1985 although government policy has been resisted, especially in rural areas. Reports have spoken of forced abortion should a mother conceive a second time, forced sterilisation and even female infanticide if the firstborn was a girl (therefore considered likely to work less usefully in the fields) in the hope of a later child being a boy. In 1987 the government relaxed its policy to 1.7 children per family in 'response to intermittent outrage about cases of coercion and brutality in implementing population goals'.

Singapore

The government introduced a massive family planning scheme in the late 1960s. The objectives of this were:

1 To establish family planning clinics and to provide contraceptives at minimal charges.

2 To advertise through the media the need for and advantages of smaller families.

3 To legislate so that under certain circumstances both abortions and sterilisation could be allowed.

4 To introduce social and economic incentives such as paid maternity leave, income tax relief, housing priority, cheaper health treatment and free education which would cease as the size of the family grew.

Lorenz curves

Lorenz curves are used to show inequalities in distributions. Population, industry and land use are three topics of interest to the geographer which show unequal distributions over a given area. Figure 13.17 illustrates the unevenness of population distribution over the world. The diagonal line represents a perfectly even distribution while the concave curve (it may be convex in other examples), illustrates the degree of concentration of population within the various continents. The greater the concavity of the slope, the greater the concentration of population (or industry, land use, etc).

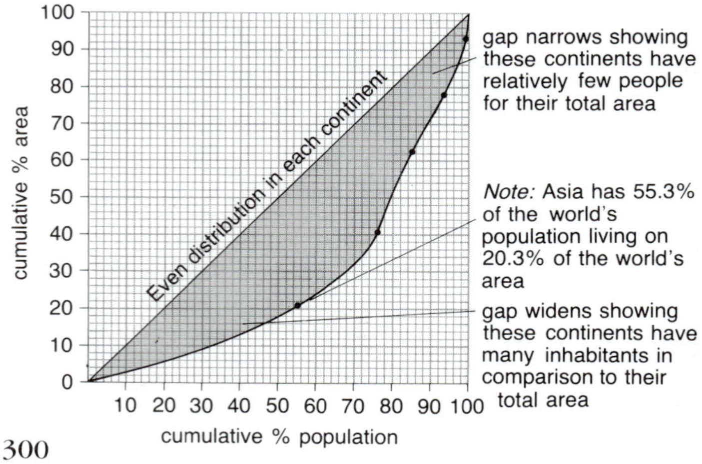

▼ Figure 13.17

Lorenz curves — the distribution of world population in 1960

gap narrows showing these continents have relatively few people for their total area

Note: Asia has 55.3% of the world's population living on 20.3% of the world's area

gap widens showing these continents have many inhabitants in comparison to their total area

Continents ranked in descending order of population 1960	Population %	Cumulative population %	Area %	Accumulative area %
Asia	55.3	55.3	20.3	20.3
Europe and USSR	21.2	76.5	20.1	40.4
Africa	9.3	85.8	22.3	62.7
Latin America	7.1	92.9	15.2	77.9
North America	6.6	99.5	15.8	93.7
Oceania	0.5	100.0	6.3	100.0

Q The table gives the estimated population totals for each continent for AD 2000.

Continent	Estimated total population in AD 2000 (millions)	Population %	Cumulative population %	Area %	Cumulative area %
Asia	3548.1			20.3	20.3
Africa	871.8			22.3	42.6
Europe and USSR	827.2			20.1	62.7
Latin America	546.7			15.2	77.9
North America	297.1			15.8	93.7
Oceania	30.1			6.3	100.0
World	6121.8				

1 Work out the percentage of world population predicted to be living in each continent in AD 2000.

2 Calculate the cumulative percentage, ranking the continents in descending order of population size.

3 Complete a Lorenz curve using Figure 13.17 for guidance.

4 Describe any differences you notice between your graph and Figure 13.17.

Changes in space: migration

Migration is a movement and in human terms usually refers to a permanent change of home. Recently it has been applied more widely to include temporary changes involving seasonal and daily movements. Migration affects the distribution of people over a given area as well as affecting the total population of a region and the population structure of a country or city. The various types of migration are not easy to classify but one means of grouping the different movements is shown in Figure 13.18.

Internal and external (international) migration

Internal migration refers to population movement within a country whereas external migration involves a movement across national boundaries between countries. External migration, unlike internal, affects the total population of a country. The **migration balance** is the difference between the number of emigrants (people who leave the country) and immigrants (newcomers arriving in the country). Countries with a net migration loss lose more through emigration than they gain by immigration and, depending upon the balance between birth and death rates, may have a declining population. Countries with a net migration gain receive more by immigration than they lose through emigration and so are likely to have an overall population increase.

Voluntary and forced

Voluntary migration is when the migrants have chosen to move because they are looking for an improved quality of life or personal freedom. Such movements are usually influenced by 'push and pull' factors (page 307). Push factors are those which cause people to leave because of pressures which make them dissatisfied with their present home while pull factors are those perceived qualities which attract people to a new settlement. When the migrant has virtually no choice but has to move from an area due to natural disasters or to economic, religious or social impositions (table, page 302) migration is said to be forced.

Times and frequency

Migration patterns include those people who may move only once in a lifetime, those who move annually or seasonally (such as the Rendille, page 400, and students), and those who move daily to work or school. Examples of these types of movement are

▼ Figure 13.18

Types of migration — how closely does each type described in this table fit in with your perception of a migrant?

Permanent	External (international)	Between countries
	(i) voluntary	West Indians to Britain
	(ii) forced	African slaves to America
	Internal	Within a country
	(i) rural depopulation	most developing countries
	(ii) urban depopulation	British conurbations
	(iii) regional	northwest to southeast of Britain
Semi-permanent	for several years	Migrant workers in France and West Germany
Seasonal	for several months	Mexican harvesters in California
Daily	commuters	southeast of England

301

given in Figure 13.18 where it can be seen that the timescales over which migration processes take place vary considerably.

Distance
People may move locally within a city or a country or they may move between countries and continents: migration takes place at a range of spatial scales.

Migration laws and a migration model

In the late 1880s, E.G. Ravenstein put forward seven 'laws of migration' based on his studies of migration within the United Kingdom. These laws made several claims.

1 Most migrants travel short distances and their numbers decrease as distance increases (distance decay, page 348).
2 Migration occurs in waves and the vacuum left as one group of people moves out will later be filled by a counter-current of people moving in.
3 The process of dispersion (emigration) is the inverse of absorption (immigration).
4 Most migrations show a two-way movement as people move in and out: net migration flows are the balance between the two movements.
5 The longer the journey the more likely it is that the migrant will end up in a major centre of industry or commerce.
6 Urban dwellers are less likely to move than their rural counterparts.
7 Females migrate more than males within their country of birth but males are more likely to move further afield.

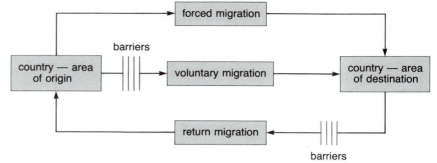

▲ Figure 13.19

A migration model (*after* Jones and Hornsby)

More recent global migration studies have largely accepted Ravenstein's 'laws', but have demonstrated some additional trends.

8 Most migrants follow a step movement which entails several small movements from the village level to a major city rather than one traumatic jump.
9 People are leaving rural areas in ever-increasing numbers.
10 People move mainly for economic reasons, e.g. jobs and the opportunity to earn more money.
11 Most migrants fall into the 20–34 age range.
12 With the exception of short journeys in developed countries, it is the male who is the more mobile. (In many societies the female is still expected to stay at home and look after the family.)
13 There are increasing numbers of migrants who are unable to find accommodation in the place they move to and so resort to living on the streets, in shanties and in refugee camps.

The table below helps to explain the model illustrated in Figure 13.19 with the use of specific examples.

Forced migration
Religious: Jews, Piligrim Fathers to New England
Wars: Muslims and Hindus in India and Pakistan, Vietnamese boat people
Political persecution: Ugandan Asians
Slaves or forced labour: Africans to SE USA
Lack of food and famine: Ethiopians into the Sudan
Natural disasters: floods, earthquakes, volcanic eruptions
Overpopulation: Chinese in SE Asia, NE Brazil
Redevelopment: British inner cities' slum clearances, Ujamaa in Tanzania
Environmental: Chernobyl, Bhopal

Voluntary migration
Jobs: Bantus into South Africa, Turks into West Germany
Higher salaries: British doctors to the USA
Tax avoidance: British pop/rock and film stars to the USA
Opening up of new areas: American colonies, Prairies, Siberia
Territorial expansion: Roman and Ottoman Empires
Trade and economic expansion: British colonies
Retirement to a warmer climate: Americans to Florida
Social amenities and services: better schools, hospitals, entertainment

Prevention of voluntary movement
Government restrictions: immigration quotas, Berlin Wall, work permits
Lack of money: unable to afford transport to and housing in new areas
Lack of skills and education
Lack of awareness of opportunities
Illness
Threat of family division and heavy family responsibilities

Reasons for return
Racial tension in new area
Earned sufficient money to return
To be reunited with family
Foreign culture proved unacceptable
Causes of initial migration removed (e.g. political or religious persecution)

Barriers to return
Insufficient money to afford transport
Standard of living lower in original area
Racial, religious or political problems in original area

▶ Figure 13.20

Percentages of urban and rural dwellers in the USA, 1870–1985

Rural and urban dwellers in the USA as a percentage of total population		
	% urban	% rural
1870	24	76
1900	40	60
1920	51	49
1940	56	44
1960	70	30
1980	76	24
1985	79	21

Internal migration in developed countries

Certain patterns of internal immigration are more characteristic of developed countries than Third World countries. The following examples have been chosen to illustrate this.

Rural–urban movement

Although rural depopulation is now a world-wide phenomenon, it has been taking place for much longer in the more developed, industrialised countries. Figure 13.20 shows the changing balance between rural and urban dwellers in the USA since 1870.

Q Why do you think the 1870 figures have been reversed in the 110 years up to 1980? Explain why this trend is still continuing.

CASE STUDY

Rural–urban movement in the USA

The 1870s was the decade after the American Civil War and the abolition of slavery. Many black farmworkers, most of whom were sharecroppers (page 385), could find no vacant land and they could not afford to buy any that might become available. Consequently, many began to move to the cities.

During the 1930s, drought and soil erosion in the Dust Bowl of the Midwest (page 150) caused many farmers to give up their land, and the economic depression of that time lowered prices for farm produce. Farmhands, who had always been poorly paid yet worked harder and much longer hours compared with their counterparts in industry, were often on yearly or shorter term contracts and so were paid off by farmers whose profits were falling. A further decline in the farm population occurred in the early 1940s when America's export markets were cut off during World War II.

During the 1950s and 1960s the mechanisation of farming and the availability of cheap oil as a fuel saw a big reduction in numbers of farmworkers. The consolidation of farms into larger units and the introduction of contract farming, e.g. farmers with combine harvesters who travel northwards in 'teams' as the cereal ripens, has meant employment for even fewer people living in rural areas, whilst the improvement in private transport has led to the decline of small service centres.

Since the 1970s there has been an increase in the number of 'suitcase farmers' — farmers who live in the city and travel out to their farms periodically. Added to this may be the limited numbers of schools, hospitals, and places of entertainment for members of the rural community, together with few jobs for females. Despite their professed love of the 'big outdoors', most Americans seem to prefer living in urban areas.

Regional movement in Britain

For over a century there has been a drift of people from the north and west of Britain to the southeast of England. The early nineteenth century was the period of the Industrial Revolution when large numbers of people moved to form large urban settlements on the coalfields of northern England, central Scotland and South Wales, and to work in the textile, steel, heavy engineering and shipbuilding industries. However, since the 1920s there has been more than a steady drift of population away from the north of Britain to the south. Some of the major reasons for this movement are listed here.

1 A decline in the farming workforce and rural population for reasons similar to those quoted in the previous section on rural–urban movements.

2 The exhaustion of supplies of raw materials (e.g. coal and iron ore).

3 The decline in the basic heavy industries such as steel, textiles and shipbuilding. Many industrial towns had relied not only on one form of industry but, in some cases, on one individual firm. With no alternative employment, those wishing to work had to move south.

4 Higher birth rates in the industrial cities meant more potential job seekers.

5 New post-war industries, which include car manufacturing, electrical engineering, food processing and, in the 1980s, micro-electronics and hi-tech industries have tended to be market orientated — they are said to be **footloose**, in the sense that they have a free choice of location in so far as they are not tied to the availability of raw materials.

6 The growth of tertiary industries has been mainly in the southeast. This includes the many new office blocks wanting a prestigious London address, tourism taking advantage of Britain's warmer south coast and the growth of government offices in London.

7 Joining the EC has meant increased job opportunities in the south and east while ports such as Glasgow, Liverpool and Bristol, which had links with the Americas, have further declined.

8 Salaries tend to be higher in the south.

9 With so much older housing, derelict land and waste tips, the quality of life is often regarded to be lower in the north, despite the beauty of its natural scenery and slower pace of life.

10 There are more social and cultural amenities in the south.

11 Communications are easier to build in the south where land is flatter and these are more developed because of the greater wealth and population size. The area is more accessible with its motorways, railways, cross-Channel ports and international airports.

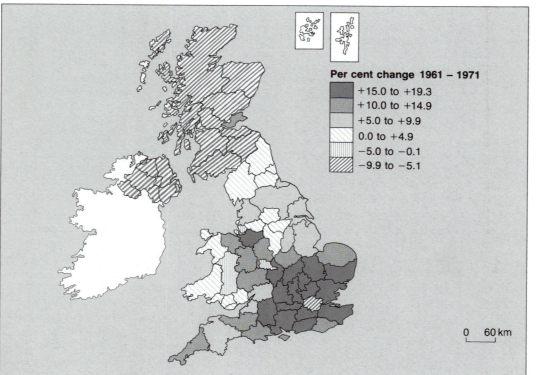

◀ Figure 13.21

Population changes in the UK 1961–1971

Per cent change 1961 – 1971

+15.0 to +19.3
+10.0 to +14.9
+5.0 to +9.9
0.0 to +4.9
−5.0 to −0.1
−9.9 to −5.1

0   60 km

Q

Using Figures 13.21 and 13.22:

1 Describe and give reasons for the changes in UK population between 1961 and 1971.

2 Describe the changes in population between 1981 and 1986.

3 Compare the changes in population between the two periods, and suggest reasons for these changes.

▶ Figure 13.22
Population changes in the UK
1981– 1986 (note that county
boundaries were altered in 1974)

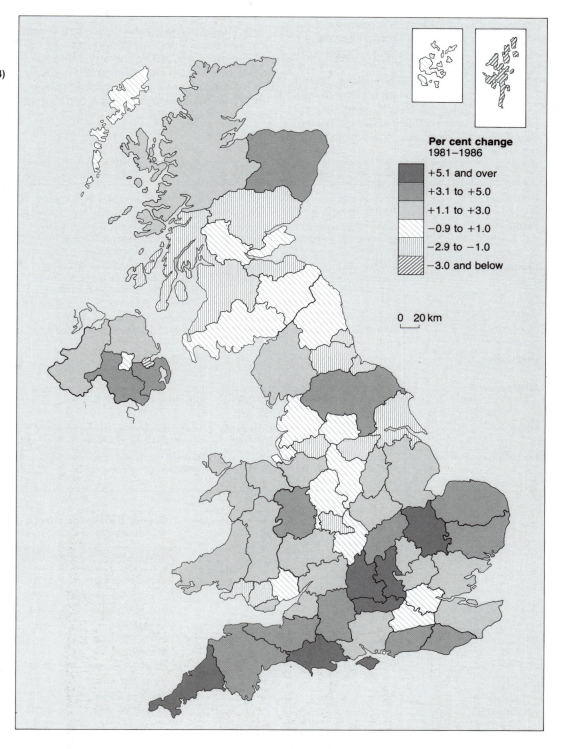

Per cent change
1981−1986

	+5.1 and over
	+3.1 to +5.0
	+1.1 to +3.0
	−0.9 to +1.0
	−2.9 to −1.0
	−3.0 and below

0   20 km

Movements within urban areas: inner city to suburbs

There has been in Britain since the 1930s a movement away from the inner cities to the suburbs, a movement accelerated by the increase in private car and public bus users together with the extension, in the London area, of the underground system. Some of the many reasons for this movement are summarised in the table below.

Two trends are apparent as a result of these movements. (Beware of stereotyping when discussing these inequalities, Framework 10, page 372.)

- Inner cities tend to be left with a higher proportion of low income families, handicapped people, the elderly, single parent families, people with few skills and limited qualifications, first-time home buyers, unemployed, recent immigrants and ethnic minorities.
- Suburbs tend to attract people with higher incomes, capable of buying their own homes. They often possess higher skills and qualifications, and tend to be parents, of whom at least one is in employment, with a growing family.

Movements between conurbations and new towns

Figure 13.23 shows how in recent years there has been a decline in population even in the outer parts of London as people have moved away into surrounding new towns, or into commuter and suburbanised villages. This trend has become characteristic of all Britain's conurbations.

	Inner city	Suburbs
Housing	Poor quality; lacking basic amenities; high density; overcrowding	Modern; high quality; with amenities; low density
Traffic	Congestion; noise and air pollution narrow, unplanned streets; parking problems	Less congestion and pollution; wider, well-planned road system; close to motorways and ring roads
Industry	Decline in older secondary industries; cramped sites with poor access on expensive land	Growth of modern industrial estates footloose and service industries hyper-markets, office blocks and hotels on spacious sites
Jobs	High unemployment; lesser skilled jobs in traditional industries	Lower unemployment in cleaner environments; often more skilled jobs in newer type industries
Open space	Limited parks and gardens	Individual gardens; more, larger parks nearer countryside
Environment	Noise and air pollution from traffic and industry; derelict land and buildings higher crime rate; vandalism	Cleaner; less noise and air pollution lower crime rate; less vandalism
Social	Fewer, older services, e.g. schools and hospitals; ethnic and racial problems	Newer and more services; fewer ethnic and racial problems
Planning and investment	Often wholescale redevelopment/clearances; limited planning and investment before 1988	Planned, controlled development; public and private investment

Some causes of migration from the inner cities to the suburbs

▼ Figure 13.23

a Population changes in London 1801–1986

b Annual percentage change in the population of British conurbations

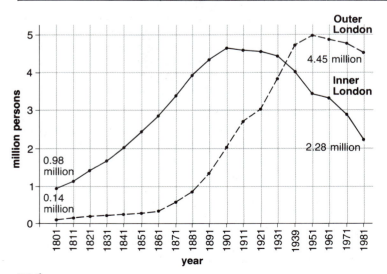

Annual percentage change of population in conurbations (Britain)

Conurbation	1961 – 1971	1971 – 1981	1981 – 1985
Greater London	−0.70	−1.04	−0.14
Inner London	−1.41	−1.92	+0.15
Outer London	−0.18	−0.47	+0.25
Greater Manchester	+0.05	−0.47	−0.34
Merseyside	−0.37	−0.91	−0.68
South Yorkshire	+0.15	−0.14	−0.26
Tyne and Wear	−0.26	−0.59	−0.33
West Midlands	+0.22	−0.53	−0.29
West Yorkshire	+0.31	−0.15	−0.17
Glasgow City	−1.48	−2.46	−0.96

Q Why has there been a movement out of conurbations to new towns in Britain?

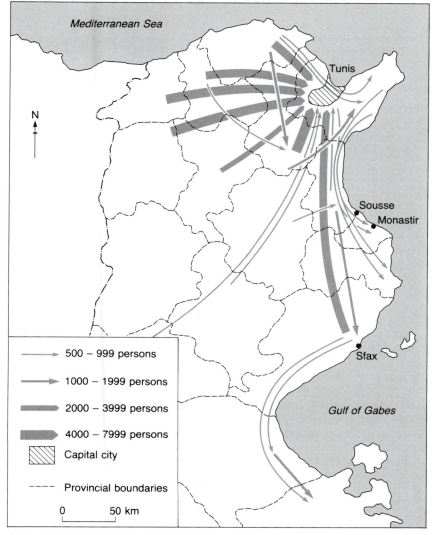

▲ Figure 13.24
Migration patterns in Tunisia

Internal migration in Third World countries

Rural–urban movements

Many large cities in Third World countries are growing at a rate of more than 20 per cent every decade. This movement is partly accounted for by rural 'push' and partly by urban 'pull' factors.

Push factors are those which force or encourage people to leave the countryside. Many families do not own their own land, or their land may have been repeatedly divided by inheritance laws until the plots are too small to support a family. Food shortages develop from the agricultural output being too low for the population of an area, or from

crop failure. Crop failure may be the result of overcropping, overgrazing (page 234), or natural disasters such as drought (Sahel countries), floods (Bangladesh), hurricanes (Caribbean) or earth movements (Andean countries). Elsewhere, farmers are encouraged to produce cash crops for export to help their country's national economy instead of growing sufficient edible food for themselves. Mechanisation reduces the number of farmers needed while high rates of natural increase may lead to overpopulation (page 317). Some people may move because of a lack of services (e.g. schools, hospitals, water supply) or be forced to move by the activity of major industrial concerns (Amerindians in the Brazilian rainforests).

Pull factors are those which encourage people to move to the cities. People in many rural communities may have a perception of the city which in reality does not exist. People migrate to the cities hoping for better housing, better job prospects, more reliable sources of food, better services and the attraction of the 'bright lights'. While it is usually true that in most countries more money is spent on the urban areas — that is where the people who allocate the money live themselves — the present rate of urban growth far exceeds the amount of money available to accommodate all the new arrivals. Recent studies seem to confirm that many migrants make a stepped movement from their rural village first to small towns, then to larger cities and finally to the major city.

Figure 13.24 shows migration patterns in Tunisia. There are several points to notice.

- There is a greater movement of rural than of urban dwellers.
- Most people move to Tunis, the capital city.
- Most migrants tend to travel short distances: relatively few make long journeys (distance decay factor, page 348).
- Most move from rural inland, desert areas to urban coastal areas.
- A few move from Tunis to coastal towns, such as the new holiday resorts of Sousse and Monastir.
- Very few migrants return to rural districts.
- There is evidence of a twofold movement into and out of Tunis and Sfax.

Q How closely do the patterns shown in Figure 13.24 correspond to those suggested by Ravenstein (page 302)?

Political resettling

National governments may direct, control or enforce movement as a result of decisions which they believe to be in the country's best interests. The following are examples of migration for political reasons.

1 **Tanzania** When the country became independent from British colonial rule in 1961, over 90 per cent of its population lived in dispersed farmsteads. In 1967, the Tanzanian government introduced a scheme to encourage people to move to *ujamaa* or family villages. The hope was that these new communities, where work and wealth were shared, would lead to self-reliance and co-operation within the local population. Between 1970 and 1976 the number of villages increased from just under 2000 to over 6200. However, although most of those who had to move had to migrate only short distances, there was considerable resentment and resistance to the plan.

2 **Brasilia** For decades, concern had been expressed in Brazil at the relative richness and rapidly growing population of the southeast of the country around Rio de Janeiro and São Paulo in comparison with the inland states (Figure 13.4). In 1952, Congress agreed to move the capital 1200 km inland from Rio de Janeiro, where most politicians preferred to live, to Brasilia which at that time was an uninhabited site in a savanna landscape, in an attempt to open up the central areas of the country. Building began in 1957 and Brasilia was officially inauguratted as capital in 1960. The population of the city had already exceeded one million by 1986, the figure initially forecast for the year AD 2000. Despite this growth, Brasilia remains a bureaucratic city with many of its 'weekly' inhabitants commuting back to the older centres for weekends.

3 **Brazil** and **South Africa** Here the poorer members of the community have been forced to live on reservations or homelands. The Amerindians of the Amazon Basin have been forced to move as the rainforests have been cleared. In South Africa black people have been forced on to the least favourable land for settlement as a result of the white government's apartheid policy.

External migration

Refugees

The United Nations definition of a refugee is 'a person who cannot return to his or her own country because of a well-founded fear of persecution for reasons of race, religion, nationality, political association or social grouping.

Before World War Two the majority of refugees tended to become assimilated in their new host country but in the last 50 years, the number of permanent refugees has risen rapidly to approximately 14 million in 1989, according to UN estimates. The first, apparently insoluble, refugee problem was the setting up of Palestinian Arab camps following the creation of the state of Israel in 1948. The refugee problem has intensified over 30 years of conflict in southeast Asia, e.g. Vietnam and Afghanistan, and more recently by food shortages and political unrest in much of Africa.

Half of the world's refugees are children of school age; most adult refugees are female and four-fifths are in the Third World. These countries have fewest resources to deal with the problem. Refugees usually live in extreme poverty and lack food, shelter, clothing, education and medical care. They rarely have citizenship, few (if any) civil, legal or basic human rights, there is little prospect of returning home and the long periods spent in camps mean they often lose their sense of identity and purpose, e.g. Vietnamese boat people in Hong Kong, 1990.

Semi-permanent movements

Perhaps because economic development has taken place at different rates in different countries, supplies of and demand for labour are uneven — due to improvements in transport, there has been an increase in the number of people who move from one country to another looking for work. For a century and a half, the UK has received Irish workers and, especially in the 1950s, West Indians. The South African economy depends largely upon black labour from adjacent nation states.

CASE STUDY

Turks in West Germany

Sakaltutan is a village of 900 inhabitants in central Turkey. Until recently it was a poor, isolated settlement dependent upon agriculture. With a high birth rate and limited resources the village had become overpopulated. There were too many males to work on the land — women were not expected to work outside the home — and the demand for craftsmen was limited.

When an all-weather road was built this encouraged the sale of surplus produce in the nearby large towns and led to an increase in mechanisation and a further decline in the need for agricultural labourers. The village school was expanded from one teacher to seven and farmers were able to obtain advice as to how to increase output. The result was a growth in the aspirations of the villagers. Outward migration followed, to the larger cities of Adana and Ankara or overseas to West Germany.

Pforzheim is an industrial town near Stuttgart in West Germany. Like other west European towns it had to be rebuilt after 1945 at a time when there were more job vacancies than workers. The extra labour needed was obtained from the poorer parts of southern Europe and the Middle East. Many of these 'guest workers' or *Gastarbeiter* initially went into agriculture as many were originally farmers, but they soon turned to the relatively better paid jobs in factories and the construction industry. These jobs were not taken by the local Germans because they were dirty, unskilled, poorly paid and often demanded long and unsociable hours.

At one stage there were 15 families from Sakaltutan living in Pforzheim. The first arrivals were males in their twenties, all of whom had had some education and were skilled at a craft, making them acceptable in the local car factory or in construction. In an attempt to earn as much money as possible they often took accommodation in poorly equipped company hostels, missed meals and used public transport to reach work.

Although, by Turkish standards, the migrants were earning high salaries, they found the West German cost of living much higher and what was intended to be a short working stay became more permanent. In time, the families of *Gastarbeiters* arrived and by the mid 1980s over 3 per cent of the West German workforce was Turkish.

This movement of labour has advantages and disadvantages to Sakaltutan in Turkey, the losing village/country, and to Pforzheim in West Germany, the receiving town/country. These have been summarised as a table.

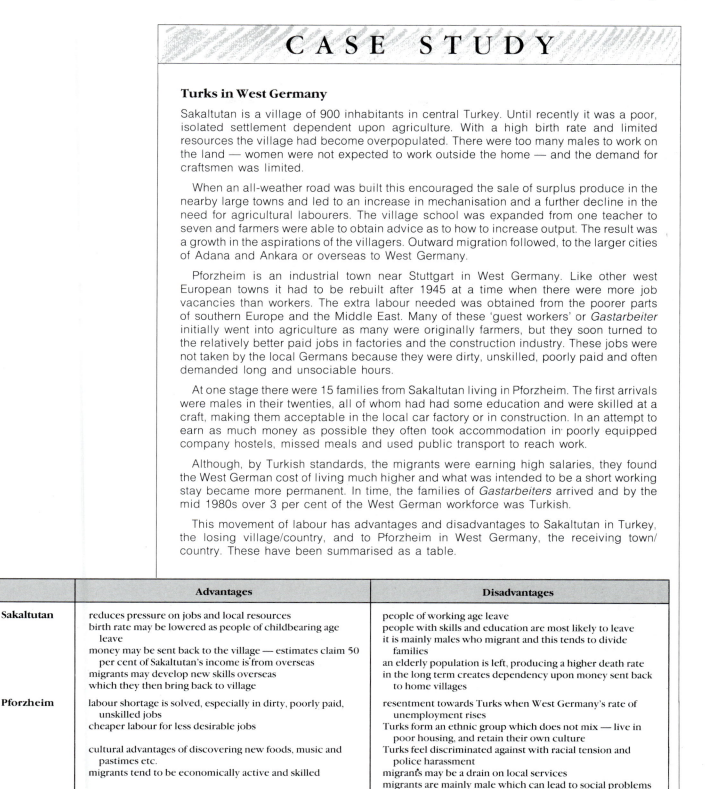

	Advantages	Disadvantages
Sakaltutan	reduces pressure on jobs and local resources birth rate may be lowered as people of childbearing age leave money may be sent back to the village — estimates claim 50 per cent of Sakaltutan's income is from overseas migrants may develop new skills overseas which they then bring back to village	people of working age leave people with skills and education are most likely to leave it is mainly males who migrant and this tends to divide families an elderly population is left, producing a higher death rate in the long term creates dependency upon money sent back to home villages
Pforzheim	labour shortage is solved, especially in dirty, poorly paid, unskilled jobs cheaper labour for less desirable jobs cultural advantages of discovering new foods, music and pastimes etc. migrants tend to be economically active and skilled	resentment towards Turks when West Germany's rate of unemployment rises Turks form an ethnic group which does not mix — live in poor housing, and retain their own culture Turks feel discriminated against with racial tension and police harassment migrants may be a drain on local services migrants are mainly male which can lead to social problems

In 1973 the West German government imposed a ban on the recruitment of foreign workers though Turks still arrived to make family reunions. When, in 1980, grants were offered to Turks wishing to return home very few took advantage of the offer and even fewer took out German citizenship. However, as tension between the two communities has continued to increase, Turks from villages like Salkatutan are now turning to Saudi Arabia and Libya as places to work and to live.

Q

Figure 13.25 shows some major world migrations since 1970.

1 Attempt to explain the international flow of migrants since 1970 by referring to countries which have experienced **a** most emigration and **b** most immigration.

2 Suggest reasons for the principal international migrations since 1970. (You should try to identify patterns of movement using specific countries as examples, rather than trying to identify each country in turn.)

3 Discuss the cultural, economic and social benefits and problems that have resulted from either
   a international migrations of population, or
   b the movement of migrant labour.

4 Population pyramids for West Germany and Turkey are shown in Figure 13.26. What effect has migration had upon the pyramids (and population structure) of these two countries?

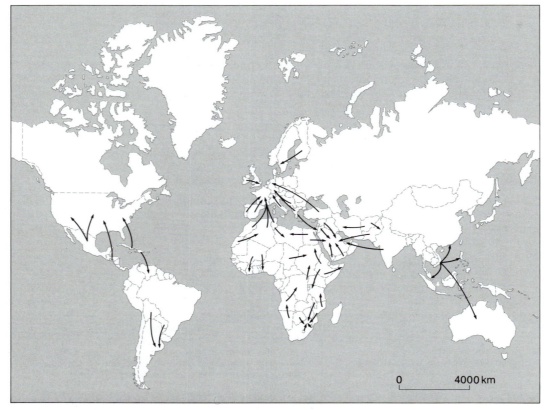

◄ Figure 13.25
Directions of some major international migrations since 1970

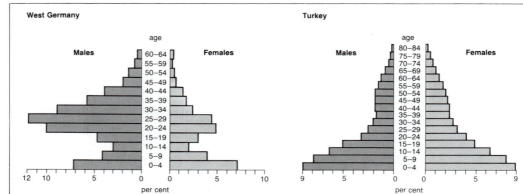

◄ Figure 13.26
Population pyramids for West Germany and Turkey

Multicultural societies

This is often a sensitive and emotive issue. Attempts to explain terms are not intended to cause insult or resentment.

The latest scientific research claims that humans evolved in central Africa about 200 000 years ago and began, 100 000 years later, to migrate to other parts of the world. This common origin, identified by the study of genes, shows that humans are genetically homogeneous to a degree unparalleled in the animal kingdom.

Previous scientific opinion suggested that the many peoples of the modern world had descended from three main races. These were the Negroid, Mongoloid and Caucasoid. The dictionary definition of *race* is 'a group of people having their own inherited characteristics distinguishing them from people of other races', e.g. colour of skin and physical features. In reality, often because of intermarriage, the distinction between races is now so blurred that the word 'race' has little significant scientific value. Today, while colour still remains the most obvious visible characteristic, groups of people differ from one another in religion, language, nationality, and culture. These differences have led to the identification of many ethnic groups.

What criteria do members of various ethnic groups prefer to use when identifying themselves?

- **Colour of skin** Whereas people of 'European' stock have long accepted being 'white' it is only in more recent years that, in Britain, people from Africa and the Caribbean have preferred to be known collectively as 'black'. The 1971 census in the United Kingdom divided immigrants born in Commonwealth countries into the Old (white) and New (black) Commonwealth.
- **Place of birth** (nationality) *The Annual Abstract of Statistics* for the UK lists immigrants under the heading 'country of last residence' — thus avoiding a reference to colour. Most groups of people, for example in the USA, are identified by their place of birth, or that of their ancestors and are known as Chinese, Mexican, Puerto Rican, etc. There is currently a major movement in the USA (and to a lesser extent in the UK) by blacks, the only group in the country not identified by place of origin, to be referred to as African-Americans. Will black people in the UK eventually prefer to be known as African-Caribbean, African-British or another term not yet coined?
- **Religion** Other ethnic groups prefer to be linked with, and are easily recognised by, their religion, e.g. Jews, Sikhs, Hindus and Muslims.

The following question on ethnic origin is to be included in the 1991 Census in Britain and asks respondents to identify themselves for the first time in a British census by ethnic group. It was tested in April 1989 to test people's reactions and to find out if the wording is appropriate and acceptable.

Ethnic groups		
White	☐	1
Black-Caribbean	☐	2
Black-Africa	☐	3
Black-other (please describe)	☐	4
Indian	☐	5
Pakistani	☐	6
Bangladeshi	☐	7
Chinese	☐	8
Any other group (please describe)		

If the person is descended from more than one ethnic or racial group, please tick the group to which he/she considers he/she belongs.

The migrations of different ethnic groups have led to the creation of multicultural societies in many parts of the world. In most countries there is at least one minority group. While such a group may be able to live in peace and harmony with the majority group, unfortunately it is more likely that there will be prejudice and discrimination leading to tensions and conflict. Four multicultural countries with differing levels of integration and ethnic tension are: South Africa, the USA, Brazil and Singapore. Remember, though, that when we look at these countries from a distance we can rarely appreciate the feelings generated by, or the successes of, different state or government policies.

South Africa

The source of South Africa's difficulties lies in the fact that 23 million Blacks are politically, economically and socially dominated by 5 million Whites. The country's racial policy of apartheid meaning 'living apart', is further complicated by the presence of 3 million Coloureds and 1 million Asians.

The white population is a mixture of Dutch, who call themselves Afrikaners or Boers which is the Dutch word for farmers, and British and Germans who arrived later. These two groups are divided among themselves: the Boer community is the staunchest advocate of white supremacy. The Blacks are mainly Bantu who moved southwards into South Africa shortly after the arrival of the Dutch. The Coloureds are a minority group resulting from mixed marriages between the early white settlers (usually Dutch) and the indigenous peoples (mainly Hottentots and Bushmen). The Asian population arrived later, after 1860, and came mainly from India.

The situation today is that Whites are perceived as the 'first class' citizens. Coloureds and Indians enjoy some rights but in comparison to the Whites are regarded as 'second class'. Blacks have virtually no rights outside their 'homelands', the territories where they have been forced to live, and are treated as 'third class' citizens.

In 1948 the National Party came to power. This new, white government, to allay the fears resulting from the fact that Blacks outnumbered Whites, legalised the apartheid system in the hope that segregation of the various racial groups would solve the problems.

What did the apartheid laws seek to achieve?

Housing The Group Areas Act ensured that white, Coloured and Asian communities lived in different parts of a city, and that Blacks lived in 'townships'. Figure 13.28 shows the segregated residential areas in Johannesburg and Cape Town. Since then, Blacks have been forced away from cities to live in one of the ten designated reserves or homelands, where the environmental advantages are minimal (conditions of drought, poor soils, and a lack of raw materials prevail). Blacks are allowed to live outside these areas, for example in cities, only if they have lived there since birth or have worked there for the same employer continuously for ten years. However, most Blacks living in the homelands are employed on one year contracts so that they are prevented from gaining residential rights.

◀ Figure 13.27
A shanty town with corrugated iron roofs

The homelands take up 13 per cent of South Africa's land, hold 72 per cent of its total population and produce 3 per cent of the country's wealth.

Life in the townships is no less difficult. The townships have been created far away from white residential areas, so those Blacks finding jobs in cities have long and expensive journeys to work. Many of the original shanty towns have been bulldozed and replaced by rows of identical single storey houses with four rooms and with toilets in the backyards: only 20 per cent of the houses have electricity. Corrugated iron roofs make the buildings hot in summer and cold in winter. The settlements lack infrastructure and services and, because the population is growing faster than the rate of house construction, are surrounded by vast squatter camps (page 374). The two best-known townships are Soweto in Johannesburg, which has 1.5 million inhabitants, and Crossroads in Cape Town.

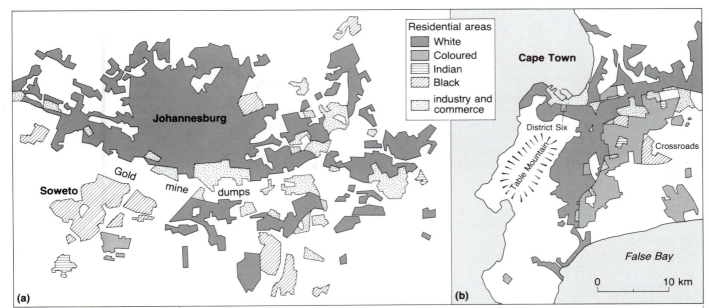

▲ Figure 13.28
Segregated residential areas in
South Africa
a Johannesburg
b Cape Town

► Figure 13.29
Housing in the Soweto
township, South Africa

Employment Blacks are severely restricted in mobility and type of job. A man must return to his homeland in order to apply for a job. If successful, he will be given a contract to work in 'white' South Africa for eleven months after which he must return to his homeland if he wishes to renew the contract. As described above, this system prevents Blacks from becoming permanent residents in the city. All Blacks had to carry a passbook and if caught without one on their person were imprisoned. This repressive law was relaxed in 1986.

Political rights Blacks do not have a vote in general elections although they can elect members for their township council or homeland committee. Coloureds and Asians have members in the House of Assembly (the national government) but as numbers of representatives are fixed they can be out-voted by the Whites. The main black party, the African National Congress, or ANC, is at present (1990) banned from official political activity.

Amenities and education Schooling is free and compulsory for all Whites and Asians but not for Coloureds or Blacks. Schools and universities are segregated by law as are many amenities. For example, Whites have their own buses, sections on trains, beaches, restaurants and places of entertainment, although more recently these laws have not always been fully enforced. There are few cultural links between racial groups though sporting integration began to increase in 1989.

CASE STUDY

The United States of America

Americans have long prided themselves that their country is a 'melting pot' in which peoples of all ethnic groups could be assimilated into one nation. Yet problems do exist. The indigenous Indian populations have been granted reservations, often in areas lacking resources, although they are not forced to live on them. However, it is usually only on reservations occupied by tribes such as the Navajo and Hopi that traditional Amerindian culture still flourishes. Indians who drift to towns are often the focus of social problems.

The African-Americans, released from slavery after the Civil War, could not find land on which to farm. They too moved to large urban centres where they congregated in inner city 'ghettos' (Chicago 15.6). Despite some improvements following the Civil Rights movement of the 1960s, this ethnic group remains socially and economically deprived in comparison with most whites.

Despite the US claim that it has an 'open door' policy, strong restrictive laws have frequently been imposed as a barrier to immigration (Figure 13.19). This policy has been noticeable against Chinese in the 1920s, Japanese during World War Two, Mexicans throughout the 1980s and, most recently, illegal migrants from Central America. Many immigrant groups still identify themselves with their 'home' country and its culture, living and marrying within their own ethnic group, e.g. Puerto Ricans in New York, and congregating to form ethnic areas, e.g. Los Angeles has its Chinatown, Japantown, Koreatown and Filipinotown.

Brazil

Most of the inhabitants of Brazil, though being of almost every colour of skin conceivable, regard themselves as Brazilians, and the country rightly claims that it has little racial discrimination or prejudice. The Census Department does, however, recognise the following divisions based upon colour.

1 Whites (*Branco*: anyone with a *café au lait* or lighter colour skin. (The USA and South Africa would regard *café au lait* as black or Coloured.) Many of the European migrants come from Portugal (the original colonists), Italy, Germany and Spain.

2 Mulatto (*Pardo*): darker skins but with a discernible trace of European ancestry. They are the result of mixed marriages or 'liaisons' between the early Portuguese male settlers and either female Indians or African slaves. There is pride rather than prejudice in coming from two racial backgrounds.

3 Blacks (*Preto*): those of pure African descent.

4 Orientals (*Amarelo*): those who have more recently emigrated from south and east Asia.

5 Amerindians: a continually declining, yet still distinctive, indigenous group.

All these groups mix freely, especially at football matches in carnivals and on the beach. Yet despite the lack of racial tension there tends to be a correlation between colour and social status and jobs. Walking into a hotel on arrival in Rio it is apparent that the baggage carriers are black, hotel porters a slightly lighter colour and the receptionists and cashiers *café au lait*.

In the army, officers are usually white and the ranks black or mulatto. Similarly, the lighter the colour of skin, the more likely it is that a person will become a doctor, bank manager, solicitor or airline pilot. Yet in the author's experience two indian tourist guides, one at Manaus in the Amazon rainforest and the other at Iguaç on the border with Paraguay, both stated that blacks and Amerindians had the opportunity of reaching the top in Brazil but preferred to avoid the stresses of modern life.

CASE STUDY

Singapore: racial and religious harmony

The three main races of Singapore have separate religions yet each is completely tolerant of the others and some people even celebrate all three 'New Years'. Although in 1988 there was still a Chinatown, restricted to ten streets, Arab Street (four streets) and Little India (six streets) the Singapore government has pulled down many old houses in these areas. Ethnic concentrations have been broken up and 86 per cent of Singaporeans now live in modern high rise flats in new towns (page 377). All races, religions and income groups live together in what appears to be a most successful attempt to create a national unity — a unity best seen on Independence Day.

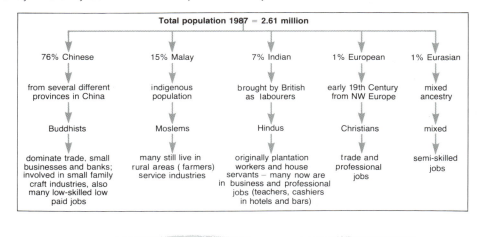

▶ Figure 13.30
Racial and religious groups in Singapore

Daily migration: commuting

A commuter is a person who lives in one community and works in another. There are two types of commuting.

1 **Rural–urban** is where the commuter lives in a small town or village and travels to work in a larger town or city. There is rarely much movement in the reverse direction. The **commuter village** may also be referred to as a **dormitory** or a **suburbanised** village.

2 **Intra-urban** movement is where people who live in the suburbs travel into the city centre for work. This now includes inhabitants of inner city areas who have to travel to edge-of-city industrial estates.

Commuter hinterlands

A **commuter hinterland**, or **urban field**, is the area surrounding a large town or city where the workforce lives. Patterns of commuting are likely to develop in some of the following situations.

- Hinterlands are larger where communications are fast and reliable, e.g. the London Underground, and where public transport is developed and private car ownership is high, e.g. SE England.
- Where new housing is a long way from the older industrial areas of a city or from the central business district (CBD), e.g. the new towns in central Scotland.
- Where there is a city or conurbation with plenty of jobs, especially in service industries, e.g. London.
- Where there is no rival urban centre within easy reach, e.g. Plymouth.
- Areas of higher salaries may enable people to pay the transport costs.
- Where people feel that the need to live in a cleaner environment outweighs the disadvantages of time and cost of travel to work, e.g. living in the Peak District and working in Sheffield or Manchester.
- In areas of high housing costs where younger people may find they are forced to look for cheaper housing further away from their work.
- A shorter working day allowing more time in which people can travel.
- The more elderly members of the workforce who buy a home in the country or near to the coast and commute until they retire, e.g. Sussex coast.
- Those who find themselves living in an area where there have been severe job losses which force them to look for work in other towns, e.g. Inhabitants of Cleveland working in SE England.

What problems arise from commuting?

Most problems occur in or near to city centres. There is often traffic congestion as cars, buses and delivery vehicles all focus on the CBD during the peak at morning and evening rush hours. Many older cities have narrow, unplanned roads in the centre or in the inner area surrounding the CBD. The problems of where to park the car and the cost of leaving it for a day cause frustration and expense. The volume of traffic increases accident potential and heightens levels of noise and air pollution. To try to relieve congestion wider roads are built, but this may mean the demolition of shops and houses.

In large cities such as London, New York and Tokyo, many commuters may take up to two hours travelling each way, which leaves them tired and with little free time in the evenings. The problem is not confined to cities in developed countries. In Third World countries, most jobs tend to be in the largest or capital city. As these settlements are expanding rapidly, anyone living on the outskirts may have many kilometres to travel on very congested roads with poor public transport systems — the diameter of São Paulo was 60 km in 1990.

Job ratios

This is a method of quantifying the relationship between where people work and where they live. The ratio is expressed by the following formula:

$$\frac{\text{People working in the area (residents and commuters)}}{\text{Total number of employed people living in the area}} \times \frac{100}{1}$$

If the ratio is over 100 it means that there are more commuters travelling into the area than people who already live and work there; if the ratio is less than 100 it means that there are fewer jobs available than there are people living in the area, so people have to travel to find work.

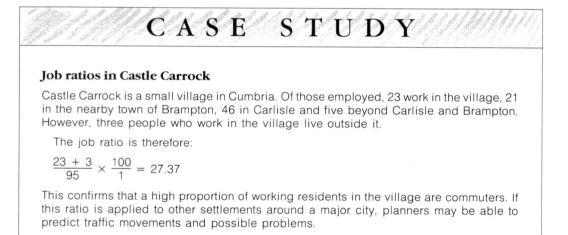

CASE STUDY

Job ratios in Castle Carrock

Castle Carrock is a small village in Cumbria. Of those employed, 23 work in the village, 21 in the nearby town of Brampton, 46 in Carlisle and five beyond Carlisle and Brampton. However, three people who work in the village live outside it.

The job ratio is therefore:

$$\frac{23 + 3}{95} \times \frac{100}{1} = 27.37$$

This confirms that a high proportion of working residents in the village are commuters. If this ratio is applied to other settlements around a major city, planners may be able to predict traffic movements and possible problems.

Figure 13.31 shows job ratios for the boroughs of Greater London.

Describe and account for the differences in job ratios over Greater London.

In thinking about this question it may be helpful to consider the following:

a Name the two boroughs with the highest job ratios and try to explain these ratios by referring to the presence of tertiary industry (offices, shops, tourism, administration and entertainment) and the lack of housing (with the exception of luxury flats in the Barbican) due to high land values.

b The three inner city boroughs north of the river adjacent to the City were the location for the traditional secondary industries of London. Give specific examples of these industries and name the boroughs.

c Explain the high density of housing in the four inner city boroughs along the south bank of the Thames and the relative lack of jobs in these areas?

d Give three examples of outer suburbs with a low job ratio. Give reasons for their low job ratios with reference to the types of residential area and the fact that industry tends to be found along the main lines of communication.

e Name the two outer western suburbs which have a high job ratio, and suggest why they are anomalous. (Communications may be an important clue.)

▶ Figure 13.31
Job ratio figures for Greater London

f How might job ratio figures for some of the surrounding dormitory and new towns (name two of each) have helped to show the scale of commuting from those settlements.

Optimum, over and under population

Demography is the study of population.

In theory, the **optimum population** of an area is the number of people which, when working with all the available resources, will produce the highest per capita economic return, i.e. the highest standard of living and quality of life. If the size of the population increases or decreases from the optimum then the output per capita, or standard of living, will fall. This concept is of a dynamic situation: it changes with time as technology improves, as population totals and structure change (age and sex ratios, e.g. population pyramids, page 295) and as new raw materials are discovered to replace old ones which are exhausted or whose values change over a period of time.

Standard of living is the interaction between physical and human resources and can be expressed in the following formula.

$$\text{Standard of living} = \frac{\text{Natural resources} \begin{cases} \text{minerals} \\ \text{energy} \\ \text{soils etc} \end{cases} \times \text{Technology}}{\text{Population}}$$

Overpopulation is when there are too many people relative to the resources and technology locally available to maintain an 'adequate' standard of living. Bangladesh, Ethiopia and parts of Brazil and India are often said to be overpopulated as they have insufficient food, minerals, and energy resources; they suffer from natural disasters such as drought and famine; and they are characterised by low incomes, poverty, poor living conditions and often a high level of emigration. In the case of Bangladesh (Case Study, page 318), where the population density increased from 282 people per km^2 in 1950, to 704 in 1985, it is easier to appreciate the problem of 'too many people' than in the case of the northeast of Brazil where the density is less than two people per km^2.

Underpopulation is when there are far more resources in an area, e.g. food production, energy and minerals, than can be used by the number of people living there. Canada, with a mid-1980s total of 25 million, could theoretically double its population by AD 2000 and still maintain its standard of living (Case Study, page 318). Countries like Canada and Australia can export their surplus food, energy and mineral resources and they have high incomes, good living conditions, high levels of technology and high levels of immigration. It is probable that standards of living would rise, through increased production and exploitation of resources, if population were to increase.

C A S E   S T U D Y

Is Bangladesh overpopulated?

Bangladesh has a high population density of 704 per km². It has a high birth rate (47.4) and a declining death rate (from 28 in 1970 to 19.5 in 1985). This has resulted in a high and accelerating rate of natural increase, from 1.59 per cent in 1950 to 2.74 per cent in 1985 (birth and death rates, page 290). Over 20 per cent of the population is under nine years of age. The gross domestic product of 158 US$ per capita is very low.

Most of the land is a delta and consequently is prone to frequent and severe natural disasters. Flooding of the Ganges and Brahmaputra rivers is due partly to the monsoon rains but is mainly a result of deforestation in the Himalayas (page 425). Flood damage is also caused by tropical cyclones in the Bay of Bengal. Most (71 per cent) of the inhabitants are farmers and 89 per cent live in rural communities. There is a shortage of industry, services and raw materials (Bangladesh has no energy or mineral resources of note), and the transport network is poorly developed. The low level of literacy has led to limited internal innovation and a lack of capital has meant that the country can ill afford to buy technical skills from overseas.

◄ Figure 13.32
Population pressure has led to overcrowding in Bangladesh

Is Canada underpopulated?

Canada's population density is low — three people per km². It has low birth and death rates of 14.9 and 7.6 respectively giving a rate of natural increase of only 1.01 per cent a year. Less than 15 per cent of the population is aged under nine. The gross domestic product per capita is very high at US$ 13034. Natural disasters are rare. Relatively few (4 per cent) of the inhabitants are farmers or live in rural areas (24 per cent). Canada has developed industries, services, energy supplies, mineral resources (of which there are large reserves) and an effective transport network. The high level of literacy and national wealth has enabled the country to develop its own technology and to import modern innovations.

However, making comparisons on a global scale suggests that there does not seem to be a direct correlation between population density and over/under population. Northeast Brazil is 'overpopulated' with two people per km² whereas California, despite water problems and pollution, is 'underpopulated' with over 500 per km². Similarly, population density is not necessarily related to gross domestic product (GDP) per capita. The Netherlands and West Germany both have a high GDP per capita and high population density; Canada and Australia have a high GDP per capita and low population density; Bangladesh and Puerto Rico have a low GDP per capita and a high population density; the Sudan and Bolivia have a low GDP per capita and low population density.

> ***Carrying capacity*** *is the largest population that the environment of a particular area can carry or support.*

The balance of resources within a country may be uneven, e.g. a country may have a population too great for one resource, e.g. energy, yet too small to use fully a second, e.g. food supply. The relationships between population and resources are highly complex and the terms 'over' and 'under' population must therefore be handled with extreme care.

Theories relating to world population and food supply

Malthus

Thomas Malthus was a British demographer who believed that there was a finite optimum population size in relation to food supply and that any increase in population beyond that point would lead to a decline in living standards and to 'war, famine and disease'. He published his views in 1798 and although, fortunately, many of his pessimistic predictions have not come to pass, they form an interesting theory and provide a possible warning for the future. His theory was based on two principles.

1 Human population, if unchecked, grows at a geometric or **exponential** rate, i.e. $1 \rightarrow 2 \rightarrow 4 \rightarrow 8 \rightarrow 16 \rightarrow 32$, etc.
2 Food supply, at best, only increases at an **arithmetic** rate, i.e. $1 \rightarrow 2 \rightarrow 3 \rightarrow 4 \rightarrow 5 \rightarrow 6$, etc. Malthus claimed that this was due to a shortage of land or that yields from a given field could not go on increasing forever.

Although he did not try to explain why population and food supply grew at these rates Malthus did demonstrate that any rise in population, however small, would mean that eventually population would exceed increases in food supply. This is shown in Figure 13.33 where the exponential curve intersects the arithmetic curve. Malthus therefore suggested that after five years, the ratio of population to food supply would increase by 16:5, and after six years by 32:6.

He suggested that once a ceiling had been reached, further growth in population would be curbed by negative (preventive) or by positive checks.

Preventive (or **negative**) checks were methods of limiting population growth and included abstinence from or a postponement of marriage which would lower the fertility rate. Malthus noted a correlation between wheat prices and marriage rates (remember this was the late eighteenth century): as food became more expensive, fewer people got married.

Positive checks were ways in which the population would be reduced in size by such events as famine, disease and war, all of which would increase the mortality rate and reduce life expectancy.

The carrying capacity of the environment
The concept of a population ceiling, first suggested by Malthus, is of a saturation level where population equals the **carrying capacity** of the local environment. This may refer to plants or animals in an ecosystem (Chapter 11) as well as to human populations. There are three models which portray what might happen as a population, growing exponentially, approaches carrying capacity (Figure 13.34).

1 The rate of increase may be unchanged until the ceiling is reached, at which point the increase drops to zero. This highly unlikely situation is unsupported by evidence from either human or animal populations.
2 Here, more realistically, the population increase begins to taper off as the carrying capacity is approached, and then to level off when the ceiling is reached. It is claimed that populations which are large in size, have long lives and low fertility rates conform to this 'S' curve pattern.

▶ **Figure 13.33**
Relationships between population growth and food supply (*after* Malthus)

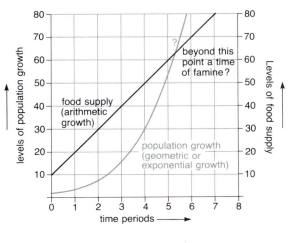

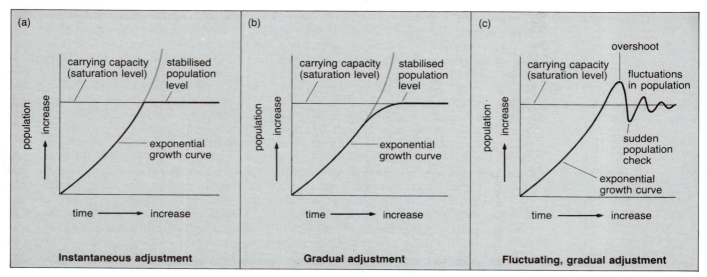

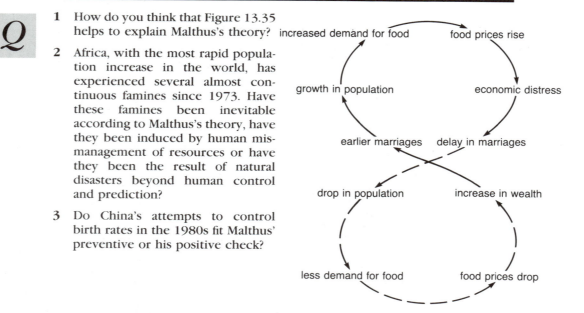

3   In this instance the rapid rise in population overshoots the carrying capacity, resulting in a sudden check which causes a dramatic fall in the total population, e.g. famine and reduced birth rates. After this, the population recovers and fluctuates around — eventually settling down at — the carrying capacity. This '𝒥' curve appears more applicable to populations which are small in number, have short lives and high fertility levels.

Boserup

Boserup was a Danish economist who in 1965 offered an alternative theory. Whereas Malthus claimed that food supply was the main limit to population size, she asserted that in a pre-industrial society, an increase in population was likely to stimulate a change in agrarian technology and result in increased food production, i.e. 'necessity is the mother of invention'. Boserup studied various land use systems based upon differences in intensity of production and frequency of cropping, from shifting cultivation in the tropical rainforests to multiple cropping in SE Asia. She suggested that as population increased, farming became more intensive with the innovation and introduction of new methods and technology.

The Club of Rome (1972)

An international team studying world problems predicted that if present trends in population and resource utilisation were to continue, a sudden decline in growth could occur within a hundred years. They suggested plans for global equilibrium, e.g. an even population growth, an even economic development and an even use of resources.

▲ Figure 13.34
Three models illustrating the relationship between an exponentially growing population and an environment with a limited carrying capacity
a Instantaneous adjustment
b Gradual adjustment — the 'S' curve
c Fluctuating, gradual adjustment — the 'J' curve

▼ Figure 13.35
Population growth and population checks

Q

1   How do you think that Figure 13.35 helps to explain Malthus's theory?

2   Africa, with the most rapid population increase in the world, has experienced several almost continuous famines since 1973. Have these famines been inevitable according to Malthus's theory, have they been induced by human mismanagement of resources or have they been the result of natural disasters beyond human control and prediction?

3   Do China's attempts to control birth rates in the 1980s fit Malthus' preventive or his positive check?

How may human and physical resources affect development?

HUMAN AND ECONOMIC RESOURCES	
Capital	Uneven distribution of wealth, money for development of resources and services, vicious circle of poverty
Labour supply	Quality and quantity of workforce, level of education (literacy, skills and organisation), health
Social factors	Class structures (e.g. caste systems), attitude to women, religious and cultural attitudes, land ownership
Population	Natural increase, over or underpopulation (in terms of people and animals), migration
Political	Colonialism, state control and government policies, socialism/capitalism
Transport	Differences in types and networks
PHYSICAL RESOURCES	
Raw materials	Soils, minerals, energy supplies, water supplies
Climate	Temperature (extremes, length of growing season), rainfall (seasonality, reliability, intensity, drought)
Soils	Productivity, erosion, leaching
Hazards	Natural (e.g. earthquakes, floods, drought, landslides, typhoons, hurricanes), human induced (e.g. deforestation, soil erosion, salinisation)

Population, resources and development

A number of problems or conflicts frequently arise as people in different parts of the world attempt to manage their environments and utilise unevenly distributed resources to satisfy human needs.

Resources

These can be conveniently divided or classified into human and physical with the latter being subdivided into finite and renewable.

- **Human resources** include the numbers, skills, abilities and wealth of the inhabitants of an area.
- **Physical resources** are the natural materials obtained from the environment.

Finite resources are non-renewable: they cannot be replaced once they have been used, other than over a long period of geological time. Minerals and fossil fuels come into this category and modern technological advances have accelerated the use of many of them. Today the USA consumes resources at seven times the world average per capita. What will be the situation if Third World countries manage to become more developed? In the last 50 years, human economic development has used more minerals and fossil fuels than in the rest of historical time.

Renewable resources can be used more than once and include soils, biotic resources (e.g. forests) and 'flow' resources (e.g. tides, waves, wind and solar energy). While the technology to use flow resources on a large scale has yet to be developed, many soils and biotic resources are being 'overused'.

Population and food supply

It is almost 200 years since Malthus expressed his fears that world population would outstrip food supply. Today, perhaps surprisingly, as famine is still prevalent in much of Africa, there is enough food for everyone in the world. However, there are two major problems:

1. massive food surpluses in North America and within the EC;
2. food shortages in many Third World countries.

This uneven distribution is reflected in Figure 13.36 which shows the variation in calorie intake throughout the world. Dieticians claim that the average adult in temperate latitudes requires 2600 kilocalories a day compared with 2300 kilocalories for someone living within the tropics. The Food and Agriculture Organisation (FAO) reports that the actual average intake for the developed world is 3300 kilocalories but only 2200 kilocalories for Third World countries. However, calorie intake alone is not an accurate guide to the quality of the diet as it fails to differentiate between carbohydrate, fat and protein. It is imbalances within the diet as much as low calorie intake which lead to malnutrition. The FAO estimates that over 25 per cent of the population in the Third World is suffering from insufficient food and over 60 per cent from inadequate diet. Each year, 45 million people die from hunger and hunger-related diseases — almost half of these are children. Even when malnutrition does not directly cause death, it reduces the capacity

321

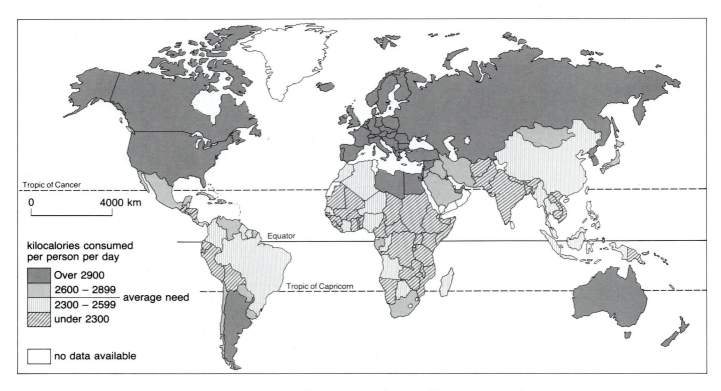

▲ Figure 13.36

World food supply — average
kilocalories consumed per
person per day, by country

to work and increases susceptibility to other diseases.

Trends in food supply

With the exception of 1973 when there was a global rainfall deficit, world food output has increased more rapidly than population since the early 1950s. This output has recently been increasing more rapidly in the developing world than in the developed (albeit from a much lower base), although in several African countries there has been a decline in food output per person (Figure 13.37). The FAO announced in 1986 that there was sufficient food to feed everyone in the world but that its distribution was uneven and that its cost was rising. In 1986 there was sufficient to give every person in the world 1 kg of food every day. However, disparities on a continental scale meant that there was 5 kg per person for North America, 3.5 kg for Oceania, 2 kg for western Europe, 1 kg for Latin America and SE Asia but less than 0.5 kg for Africa.

Problems of food surpluses

The surpluses in the EC and North America have tended to depress world prices of raw foodstuffs, to force farmers in Third World countries out of business, which accelerates rural–urban migration, and to contribute to a decline of production in some Third World countries. The increase in output from the developed countries is a combination of biological innovations (the introduction of various new seeds and the use of fertilisers)

and economic factors (farmers are often guaranteed a fixed price or subsidy). By 1986 the EC had built up a butter 'mountain' of 1.5 million tonnes, and a powdered skimmed milk surplus of 1 million tonnes. The 32 million tonnes of cereal surplus of that year had grown from a figure of 13 million in 1980. It was predicted to rise to 89 million in 1990. A reduction in subsidies saw an appreciable decrease in these 'stores' in 1989.

Causes of food shortages in sub-Saharan Africa

The population of sub-Saharan Africa is growing faster than anywhere else in the world. The high birth rates and falling death rates mean more mouths to feed. With 71 per cent of the labour force in agriculture

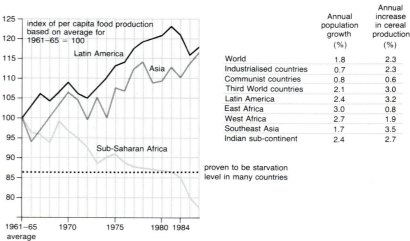

	Annual population growth (%)	Annual increase in cereal production (%)
World	1.8	2.3
Industrialised countries	0.7	2.3
Communist countries	0.8	0.6
Third World countries	2.1	3.0
Latin America	2.4	3.2
East Africa	3.0	0.8
West Africa	2.7	1.9
Southeast Asia	1.7	3.5
Indian sub-continent	2.4	2.7

and 77 per cent of the population living in rural areas, the income, nutrition and health of most Africans is closely tied to farming. In an area where the use of new seeds, fertilisers, machinery and irrigation is the lowest in the world, agriculture is almost wholly reliant upon an environment which is not naturally favourable. The soils in many areas have fertility constraints, low water holding capacity and are vulnerable to erosion. High evapotranspiration rates harm crops, as do the unreliable rains which may cause flooding one year and then fail for several years. While the periods of water budget deficiency (drought) are getting longer and more frequent, experts argue as to whether this is part of a natural climatic cycle, a consequence of deforestation which has reduced the amount of moisture in the atmosphere or of the greenhouse effect.

Traditional farming methods such as shifting cultivation and nomadic pastoralism use land for a limited period before abandoning it for several years but later returning to it. With increases in population, fallow periods have been reduced and land has been overgrazed. This, together with the destruction of forests for fuelwood, has allowed accelerated erosion. Efforts to increase food production have been impeded by a lack of money for fertilisers, seeds and tools, and even when overseas financial aid has been given it has in some instances been directed towards unsuitable projects (Case Study, Senegal). Such unsuitable schemes include promoting monoculture (page 235), increasing cattle herds on marginal land and ploughing soils which should receive only minimal disturbance if equilibrium is to be retained. Financial aid schemes from overseas may create problems as the recipient country is likely to fall into debt, while the doner expects crops to be grown for export rather than for consumption by the local population. To add to the difficulties faced in some countries there is civil war and administrative corruption both of which interrupt farming and the distribution of relief supplies.

Q

Figure 13.38, page 324, shows some of the causes of famine which affected Bangladesh in the mid-1980s.

1 In what ways is famine in that country beyond human control?

2 How is the problem aggravated by social and economic factors?

3 What, if anything, can be done to overcome the problems of famine in the short term and the long term?

CASE STUDY

Failures of aid in Senegal

Bread is the staple diet in this West African country. Although millet is grown locally and can be used for flour, while Senegal was a French colony the population grew to prefer bread made from wheat which has to be imported. Groundnuts were also grown as a cash crop for the European market but recently overseas demand has fallen and caused a loss of income. In the early 1970s, a Dutch firm began to grow top quality fruit and vegetables on irrigated land for the northwest European market. Even before the company pulled out in 1979 there had been no domestic benefits.

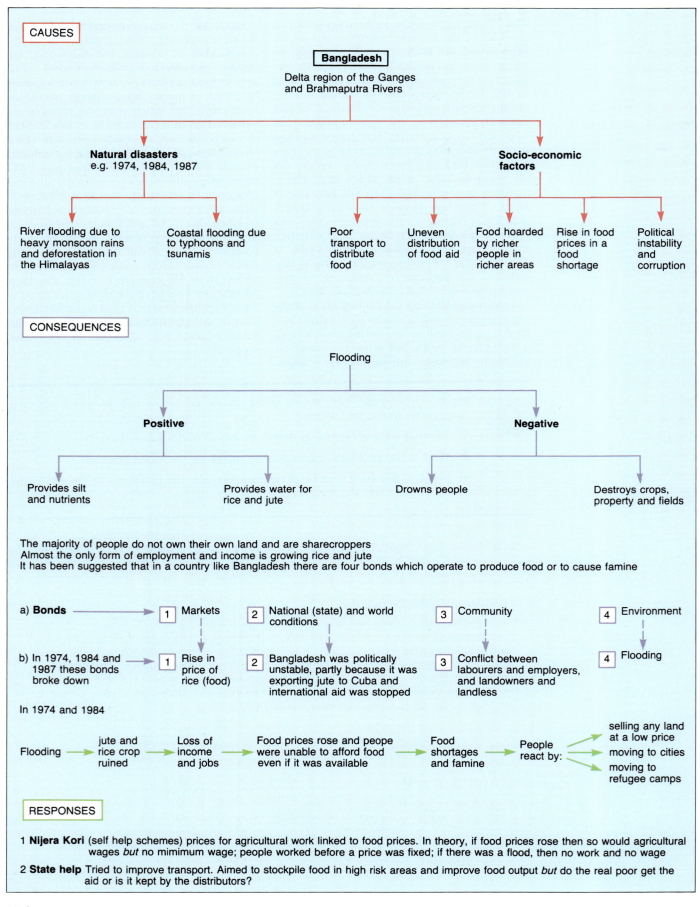

CAUSES

Bangladesh

Delta region of the Ganges and Brahmaputra Rivers

Natural disasters
e.g. 1974, 1984, 1987

Socio-economic factors

River flooding due to heavy monsoon rains and deforestation in the Himalayas

Coastal flooding due to typhoons and tsunamis

Poor transport to distribute food

Uneven distribution of food aid

Food hoarded by richer people in richer areas

Rise in food prices in a food shortage

Political instability and corruption

CONSEQUENCES

Flooding

Positive

Negative

Provides silt and nutrients

Provides water for rice and jute

Drowns people

Destroys crops, property and fields

The majority of people do not own their own land and are sharecroppers
Almost the only form of employment and income is growing rice and jute
It has been suggested that in a country like Bangladesh there are four bonds which operate to produce food or to cause famine

a) **Bonds** ────▶ 1 Markets   2 National (state) and world conditions   3 Community   4 Environment

b) In 1974, 1984 and 1987 these bonds broke down ───▶ 1 Rise in price of rice (food)   2 Bangladesh was politically unstable, partly because it was exporting jute to Cuba and international aid was stopped   3 Conflict between labourers and employers, and landowners and landless   4 Flooding

In 1974 and 1984

Flooding ──▶ jute and rice crop ruined ──▶ Loss of income and jobs ──▶ Food prices rose and peope were unable to afford food even if it was available ──▶ Food shortages and famine ──▶ People react by: ──▶ selling any land at a low price / moving to cities / moving to refugee camps

RESPONSES

1 **Nijera Kori** (self help schemes) prices for agricultural work linked to food prices. In theory, if food prices rose then so would agricultural wages *but* no mimimum wage; people worked before a price was fixed; if there was a flood, then no work and no wage

2 **State help** Tried to improve transport. Aimed to stockpile food in high risk areas and improve food output *but* do the real poor get the aid or is it kept by the distributors?

◀ **Figure 13.38**

Famine and food supply in Bangladesh

The concept of economic development

Definition of terms

Terms such as 'developed' and 'developing' have been used for several decades to indicate the economic conditions of a group of people or a country. More recently the term 'developing' has been regarded as a stigma and is being replaced by the 'South' (*Brandt Report*, 1980) and, with increasing popularity, the 'Third World'. A growing realisation and appreciation that poverty is a relative, not an absolute, concept has led to the introduction of the terms 'more developed countries' (MDCs) and 'less developed countries' (LDCs).

However, all these definitions (summarised in Figure 13.39) have been based upon, and have over-emphasised, economic growth. To those living in a western, industrialised society, economic development tends to be synonymous with wealth, i.e. a country's material standard of living. This is measured as the **gross domestic product** (GDP) per capita by dividing the money value of all the goods and services produced in a country by its total population. When trade figures for 'invisibles' (mostly financial services and deals) are included, the term **gross national product** (GNP) per capita is used. In the mid-1980s it was estimated that 80 per cent of the world's wealth was owned by the 20 per cent of its inhabitants who lived in developed countries, leaving only 20 per cent to those 80 per cent of its inhabitants in the Third World. In 1981 the UN estimated that over half of the world's population earned under US$200 a year.

For consistency, the terms 'developed' and 'Third World' are usually used in this book.

A number of definitions, usually involving cultural development and social well-being, have been suggested as alternatives to those based upon economic criteria, e.g. they emphasise 'quality of life' in contrast to 'standard of living'. The Overseas Development Council (ODC) has introduced the term **physical quality of life index** or PQLI.

Criteria for measuring development

As suggested above, these fall into two main categories:

1 Economic development, based on GDP figures (used by EC countries) or GNP (used by the UN and the USA).
2 Cultural, social and welfare criteria including the PQLI described.

1 Economic wealth

Using GDP per capita figures, which two countries would you consider to be the world's 'richest' and 'poorest'? You may have assumed the richest to be the USA — in fact in 1987 it was third, after Liechtenstein and the United Arab Emirates (UAE). You may have perceived Ethiopia to be the poorest yet its GDP per capita in 1987 was greater than that of Bhutan, Cambodia, Laos, Vietnam and Afghanistan.

Although GDP figures are easier to measure and to obtain than other development indicators such as social well-being, there are limitations to their use and validity. GDP figures are more accurate in countries which have many economic transactions and where trade in goods, services and labour can be measured as they pass through a marketplace — hence the term 'market economies'. Where markets are less well developed, trading is done informally or through bartering and where much production is made in the home for personal subsistence, GDP figures are less reliable. Countries at war cannot provide data. Centrally planned, socialist economies operate in a largely non-market economy, play a relatively small role in international trade and include few services, so their GDP figures are difficult to calculate and interpret.

Comparison of GDP requires the use of a single currency — generally US dollars — but currency exchange rates fluctuate. The size and growth of GDP may prove to be poor long term economic indicators and do not take human and natural resources into consideration. GDP per capita is a crude average and hides extremes and uneven distribution of income between regions and across socio-economic groups, especially in

▼ **Figure 13.39**

Terms used in relation to economic development

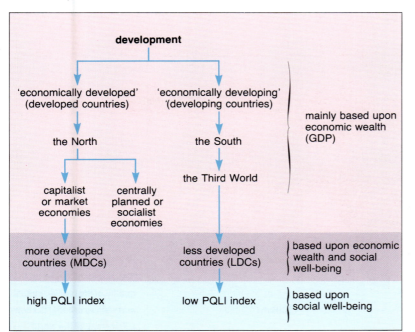

Third World countries where there may be very few immensely wealthy people while the majority live at subsistence level. Despite these limitations, GDP is still regarded as a relatively good indicator of development (Figure 13.40).

2 Social, cultural and welfare criteria

The PQLI is the average of three characteristics: literacy, life expectancy and infant mortality. Each is given an index scale ranging from 0 to 100:

- Literacy rates of zero and 100 per cent would be scaled as 0 and 100 respectively.
- Sierra Leone, which has the world's shortest life expectancy at 35 years, is scaled 0; Norway, with the longest expectancy of 77 years, is given an index of 100.
- Cambodia, with the world's highest infant mortality rate of 260/1000 has an index of 0; Sweden, with only 8/1000, is 100.

The resultant PQLI map of the world is shown in Figure 13.41.

There are other criteria which may be used to measure the quality of life as an indicator of the level of development. Several are linked to population, e.g. birth rates, natural increase and percentage of the population aged under fifteen. Death and infant mortality rates may also reflect nutrition, health and medical care.

In Third World countries the prevalence of disease may result from an inadequate diet, a lack of clean water and poor sanitation — a situation often aggravated by the limited numbers of doctors and hospital beds per person. The majority of people live in rural areas and are dependent upon farming. Only a small percentage of the population work in manufacturing or service industries and the level of energy consumption is low. Many jobs are informal (Figure 17.31) and at a

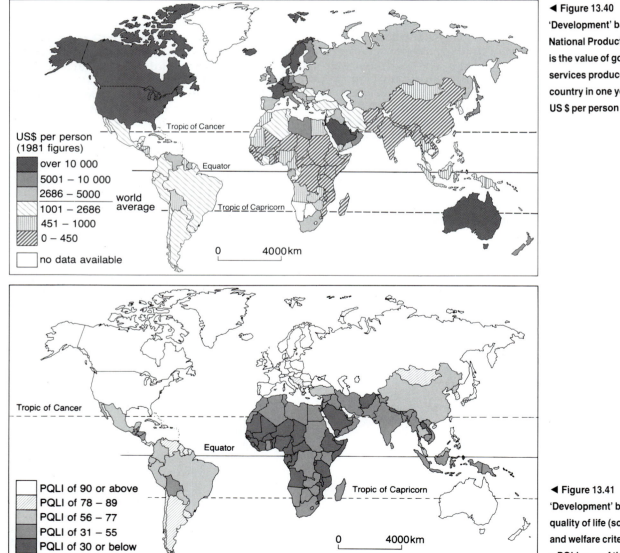

◄ Figure 13.40
'Development' based on Gross National Product (GNP) — GNP is the value of goods and services produced by each country in one year, measured in US $ per person

US$ per person (1981 figures)
- over 10 000
- 5001 – 10 000
- 2686 – 5000
- 1001 – 2686   world average
- 451 – 1000
- 0 – 450
- no data available

0   4000km

◄ Figure 13.41
'Development' based on the quality of life (social, cultural and welfare criteria) is shown in a PQLI map of the world

- PQLI of 90 or above
- PQLI of 78 – 89
- PQLI of 56 – 77
- PQLI of 31 – 55
- PQLI of 30 or below

0   4000km

subsistence level. Third World countries often import manufactured goods, energy supplies and sometimes even foodstuffs, especially grain. In return they may export raw materials for processing in the developed world, accumulate a trade deficit and get increasingly into debt. Internal trade tends to be limited. The density of communication networks, circulation of newspapers and numbers of cars, tractors, telephones and television sets per household or per capita are also associated with development.

Correlation and development

In the previous section it was suggested that there was a correlation between certain criteria and the level of development. Correlation is used in this sense to describe the degree of association between two sets of data. This relationship may be shown graphically by means of a scattergraph. This involves the drawing of two axes — the horizontal or 'x' axis and the vertical or 'y' axis. Usually one variable to be plotted is dependent upon the second variable. It is conventional to plot the **independent** variable on the x axis and the **dependent** variable on the y axis.

Figure 13.42 shows two relationships, one from physical geography and the other from human and economic. The physical example shows rainfall as the independent variable with runoff being dependent upon rainfall. The human-economic example shows GDP as the independent variable and energy consumption per capita to be dependent upon this measure of the wealth of the country.

The data is plotted against the scales of both axes. The degree of correlation is estimated by the closeness of these points to a **best fit** line. This line is usually drawn by eye and shows any trend in the pattern indicated by the location of the various points. One or two points, **residuals**, may lie well beyond the best fit line and, being anomalous, may be ignored at this stage. (Later it may be relevant to try to account for these anamolies or exceptions.)

The best fit line may be drawn as a straight line (on an arithmetic scale) or a smooth curve (on log or semi-log scales). If all of the points fit exactly on the best fit line, there is perfect correlation between the two variables. However, most points at best will lie close to and on either side of the drawn line. A **positive correlation** is where both variables increase, i.e. the best fit line rises from the bottom left towards the top right (Figure 13.43a and b). A **negative correlation** occurs where the independent variable increases as the dependent variable decreases, i.e. the best fit line falls from the top left to the bottom right (Figure 13.44d and e). In some instances the arrangement of the points makes it impossible to draw in a line, in which case the inference is that there is no correlation between the two sets of data chosen (Figure 13.43c). In the event of one, or both, of the variables having a wide range of values then it may be advisable to use a logarithmic scale (page 52).

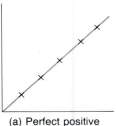

▶ Figure 13.42
Plotting the dependent and independent variable to show positive and negative correlations

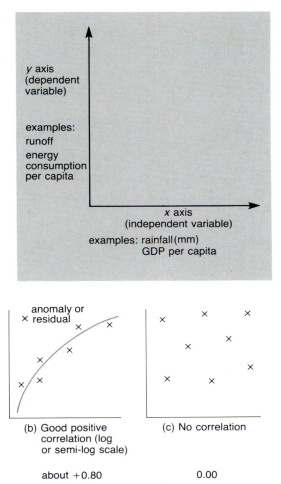

▼ Figure 13.43
Types of correlation and their associated Spearman Rank co-efficients

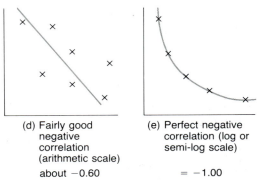

(a) Perfect positive correlation (arithmetic scale)

= + 1.00

(b) Good positive correlation (log or semi-log scale)

about +0.80

(c) No correlation

0.00

(d) Fairly good negative correlation (arithmetic scale)

about −0.60

(e) Perfect negative correlation (log or semi-log scale)

= −1.00

Q

The table in Figure 13.44 lists the GDP and energy consumption per capita for 15 selected countries.

1 Using only the information in these two columns, draw a scattergraph.

2 Attempt to draw in the best fit line.

3 Does there appear to be a strong or a weak correlation between GDP and energy consumption?

4 Is this correlation positive or negative?

5 Explain any relationship in words.

It is possible to use an appropriate statistical test to see if there is any correlation and to quantify the relationship. Of course, this is only worth doing if your scattergraph indicated the possibility of such a correlation.

	GDP per capita (US $)	Rank	Energy consumption per capita (kg coal-equivalent)	Rank	d	d²	Birth rate per 1000	Rank	d	d²
Switzerland	8246	1	3642	5	4	16	12.0			
West Germany	6451	2	5345	2	0	0	9.8			
Norway	6400	3	4607	3	0	0	13.3			
France	5859	4	3944	4	0	0	13.6			
Japan	4485	5	3622	7	2	4	16.3			
New Zealand	3663	6	3111	8	2	4	18.5			
USSR	2600	7	5546	1	−6	36	18.5			
Hungary	2100	8	3624	6	−2	4	17.5			
Argentina	1900	9	1754	9	0	0	22.9			
Turkey	980	10	630	11	1	1	39.6			
Colombia	540	11	671	10	−1	1	30.0			
Zambia	390	12	504	12	0	0	51.5			
Egypt	310	13	405	13	0	0	35.5			
Kenya	220	14	174	15	1	1	48.7			
India	132	15	221	14	−1	1	34.6			
(Source: JMB, 1982)						$\Sigma d^2 = 68$				$\Sigma d^2 =$

The Spearman rank correlation coefficient

This is a statistical measure to show the strength of a relationship between two variables. Figure 13.44 lists the GDP per capita for 15 selected countries. Fifteen is the minimum number needed in a sample for the Spearman rank rule to be valid. The first stage is to see if there is any correlation between the GDP and the energy consumption per capita. This can be done using the following steps.

1 Rank both sets of data. This has already been done in Figure 13.44. Notice that the highest value is ranked first. Had there been two or three countries with the same data then they would have been given equal ranking, e.g. rank order: 1, 2, 3.5=, 3.5=, 5, 7=, 7=,7=,9, 10.

2 Calculate the difference, or d, between the two rankings. Note that it is possible to get a negative answer.

3 Calculate d^2, to eliminate the negative values.

4 Add up, or sum, all of the d^2 values (in this example the answer is 68).

5 You are now in a position to calculate the correlation coefficient, or r, by using the formula:

$$r = 1 - \frac{6\Sigma d^2}{n^3 - n}$$

▲ Figure 13.44
Spearman Rank data for GDP and energy consumption for different countries

where d^2 is the sum of the squares of the differences in rank of the variables, and n is the number in the sample.

So in our example it follows that:

$$r = 1 - \frac{6 \times 68}{3375 - 15}$$

$$r = 1 - \frac{408}{3360}$$

$r = 1 - 0.12$ (then do not forget the final subtraction)

$r = 0.88 \qquad$ (it is usual to give an answer correct to two decimal places)

In this example there is a strong, positive correlation (a perfect positive correlation is 1.00) between the GDP and the energy consumption per capita.

Q Figure 13.44 also lists figures for the birth rates of the 15 selected countries. Using the Spearman method of rank correlation, calculate the correlation coefficient between GDP per capita and the birth rate. (This time you should find that your answer shows a very good negative correlation.)

So although the closer r is to $+1$ or -1 the stronger the likely correlation, there is a danger in jumping to quick conclusions. It is possible that the relationship described has occurred by chance. Figure 13.45 is a graph used to test the **significance** of the relationship. Note that the correlation coefficient r is plotted on the y axis and the **degrees of freedom** on the x axis. Degrees of freedom is the number of pairs in the sample minus two.

Using the correlation coefficient between GDP per capita and energy consumption per capita, which we have worked out to be 0.88, then we can read off 0.88 on the vertical scale and 13 (i.e. 15 in sample minus 2) on the horizontal. We can see that the reading lies on the 0.1 per cent significance level curve. This means that we can say with 99.9 per cent confidence that the correlation has not occurred by chance. The graph also shows that if the correlation falls below the 5 per cent significance level curve then we can only say with less than 95 per cent confidence that the correlation has not occurred by chance.

Below this point the correlation or hypothesis is rejected in terms of statistical significance, i.e. there is too great a likelihood that the correlation has occurred by chance for it to be meaningful.

Even if there is a significant correlation, the result does not prove that there is necessarily a *causal* relationship between variables. It cannot be assumed that a change in data A causes a change in data B. Further investigation is necessary to establish this.

▼ **Figure 13.45**
The significance of the Spearman Rank correlation — coefficients and degrees of freedom

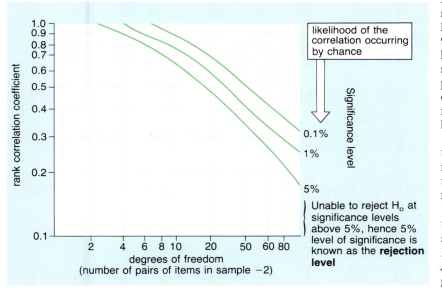

References

Annual Abstract of Statistics 1988, HMSO, 1988.

An Introduction to Population Geography, W.F. Hornby and M. Jones, Cambridge University Press, 1980.

Famine and Flood in Bangladesh, Open University/BBC Television (plus booklet), 1983.

Human Geography, M.G. Bradford and W.A. Kent, Oxford University Press, 1977.

Human Geography, K. Briggs, Hodder & Stoughton, 1982.

Human Geography, Patrick J. McBride, Blackie, 1980.

Landstat Development Project, Dr. R. Dutton, L.D.P., 1987.

North–South: a Programme for Survival, Report of the Brandt Commission, Pan Books, 1980.

Pattern and Process in Human Geography, Vincent Tidswell, U.T.P., 1976.

Small Area Statistics, HMSO, 1983.

The New Geographical Digest, George Philip & Sons, 1986.

The Third World, M. Barke and G. O'Hare, Oliver & Boyd, 1984.

The World, David Waugh, Thomas Nelson, 1987.

Times Atlas of World History, Times Books, 1979.

Turkish Migration/Open University, BBC Television (plus booklet), 1983.

CHAPTER 14

Settlement

'The largest single step in the ascent of man is the change from nomad to village agriculture.'

The Ascent of Man, J. Bronowski, BBC 1973

Early settlement

About 8000 BC, at the end of the last ice age, the world's population consisted of small bands of hunters and collectors living in subtropical lands at a subsistence level. These groups of people, who were usually migratory, could support themselves only if everyone was almost continually involved in the search for food. At this time two major technological changes, known as the 'Neolithic revolution', turned the migratory hunter-collector into a sedentary farmer. The first was the domestication of animals (sheep and cattle) and the second the introduction and cultivation of new strains of cereals (wheat, rice and maize). This gradually led to food surpluses and enabled an increasing proportion of the community to specialise in non-farming tasks.

The evolution in farming appears to have taken place independently, but at about the same time, in three river basins: the Tigris-Euphrates in Mesopotamia, the Nile and the Indus (Figure 14.1). These areas had similar natural advantages:

- flat flood plains next to large rivers
- rich, fertile silt deposited by the rivers during times of flood
- a dry climate which maintained soil fertility as there was limited leaching (though these areas were more moist than they are today)
- a warm subtropical climate
- a permanent water supply from the rivers for domestic use and, as farming developed, for irrigation
- a dry climate allowing mud from the rivers to be used to build houses.

By 3000 BC larger towns and urban centres had developed with an increasingly wider range of functions. Administrators were needed to organise the collection of crops and distribution of food supplies; traders exchanged surplus goods with other urban centres; and early planners introduced irrigation systems. Many administrative functions were under the control of a priesthood or a secular ruler but in either

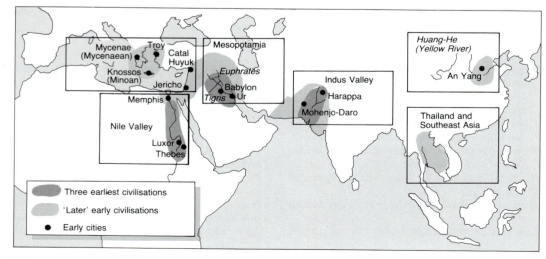

◄ Figure 14.1

Civilisations and cities before 1500 BC

case centred around a temple. Craftsmen were required to make farming equipment and household articles — the oldest known pottery and woven textiles were found at Catar Huyuk in present day Turkey — and copper and bronze were being worked by 3000 BC. As towns continued to grow it became necessary to have a legal system and an army to defend the settlement. Although there is divergence of opinion over exact dates, Figure 14.2 gives a chronological sequence of early settlements.

Site and situation of early settlements

Site describes the characteristics of the actual point at which a settlement is located and was of major importance in the early establishment and growth of a village or town. **Situation** describes where a place is in relation to its surroundings (e.g. neighbouring settlements, rivers and uplands). Situation, along with human and political factors, determined whether or not that settlement remained small or grew into a larger town or city.

Early settlements, similar, in most parts of the world, to those previously described in the Nile, Euphrates and Indus valleys, developed in a rural economy which aimed at self sufficiency, largely because transport systems were limited. The most significant factors in determining the site of a village included those listed below, but remember that several factors would usually operate together when a choice of site for settlement was being made.

- **Wet point sites** Without water there can be no life — human, animal or plant — and so an accessible source of water was essential. Water is needed daily throughout the year and is heavy to carry any distance. In earlier times, rivers were sufficiently clean to give a safe, permanent supply. In lowland Britain, many early villages were located at springs which formed at the foot of a chalk or limestone escarpment (page 164). In regions lacking rainfall, people lived where the water table was sufficiently near to the surface to allow wells to be constructed, e.g. a desert oasis.

- **Dry point sites** Elsewhere, the problem may have been too much water. In the English Fenlands or on coastal-marshes, villages were built on mounds which

▶ Figure 14.2

A chronology of early settlement

Approximate date BC		Near East	Tigris–Euphrates / Nile / Indus	Rest of world
9000			Hunters and collectors	
8000	8500		First domestic animals and cereals	Northern Europe recovering from the last ice age
	8300		Jericho first walled city	
7000				
6000	6250		Catal Huyuk — first pottery and woven textiles, became largest city in world	
5000	5500		Growth of villages in Mesopotamia	
			Growth of many villages in Nile and Indus valley	
	5000		Early methods of irrigation	Rice cultivation in SE Asia
4000			Bronze casting	
3000	3500		Invention of the wheel and plough in Mesopotamia and the sail in Egypt	First Chinese city
	3000		Cities in Mesopotamia	First crops grown in central Africa, bronze worked in Thailand
2000	2600		Pyramids	
	2000		Minoan civilisation in Crete	Metal working in the Andes
1000	1600		Mycenaean civilisation in Greece	

formed natural islands, e.g. Ely. Other settlements were built on river terraces (page 70) which were above the level of flooding and, in some cases, avoided the diseases which spread from stagnant water. In the Netherlands, farms are found along the raised canal banks.

■ **Building materials** Materials were heavy and bulky to move and transport was poorly developed so it was important to build settlements close to a supply of stone, wood or clay.

■ **Food supply** The ideal location was in an area which was suitable both for the rearing of animals and the growing of crops.

■ **Soils** Fertility of soils determined the quality and quantity of farm produce, the best soils being deep loams and the poorest sands and gravels.

■ **Relief** Flat, low-lying land, e.g. the North German Plain, was easier to build on than steeper, higher ground.

■ **Fuel supply** Even tropical areas need fuel for cooking purposes as well as for warmth during colder nights. In many early settlements, firewood was the main source — as it still is in the least developed countries such as those in the Sahel.

■ **Defence** Protection against surrounding tribes was often essential. Jericho is the oldest city known to have had walls — it was built over 10 000 years ago in about 8350 BC. In Britain the two best types of defensive site were when the settlement was surrounded on three sides by water (e.g. Durham) or when built upon a hill commanding views over the surrounding countryside (e.g. Edinburgh). The latter type may have had problems with water supply.

■ **Nodal points** These included places where several valleys met to form a **route centre** (e.g. Carlisle, Paris) or which commanded communicating routes through hills (e.g. Dorking, Carcassonne).

■ **Bridging points** Settlements have tended to grow where routes had to cross rivers, initially as fords (e.g. Oxford) and later by means of bridges. Of greater significance for trade and transport was the lowest bridging point before a river entered the sea (e.g. Newcastle upon Tyne).

■ **Shelter and aspect** In Britain a south-facing slope has the advantage of being protected from the cold northerly winds and gains the maximum benefit from the sun's warmth (e.g. Torquay).

Whereas the factors listed above were all natural, today the choice of site for a new settlement is usually for political reasons, e.g. Israeli towns in the Gaza Strip, or for social (many of Britain's new towns), or economic considerations, e.g. towns on the Dutch Polders.

Q Figure 14.3 shows the original site of London and Figure 14.4 shows spring-line villages in SE England.

1 List as many advantages as you can for the initial growth of the two settlements.

2 What are the advantages of the five sites shown in Figure 14.5?

▼ Figure 14.3

The site of London

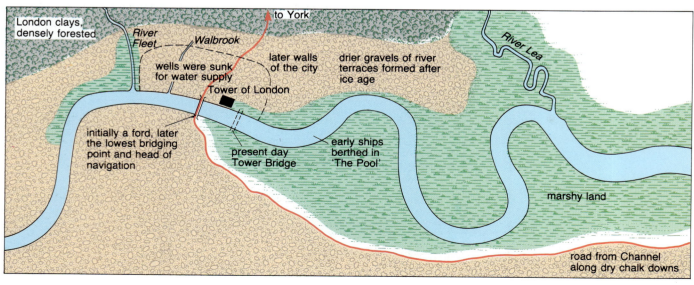

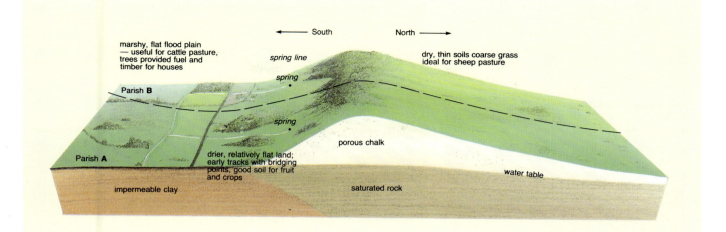

marshy, flat flood plain
— useful for cattle pasture,
trees provided fuel and
timber for houses

spring line

spring

Parish **B**

spring

Parish **A**

drier, relatively flat land;
early tracks with bridging
points, good soil for fruit
and crops

impermeable clay

← South   North →

dry, thin soils coarse grass
ideal for sheep pasture

porous chalk

water table

saturated rock

▲ **Figure 14.4**
**Spring-line villages — the North
Downs, England**

▶ ▼ **Figure 14.5**
**Five different sites for early
settlement**
a *right* **A defensive settlement
— Masada, Israel**
b *below left* **An oasis —
Morocco**
c *below right* **In the loop of a
meander — Durham, England**
d **page 334**, *top left* **An island in
a river — Paris, France**

e **page 344**, *top right* **Spring-line
settlement — Fulking
escarpment, Sussex, England**

Functions of settlements

As early settlements grew in size, each one tended to develop a specific function or functions. The **function** of a town relates to its economic and social development and refers to its main activities. There are problems in defining and determining a town's main function and, often due to a lack of data, e.g. employment and/or income figures, subjective decisions are made. As settlements have a great diversity then it is an advantage to try to group together those with a similar function. Any classification should, however, be consistent, relatively easy (convenient) to use and yet not too oversimplified nor generalised (Framework 3, page 130). The table lists categories for classifications put forward by Houston, Money, Harris and Nelson.

Classification of settlements based on function

J. Houston	D. Money	C. Harris	H. Nelson
Industrial	Industrial	Industrial	Manufacturing
Commercial	Commercial	Retail	Retail
Service	Cultural	Learning	Professional
Administrative	Communication	Transport	Transport
	Resort	Resort	Personal Service
	Administration	Political	Public Administration
	Mining	Wholesale	Wholesale
	Defence/Military	Mining	Financial
	Religious	Diverse	Mining
			Diverse

Of these, it could be argued that Houston's is too generalised to be useful, whereas Nelson's is the most consistent and precise as it was based upon census figures for the employment of people in each town or city. It should be realised though that some settlements grew based upon an activity which no longer exists or where its original function has changed over *time*, e.g. a Cornish fishing village may now be a holiday resort. Further, functions of towns in the developed world may be different from those in Third World countries, meaning that there is a difference over *space*.

One of the most important points to remember is that today, especially in developed countries, most towns and cities tend to be multifunctional and have numerous functions although one or two may predominate.

An alternative classification of settlement

Rural	19th Century (Developed)	20th Century (Developed)	20th Century (Third World)
Market and agricultural Route centre Small service town Defensive Dormitory and overspill	Mining Manufacturing Route centre Religious Trade/commerce	Administration Manufacturing Route centre Service Commercial Cultural/religious Resort/recreation Residential New towns	Administration Market and agricultural Route centre or port Mining Commercial Service

1 Explain what is meant by, and name an example of each of the following: market town, mining town, manufacturing/industrial town, route centre, administration town, commercial/retail settlement cultural and religious centre, resort, dormitory and overspill town.

2 What do you consider to be the merits and disadvantages of the four classifications listed in the tables above?

3 Can you devise your own classification of settlements? Ideally, it should not be too generalised nor too detailed but it will be valid providing you can *justify* your choice.

Differences between rural and urban settlement

Figure 14.6 shows the commonly accepted types of settlement, but hides the divergence of opinion as to how and where to draw the borders between each type. Several methods have been used to try to define the difference between a village or rural settlement and a town or urban settlement.

- **Size** There is a wide discrepancy of views over the minimum size of population required to enable a settlement to be termed a town, e.g. in Denmark it is considered to be 250 people, in France 2000, in the USA 2500, in Spain 10 000 and in Japan 30 000. In India, where many villages are larger than British towns, a figure of less than 25 per cent engaged in agriculture is taken to be the dividing point.
- **Occupations** Rural settlement may be defined as those areas where most of the workforce is farmers or is engaged in other primary activities: urban areas are where the workforce is employed in secondary and service industries.
- **Services** (range of functions) Service provision such as schools, hospitals, shops, public transport and banks are usually absent or limited in rural areas.
- **Land use** In rural areas, settlements are widely spaced with open land between adjacent villages. Within each village there may be individual farms as well as residential areas and possibly small scale industry. In urban areas, settlements are often packed closely together and within towns there is a greater mixture of land use with residential, industrial, services and open space provision.
- **Age structure** Rural settlements, especially those in more remote areas, may have more inhabitants in the over 65 age group. Urban areas are more likely to have most of their inhabitants within the economically active age group (page 296) or under secondary school age.

It is becoming increasingly difficult to differentiate between villages and towns as urban areas spread outwards into the rural fringe.

▼ Figure 14.6

A hierarchy of settlement types

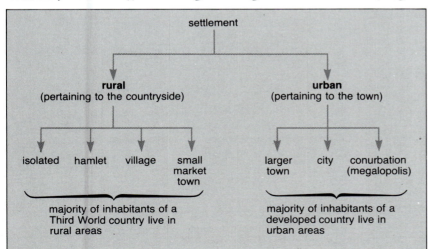

Rural settlement

Morphology

Recently geographers have become increasingly interested in the **morphology**, the pattern or shape, of villages and small towns rather than emphasising their particular function. Although village shapes vary spatially in Britain and across the world, it is possible to recognise seven types. (Remember that, as in other classifications, some geographers may identify more or fewer than seven categories.) Examples of each type are shown in Figure 14.7.

1 Isolated This usually refers to an individual building found in an area of extreme physical difficulty where the natural resources are insufficient to maintain more than a few inhabitants, e.g. the Amazon rainforests where tribes live in a communal home called a maloca. Isolated houses may also be found in planned pioneer areas as on the Canadian Prairies where the land was divided into small squares each with its own farmbuildings.

2 Dispersed In this grouping, farmsteads are scattered, as in the Scottish Highlands where the community may consist of individual crofts dispersed along a road or raised beach (page 133). Each group may be separated from the next by 2 or 3 km of open space or farmland. Hamlets are common in rural areas of northern Britain, on the North German Plain (where their name *Urweiler* means 'primeval hamlet') and in sub-Saharan Africa.

3 Nucleated These are common in many rural parts of the world where buildings were grouped together for defensive or economic purposes. In Britain they were found every 3 or 4 km so that there was enough land around the hamlet or village to grow crops and rear animals to be self

▶▼ Figure 14.7

The morphology of villages

a *below left* Dispersed — Scottish Crofts

b *below right* Nucleated — a village on Dartmoor

c *bottom left* Dispersed — a hamlet in the Bolivian Andes

d *bottom right* Linear or ribbon — Combe Martin, Devon

e *opposite* Ring village — Kraito in the Amazon

f *opposite, far right* Planned settlement — Milton Keynes

sufficient. Clusters of buildings are often found at crossroads or 'T' junctions. The majority of villages in India fall into this category. More recently, villages in areas of guerilla warfare have become increasingly compact in size, e.g. *kampongs* in Malaysia during the communist offensive of the 1950s.

4 Loose-knit These villages are similar to nucleated ones except that the buildings are more spread out, possibly along the routes leading from a crossroads.

5 Linear In many parts of the world settlement has become strung out along a main line of communciation or along a confined river valley. In Britain, linear settlement occurred along main roads following the availability of the private car and development of public transport, as well as along the narrow coalmining valleys of South Wales, the raised beaches of western Scotland and river terraces above the level of flooding. In areas as far apart as the Netherlands and Thailand, houses have been built alongside canals and waterways.

6 Ring and 'green' villages Ring villages are found in Southern Africa where *kraals* were built surrounding a open central area in which the tribal meetings and communal life took place. In northern England, villages were built around a central green, probably so that livestock could be protected during border raids.

7 Planned Historically, most villages developed a random shape because of their origins in pre-planning days. More recently, suburbanised villages surrounding large urban areas in, for example, Britain and the Netherlands, have expanded with small and often crescent shaped estates which house commuters to the larger settlements.

If you study maps of villages it is very likely that you will find many settlements with a mixture of the above shapes. It is possible that a British village may have a nucleated centre, a planned estate on its edges and a linear pattern extending along the road leading to the nearest large town.

Dispersed and nucleated rural settlement

Whether settlement is dispersed or nucleated depends upon local physical conditions; economic factors such as the time and distance between places; and social factors which include who owns the land and how the people of the area live and work on it. (Definitions are in the margin.)

Why are some settlements dispersed?
The more extreme the physical conditions and possible hardship of an area, the more the settlement is likely to be dispersed, e.g. in places which are very cold, wet, dry, rocky, high or steep. Similarly, dispersed settlement develops in areas where natural resources are limited and insufficient to support many people. This lack of resources could include a limited water supply, as on the Carboniferous limestone outcrops of the Pennines; forested areas of the Canadian Shield and the Amazon rainforests; and marginal farmland as in the Scottish Highlands and the Sahel countries, where farmers are usually pastoralists and the number of animals is limited by the quality and quantity of available grass. Areas with physical difficulties are less likely to have good transport networks which in turn discourages further settlement.

Forms of land tenure can also result in dispersed dwellings, especially in those parts of the world where inheritance laws have meant that the farm is successively divided between several sons. Similar patterns, though with larger farm units, can be found in pioneer areas such as the Canadian Prairies and the Dutch Polders. In more favourable parts of Prairies, land was divided into rectangles

Dispersed settlement is when there are isolated dwellings or small scattered hamlets where the buildings are spread out. Farms are usually away from the village.

Nucleated settlement consists of hamlets, villages and small market towns where buildings are grouped or clustered closely together. Individual farms may be incorporated within an actual hamlet or village.

measuring one square mile which in turn was subdivided equally into four individual farms. The Enclosure Acts in eighteenth-century Britain led to the enclosing of medieval open fields and the enlarging and consolidation of farms to give the now familiar hedgerows and walls and a more dispersed settlement pattern.

The 'agrarian revolution' in Britain in the eighteenth century was an acknowledgement that the open field system, in which strips were owned individually but the crops and animals were controlled by the community, was wasteful. It was replaced by enclosing several fields which were owned by a farmer who became responsible for all the decisions affecting that farm.

Two other changes about the same time increased the incidence of dispersed settlement. The first was the growth of large estates belonging to wealthy landowners. The second was the uphill extension of farming in hilly areas — initially in the eighteenth century as enclosures pushed upslope, and thereafter in the nineteenth century — to produce the extra food needed to feed the rapidly growing urban areas. Much moorland in the Pennines was walled and fenland areas previously limited in value were farmed, though often drainage was required first. Areas of downland were also put under the plough. This led to an increase in the number of people living in both towns and villages with the result that the latter may have become overpopulated. As the numbers of people became too great for the resources available, some members of the community moved to more isolated areas.

Settlement was more likely to develop a dispersed pattern where there was less risk of war or civil unrest because there was less need for people to group together for protection.

Why are some settlements nucleated?
The majority of humans have always preferred to live together in groups, as witnessed by the cities of ancient Mesopotamia and Egypt, through history up to the present day conurbations and cities with more than one million inhabitants (page 350). One major reason for people to nucleate has been a limited or excess water supply in an area. This includes settlements around springs as at the foot of chalk escarpments in southern England (Figure 14.4) and waterholes and oases in the desert, or on mounds in marshy fenland regions or on river terraces above the level of flooding (Figure 14.3). A second cause was the need to group together for defence and protection. Examples of defensive settlements include living in walled cities on relatively flat plains, e.g. Jericho and York; behind stockades as in African *kraals*; in hill-top villages in southern Italy and Greece; or in meander loops taking advantage of a natural water barrier, e.g. Durham (Figure 14.5c).

In Anglo-Saxon Britain, when many villages had their origin, the feudal open field system of farming encouraged nucleation as the local lord could better supervise his serfs if they were clustered around him. By living in the village the serf was probably more equidistant from his fragmented strips of farmland. Today, the more intensive the nature of farming, the more nucleated settlements tend to be. In the twentieth century people like to be as near as possible to services — the larger and more nucleated the village the more likely it is to have a wide range of services, e.g. a primary school, shops and public house.

Transport and routeways have always had a major influence on the clustering of dwellings. Buildings tend to be grouped together at crossroads and 'T' junctions; at gaps in hills; at bridging points and along main roads, waterways and railways. Compact settlement patterns are also found in areas with an important local resource, such as in the Durham coalmining or North Wales slate quarry villages, or where there was an abundance of building materials. More recently, governments have encouraged new, nucleated settlements in an attempt to achieve large scale self-sufficiency. Examples may be found as far afield as the Soviet collective farm, the Chinese commune, the Tanzanian *ujamaa* and the Israeli *kibbutz*.

1 Briefly describe the differences between a dispersed and a nucleated settlement pattern.

2 Illustrating your answer with specific examples from any part of the world, explain why some rural settlement patterns are dispersed whereas others are nucleated.

3 Why have some rural settlement patterns changed in the last few decades?

▶ Figure 14.9
Suburbanised villages

Changes in rural settlement in Britain

Within the British Isles there are areas where the rural population is increasing, resulting in a changing size, morphology and function of villages. In contrast, usually in more remote areas, there is rural depopulation. Figure 14.8 shows that there is some relationship between the type and rate of change of a rural settlement and its distance from a large urban area.

Accessibility to urban centres

As public and private transport improved during the interwar period (1919–1939), British cities expanded into the surrounding countryside at a rapid and uncontrolled rate. In an attempt to prevent this urban sprawl a **green belt** was created around London by the 1947 Town and Country Planning Act. The concept of a green belt, later established around most of Britain's conurbations, was to restrict the erection of houses and other buildings and to preserve and conserve areas of countryside for farming and recreational purposes. Beyond the green belt, **new towns** and **overspill towns** were built, initially to accommodate new arrivals to the nearby city and later those forced to leave it under slum clearance and other redevelopment schemes. These new settlements were designed to become self-supporting both economically and socially, and although in rural areas, they developed urban characteristics and functions.

Meanwhile, also beyond the green belt, uncontrolled growth continued in many small villages, despite the 1968 Town and Country Planning Act. Referred to during the interwar period as **dormitory** or **commuter** villages (page 315), these settlements have increasingly adopted some of the characteristics of nearby urban areas and have been termed **suburbanised villages**. Some of the changes evident in a village which has been suburbanised are listed in Figure 14.9.

▼ Figure 14.8
Rural settlements and distance from large urban areas

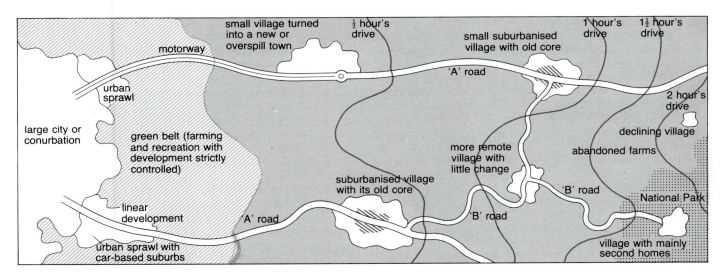

Characteristic	Original village	Suburbanised village
Housing	Detached, stone built houses with slate/thatch roofs. Some farms. Most over 100 years old. Barns.	New, mainly detached or semis. Renovated barns or cottages. Expensive planned estates, garages.
Inhabitants	Farming and primary jobs. Labouring/manual groups.	Professionals/executives, commuters. Wealthy with young families or retired.
Transport	Bus service, some cars, narrow/winding roads.	Decline in bus service as most families have one or two cars. Better roads.
Services	Village shop. Small junior school. Public house. Village hall.	More shops, enlarged school, modern public houses/restaurants.
Social	Small, close-knit community.	Local community swamped. Village may be deserted during day.
Environment	Quiet, relatively pollution-free.	More noise and risk of more pollution. Loss of farmland/open space.

Less accessible settlements

These villages are further in distance from or have poorer transport links to the nearest city, i.e. they are beyond commuting range. This makes the journey longer in time, more expensive and less convenient. Though these villages may be relatively stable in size their social and economic make-up is changing. Many in the younger age groups move out, pushed by a shortage of jobs and social life. They are replaced by retired people seeking quietness and a pleasant environment though often not realising that the rural areas lack many of the services required by the elderly such as shops, buses, doctors, and libraries.

Villages in National Parks or other areas of attractive scenery in upland or coastal areas are being changed by the increased popularity of **second** or **holiday homes** (Figure 14.10). The more wealthy urban dwellers, seeking relaxation away from the stress of their local working and living environment, buy vacant properties in villages. While this may bring trade to the local shop and improve the quality of some buildings, it means that the local inhabitants cannot afford the inflated house prices and many properties may stand empty for much of the year.

Remote areas

These areas suffer from a population loss which leaves houses empty and villages decreasing in size. Resultant problems include the lack of job opportunities, few services and poor transport facilities. Employment is often limited to the shrinking primary industries which tend to be low-paid and lacking in future prospects. The cost of providing services to remote areas is high and there is often insufficient demand to keep the local shop or village school open. With fewer inhabitants to use public transport, bus services may decline or stop altogether, forcing people to move to more accessible areas.

Measuring settlement patterns

Nearest neighbour analysis

Settlements often appear on maps as dots. Dot distributions are commonly used in geography, yet their patterns are difficult to describe. Sometimes patterns are obvious, such as when settlement is extremely nucleated or dispersed (Figure 14.11), but usually any description is subjective. This is because, in reality, the pattern lies between those two extremes. One way in which the pattern can be measured objectively is by using nearest neighbour analysis.

This technique was devised by a botanist who wished to describe patterns of plant distributions. It can be used to identify a tendency towards nucleation (clustering) or dispersion for settlements, shops, industry, etc. as well as plants. Nearest neighbour analysis gives a precision which enables one region to be compared with another and changes in distribution to be compared over a period of time, but it *is* only a technique and it does *not* offer any explanation of patterns.

The formula used in nearest neighbour analysis produces a figure (expressed as Rn) which measures the extent to which a particular pattern is clustered (nucleated), random, or regular (uniform) (Figure 14.11). **Clustered** is when all the dots occur very close to the same point. An example of this in Britain is on coalfields where mining villages tended to coalesce. In an extreme case, Rn would be zero.

Random distribution, where there is no pattern at all, has an Rn value of 1.0. The usual pattern for settlement is one which is predominantly random with a tendency either towards clustering or regularity.

Regular is a perfectly uniform pattern which, if ever found in reality, would have an Rn value of 2.15. This would mean that each dot was equidistant from all of its neighbours. The closest example of this in Britain is the distribution of market towns in East Anglia.

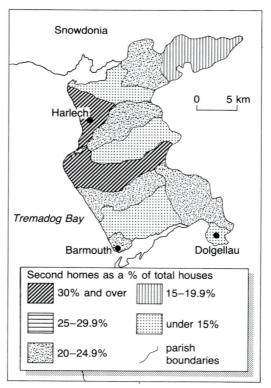

◄ Figure 14.10

Second homes as a percentage of the total number of houses in part of North Wales

▶ Figure 14.11

Nearest neighbour values (*Rn*)

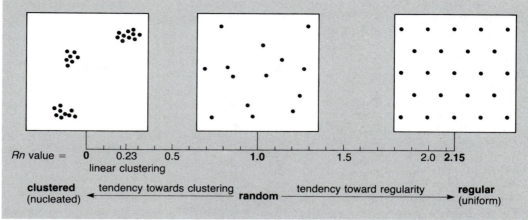

Rn value = 0   0.23   0.5   1.0   1.5   2.0 2.15
             linear clustering

clustered   ←— tendency towards clustering — **random** — tendency toward regularity —→   **regular**
(nucleated)   (uniform)

Using the technique of nearest neighbour analysis

Figure 14.12 shows settlements in northeast Warwickshire and southwest Leicestershire, a part of the English Midlands where it might be expected that there would be evidence of regularity in the distribution.

1 A total of 30 settlements were located — this is the minimum recommended for a nearest neighbour analysis. Each settlement was given a number.

2 The nearest neighbour formula was applied.

This formula is $Rn = 2\bar{d} \sqrt{\dfrac{n}{A}}$

where:

Rn = the description of the distribution
$\bar{d}$ = the mean distance between the nearest-neighbours (km)
n = the number of points (settlements) in the study area
A = the area under study (km²).

3 To find $\bar{d}$, measure the straight line distance between each settlement and its nearest neighbour, e.g. settlement 1 to 2, settlement 2 to 1, settlement 3 to 4 and so on. (It is permissible to refer back to previous points.) One point may have more than one nearest neighbour (e.g. settlement 8) and two points may be each others' nearest neighbour (e.g. settlements 1 and 2). In this example, the mean distance between all of the pairs of nearest neighbours was 1.72 km, i.e. the total distance between each pair, which was 51.7 km, divided by the number of points, which was 30.

4 Find the total area of the map, i.e. 15 km × 12 km = 180 km².

5 Calculate the nearest neighbour statistic, Rn, by substituting the formula. This has been done in Figure 14.12 and gives an Rn value of 1.41.

6 Using this Rn value, refer back to Figure 14.11 to determine how clustered or regular the pattern is. A value of 1.41

▼ Figure 14.12

Nearest neighbour analysis — a worked example for part of northeast Warwickshire and southwest Leicestershire

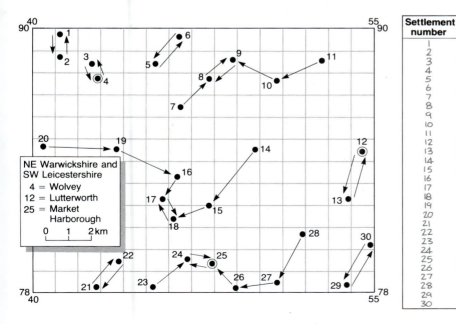

NE Warwickshire and
SW Leicestershire

4 = Wolvey
12 = Lutterworth
25 = Market
      Harborough

0   1   2km

Settlement number	Nearest neighbour	Distance (km)
1	2	1·0
2	1	1·0
3	4	0·6
4	3	0·6
5	6	1·6
6	5	1·6
7	8	1·8
8	9	1·3
9	8	1·3
10	9	2·1
11	10	2·2
12	13	2·2
13	12	2·2
14	15	3·3
15	18	1·7
16	17	1·3
17	18	1·0
18	17	1·0
19	16	3·0
20	19	3·2
21	22	1·6
22	21	1·6
23	24	2·1
24	25	1·1
25	24	1·1
26	25	1·5
27	26	1·8
28	27	2·5
29	30	2·2
30	29	2·2
		Σ 51·7

nearest neighbour formula

$Rn = 2\bar{d} \sqrt{\dfrac{n}{A}}$

In this example:

$Rn = 2 \times 1·72 \sqrt{\dfrac{30}{180}}$

$Rn = 3·44 \sqrt{0·17}$

$Rn = 3·44 \times 0·41$

$Rn = 1·41$

shows that there is a fairly strong tendency towards a regular pattern of settlement — which is what we had assumed might be the case for an area in the English Midlands.

7   However, there is a possibility that this pattern has occurred by chance. Referring to figure 14.13, it is apparent that the values of *Rn* must lie outside the shaded area before a distribution of clustering or regularity can be accepted as significant. Values lying in the shaded area at the 95 per cent probability level show a random distribution. (*Note:* with fewer than 30 settlements it becomes increasingly difficult to say with any confidence that the distribution is clustered or regular.) By reading this graph it confirms that our *Rn* value of 1.41 has a significant element of regularity.

How can the nearest neighbour statistic be used to compare two or more distributions? Figure 14.13 shows the *Rn* value for three areas in England, including that for the English Midlands. The *Rn* statistic for part of East Anglia has been calculated to be 1.57 which shows that the area has a more pronounced pattern of regularity than the Midlands. An *Rn* value of 0.61 for part of the Durham coalfield indicates that it has a significant tendency towards a clustered distribution.

Limitations and problems

As noted earlier, nearest neighbour analysis is a useful statistical technique but it has to be used with care. For example, the following must be taken into account.

1   The size of the area chosen is critical. Comparisons will be valid only if each area chosen is of the same size.

2   The area chosen should not be too large, as this lowers the *Rn* value (i.e. it exaggerates the degree of clustering), or too small, as this increases the *Rn* value (i.e. it exaggerates the level of regularity).

3   Distortion is likely to occur in valleys, where nearest neighbours may be separated by a river, or where spring line settlements are found in a linear pattern, as at the foot of a scarp slope (page 164).

4   How large are the settlements to be included? Are hamlets acceptable or is the village to be the smallest size? If so, when is a hamlet large enough to be called a village?

5   There may be difficulty in determining the centre of a settlement for measurement purposes, especially if it has a linear or a loose-knit morphology.

6   The boundary of an area is significant. If the area is a small island or lies on an outcrop of a particular rock then there is little problem but if, as in Figure 14.12, the area is part of a larger region then the boundaries must have been chosen arbitrarily (in this instance by predetermined grid lines). In this case it is likely that the nearest neighbour to some of the points (e.g. number 20) will be off the map. There is disagreement as to whether those points nearest to the boundary of the map should be included or not, but perhaps of more importance is the need to be consistent in approach and to be aware of the problems and limitations.

Despite these limitations, nearest neighbour analysis forms a useful basis for further investigation into why any clustering or regularity of settlement has taken place.

The rank–size rule

This is an attempt to find a numerical relationship between the population size of settlements within a country or county. The rule states that **the size of settlements is inversely proportional to their rank**. Settlements are ranked in descending order of population size with the largest or **primate** city placed first. The assumption is that the second largest city will have a population half that of the primate, the third largest city a population one-third of the primate, the fourth largest one-quarter of the primate, and so on.

The rank–size rule is expressed by the formula: $Pn = \dfrac{Pl}{n \ (\text{or } R)}$

where:

Pn = the population of the city
Pl = the population of the largest (primate) city
n (or R) = the rank–size of the city.

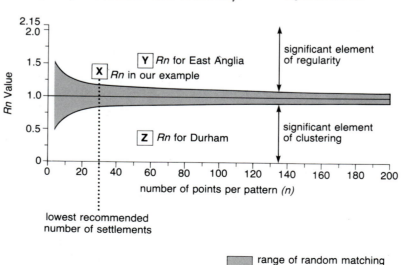

▼ Figure 14.13
Interpretation of the *Rn* statistic
— significant values

significant element of regularity

significant element of clustering

lowest recommended number of settlements

range of random matching (at the 95% probability level)

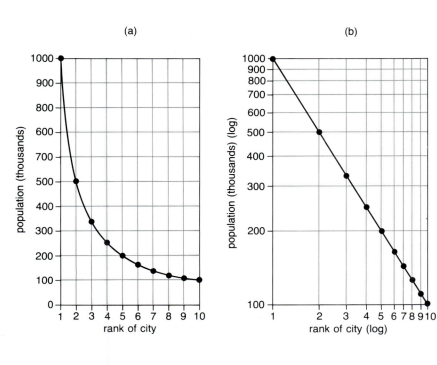

(a)

(b)

For example, if the largest city has a population of 1 000 000 then:

the second largest city will be 1 000 000 divided by 2 = 500 000;

the third largest city will be 1 000 000 divided by 3 = 333 333;

the fourth largest city will be 1 000 000 divided by 4 = 250 000.

If that perfect negative relationship actually occurred then it would produce a steeply downward-sloping, smooth, concave curve on an arithmetic graph (Figure 14.14a). However, it is more usual to plot the rank–size distribution on a logarithmic scale, in which case the perfect relationship would appear as a straight line sloping downwards at an angle of 45° (Figure 14.14b).

In reality it is rare to find a close correlation between the city-size of a country and the rank–size rule. Yet graphs drawn to show this rule may indicate that in some countries the primate city is very many times larger than the second and that in other countries there are two very large primate cities — a situation referred to as a **binary distribution**.

▲ Figure 14.14

The rank-size rule

Figure 14.15 shows the rank–size graph for Brazil. Below it are the actual populations for the ten largest cities in four other countries.

Rank	City	Actual	Estimated
1	Sâo Paulo	8493	
2	Rio de Janeiro	5091	4246
3	Belo Horizonte	1781	2831
4	Salvador	1502	2123
5	Fortaleza	1300	1698
6	Recifé	1204	1415
7	Brasilia	1177	1213
8	Porto Alegre	1125	1061
9	Nova Iguaçu	1095	943
10	Curitiba	1025	849

Brazil 1980
(thousands)

Rank	USA 1984	Actual	Italy 1983	Actual	Argentina 1980	Actual	Australia 1983	Actual
1	New York	17807	Rome	2831	Buenos Aires	9927	Sydney	3335
2	Los Angeles	12373	Milan	1561	Cordoba	982	Melbourne	2865
3	Chicago	8035	Naples	1209	Rosario	955	Brisbane	1138
4	Philadelphia	5755	Turin	1069	Mendoza	597	Adelaide	971
5	San Francisco	5685	Genoa	747	La Plata	560	Perth	969
6	Detroit	4577	Palermo	712	Tucuman	497	Newcastle	414
7	Boston	4027	Bologna	448	Mar Del Plata	407	Canberra	256
8	Houston	3566	Florence	441	Sante Fe	287	Wollongong	235
9	Washington	3429	Catania	380	San Juan	280	Hobart	174
10	Dallas	3348	Bari	370	Salta	260	Geelong	143

▲ Figure 14.15

1 For each of these countries work out the estimated population based on the rank–size rule, and then plot these on a logarithmic graph.

2 Describe the resultant relationship between actual city-size and the estimated rank–size.

3 The 1984 figures for the USA (given above) have a less close correlation to the rank–size rule than did the 1971 census figures. Can you suggest reasons for the change between 1971 and 1984?

Variation from the rank–size rule

Primate distribution (urban primacy) is where the largest city is many times larger than the second largest. Montevideo (Uruguay) is seventeen times larger and Lima (Peru) is twelve times larger than the second city. (Notice Buenos Aires in Figure 14.15.)

Binary distribution is where there are two very large cities of almost equal size: one may be the capital and the other the chief port or major industrial centre. Examples of binary distribution include Spain (Madrid and Barcelona) and Ecuador (Quito and Guayaquil). (Notice Brazil in Figure 14.15.)

It has been suggested (though there are many exceptions) that the rank–size rule is more likely to operate if the country is developed; has been urbanised for a long time; is large in size; and has a complex and stable economic and political organisation. In contrast, primate distribution is more likely to be found (also with exceptions, e.g. France and Austria) in countries which are small in size; less developed; former colonies of European countries; only recently urbanised and which have experienced recent changes in political boundaries.

Two schools of thought exist concerning the causes of variations in urban primacy. One suggests that as a city begins to dominate a country it attracts people, trade, industry and services at an increasingly rapid rate and at the expense of rival cities. The other claims that as a country becomes more urbanised and industrialised then growth of several cities tends to be stimulated, thus reducing the importance of the primate city — several older cities in developed countries are experiencing urban depopulation.

Central place theory

A **central place** is a settlement which provides goods and services. It may vary in size from a small village to a conurbation or primate city (Figure 14.16) and forms a link in a hierarchy. The area around each settlement which comes under its economic, social and political influence is referred to as either its **sphere of influence** or its **urban field**. The extent of the sphere of influence will depend upon the spacings, size and functions of the surrounding central places.

Functional hierarchies

Four generalisations may be made regarding the spacings, size and functions of settlements.

1 The larger settlements are in size the fewer in number they will be, i.e. there are many small villages but relatively few large cities.
2 The larger settlements grow in size, the greater the distance between them, i.e. villages are usually found close together while cities are spaced much further apart.
3 As a settlement increases in size the range and number of its functions will increase (Figure 14.16).
4 As a settlement increases in size the number of higher order services will also increase (see table).

Size, spacing and functions of settlements

Central place	Population	Distance apart (km)	Service area (km²)	Functions (services)
Hamlet	10–20	2	—	perhaps none
Village	→ 1 000	7	45	church, post office, shop, junior school
Small town	→ 20 000	21	415	shops, churches, senior school, bank doctor
Large town	→ 100 000	35	1 200	shopping centre, small hospital, banks senior schools
City	→ 500 000	100	12 000	shopping complex, cathedral, large hospital, football team, large bus and rail station, cinemas, theatre
Conurbation	> 1 million	200	35 000	shopping complexes, several CBDs
Capital or primate city	several million	—	whole country	government offices, all other functions

Note: the distances and service areas have been taken from Christaller's work in southern Germany (1933) with, in some cases, a rounding off of figures for simplicity. The population figures and functions are more applicable to the UK and the present time. Population, distances and service areas vary between and within countries and should be taken as comparative and approximate rather than absolute.

▶ Figure 14.16
Settlement hierarchy — the relationship between size and functions

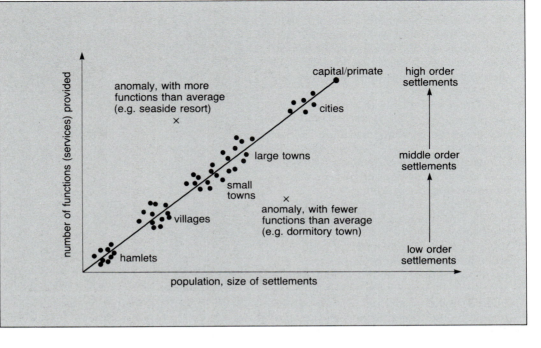

Central place functions are activities, mainly within the tertiary sector, that market goods and services from central places for the benefit of local customers and clients drawn from a wider hinterland.

Range and threshold of central place functions

The **range** of a service is the maximum distance that people are prepared to travel to use a particular good or service. It is dependent upon the value of the good, the length of the journey, and the frequency that the service is needed. People are not prepared to travel as far to buy a newspaper which they need daily than to buy furniture which they might purchase only once every several years.

The **threshold** is the minimum number of people required to support a particular service. As a rule, the more specialised the service, the greater the number of people needed. It is suggested that in the UK about 300 people are necessary for a village shop, 500 for a primary school, 2500 for a doctor, 10 000 for a senior school or a small chemist shop, 25 000 for a shoe shop, 50 000 for a small Marks and Spencer, 60 000 for a medium-sized Sainsburys, 100 000 for a large John Lewis and over one million for a university.

Delimiting the sphere of influence

The size of the sphere of influence will depend upon the answers to the following questions.
1 Is the surrounding land flat or are there water and relief barriers?
2 How good are local communications?
3 Are there any political barriers?
4 How densely populated and affluent is the surrounding area?
5 Are there rival urban areas?

The sphere of influence of a settlement may be determined by choosing criteria such as the area served by the local newspaper, hospital, main furniture shop or superstore, bank, senior school and local government offices.

Christaller's model of central places

Walter Christaller was a German who, in 1933, published a book in which he attempted to demonstrate a sense of order in the spacings and functions of settlements. He suggested that there was a pattern in the distribution and location of settlements of different sizes and also in the ways in which they provided services to the inhabitants living within their sphere of influence. Regardless of the level of service provided, he termed each settlement a **central place**. Although Christaller's **central place theory** was based upon investigations in southern Germany, and it was not translated into English until 1966, his work has contributed a great deal to the search for order in the study of settlements.

The two principles underlying Christaller's theory were the **range** and the **threshold** of a goods and services. He made a set of assumptions which were similar to those of two earlier German economists, von Thünen in his agricultural land use model (page 387) and Weber in his industrial location theory (page 434).

These assumptions were as follows.
- There was an unbounded isotrophic (flat) plain so that transport was equally easy and cheap in all directions. Transport costs were proportional to distance from the central place and there was only one form of transport.
- Population was evenly distributed across the plain.
- Resources were evenly distributed across the plain.
- Goods and services were always obtained from the nearest central place so as to minimise distance travelled.
- All customers had the same purchasing power (income) and made similar demands for goods.
- Some central places offered only low order goods, for which people were not prepared to travel far, and so had a small sphere of influence. Other central places offered higher order goods, for which people would travel further, and so they had much larger spheres of influence (page 344). The higher order central places provided both higher and lower order goods.
- No excess profit would be made by any one central place, but each would locate as far away as possible from a rival to maximise profits.

The ideal shape for the sphere of influence of a central place is circular, as then the distance from it to all points on the boundary is equal. However, when circles are drawn they produce a series of trade areas where some parts are left unserved by any central place (Figure 14.17a) while others are served by more than one central place (Figure 14.17b).

If the boundary is drawn midway through the overlapping trade areas this gives a series of hexagons (Figure 14.17c). A hexagon is almost as efficient as a circle in terms of accessibility from all points of the plain and is considerably more efficient than a square or triangle (Figure 14.17d). A hexagonal pattern also produces the ideal shape for the location of central places with different levels of functions — the village, town and city of Christaller's hierarchy. Figure 14.18 shows a large trade area for the third order central place, a smaller trade area for the six second order central places, and even smaller trade areas for the 24 first order central places.

By arranging the hexagons in different ways Christaller was able to produce three different patterns of service or trading areas. He called these $k = 3$, $k = 4$ and $k = 7$, where k is the number of places dependent upon the highest order central place.

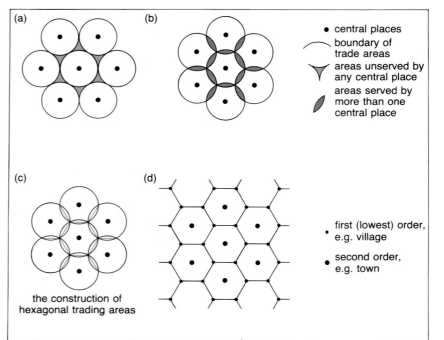

▲ Figure 14.17
Theoretical shapes of spheres of influence around settlements (*after* Christaller)

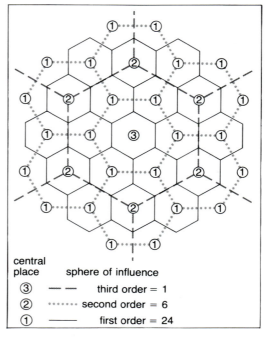

central place | sphere of influence
③ — — — third order = 1
② ·········· second order = 6
① ———— first order = 24

◀ Figure 14.18
Christaller's central places and spheres of influence

The following should be noted at this point. Where $k = 3$, the trade area of the third (i.e. highest) order central place is three times the area of the second order central place, which in turn is three times larger than the trade area of the first (lowest) order central place.

Where $k = 4$, the trade area of the third order central place is four times the area of the second order central place, which is four times larger than the trade area of the first order central place.

Where $k = 7$, the trade area of each order is seven times greater than the order beneath it.

$k = 3$

The arrangement of the hexagons in this case is the same as given in Figure 14.18 and the explanation of how $k = 3$ is reached is shown in Figure 14.19. In Figure 14.19:

A = the central place or third order settlement,

B, C, D, E, F and G are six second order settlements surrounding A,

U, V, W, X, Y and Z are some of the 24 first order settlements which lie between A and the second order settlements.

It is assumed that one-third of the inhabitants of Y will go to A to shop, one-third to D and one-third to E. Similarly, one-third of people living at X will shop at A, one-third at D and one-third at C. This means that A will take one-third of the customers from each of U, V, W, X, Y and Z ($6 \times \frac{1}{3} = 2$) plus all of its own customers (1). In total, A therefore serves the equivalent of three central places ($2 + 1$).

Christaller based the $k = 3$ pattern on a **marketing principle**, i.e. it maximises the number of central places and so brings the supply of higher order goods and services as near as possible to all of the dependent settlements and therefore to the inhabitants of its trade area.

$k = 4$

In this case, the size of the hexagon is slightly larger and it has been re-oriented (Figure 14.20). The first order settlements, again labelled U, V, W, X, Y and Z, are now located at the mid-point of the sides of the hexagon instead of at the apex as in $k = 3$. Customers from Y now have a choice of only two markets, A and N, and it is assumed that half of those customers will go to A and half to N. Similarly, half of the customers from X will go to A and the other half to M. So A will now take half of the customers from each of the six settlements at U, V, W, X, Y and Z ($6 \times \frac{1}{2} = 3$) plus all of its own customers (1) to serve the equivalent of four central places ($3 + 1$).

This pattern is based on a **traffic principle**, whereby travel between two centres is made as easy and cheap as possible. The central places are located so that the maximum number may lie on routes between the larger settlements.

$k = 7$

Here the pattern shows the same high order central place, A, but here all of the lower order settlements, U, V, W, X, Y and Z lie within the hexagon or trade area (Figure 14.21). In this case all of the customers from the six smaller settlements will go to A ($6 \times 1 = 6$) together with all of the inhabitants of A (1). This means that A serves seven central places ($6 + 1$). As this system makes it efficient to organise or control several places, and as the loyalties of the inhabitants of the lower order settlements to a higher one are not divided, it is referred to as the **administrative principle**.

▶ **Figure 14.19**
Christaller's $k = 3$

▼ **Figure 14.20**
Christaller's $k = 4$

▶ **Figure 14.21**, *below right*
Christaller's $k = 7$

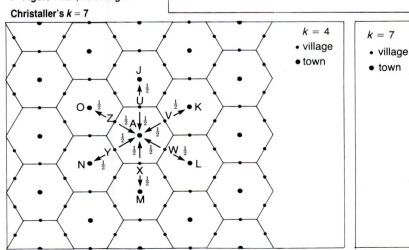

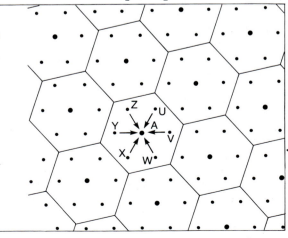

Why can no perfect example of Christaller's model be found in the real world? The answer lies mainly in the hypothetical basis from which he devised his model.

- Large areas of flat (isotropic) land rarely exist and the presence of relief barriers or routes along valleys means that transport is channelled in certain directions. There is more than one form of transport, costs are not proportional to distance and both systems and types of transport have changed since Christaller wrote in 1933.
- Neither people nor resources are evenly distributed.
- People do not always go to the nearest central place, e.g. they may travel much further to a new edge-of-city hypermarket.
- People do not all have the same purchasing power nor make equal demands.
- Governments often have control over the location of industry and of new towns.
- Perfect competition is unreal and some firms make greater profits than others.
- Christaller saw each central place as having a particular function whereas in reality, places may have several functions which change over time.
- Christaller's model does not really fit industrial areas, although there is some correlation with farming areas such as East Anglia and the Canadian Prairies.

Christaller has, however, provided us with a model with which we can test the real world, and without which we would have to make subjective descriptions about the spacings of settlements. His theories have helped geographers and planners to locate new services such as retail outlets and roads.

Interaction or gravity models

These models aim to predict the degree of interaction between two places. They are derived from Newton's law of gravity which states that **any two bodies attract one another with a force that is proportional to the product of their masses and inversely proportional to the square of the distance between them**. When used geographically, the words 'bodies' and 'masses' are replaced by 'towns' and 'population' respectively.

The interaction model in geography therefore suggests that as the size of one, or both, of the towns increases then there will also be an increase in movement between them. The further apart the two towns are, the less the movement there will be between them. This phenomenon is known as **distance decay**.

The model can be used for several purposes.
1 To estimate traffic flows between two places (page 476).

2 To predict migration between two areas.
3 To estimate the number of people likely to use one central place, e.g. a shopping area, in preference to a rival central place.

It can also be used to determine the sphere of influence of each central place by estimating where the **breaking point** between two settlements will be, i.e. the point at which customers find it preferable because of distance, time and expense considerations to travel to one centre rather than the other.

Reilly's law of retail gravitation (1931)

Reilly's interaction breaking point is a method used to draw boundary lines showing the limits of the trading areas of two adjacent towns or shopping centres. Unlike Christaller, Reilly suggested that there were no fixed trade areas, that these areas could vary in size and shape, and that they could overlap. His theory states that **two centres attract trade from intermediate places in direct proportion to the size of the centres and in inverse proportion to the square of the distances from the two centres to the intermediate place**.

This can be expressed by the formula:

$$Db = \frac{Dab}{1 + \sqrt{\frac{Pa}{Pb}}}$$

or similarly $djk = \dfrac{dij}{1 + \sqrt{\dfrac{Pi}{Pj}}}$

where:
Db (or djk) = the breaking point between towns A and B
Dab (or dij) = the distance (or time) between towns A and B
Pa (or Pi) = the population of town A (the larger town)
Pb (or Pj) = the population of town B (the smaller town)

Taking as an example Grimsby–Cleethorpes which has a population of 131 000 and Lincoln, 71 km away, with a population of 75 000 then the formula can be written as:

$$Db = \frac{71}{1 + \sqrt{\frac{131\,000}{75\,000}}}$$

which means $Db = \dfrac{71}{1 + 1.32}$

Therefore $Db = 30.58$,
i.e. the breaking point is 30.58 km from Lincoln (town B) and 40.42 km from Grimsby–Cleethorpes (town A). This is shown in Figure 14.22.

348

▶ Figure 14.22

Reilly's breaking point between settlements of different sizes

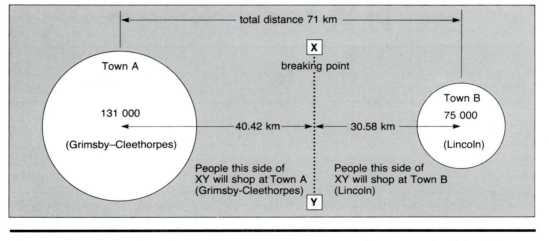

total distance 71 km

X

breaking point

Town A

131 000

(Grimsby–Cleethorpes)

40.42 km

30.58 km

Town B

75 000

(Lincoln)

People this side of XY will shop at Town A (Grimsby-Cleethorpes)

People this side of XY will shop at Town B (Lincoln)

Y

Q Scunthorpe has a population of 71 000 and is 59 km from Grimsby–Cleethorpes and 56 km from Lincoln. What is the breaking point between Scunthorpe and each of the other two settlements?

Limitations of Reilly's model

As with other models it is based on assumptions which are not always applicable to the real world.

In this case the assumptions are:

■ The larger the town the stronger its attraction.

■ People shop in a logical way, seeking the centre which is nearest to them in terms of time and distance.

These assumptions may not always be true.

■ There may be traffic congestion on the way to the larger town and car parking may be more difficult and expensive.

■ The smaller town may have fewer but better quality shops.

■ The smaller centre may be cleaner, more modern, safer and less congested.

■ The smaller town may advertise its services more effectively.

A variation on Reilly's law of retail gravitation

Like central place theory, Reilly's law seems to fit rural areas better than closely packed, densely populated urban areas. A variation on Reilly's law is based upon the drawing power of shopping centres (i.e. the number and type of shops in each) rather than distance between the two towns.

This version of the gravity model has the formula:

$$Db = \frac{Dab}{1 + \sqrt{\dfrac{Sa}{Sb}}}$$

where: Sa = the number of shops in town A.
Sb = the number of shops in town B.

Referring to our original example, suppose Grimsby–Cleethorpes has 800 shops and Lincoln has 300 shops.

The formula could then be written:

$$Db = \frac{71}{1 + \sqrt{\dfrac{800}{300}}}$$

Therefore $Db = 27$

This means that out of every 71 shoppers, 44 would go to Grimsby–Cleethorpes and 27 to Lincoln.

Measuring settlement patterns: conclusion

Nearest neighbour analysis, the rank–size rule, Christaller's central place theory and the interaction models are all theories which are difficult to observe in the real world. Their value lies in the fact that they form hypotheses against which reality can be tested — provided you do not seek to *make* reality fit them! (Hypothesis testing, Framework 6.) When theory and reality diverge the geographer can search for an explanation for the differences. An important component of these theories is that they aim to find order in spatial distributions.

References

Human Geography: Concepts and Applications, K. Briggs, Hodder & Stoughton, 1982.

Human Geography: Theories and their Applications, M.G. Bradford and W.A. Kent, Oxford University Press, 1977.

Statistics in Geography for 'A' level Students, J. Wilson, Schofield & Sims, 1984.

Times Atlas of World History, Times Books, 1979.

CHAPTER 15

Urbanisation

'The invasion from the countryside . . . is overwhelming the ability of city planners and governments to provide affordable land, water, sanitation, transport, building materials and food for the urban poor. Cities such as Bangkok, Bogota, Bombay, Cairo, Delhi, Lagos and Manila each have over one million people living in illegally developed squatter settlements or shanty towns.'

Only One Earth, Lloyd Timberlake, BBC Enterprises Ltd 1987

Urbanisation is defined as the process 'whereby an increasing proportion of a nation's or region's population live in urban areas'. There is not, however, any global agreement as to what constitutes an 'urban area'. Although there are still more rural dwellers in the world than those living in urban areas, in the last century in developed countries and during the last 30 years in Third World countries there has been an unprecedented growth in the number and size of large cities (Figure 15.1).

On a world scale, rapid urbanisation has occurred twice in time and space.

1 In the nineteenth century in developed countries when industrialisation led to a demand for labour in mining and manufacturing centres. Urbanisation was therefore a consequence of development (page 446).

2 Since 1950 in Third World countries, caused by the twin processes of migra-

Number of cities with populations of more than:			
	1 million	5 million	10 million
1850	2	0	0
1900	11	0	0
1950	70	6	0
1985	273	29	10

tion from rural areas (page 307) and the high natural increase in population — the latter resulting from high birth rates and falling death rates (page 290). This trend, in many cases following independence from colonial rule, has been the result of better job opportunities and service provision in a few large cities. Urbanisation is an integral and fundamental part of development. Figure 15.2 on page 352 shows national levels of urbanisation in 1985.

Percentage of world population living in urban areas				
Area	**1950**	**1970**	**1985**	**2000(est)**
World	29.2	37.1	41.0	46.6
More developed regions	53.8	66.6	71.5	74.4
Less developed regions	17.0	25.4	31.2	39.3
Africa	15.7	22.5	29.7	39.1
Latin America	41.0	57.4	68.9	76.8
North America	63.9	73.8	74.1	74.9
Eastern Asia	16.8	26.9	28.6	32.8
South Asia	16.1	21.3	27.7	36.5
Europe	56.3	66.7	71.6	75.1
Oceania	61.3	70.8	71.0	71.3
USSR	39.3	56.7	65.6	70.7

Note: the UN prediction for the world in 2025 is 60%

◄ Figure 15.1
The proportion of world population living in urban areas

Simultaneous with urbanisation has been the growth of large cities. The 'million city' is now commonplace, even in relatively rural and less developed areas, and the number of cities exceeding five and even ten million is increasing.

Also changing is the spatial distribution of these 'million cities'. Prior to 1940, the majority of cities with over one million inhabitants were found in the temperate latitudes of the northern hemisphere, i.e. 'the North' (page 325). Since 1940, there has been a dramatic increase in the proportion of 'million cities' found within the tropics and in the Third World countries, i.e. 'the South'. Whereas in 1940 71 per cent of these cities lay north of 40°N, by 1980 this figure had fallen to 29 per cent. In the 60 years between 1920 and 1980 the mean latitude of these largest cities had altered from 44°30' to 33°27'. The 1985 distribution of 'million cities' is shown in Figure 15.3 overleaf.

There has been some indication in the 1980s that the speed of growth of these large cities has begun to slow down. Figure 15.4 overleaf gives the 1985 estimated size of the world's twelve largest cities as predicted in 1980, and also the actual size of the 'top twelve' in 1985.
This table shows several trends.

1 The fastest growing cities were in the Japan–Korea region.
2 Most of the fastest growing cities were in the Third World countries, although their increase was less rapid than had initially been estimated. Cities in the least developed countries had the greatest increase.
3 China proved to be an anomaly because of its birth control policy, e.g. Shanghai in 1985 was only the twenty-first largest city with 6.7 million instead of being the fifth largest with 14.3 million previously predicted.
4 Cities in North America and western Europe were showing a decrease in overall size. This trend towards **counter-urbanisation** was due to improved transport, a decline in the heavy industries which had been concentrated in a few areas, and a greater development of services which favoured edge-of-city environments.

However, the table does not show differences in the population density between cities. Of the largest 85 cities in the world, the ten with the lowest population density are in the developed countries (nine are in North America) while the ten with the highest densities are in the developing countries (headed, in rank order, by Hong Kong, Lagos, Jakarta, Bombay, and Ho Chi Minh City). Despite this popular image that the largest cities in Third World countries are growing so rapidly, it should be remembered that over one-third of the inhabitants, especially in India and China, still live in smaller towns of between 20 000 and 100 000 people.

Models of urban structure

As cities have grown in area and population in the twentieth century, geographers and sociologists have tried to identify and to explain urban spatial patterns and variations in their structure.

1 Burgess, 1924

Burgess attempted to identify areas within Chicago. The basis for his model was the outward expansion of the city and the socio-economic groupings of its inhabitants. He used the then current concepts of plant ecology as had been presented by the Universty of Chicago and applied ecological terms such as **invasion**, **competition**, **dominance** and **succession** (Chapter 11) to groups of people (Figure 15.6).

Basic assumptions
Although the main aim of his model was to describe residential structures of a city, subsequently geographers have presumed that Burgess made certain assumptions. These included:
- The city was built upon a flat, isotrophic surface with equal advantages in all directions, i.e. he ignored morphological features such as river valleys.
- Transport systems were of limited significance being equally easy, rapid and cheap in every direction.
- Land values were highest in the centre of the city and declined with distance outwards to give a zoning of urban functions and land use.
- The oldest buildings were in or close to the city centre with progressively newer ones towards the city boundary.
- Cities contained a variety of well-defined socio-economic and ethnic areas.
- The poorer classes had to live near to the city centre and places of work as they could not afford transport or expensive housing
- There were no concentrations of heavy industry.

Burgess's concentric zones
Figure 15.5 on page 353 shows five concentric zones.

1 The **Central business district** contains the major shops and offices and is the centre for commerce, entertainment and the focus for transport routes.

351

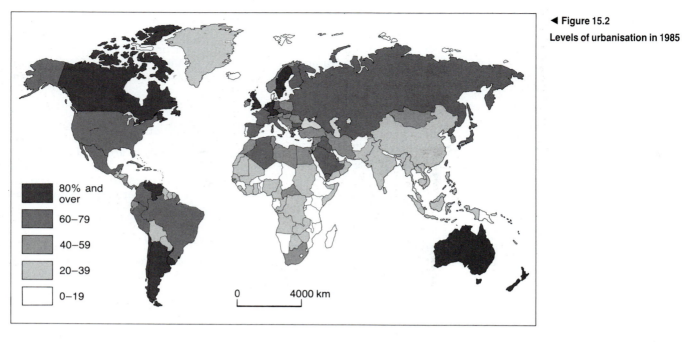

◄ Figure 15.2
Levels of urbanisation in 1985

Legend:
- 80% and over
- 60–79
- 40–59
- 20–39
- 0–19

0   4000 km

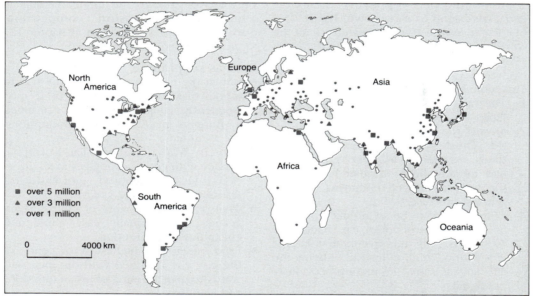

◄ Figure 15.3
Distribution of the world's
'million cities'

- ■ over 5 million
- ▲ over 3 million
- • over 1 million

0   4000 km

1985 estimate made in 1980 (millions)			1985, actual (millions)	
1	Tokyo	23.0	Tokyo	25.4
2	New York	18.0	Mexico City	16.9
3	Mexico City	17.9	São Paulo	14.9
4	São Paulo	16.8	New York	14.6
5	Shanghai	14.3	Seoul	13.7
6	Los Angeles	13.7	Osaka/Kobe	13.6
7	Calcutta	12.1	Buenos Aires	10.8
8	Bombay	12.1	Calcutta	10.5
9	Beijing	12.0	Bombay	10.1
10	Buenos Aires	11.7	Rio de Janeiro	10.1
11	Rio de Janeiro	11.4	Moscow	9.9
12	Seoul	11.2	Los Angeles	9.6

◄ Figure 15.4
Estimated (1980) and actual
(1985) size of the world's largest
cities

► Figure 15.5

The Burgess concentric model of an urban area

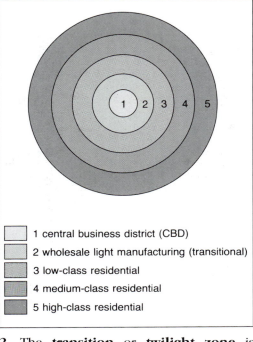

1 central business district (CBD)
2 wholesale light manufacturing (transitional)
3 low-class residential
4 medium-class residential
5 high-class residential

▼ Figure 15.6

Urban areas of Chicago (*after* Burgess)

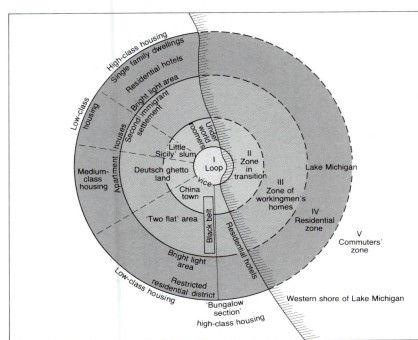

2 The **transition** or **twilight zone** is where the oldest housing is either deteriorating into slum property or being 'invaded' by light industry. The inhabitants tend to be of poorer social groups and first generation immigrants.

3 Areas of **low class housing** are occupied by those who have 'escaped' from zone **2**, or by second generation immigrants who work in nearby factories. They are compelled to live near to their place of work to reduce travelling costs and rent. In modern Britain these zones are equated with the inner cities.

4 **Medium class housing** of higher quality which, in present day Britain, would include interwar private semi-detached and council estates.

5 **High class housing** occupied by people who can afford the expensive properties and the high cost of commuting. This zone also includes the commuter or suburbanised villages beyond the city boundary — though there were few commuters when Burgess produced his model in 1924.

Criticisms of Burgess' model

Despite the advantage of its simplicity, and his own admission that it was specific to one city and to one period of time, Burgess' model has been criticised — in some instances on grounds which did not exist when it was produced. It has been tested against different examples in reality and the following flaws have been pointed out.

■ Zones are never, in reality, as clear cut as in the model and the following flows (Chicago, Figure 15.6).
■ Zones usually contain more than one type of land use.
■ Insufficient importance was attached to industry.
■ Cities are rarely built in flat plains.
■ Insufficient weighting was given to transport, though in 1924 there was limited public transport and few private cars.
■ The oldest housing is not always in or near the city centre, though in Burgess' time there had been little CBD modernisation or inner city redevelopment.
■ In 1924 there were no edge-of-city shopping centres or trading estates.
■ The model did not consider characteristics of cities in the Third World.

2 Hoyt, 1939

Hoyt's model was based on the mapping of eight housing variables for 142 cities in the USA. He tried to account for changes in, and the distribution of, residential patterns.

Basic assumptions

Hoyt made the same implicit assumptions as had Burgess (page 351), with the addition of three new factors.

■ Wealthy people, who could afford the highest rates, chose the best sites, i.e. competition based on 'ability to pay' resolved land use conflicts.
■ Wealthy residents could afford private cars or public transport, so lived further from industry and nearer to main roads.
■ Similar land uses attracted other similar land uses, concentrating a function in a particular area and repelling others. This 'attract and repel' process led to 'sector' development.

Hoyt's sector model

Hoyt suggested that areas of highest rent tended to be alongside main lines of communication and that the city grew in a series of wedges (Figure 15.7). He also claimed that once an area had developed a distinctive land use it tended to retain that land use as the city extended outwards, e.g. if an area north of the CBD was one of low class housing in the nineteenth century, then the northern suburbs of that city in the late twentieth century would be most likely to consist of low class estates.

Criticisms of Hoyt's model

Many criticisms are similar to those of Burgess, e.g. cities are not always built on flat land, zones do not always have distinct boundaries and often contain more than one type of land use (Calgary, Figure 15.8).

- Hoyt based his model on housing and neglected other land uses.
- Areas of low cost housing do occur beside main roads near to the boundaries of most cities.
- It was assumed that there were no planning laws or restrictions.
- Only cities in the USA were studied.

This model was also put forward before the rapid growth of the car-based suburb, the swallowing up of small villages by urban growth, the redevelopment of inner city areas and the relocation of shopping, industry and office accommodation on edge-of-city sites.

3   Mann, 1965

Mann tried to apply the Burgess and Hoyt models to three industrial towns in northern England: Huddersfield, Nottingham and Sheffield. His compromise model (Figure 15.9) combined the ideas of Burgess' concentric zones and Hoyt's sectors. Mann assumed that because the prevailing winds in northern Britain blow from the west, then the high class housing would be in the west and industry, with its smoke, would be located to the east of the CBD. His conclusions are summarised as follows.

- The twilight zone was not concentric to the CBD but lay on the side of the city which led to the more wealthy residential areas.
- Heavy industry was found in sectors along main lines of communications.
- Low class housing should be called the 'zone of older housing'.
- Higher class or, in Hoyt's terms, 'modern' housing was usually found away from industry and smoke.

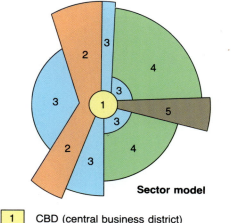

◄ Figure 15.7
The Hoyt sector model

Sector model

1	CBD (central business district)
2	wholesale light manufacturing (transitional)
3	low-class residential
4	medium-class residential
5	high-class residential

Figure 15.10 shows how Robson (1975) applied Mann's model to another northern industrial town, Sunderland. Although Mann's model is based on a small sample of three industrial towns, it does show that there are a variety of approaches to the study of urban structures.

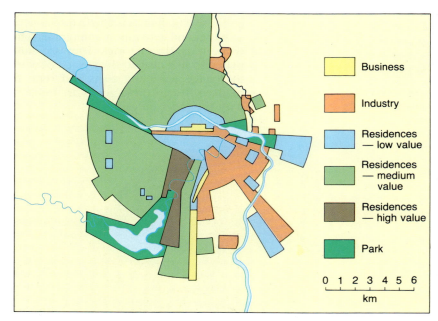

	Business
	Industry
	Residences — low value
	Residences — medium value
	Residences — high value
	Park

0 1 2 3 4 5 6
km

Ullman and Harris, 1945

Ullman and Harris set out to produce a more realistic model than those of Burgess and Hoyt but consequently ended with one that was more complex (Figure 15.11) — and more complex models may become descriptive rather than predictive if they match reality too closely in a specific example (Framework 8, page 293).

▲ Figure 15.8
Urban areas of Calgary, 1961 (*after* Hoyt). Calgary (Canada) is the standard example of a city exhibiting the characteristics of Hoyt's model (Case Study, page 386)

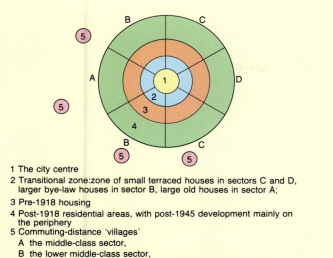

1 The city centre
2 Transitional zone:zone of small terraced houses in sectors C and D, larger bye-law houses in sector B, large old houses in sector A;
3 Pre-1918 housing
4 Post-1918 residential areas, with post-1945 development mainly on the periphery
5 Commuting-distance 'villages'
   A the middle-class sector,
   B the lower middle-class sector,
   C the working-class sector and main municipal housing areas,
   D industry and lowest working-class sector.

▲ Figure 15.9
The Mann model of urban structure

▶ Figure 15.10, *above right*
Urban areas of Sunderland, England (*after* Mann)

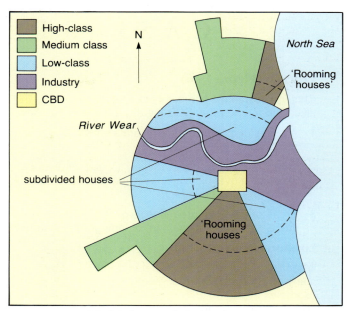

High-class
Medium class
Low-class
Industry
CBD

River Wear

subdivided houses

North Sea

'Rooming houses'

'Rooming houses'

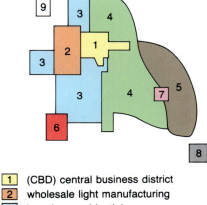

1 (CBD) central business district
2 wholesale light manufacturing
3 low-class residential
4 medium-class residential
5 high-class residential
6 heavy manufacturing
7 outlying business district
8 residential suburb
9 industrial suburb

▶ Figure 15.11
The Ullman–Harris multiple nuclei model

Basic assumptions

- Modern cities have a more complex structure than that suggested by Burgess and Hoyt.
- Cities do not grow from one CBD but from several independent nuclei.
- Each **nucleus** acts as a **growth point**, and probably has a function different from other nuclei within that city, e.g. administrative, retail or transport.
- In time there will be an outward growth from each nucleus until they merge as one large urban centre.
- If the city becomes too large and congested then some functions may be dispersed to new nuclei, e.g. an edge-of-city shopping centre or around a new airport.

Multiple nuclei developed as a response to the need for maximum accessibility to a centre, to keep certain types of land use apart, to differences in land values and, more recently, to decentralise.

Models of urban structure: conclusions

It must remembered that the four models described were put forward to try to explain differences in structure within cities in the developed world. If you have to study your local town or city you must avoid the temptation of saying that it *fits* one of those models — at best it will show characteristics of one or possibly two. Each city is unique and will have its own structure — a pattern not necessarily derived according to any previously devised model (Framework 8).

The land value model or bid–rent theory

This model of urban structure is similar to von Thünen's ideas of rural land use (Chapter 16) as it is based on locational rent. The main assumption is that in a free market the highest bidder will obtain the use of the land. The highest bidder is likely to be the one who can obtain the maximum profit from that site and so can pay the highest rent. Competition for land is keenest at the centre of the city. Figure 15.13 overleaf shows the locational rent that three different land users are prepared to pay for land at various distances from the city centre.

CASE STUDY

Chicago (Figure 15.6)

Chicago lies on the shores of Lake Michigan and its CBD, known as the 'Loop', faces the lake. Surrounding this CBD the city's housing has developed in a distinctive pattern.

1 A series of income sectors radiate outwards, e.g. higher class housing extends to the northwest.

2 Three concentric zones linked to age and family size, e.g. people in their early twenties or over 60 live close to the CBD while families with young children tend to live nearer the city boundary.

3 Areas of ethnic segregation, e.g. Chinatown and the black belt. Burgess suggested that a centrifugal movement of immigrants occurred over a period of time.

London (Figure 15.12)

When Ullman and Harris' model is applied to London (Figure 15.12) it can be seen that many of the nuclei were originally separate settlements. This is shown in the following examples.

■ the three specialist business districts of the City (finance), Westminster (administration) and the West End (retail and entertainment).

■ Adjacent to main line rail terminals (e.g. King's Cross and Paddington) or near to airports (e.g. Heathrow).

■ Suburban centres with their own CBDs, e.g. Richmond and Croydon.

■ Areas of heavy industry associated with the docks and the Lea Valley.

■ Lighter manufacturing centres in the inner areas of east London or the edge-of-city trading estates as at Slough.

■ Commuter settlements as at Wimbledon, Finchley and Bromley.

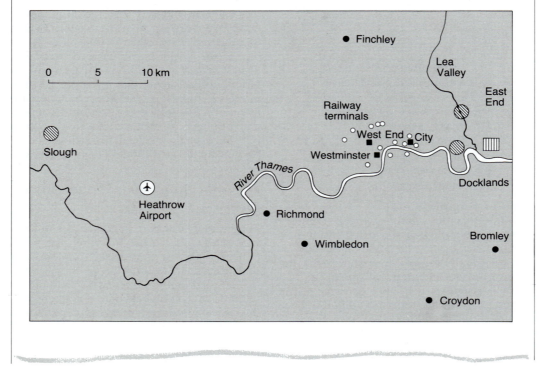

◀ Figure 15.12
London — examples of multiple nuclei using Ullman and Harris' model

The most expensive or 'prime site' in most cities is the CBD mainly because of its accessibility and the shortage of space there. Shops, especially department stores, conduct their business using a relatively small amount of groundspace and due to their high rate of sales and turnover they can bid a high price for the land (which they try to compensate by building upwards and by using the land intensively). The most valuable site within the CBD is called the **peak land value intersection** or PLVI —

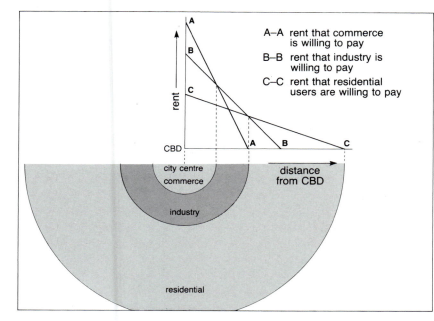

▲ Figure 15.13

Bid-rent curves

often a Marks and Spencer is at the peak land value site! Competing with retailers are the offices which also rely upon good transport systems and a proximity to other commercial buildings.

Away from the CBD land rapidly becomes less attractive for commercial activities — as indicated by the steep angle of the bid−rent curve in Figure 15.13. Industry, partly because it takes up more space and uses it

less intensively, bids for land that is less valuable than that prized by shops and offices. Residential land, which has the flattest of the three bid−rent curves, is found further out from the city centres where the land values have decreased as there is less competition for land. Individual householders cannot afford to pay the same rents as shopkeepers and industrialists. Figure 15.14 shows the predicted land use pattern when land values decrease rapidly and at a constant rate from the city centre. The resultant pattern is similar to that suggested by Burgess. It partly accounts for why inner city areas have a higher population density than places nearer to the city boundary (Figure 13.6).

One basis of this model is 'the more accessible the site, the higher its land value'. Rents will therefore be greater along main routes leading out of the city and along outer ring roads. Where two of these routes cross there may be a secondary or subsidiary land value peak (Figure 15.15) overleaf. The land use at this point is likely to be a small suburban shopping parade or a small industrial estate. The 'retail revolution' of the 1980s (page 361) has altered this pattern as massive edge-of-city shopping centres have been developed, e.g. MetroCentre in Gateshead and Brent Cross in north London. Similarly, large industrial estates are located near to motorway interchanges.

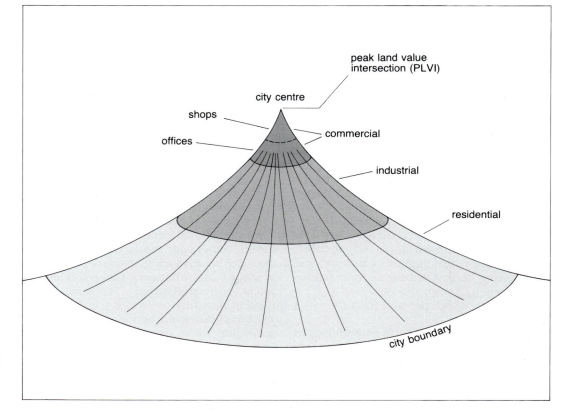

▶ Figure 15.14

Urban land use patterns based on land values

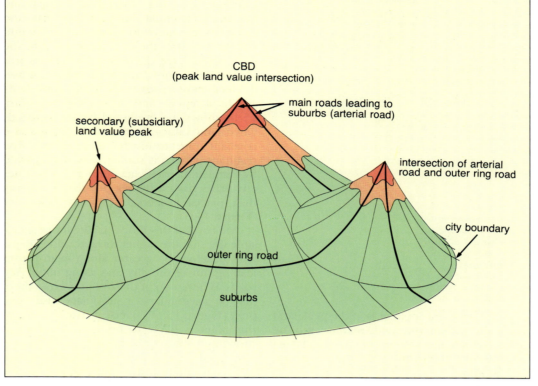

CBD
(peak land value intersection)

main roads leading to
suburbs (arterial road)

secondary (subsidiary)
land value peak

intersection of arterial
road and outer ring road

city boundary

outer ring road

suburbs

Functional zones within a city

The central business district (CBD)

The CBD is regarded as the centre for retailing and service industries (e.g. banks and offices). It contains the principal commercial streets and main public buildings and so forms the core of a city's business and commercial activities — although, as already seen in the case of London, very large cities may have more than one CBD.

How can the CBD be delimited?

This is a practical exercise which might be carried out in your local town. Several methods have been suggested, based on the pioneer work of Murphy and Vance in North America, but no single one appears to be completely satisfactory.

Characteristics of the CBD are listed below.

■ It contains the major retailing outlets. The principal department stores and specialist shops with the highest turnover and requiring largest threshold populations compete for the prime sites.

■ It contains a high proportion of the city's main offices.

■ A concentration of the tallest buildings in the city or highest vertical growth (more typical of North American cities), which is due to high rents resulting from competition for land (Figure 15.13, page 357).

■ A concentration of the greatest number of pedestrians.

■ A concentration of the greatest volume of traffic. The city centre grew at the meeting point of the major lines of communication into the city. Much of present day planning aims to limit the number of vehicles entering this zone.

■ It has the highest land values in the city, including the PLVI (**peak land value intersection**).

■ It is constantly undergoing change. New shopping centres, taller office blocks and new traffic schemes seem to be announced daily. Even the grandiose schemes of the early 1960s are now viewed as out of date and unfashionable taste, as in the case of Birmingham's Bull Ring and London's Paternoster Square at St. Paul's. Many of these are to be demolished and rebuilt.

Recent studies have shown the CBD of many cities to be advancing in some directions (**zone of assimilation**) and retreating in others (**zone of discard**). The zone of assimilation is usually towards the higher status residential districts whereas the zone of discard tends to be nearer the industrial and poorer quality residential areas. There is also a significant trend for retailing to be static, or even declining (possibly because of competition from out-of-town developments),

*Shops and offices are said to be **central** or **CBD functional** elements. The land uses remaining — residential property, government and public buildings, churches and educational establishments, wholesaling and storage, industry and vacant land — are classed as **non-central** or **non-CBD functional** elements.*

while offices, banks, insurance etc. are increasing in terms of space taken and income and employment generated.

How can these characteristics be mapped?

Taking the seven characteristics listed above, the following are suggested methods which might be used in fieldwork study.

- **Land use mapping of shops**
  1 Plot the location of all the shops: where the ratio of shops to other properties is more than 1:3, then count that as being within the CBD. This is a straightforward exercise using the (UK) Census of Information definition that over 33 per cent of buildings in the CBD are connected with retailing.
  2 An alternative method is to include all shops which are less than 100 m (or any distance which you judge to be suitable) from another shop as being within the CBD. This may produce a central 'core' and several smaller groupings.
  3 A third possibility might be to take the mean frontage (in metres) of, for example, the middle five buildings or shop units in a block. Shop frontages are likely to be greatest near to the PLVI as this is where most department stores will be located.

- **Land use mapping of offices** The first method described above could be repeated using offices instead of shops and using a ratio of 1:10. This recognises that at ground floor level offices are less numerous than are shops. This would include banks and building societies.

- **Height of buildings** Another simple technique is to plot the height of individual buildings or the mean of a group of buildings in the centre of a block. Most cities tend to have a sharp decline in building height at the edge of the CBD.

- **Accessibility** This is suggested as a group activity to count the number of pedestrians passing a given point over a certain time. The greater the number of sites (ideally chosen by using random numbers (sampling, page 138)) the greater the accuracy of the survey. Define a pedestrian as someone of school age and over, entering, leaving or passing a shop on your side of the street. Any of these criteria can be altered as long as they are applied by the whole group.

- **Accessibility to traffic** This is a similar survey to pedestrian accessibility except that vehicles are counted. Make sure all groups have the same definition of a vehicle e.g. do you include a bicycle or a pram?

- **Land values** These might be expected to decline outwards at a uniform rate which would give no 'natural breaks' by which to delimit the area. However, providing rateable values can be obtained (try the rates office) and there is the time to process them, then this is often a good indicator of CBD extent. To overcome problems of variations in the floor space of different buildings (a Marks and Spencer store will cover more space than a shoe shop) you will need to work out the rateable index (*RI*).

$$RI = \frac{\text{gross rateable value of the property}}{\text{ground floor area of the property}}$$

It may then be useful to take the PLVI point and call this 100 per cent, and then convert the rateable index for all other properties as a percentage of the PLVI. It has been suggested than a figure of five per cent delimits the CBD for a North American urban area and 20 per cent for a British city.

- **Changing land use and functions** This is a mapwork exercise using an old map of the central area (shopping maps are produced by GOAD plans) and superimposing onto it the present day changes. Look for evidence of zones of assimilation and discard.

- **Central business index** This is probably the best method as it involves a combination of land use characteristics, building height and land values. The problem is in obtaining the necessary data which is:
  1 the total floor area of all central or CBD functions
  2 the total ground floor area
  3 the total floor area (i.e. includes all upstair floor area as well as the ground floor).

You may laboriously work this out from a large scale plan, or choose to compromise by taking the mean of a sample of buildings in each block. From this data two indices can be used.

The **central business height index** or CBHI which is expressed as:

$$CBHI = \frac{\text{total floor area of all CBD functions}}{\text{total ground floor area}}$$

The **central business intensity index** or CBII which is expressed as:

$$CBII = \frac{\text{total floor area of all CBD functions}}{\text{total floor area}} \times \frac{100}{1}$$

To be considered part of the CBD, the CBHI of a plot should be over 1.0 and the CBII over 50 per cent (Figure 15.16) — though some studies identify a central core within the CBD with a CBHI over 4.0 and a CBII over 80 per cent.

Plotting the data

Careful consideration should be given as to which cartographic technique is best applied to each set of collected data. You may wish to use one or several of the following: land use map, isolines, choropleth, flow graphs, histograms, bar graphs, scattergraphs and transects. Alternatively you may be able to devise a technique of your own. You may save time and produce results which are easier to compare by using tracing overlays.

Delimiting the CBD: conclusions

If you have carried out a survey of the CBD in your town, your report might include comments on the following questions.

1 What problems did you encounter in collecting and refining the data?
2 Which of the methods used in collecting the data appeared to give the most, and the least, accurate delimitation of the CBD?
3 In your town was there an obvious CBD; did you find an inner and an outer core?

(Figure 15.17). Was there evidence of zones of assimilation and discard? Were there any specific functional zones other than shops and offices? Was the area of the CBD similar to your **mental map** (your preconceived picture) of its limits?
4 What refinements to the techniques used would you make if you had to repeat this task in another urban area?

Shopping

There are two main types of shop.
1 Those selling **convenience** or **low order** goods which are bought frequently, usually daily, but are not sufficiently high in value to attract customers from further than the immediate catchment area, e.g. newsagents and small chain stores.
2 Those selling **comparison** or **high order** items which are purchased less frequently but which need a much higher threshold population, e.g. department store and specialist shop.

The frequency of visit, combined with accessibility and the cost of rents (Figure 15.14), determines the preferential location of different shops (Figure 15.18, page 362).

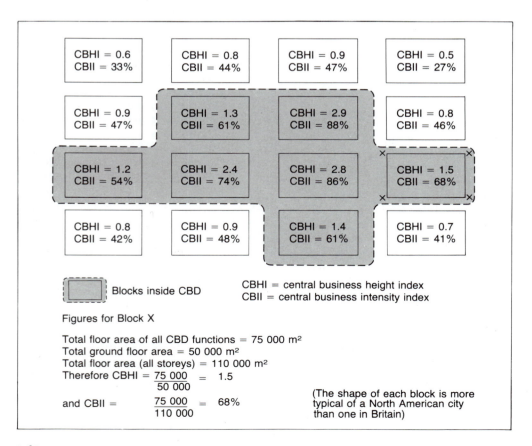

CBHI = 0.6, CBII = 33% CBHI = 0.8, CBII = 44% CBHI = 0.9, CBII = 47% CBHI = 0.5, CBII = 27%

CBHI = 0.9, CBII = 47% CBHI = 1.3, CBII = 61% CBHI = 2.9, CBII = 88% CBHI = 0.8, CBII = 46%

CBHI = 1.2, CBII = 54% CBHI = 2.4, CBII = 74% CBHI = 2.8, CBII = 86% CBHI = 1.5, CBII = 68%

CBHI = 0.8, CBII = 42% CBHI = 0.9, CBII = 48% CBHI = 1.4, CBII = 61% CBHI = 0.7, CBII = 41%

Blocks inside CBD

CBHI = central business height index
CBII = central business intensity index

Figures for Block X

Total floor area of all CBD functions = 75 000 m²
Total ground floor area = 50 000 m²
Total floor area (all storeys) = 110 000 m²

Therefore $\text{CBHI} = \dfrac{75\ 000}{50\ 000} = 1.5$

and $\text{CBII} = \dfrac{75\ 000}{110\ 000} = 68\%$

(The shape of each block is more typical of a North American city than one in Britain)

◀ Figure 15.16

The central business index

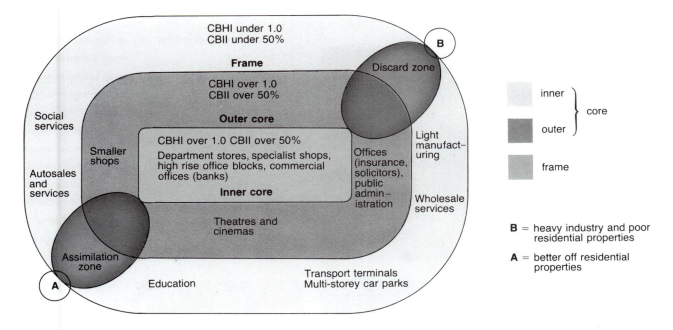

▲ Figure 15.17

The core and frame concept for the CBD

Convenience shops are commonly located in housing estates and neighbourhood units so as to be within easy reach of their customers — often within walking distance. With a lower turnover of goods than retail units in the CBD they may have to charge higher prices but their rent and rates are lower. Ideally they are located along suburban arterial roads or at a crossroads for easier access and possibly to encourage impulse buying by motorists driving into the CBD.

Convenience shops are found in suburban shopping parades (Figure 15.18c) where the inhabitants live a long way from the central shopping area, in inner cities where the corner shop (Figure 15.18b) caters for a population which cannot afford high transport costs, and along side streets in the CBD where they take advantage of lower rents to provide daily essentials for those who work in the city centre.

Comparison shops need a large threshold population and therefore have to attract people from the whole urban area and beyond. As they bid for a central location they must have a high turnover in order to pay the high rents. This central area has traditionally afforded the greatest accessibility for shoppers with public transport competing with the private motorist. Today congestion and other traffic problems have meant that urban access is a more complex concept than in the past and has resulted in changes in retail locations. Many city centres are having to undergo constant change to try to attract more customers (e.g. Peterborough's Queensgate, Newcastle's Eldon Square and Birmingham's Bull Ring) or in

some cases just to maintain previous levels. These modern, integrated centres with traffic-free, covered malls (Figure 15.18a) differ considerably from the traditional CBD and increasingly have to compete with extensive edge-of-city shopping complexes (Figure 15.18d). While department stores and specialist shops dominate in CBD centres, smaller specialist shops may occupy ground floors of the increasing number of office blocks. Some comparison shops may locate in more affluent suburbs.

Out-of-town shopping centres

Although a relatively new concept in the UK, it is estimated that out-of-town complexes will number over 800 by the early 1990s. These take advantage of economies of scale, lower rents and a pleasant, planned shopping environment. Hypermarkets and superstores build on cheap land at or beyond the city margins, allowing them to buy space for their immediate use, for possible future expansion and for large car parking areas. They aim to capture a high threshold population, especially one which is wealthy and mobile. As a consequence these centres seek locations next to main roads and, ideally, motorway interchanges, to facilitate access for customers and delivery drivers, thus avoiding some of the congestion of the more central parts of the city. Often a full range of goods — convenience and comparison — is sold under one roof so shoppers are able to complete their purchases in the warmth, in bulk, at times suitable to the working family, in a cleaner, less congested environment and increasingly with the additional amenities of cafés, cinemas and children's play areas.

Figure 15.18 Functional zones in a city

A   Undercover shopping in the CBD (Singapore)

B   An inner city corner shop (Colliers Wood, London)

C   A suburban shopping parade

D   Edge-of-city sites are popular for large, new 'superstores' or 'hypermarkets' where there is plenty of space for parking (Wimbledon, London)

E   Inner city housing is often a mixture of 'modern' high rise development and nineteenth-century terraces — many older areas have undergone 'renewal' or renovation (Kennington Oval, London)

F   High tower blocks have often replaced former terraced housing (Brentford, London)

G   Low density, semi-detached housing is characteristic of inter-war residential areas

H   Post-1960s private housing

1   RETAIL ZONES ▲

2   RESIDENTIAL ZONES ▲

developments have tended to be in the form of suburban estates on green field sites on city boundaries (Bramcote, Nottingham)

I Post 1960s council housing has been mainly low rise where sufficient open space was available.

J Space is at a premium in the CBD so office blocks are built upwards to maximise the use of groundspace (Hong Kong)

K Nineteenth-century industry developed alongside terraced housing in inner city areas (Ebbw Vale, South Wales)

L In the 1950s and 1960s edge-of-city land was often used for building public amenities such as schools and hospitals which required large amounts of space (Harlow, Essex)

M Modern industrial estates seek the cheaper land with good road access found at the edge of built-up urban areas (Stevenage)

▲

Inner city

◄ **Inter-war suburbs** ►

Edge-of-city

▼

2 RESIDENTIAL ZONES ▲

3 OTHER LAND USES ▲

CASE STUDY

Eldon Square, Newcastle upon Tyne

When the Eldon Square shopping complex opened in 1976 it was the largest undercover centre in Europe. New air-conditioned malls, lined with 'anchor stores' and specialist shops, were integrated with existing shops and streets. Although the centre was traffic-free, the addition of two Metro stations, a new bus station and two multi-storey car parks made it highly accessible. Within weeks of its opening, pedestrian flows of 17 000 per hour were recorded at peak times and customers were prepared to travel up to 150 km to shop there.

Ten years later, the centre was threatened by competition from the newly opened MetroCentre shopping complex built on the outskirts of Gateshead, only some 8 km distant. Eldon Square had to adapt quickly. Day-lighting was introduced to alleviate the feeling of claustrophobia expressed by many customers. In keeping with modern thinking, a series of 'theatrical sets' were set up which could be used, demolished and replaced as new trends come into fashion. At the centre of the scheme is the Rotunda, with escalators linking three new retail levels, and the newly created Eldon Gardens with plants and fountains. Despite the success of the MetroCentre, and fears to the contrary, the numbers of shoppers to Eldon Square (and the amounts of money which they spend) have continued so far to increase.

MetroCentre, Gateshead

John Hall, whose brainchild the MetroCenter is, was a pioneer among modern entrepreneurs in recognising that shopping could become part of a wider leisure experience as people's mobility, disposable income and leisure time (for those with jobs) increased. The MetroCentre, taking advantage of tax concessions by locating in a government Enterprise Zone (page 369), was the prototype of a new shopping development concept in Britain. Its success was assured after Marks & Spencer opted to make the MetroCentre its first out-of-town location. There are now over 200 retail units set in a pleasant shopping environment (Figure 15.19) with wide tree-lined malls, air conditioning, 1 km^2 of glazed roof to let in natural light (supplemented by modern lighting in 'old world' lamps), numerous seats for relaxing, window boxes, hot air balloons, escalators and lifts for the disabled. A street market atmosphere has been created by traders selling from barrows and there are over 40 eating places offering all kinds of cuisine from Mexican to Chinese.

◄ Figure 15.19
Aerial view of the MetroCentre, Gateshead — to the left is the Western bypass providing road access and to the right is the Newcastle to Carlisle railway with the covered way connecting the new railway station to the Centre

Leisure is a vital part of the scheme. There is a 'children's village' and crèche, a ten-screen cinema, a 'space city' for computer and space enthusiasts, a covered 'fantasy-land' with all the attractions of the fair without the worries of the British climate and theme areas such as 'Antiques Village' and 'The Roman Forum'. In the pipeline are a 150-bedroom hotel and leisure centre, an office block and an exhibition centre. There are 10 000 free car parking spaces, 100 buses per hour, and 69 trains daily — the MetroCentre has its own bus and railway station.

▶ **Figure 15.20**
Inside the MetroCentre

▼ **Figure 15.21**
Land use in a city — transect from the CBD to the city boundary

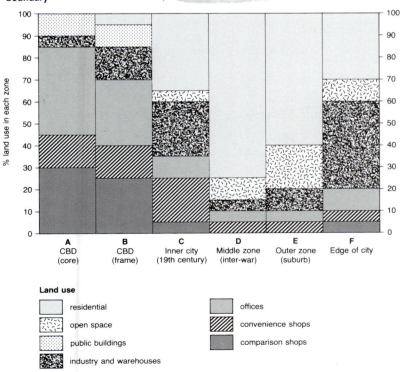

Offices

Offices use land intensively and often compete with shops for sites in the CBD (Figures 15.16, 15.18j). In addition to building their own tower blocks, businesses requiring offices also locate above main shops or along side streets running off the main shopping thoroughfares where rents are marginally lower. Offices often locate in CBDs for prestige reasons, for ease of access for their clients and staff, to be near **functional links** (banks, insurance and entertainment) and sources of data and information. Banks can afford prime corner sites, while building societies and estate agents vie for high visibility locations.

Other offices take advantage of the cheaper land values found in the frame (Figure 15.17) surrounding the core of the city, or in the twilight zone. Doctors, dentists, accountants and solicitors number amongst those utilising renovated buildings. A recent trend has been for office activity to move to edge-of-city locations for similar reasons as retail outlets. Modern methods of transferring data, the replacement of staff with computerisation, and the nearness to motorways and airports for clients have reduced the importance of a central location. The high rents, congestion and pollution of the city centre have pushed offices out into a cheaper, more pleasant working environment. Many out-of-town locations are now sited in purpose-built business or science parks (Figure 17.18, page 443).

Industry

Industry within urban areas has changed its location over time. In the early nineteenth century it was usually sited within city centres, e.g. textile firms, slaughter houses and food processing. However, as the Industrial Revolution saw the growth in size and number of factories, and later when shops began to compete for space in the city centre, industry moved centrifugally outwards into what today is the inner city. Inner city areas could provide the large quantity of unskilled labour needed for textile mills, steelworks and heavy engineering. The land was cheaper and had not yet been built upon. Factories were also located next to main lines of communications: originally, rivers and canals, then railways and finally

roads. Firms including bakeries, dairies, printing (newspapers) and furniture which have strong links with the city centre are still found here. Until the 1980s this zone had increasingly become one of industrial decline as older, traditional industries closed down and others moved to edge-of-city sites (Figure 15.18). In Britain, recent changes in government policy have led to attempts to regenerate industry in these areas through initiatives such as Enterprise Zones, derelict land grants and urban development corporations (page 369).

Most modern industry is 'light' in comparison to the last century and, producing fewer obvious pollutants, has moved to greenfield sites near to the city boundary. Trading estates and modern science parks are located here on large areas of relatively cheap land where firms have built new premises and use up-to-date equipment. The internal road systems are purpose-built for cars and lorries and linked to nearby motorway interchanges and other main roads. The wide range of skills and increasing demand for female labour are satisfied by local housing estates. Most of the industries are 'footloose' and include hi-tech, electronics, assembling, food processing and distributive firms.

Q Figure 15.21 shows how land use alters with increasing distance from the centre of a British city.

1 Account for the differences in land use between zones B and E.

2 Why does there appear to be little correlation between areas with a high residential population and areas with most shops?

3 Describe and account for the distribution of industry.

4 How does bid–rent theory help to explain the differences in land use between zones A and E?

▼ Figure 15.22
Social, economic and ethnic segregation in a city in the developed world

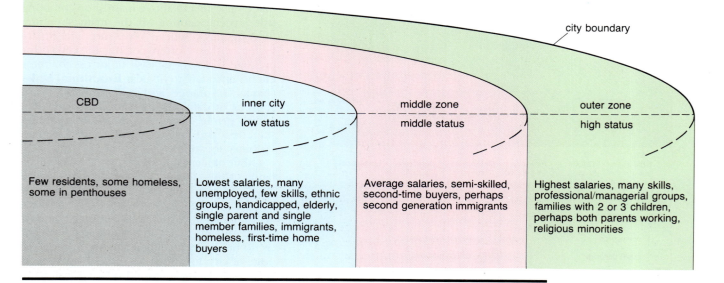

city boundary

CBD	inner city	middle zone	outer zone
	low status	middle status	high status
Few residents, some homeless, some in penthouses	Lowest salaries, many unemployed, few skills, ethnic groups, handicapped, elderly, single parent and single member families, immigrants, homeless, first-time home buyers	Average salaries, semi-skilled, second-time buyers, perhaps second generation immigrants	Highest salaries, many skills, professional/managerial groups, families with 2 or 3 children, perhaps both parents working, religious minorities

FRAMEWORK 9

Values and attitudes

In the past, some teachers, perhaps because of their religious or political opinions, felt it necessary to pass on their own values to students, while the majority tended to avoid considering the role of values in geography in an attempt to remain neutral and non-doctrinaire. Today it is being accepted, perhaps slowly and with reservations, that in order to understand the character of places and the behaviour of people in relation to their environments it is necessary to consider the motivations, values and emotions of those people involved.

The present author has tried, rightly or wrongly, to maintain a 'neutral' stance. Some would claim that what has been included in this book has been influenced by the author's own values and attitudes, e.g. a belief in the fundamental role of physical geography in an understanding of environmental problems, a preference for living in a semi-rural area rather than an inner city. Criticism could be levelled for using personal experiences as examplars. What the author has tried to do is to present readers with information in the hope that they may become more aware of their own values in relation to the behaviour of others and to enable them to discuss, with fewer prejudices and preconceptions, the foundations of their own values.

This may be illustrated with reference to the following section on inner cities. It is structured as follows:

1 **The problem of inner cities** Will these issues be seen differently by the inhabitant of an inner city area and a person living in a rural environment?

2 **The image of an inner city area** Will a description of inner city problems give a negative picture of the quality of life in those environments and in doing so perpetuate those problems, or could it help in understanding and tackling those problems? .

3 **Possible solutions to the inner city problem** Would solutions proposed by inner city residents be similar to those suggested and implemented by the government or local authority?

4 **What successes have government schemes had?** This depends upon your political views. Conservatives point out many achievements in the 1980s; Labour and other opposition parties claim that little has been done. Who, if anyone, is correct?

Issues of inner cities in Britain

Social inequality

Burgess, in his urban model, accepted as a basic assumption that there were well-defined socio-economic and ethnic areas within a city (Figures 15.5, 15.6). The segregation of different groups of people in British cities may result from differences in wealth, class, colour, religion, education and the quality of the environment. Some of the resultant social and economic inequalities within a city are summarised in Figure 15.23. While much activity in inner cities may be positive, such as London's Notting Hill Carnival and the lively multicultural society of Brixton, the more likely perception of such areas by non-residents is usually negative. To them, inner cities have an image of poverty, dirt, crime, overcrowding, unemployment, poor housing conditions and racial tension (Framework 10, page 372).

What is an inner city and what are its problems?

The widest definition is 'an area found in older cities, surrounding the CBD, where the prevailing economic, social and environmental conditions pose severe problems'. These problems vary between different inner cities — the only inner city problem common to all is, possibly, unemployment. While a description of these problems *may* present a negative picture (Framework 9, above) it is necessary to identify them if people are to begin to understand the difficulties and to offer workable solutions.

- **Economic problems** Inner cities have long suffered from a lack of investment, especially as after 1945 much money was channelled into the 'new towns', which hastened the downward spiral. Traditional industries declined and closed while those remaining shed much of their labour force due to improved technology and

(a) Liverpool 1971 — figures over 100 indicate the concentration of a problem in that area. A figure of 200 indicates the problem to be twice the average for Liverpool.

	Inner city and older council estates	Older terraced housing	Outer city and more recent council estates	Higher status owner-occupied housing
Long term unemployment	250	89	102	32
Youth unemployment	199	63	145	33
Youth job instability	302	68	112	17
Free school meals	201	74	147	29
School absenteeism	295	76	121	18
Low reading ability	210	86	131	30
Infant mortality	158	118	69	75
Infectious diseases	143	110	82	68
Children in short-term care	154	95	113	22
Children in long-term care	212	104	74	15
Delinquency	397	75	93	21
Possession orders	189	77	136	27

(b) GEAR Glasgow 1971 — Figures given are in percentages

	GEAR region	Strathclyde region
Male unemployed	20	13
Male and female unemployed	17	11
Retired persons	20	15
Household heads retired	37	25
Households with pensioners	41	29
Handicapped	29	21
Households with children 0–14	31	41
Earning under £1750 p.a.	55	39
Earning over £4126 p.a.	5	
Adults with no formal school qualifications	83	
Households with cars	16	61
Living in two rooms or less	40	18
Lacking indoor toilet	10	2
Lacking one basic amenity	18	7

◄ Figure 15.23
Symptoms of deprivation in Liverpool and Glasgow, 1971

falling demand, e.g. shipbuilding on the Tyne, Wear and Clyde and textiles in West Yorkshire and Lancashire. A subsequent decline and shift in trade affected ports such as Liverpool and London. Few new industries wish to locate in the inner city because of the environment and because the former labour force may not possess the relevant skills for new services and hi-tech industries. Between 1951 and 1981, unemployment in the inner cities rose from 33 per cent above the national average to 51 per cent.

■ **Social issues** Much inner city housing is high density (although this should not automatically mean poor quality) and pre-1914 in age. Relatively few occupants can afford to buy their homes and the houses, smoke-stained from former industrial times, may be in a poor state of repair. The small area census statistics illustrate some of the following social characteristics.

1 A lack of basic amenities. According to the 1981 census, one million dwellings lacked either a bathroom, WC or hot water.

2 Overcrowding. Many houses were built with just two rooms. Where slum clearances have taken place and housing replaced with poorly built high rise flats, other social problems have been created. (The census figures do not draw attention to the increasing number of homeless people.)

3 Higher death and infant mortality rates, a lower life expectancy and a greater incidence of illness than in other residential environments.

4 Predominance of low income, semi-skilled and manual workers.

5 Family status with a higher incidence of single parent families, elderly people, children in care and free school meals than other districts.

Many inner city areas have concentrations of ethnic minorities which may cause tension. The Scarman Report, following the 1981 riots, concluded that racial discrimination was a major issue.

■ **Environmental factors** These areas may suffer from noise and air pollution caused by heavy traffic and the few remaining factories; visual pollution in the form of derelict factories and houses, waste land, rubbish, vandalism and grafitti; and pollution of water in rivers and old canals. Although some open space has been created by clearance of slum property there is likely to be a general absence of trees and grass.

Indicators of deprivation

Deprivation is defined by the Department of the Environment, as when 'an individual's well-being falls below a level generally regarded as a reasonable minimum for Britain today'. Estimates suggest that four to six million people live under such conditions. S. Holterman (1975) used 18 chosen indicators from the 1971 census figures. He noted that while Inner London ranked highest for substandard housing, Clydeside scored as the worst area for nearly all of the other indices. Figure 15.23 shows deprivation indices for Liverpool and GEAR (Glasgow Eastern Area Renewal) based on the 1971 census figures.

Positive characteristics of inner cities

Many people make a positive choice to live in the inner cities. Proximity to the CBD provides quick and cheap access to jobs, shops and public amenities. Certain groups, such as students and the elderly, may prefer the smaller, cheaper types of accommodation. Some areas, especially those with old terraced housing, have a strong community spirit while others, where ethnic groups are present, have developed forms of international co-operation.

Government policies for the inner cities

When Margaret Thatcher was elected as Prime Minister for a third term in 1987 she stated that the regeneration of economically depressed and socially deprived inner city areas was one of the government's major objectives. When launching the government's 'Action for Cities' policy in March 1988 she said: 'The government is determined to build on a strong economy a new vitality in our inner cities. In partnership with the people and the private sector, we intend to step up the pace of renewal and regeneration to make our inner cities much better places in which to live, work and invest'.

The government specifies six aims in the inner city programme.

1 To enhance job prospects and the ability of residents to compete for them.
2 To bring land and buildings back into use.
3 To improve housing conditions.
4 To encourage private sector investment.
5 To encourage self help and improve social fabric.
6 To improve the quality of the environment.

It plans to achieve these aims through six main programmes.

1 **The Urban Programme** This gives 75 per cent grants to 57 of the most needy local authorities who are trying to tackle underlying economic, social and enviromental problems. Examples include converting part of Dock Street (Leeds) into small workshops and creating a 'Media Training Centre' in Newcastle upon Tyne.
2 **Derelict Land Grants** Reclamation schemes have included turning the redundant Edge Hill railway sidings (Liverpool) into Wavertree Technology Park; the conversion of derelict mills, railway land and old housing at Listerhills (Bradford) into a £40 million warehouse and distribution centre; and the creation of National Garden Festivals (Liverpool 1984, Stoke-on-Trent 1986, Glasgow 1988, Gateshead 1990 and Ebbw Vale 1992).
3 **Land Registers** These list unused and underused land.
4 **Enterprise Zones (EZs)** Sunderland became Britain's twenty-sixth EZ in 1989 following the closure of its last shipyard. EZs try to stimulate economic activity by lifting certain tax burdens, e.g. exemption from paying rates for ten years after the EZ is designated and 100 per cent grants for machinery and buildings, and relaxing or speeding up the application for planning permission. Gateshead's Metro-Centre (page 364) was built in an Enterprise Zone.
5 **Grants for Urban Development** The new City Grant bridges the gap between costs and value on completion, enabling investors to make a reasonable return on their capital.
6 **Urban Development Corporations (UDCs)** These now spearhead the government's attempts to regenerate areas which contain substantial amounts of derelict, unused or underused land and buildings. They aim to encourage maximum private sector investment and development. UDCs have powers to acquire, reclaim and service land and to restore buildings to effective use. They promote new industrial and housing developments and support community facilities. The first two, the London Docklands and the Merseyside Development Corporations (LDDC and MDC) were set up in 1981. Four more were created in 1986: Trafford Park in Greater Manchester; on Teesside; in the West Midlands and in Tyne and Wear. Three more were announced in 1987: Bristol, Leeds and Central Manchester. One was designated in 1988: the Lower Don Valley in Sheffield.

Q Read the case studies on page 370 before answering these questions.

1 What schemes has the government in Britain introduced to deal with inner city problems?

2 How successful do you think these schemes have been? (With the information provided, and individual research, you should be able to make your own value judgement.)

CASE STUDY

The Merseyside Development (MDC) and London Docklands Development (LDDC) Corporations: success stories . . . but for whom?

MDC The renovated Albert Dock now provides shops, restaurants, a pub, the Merseyside Maritime Museum, 114 apartments and the new 'Tate of the North' gallery of modern art. The nearby Toxteth and Harrington Docks have been turned into the Brunswick Business Park. It is hoped that the 40 ha of reclaimed waterfront will provide units for 1220 jobs by 1992 (Figure 15.24).

LDDC Success for the government? According to the Department of the Environment 21 km² of derelict docks and associated industry which had lost 10 000 jobs in the four years previous to the creation of the LDDC in 1981 had already been 'transformed' by 1988 — several major projects were still in the pipeline. In those seven years 10 000 new jobs had been created and 660 companies attracted including several newspaper corporations and Limehouse ITV studios. Local unemployment fell by 15 per cent in 1987–88. The government had invested £400 million of taxpayers' money but this was only one-tenth of that provided by the private sector. Over 8000 homes had been constructed with former local council tenants purchasing a quarter of them. The Docklands Light Railway, opened in 1987, links the City with the Canary Wharf financial centre which, when completed, will employ 50 000 people and be a 'Wall Street on the water'. Other transport improvements have been the opening of the London City Airport and the improved A13 road. Major development schemes have included the London Commodity Exchange at St. Katherine's Dock, the Free Trade Wharf and Tobacco Wharf — a Covent Garden style blend of shops, restaurants and wine bars. The proposed 'China City' will provide a commercial, cultural, tourist and trade centre in Poplar Dock (Figure 15.25).

◄ **Figure 15.24**
The Albert Docks, Liverpool (MDC)

Failure for local people? The scheme came too late to prevent many previous inhabitants from leaving the area because of slum clearances or the lack of job opportunities. Many of the new homes are extremely expensive ('Yuppy-land') which not only puts them beyond the price range of the locals but by attracting the better-off also tends to lead to a sharp rise in the cost of living in the area and the destruction of the close-knit 'Eastenders' community. Many of the new jobs are hi-tech and skilled, reducing opportunities for the local population who are relatively low-skilled. Residents are also anxious about the noise levels and safety of the airport.

▲ **Figure 15.25**
Dockland redevelopment on the Isle of Dogs, London (LDDC)

The Third World: urban models

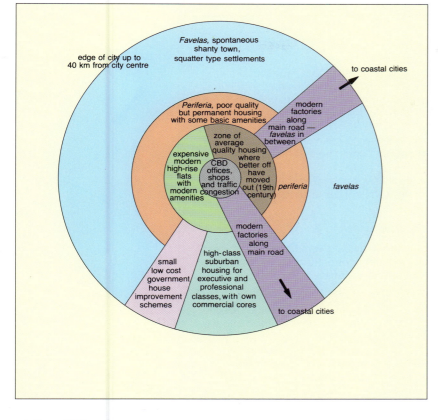

▲ Figure 15.26

Model of land use structure and residential areas in Brazilian cities (excluding Brasilia) — the zoning of housing with the more affluent living near to the CBD and the poorest further from the centre is typical of cities in the Third World

► Figure 15.27
Street dwellers in Bombay, India

Cities in the Third World, which have grown rapidly only in the last few decades, have a different structure from the older settlements in developed countries. Yet despite some observed similarities between most Third World cities, few attempts have been made to produce models to explain them. Clarke has proposed a model for African cities, McGhee for SE Asia and the author has suggested one for Brazil based subjectively on two television programmes on São Paulo and Belo Horizonte together with some limited personal fieldwork (Figure 15.26).

Functional zones in Third World cities

The **CBD** is similar to 'western' cities except that congestion and competition for space is even greater (e.g. São Paulo, Cairo and Hong Kong).

Inner zone In pre-industrial or during colonial times the wealthy landowners, merchants and administrators built large and luxurious homes around the CBD. While the condition of some of these houses may have deteriorated with time, the well-off have continued to live in this inner zone, with its proximity to the commercial centre, in new, modern high rise apartments.

Middle zone This is similar to a developed city in that it provides the 'in between' housing, except that here it is of much poorer quality. In many cases it consists of self-constructed homes to which the authorities *may* have added some of the basic infrastructure amenities such as running water, sewerage systems and electricity supply (including 'site and service' schemes, page 376).

Outer zone Unlike the western city the location of the 'lower class zone' is reversed as the quality of housing decreases rapidly with distance from the city centre. This is where migrants from the rural areas live, usually in shanty towns lacking basic amenities. Where groups of better off inhabitants have moved to the suburbs, possibly to avoid the congestion and pollution of central areas, they live together in well-guarded communities with their own commercial cores.

Industry This has either been planned or has grown spontaneously along main lines of communication leading out of the city.

FRAMEWORK 10

Stereotypes

One of several dangers which may result from making generalisations is that of creating stereotypes. Take, for example, the following unsupported, emotive comments which may not only be grossly inaccurate but which many people may consider offensive.

- All Chinese and Japanese are small. (So where did they find such tall volleyball players for the last Olympic Games?)
- Favelas are shanty settlements whose inhabitants have no chance of improving their living conditions. (Then why have some been able to benefit from self help schemes, page 376?)
- The inhabitants of a favela can only survive by a life of crime (*One*, below).
- Amazon Amerindians have not developed their environment because they are lazy and unintelligent (*Two*, below).

The following accounts are based on the author's experiences in Brazil.

One 'According to books which I had read in Britain and advice given to me by guides in São Paulo, favelas were to be avoided at all costs. Any stranger entering one was sure to lose his watch, jewellery and money and was likely to be a victim of physical violence.

With this in mind the local university made arrangements for me to be driven past several favelas so that I could take photographs. I set off in an old Volkswagen beetle converted into a taxi. On reaching the first favela, to my horror the driver turned into the settlement and we bumped along an unmade track. He kept stopping and indicating that I should take photographs. Expecting at each stop that the car would be attacked and my camera stolen, I hastily took pictures — which turned out to be over-exposed because I took them through the windscreen not daring to open the car door!

Suddenly the taxi spluttered and stopped. In one movement I had hidden the camera and was outside trying to push the car. I looked down and saw water flowing between my feet. That, according to the stereotype of a favela, would be sewage. I raised my eyes to find three well built males helping me push the car. Which one would hit me first? I smiled and they smiled. I pointed to each one in turn and called him after one of Brazil's football players and then referred to myself as Lineker. Huge smiles, big pats on the back and comments like *Ingleesh amigo* were only halted by the car re-starting. Now my stereotyped view of a favela inhabitant has been replaced and I would describe him or her as someone who is poor but very cheerful, friendly and helpful.'

Two 'I was surprised to find, on landing at Manaus airport in the middle of the Amazon rainforests, that our courier was an Amerindian. He dashed around quickly getting our party organised and our luggage collected (he certainly did not seem to be slow or lazy). As we talked, he admitted, and proved, that he was able to speak in seven languages. (Hardly the sign of someone unintelligent — how many languages can you speak?)

Later, I asked him why so few Amerindians appeared to have good jobs and why he kept talking about returning to the jungle. His reply was simple: 'to avoid hassle'. He considered that the indian lifestyle was preferable to the western one with its quest for material possessions often only achieved by winning in the 'rat race'. Had he returned to the jungle he would have rejoined his family and become a shifting cultivator living in harmony with the environment (page 401). Is that traditional way of life really less demanding of intelligence than that imposed by invading beefburger multinationals engaged in the wholesale destruction of large tracts of rainforest?'

From these examples we can see how easy it is to accept stereotypes without realising we are doing so, and also how the experience of seeing a situation for ourselves may alter the picture we have received through incomplete information from books, newspapers, television etc. An important question is whether geographers should take a role in overcoming the problems of stereotyped images (on the basis of which planning decisions, for example, may be made) by helping to provide relatively unbiased information to improve knowledge and understanding.

CASE STUDY

Calcutta

Although over 100 000 people live and sleep on Calcutta's streets, one in three inhabitants of the city live in *bustees*. These dwellings are built from wattle, with tiled roofs and mud floors, materials which are not particularly effective in combating the heavy monsoon rains. The houses, packed closely together, are separated by narrow alleys. Inside there is often only one room, no bigger than an average British bathroom. In this room the family, often up to eight in number, live, eat and sleep. Yet, despite this overcrowding, the insides of the dwellings are clean and tidy. The houses are owned by landlords who readily evict those bustee families who cannot pay the rent.

Rio's favelas (Figure 15.28)

A *favela* was a wildflower that grew on the steep *morros*, or hillsides, which surround and are found within Rio de Janeiro. Today these same morros are covered in *favelas* or shanty settlements. The houses of the *favelados*, the inhabitants, are constructed from any materials available — wood, corrugated iron and even cardboard. Some houses may have two rooms, one for living in and the other for sleeping. There is no running water, sewerage, or electricity, and very few local jobs, schools, health facilities or forms of public transport. The land upon which the favelas are built is too steep for normal houses. The most favoured sites are at the foot of the hills near to the main roads and water supply, although these may receive sewage running in open drains downhill from the more recently built homes. Often there is only one water pump for hundreds of people and those living at the top of the hill (with fine views over the tourist beaches of Copacabana and Ipanema!) need to carry water cans several times a day. When it rains, mudslides occur on the unstable slopes (page 38) and these carry away the flimsy houses (over 200 people were killed in February 1988).

Several attempts have been made to clean up the favelas or to remove them altogether. In some cases new homes have been built for the favelados, but over 40 km away, in areas lacking jobs and transport. In other examples the evicted inhabitants have simply moved back again as they had nowhere else to go.

▶ Figure 15.28
Favelas in Rio de Janeiro, Brazil
(see also Figure 13.27)

Problems resulting from rapid growth

The 'pull' and rapid growth of cities in the Third World has led to serious problems in providing housing, basic services and jobs — problems accentuated by a much wider gulf between the rich minority and the poor than exists in the western world. (Remember to be aware of the dangers of stereotyping — Framework 10 — when looking at urban environments in the Third World.)

Housing

Despite some promising initiatives, most authorities have been unable to provide adequate shelter and services and so have had to leave the majority of the poor to fend for themselves and to survive by their own efforts. Estimates suggest that one-third of the urban dwellers in Third World countries either cannot afford or cannot find accommodation which meets basic health and safety standards. Consequently they are faced with three alternatives: to sleep on pavements or in public places (Figure 15.27); to rent a single room if they have some resources; or to build themselves a shelter, possibly with the help of a local craftsman, on land which they do not own and on which they have no permission to build (Figure 15.28).

A growing number have adopted the third option and set up home in illegal 'shanty towns' where they face the constant threat of eviction or demolition of their homes. Estimates in 1981 suggested (no accurate figures are available so nobody knows how many people actually live in these squatter settlements) that 30 per cent of Rio de Janeiro's total population live in shanty towns, 25 per cent in São Paulo, 45 per cent in Mexico City, 40 per cent in Bombay, 60 per cent in Calcutta and 75 per cent in Ibadan.

In time, some squatter settlements may develop into residential areas of 'adequate' standards (The periferia, Figure 15.26). These areas may eventually have a water supply, sewerage systems, public services (e.g. refuse disposal) and occupants may obtain legal tenure of the land.

Services

Only small areas within many developing cities have running water and mains sewerage. Rubbish, dumped in the streets, is rarely collected. When heavy rains fall, especially in the monsoon countries, the drains are not adequate to carry the water away. A lack of electricity hinders industrial growth as well as affecting the material standard of living in homes. There is a shortage of schools and teachers, of hospitals, doctors and nurses. Police, fire and ambulance services are unreliable. Shops may sell only essentials, and food is exposed to heat and infection-carrying flies.

Pollution and health

Drinking water is often contaminated with sewage which may give rise to outbreaks of cholera, typhoid and dysentery. The uncollected rubbish is an ideal breeding ground for disease. Most children have worms and suffer from malnutrition as their diet lacks fresh vegetables, protein, calories and vitamins. Local industry is rarely subjected to pollution controls and so discharges waste products into the air which may cause respiratory diseases, or into water supplies. The constant struggle for survival often causes stress-related illnesses. It is not surprising that in these rapidly growing urban areas infant mortality is high and life expectancy is low.

Unemployment and underemployment

New arrivals to a city far outnumber the jobs available and a high unemployment rate is the result. As manufacturing industry is limited, full time occupations are concentrated in service industries such as the police, the army, cleaning, guards and the civil service. The majority of people who do work are in the **informal sector**, i.e. they have to find their own form of employment. This includes street trading (e.g. selling food or drinks), food processing, services (e.g. shoe-cleaning) and local crafts (e.g. making furniture and clothes). Most of these people are underemployed and live at a subsistence level because the market for their product or service is so poor.

Transport

Few Third World cities can afford elaborate public transport systems and what there is tends to be outdated and overused. The wealthy have cars but the road network is often inadequate to cope with the volume of traffic. Local traditional forms of transport (e.g. rickshaws, bullock carts, donkeys) compete with other road users.

CASE STUDY

Cairo: problems of growth

1988 figures put Cairo's population somewhere between 12 and 16 million (Framework 11). By Third World standards, the city has relatively few squatter settlements. Most newcomers disappear into the medieval centre of the old city to live in overcrowded two-roomed apartments within tall blocks of flats (Figure 15.29) or in roof-top slums (the desert climate allows roofs to be flat). There are few public services and washing hangs everywhere. Many migrants live in the 'City of the Dead', a huge muslim cemetry (Figure 15.30). People actually live within the tombs, which are often cleaner than the city apartments, but are over a kilometre from the nearest water supply. New flats are being built remarkably quickly but these are in the suburbs and their rents are unaffordable or costs of the journey to work in the city centre are too high for most people.

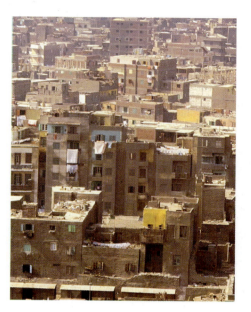

▶ Figure 15.29
Apartments in the Medieval centre of Cairo, Egypt

Cairo's narrow, unplanned streets were not built for the volume of motorised traffic which is noted for its noise and pollution. Pollution also comes from the breakdown of sewerage systems which were built in the early twentieth century, or their total absence, while the many small factories emit their wastes into the air and streets. Although small workshops appear to be everywhere, over 35 per cent of the population are without full time jobs. Workshops are set up in backyards, within houses and on rooftops. Donkey carts take rubbish to dumps on the edge of the city, providing jobs for refuse collectors and people who sort out the bottles, plastic and paper on the dump ready for recycling in local industry. The rubbish dumps, limited clean water, poor sewage, industrial pollution and high density housing all contribute to create potential health hazards.

FRAMEWORK 11

How reliable are statistics?

Accurate and reliable statistics are often difficult to obtain even for developed countries. Some of the least reliable are for population and those presented in this book, e.g. birth rates, should be used with considerable caution. It is suggested that Britain's 1981 census could have had a margin of error, even with supposedly high levels of refinement, of plus or minus 1 per cent. The size of this error increases considerably in Third World countries. For example ...

What is Cairo's population?

The following all produced estimates for 1985 (unless otherwise stated):

a The *New Geographical Digest* (George Philip) put Cairo at 6 818 000 and El Giza (the suburb surrounding the Pyramids, now considered part of Cairo) at 1 230 000 = 8 048 000
b The United States Bureau of Census = 8 595 000
c Reader's Digest (UN figures) = 9 770 000
d *Geofile* (No. 88, 1987) = 10–12 000 000
e A Granada Television programme for schools on Cairo = 15 000 000
f Official Cairo Tourist guide (1989) = 15 000 000

The first problem in obtaining accurate figures is that Cairo's most recent census was in 1976 and since then the size has had to be estimated: various organisations use different criteria. A second problem is in determining whether the statistic is for a specified city area or if it includes developments which have sprawled beyond Cairo's 'official' limits. It would be highly unlikely that every inhabitant was consulted for the census as many migrants to the city 'lose' themselves in the high density housing of the 'old town' and an estimated 500 000 live in and around the tombs of the 'City of the Dead' (Figure 15.30).

Recent estimates are based upon birth and death rates, but these too are most unreliable. The country had no consistent method of registering births until 1986 when a new law was passed in an attempt to ensure that all births took place in hospital and so would be recorded. However, methods of recording natural increase remain suspect and there is no effective means of counting migrants from rural areas.

▲ Figure 15.30
Living in the 'City of the Dead', Cairo, Egypt

Improvements and self-help schemes in Third World housing

Relocation by governments: 'formal housing'

Some of the more wealthy Third World countries, e.g. Venezuela using its oil revenue and Hong Kong and Singapore with income from trade and finance, have made considerable efforts to provide new homes to replace the squatter settlements. In most cases, high rise blocks of flats have been built on new sites as close as possible to the CBD. (Contrast the case studies from Venezuela and Singapore, pages 377, 378.)

Upgrading schemes

A policy of wholesale demolition of squatter settlements, as was attempted in Rio, is often a mistaken one. Squatters have shown that they are capable of constructing cheap accommodation for themselves but they cannot provide the essential basic services. In Latin America, and less successfully in Africa and SE Asia, governments have, albeit reluctantly, at times accepted that shanties are permanent and that it is cheaper and easier to improve those houses by adding basic amenities than it is to build new ones. Several schemes in São Paulo's *periferia* (Figure 15.26) have seen running water, main drains and electricity added to houses with street lighting and improved roads if there was any surplus money. The result over a lengthy period of time has been an upgrading of living conditions, though they are still poor, with the introduction of some shops and small scale industry.

In the Mathare Valley, 4 km east of Nairobi's CBD, the original wood and cardboard houses have been upgraded to mud and wattle and in some cases stone walls, cement floors and metal roofs. The authorities have given some help in providing fresh water, sewerage and refuse collection.

Site and service schemes

This concept, funded by the World Bank, is one of self-help and seems to be most appropriate to the poorer countries whose governments cannot afford large rehousing schemes.

In Lusaka (Zambia) one typical scheme encourages about 25 individuals to group together. They are given a standpipe and 8 ha of land. If the group digs ditches then with the money saved the authorities will lay water and drainage pipes and construct the houses. Moreover, if local craftsmen are prepared to build the shells of the houses then the group will be supplied with low-priced building materials and the extra money saved by the authorities may be used to add electricity and to tarmac the roads. In some cases a small clinic and school may be added.

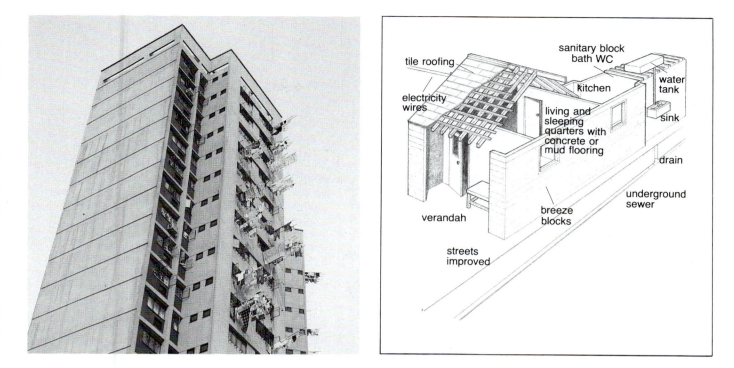

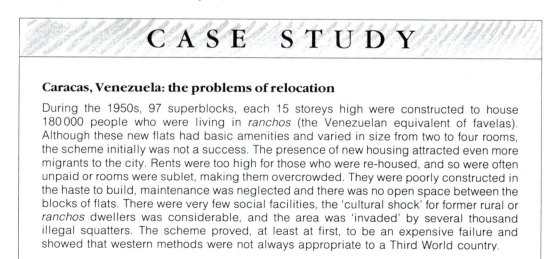

▲ Figure 15.31
Relocation housing in
Singapore

▶ Figure 15.32, *above right*
A 'sites and services' scheme,
São Paulo, Brazil

Such schemes can create a community spirit, improve the skills of local people and can result in cheap to erect accommodation. Figure 15.32 is a plan of a typical 'site and service' house. Yet their success often depends upon the motivation and skills of the local people and the use of appropriate and cheap building materials under expert guidance.

New towns

As in the developed countries, new towns have been seen in the Third World as one way of relieving overcrowding in the big cities. Although much-quoted, Brasilia, with its futuristic buildings, 'super quadras' (blocks of flats) and administrative jobs, is not typical. Cairo's problems have already been described (page 375).

Cairo is trying to overcome these by building a new sewerage system which is probably the largest public health engineering scheme in the world; organising refuse collection, constructing an underground metro system; building new roads (including a planned outer ring road); using electricity from the High Dam at Aswan; and building many high rise apartment blocks. Yet all these schemes can hardly keep pace with the population growth of Cairo and so, to date, five satellite or new towns have been begun in the desert surrounding the city. One of these is the Tenth of Ramadan (case study, page 379).

CASE STUDY

Caracas, Venezuela: the problems of relocation

During the 1950s, 97 superblocks, each 15 storeys high were constructed to house 180 000 people who were living in *ranchos* (the Venezuelan equivalent of favelas). Although these new flats had basic amenities and varied in size from two to four rooms, the scheme initially was not a success. The presence of new housing attracted even more migrants to the city. Rents were too high for those who were re-housed, and so were often unpaid or rooms were sublet, making them overcrowded. They were poorly constructed in the haste to build, maintenance was neglected and there was no open space between the blocks of flats. There were very few social facilities, the 'cultural shock' for former rural or *ranchos* dwellers was considerable, and the area was 'invaded' by several thousand illegal squatters. The scheme proved, at least at first, to be an expensive failure and showed that western methods were not always appropriate to a Third World country.

Singapore: a success story?

Faced with a large and rapidly increasing number of slum dwellers, the Singapore government set up the Housing and Development Board (HDB) in 1960. The HDB decided to build a series of new towns (there were 14 by 1989, all within 12 km of the CBD) and to construct housing units of between one and three rooms in closely packed high rise flats (Figure 15.31). These units were for low income families earning under 800$ Singapore or £266 per month (1988 figures). The rent, not including basic utilities, was 75$ Singapore (£25) per month, or 125$ Singapore (£42) per month inclusive of water and electricity. However, one-quarter of every wage earners' salary is automatically deducted and individually credited by the government into a central pension fund (CPF). Western style welfare benefits are regarded as an anti-work ethic, but Singaporeans can use their CPF capital to buy their own apartment or flat. Those earning over 800$ Singapore per month have to buy their own homes, although since 1974 the HDB have built many four and five room units for this higher income group. To buy your own home a downpayment of 20 per cent is needed and payments are made over 20 years: the mortgage rate is 3 per cent. At present, 86 per cent of Singaporeans live in government-built housing, sharing the belief of their Prime Minister that 'a population aware that street riots reduces property values is hardly likely to riot'.

The large estates are functional in design and each neighbourhood is capable of housing between 5000 and 25 000 people. Yet they contain much greenery and are well provided with amenities such as shops, schools, medical and community centres. Where several estates are in close proximity better services are provided such as department stores and entertainment facilities. The inhabitants are close enough to the city centre, the port and industrial areas for employment, and their journeys to work will be made easier by 1990 on the completion of the MRT (mass rapid transport) system.

Housing estates are free of litter and graffiti: the minimum fine for each is 500$ Singapore, or £133. The lifts are clean and almost always work. Like many other countries with high rise flats, there was a problem as the lifts were used as toilets. So the HDB put sensors in the floors which were activated by salt in the urine. The lift doors then locked automatically, an alarm was set off and the offenders were booked and later fined heavily. A problem quickly solved! Figure 15.31 shows typical flats with bamboo poles used for drying clothes.

ITDG and 'materials for shelter'

The Intermediate Technology Development Group (ITDG, page 455) is investigating, developing and promoting a range of building materials suitable for self-help schemes (Figure 15.33). One such material is fibre-concrete roofing tiles made from a mixture of cement, sand and (in Kenya) locally grown sisal (Figure 17.33). One fifth of the rice plant is inedible and in India it is burnt to produce ash which is up to 95 per cent silica in content. The silica, when mixed with lime, gives a cement good enough to construct

◄ Figure 15.33
A self-help scheme in Sri Lanka based on intermediate technology

single storey houses and irrigation channels. Elsewhere, the firing of bricks in kilns places a high demand upon the already depleted wood supplies. ITDG are supporting improvements in brick quality, more efficient kilns and substitute fuels. They are providing technical assistance in mining, quarrying and processing local raw materials which can be used for building. In India, thatch roofs may now last up to four times longer when coated with a waterproof compound of copper sulphate and cashew nut resin.

Tenth of Ramadan

This is being built 55 km ENE of Cairo on the main road to Port Said (Figure 15.34). It could eventually house over 300 000 people living in six neighbourhood units. Each unit is planned to have high rise apartment blocks with some open space and gardens, a mosque, junior school, and a local shopping centre. Nearby, though not in the units, will be industrial zones, for without provision of jobs the scheme is unlikely to succeed. After eight years, the new town has its problems. It had only grown to 30 000, partly because the rents were prohibitively high, as were the costs of travel for those still working in Cairo. The second explanation for the slow growth is that despite government loans to firms, industry has not grown quickly and so people have not been enticed to move out of the capital to obtain work.

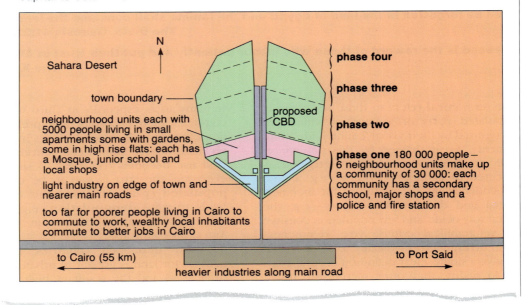

▶ Figure 15.34
Proposed layout of the new town Tenth of Ramadan, near Cairo, Egypt

References

Alternative Approaches to Development, Geography 16–19 Project, Longman, 1985.

Brazil, Phillip Vaughan-Williams, U.T.P., 1981.

Brazil: City of Newcomers, BBC Television, 1980.

Brazil: Skyscrapers and Slums, BBC Television, 1980.

Egypt: Progress in the 1980s, Geofile No. 88, Mary Glasgow Publications, 1987.

Human Geography, P.J. McBride, Blackie, 1980.

Human Geography: Theory and Applications, M.G. Bradford and W.A. Kent, Hodder & Stoughton, 1982.

Inner City Programmes 1987–1988, Department of the Environment, HMSO, 1989.

New Life for the Inner Cities, David Trippier, CPC, 1988.

North-South: A Programme for Survival, Report of the Brandt Commission, Pan, 1980.

People and Places: Cairo, ITV Granada, 1987.

Slums of Hope, Peter Lloyd, Penguin Books (Pelican), 1979.

Small Area Statistics, HMSO, 1983.

People and Places: Cairo, Granada Television, 1987.

Squatter Settlements in the Third World, Geography 16–19 Project, Longman, 1985.

The Global Housing Crisis, David Salterthwaite, CPC, 1988.

The Least Developed Countries, Geofile No. 69, Mary Glasgow Publications, 1986.

The Third World, M. Barke and G. O'Hare, Oliver & Boyd, 1984.

The World, David Waugh, Thomas Nelson, 1987.

Urban Development, Geography 16–19 Project, Longman, 1985.

Urban Planning for a Third World City: Bombay, Geography 16–19 Project, Longman, 1985.

World City Growth, J.H. Lowry, Edward Arnold, 1975.

(Many case studies to fit themes in this chapter can be found in the *Area Studies* series by David Waugh, published by Thomas Nelson.)

CHAPTER 16

Rural land use

'But of all the occupations by which gain is secured, none is better than agriculture, none more profitable, none more delightful, none more becoming to a free man.'
Cicero, *De Officiis* 1.51

'Behold, there shall come seven years of great plenty throughout all the land of Egypt: And there shall arise after them seven years of famine; and all the plenty shall be forgotten in the land of Egypt; and the famine shall consume the land ...'
The Bible, Genesis 41:29, 30

'Blessed is the reward of those who labour patiently and put their trust in Allah.'
The Koran

The term **rural** refers to those less densely populated parts of a country which are recognised by their visual 'countryside' components. Areas defined by this perception will depend upon whether attention is directed to economic criteria (e.g. a high dependence upon agriculture for income), social and demographic factors (e.g. the 'rural way of life' and low population density) or geographical criteria (e.g. remoteness from urban centres). Usually it is impossible to give a single, clear definition of rural areas as, in reality, they often merge into urban centres (e.g. the rural–urban fringe).

Although generalisations may lead to over-simplifications (Framework 7) it is useful to identify three main types of rural area.

1 Where there is relatively little demand for land, certain rural activities can be carried out on an **extensive** scale, e.g. arable farming on the Canadian Prairies and forestry on the Canadian Shield.

2 In many areas, especially in Third World countries, there is considerable pressure upon the land which results in its **intensive** use. Where human competition for land use becomes too great to sustain everyone, the area is said to be **overpopulated** (page 317) and this often leads to rural **depopulation**, e.g. the movement to urban centres in Latin American countries.

3 In many developed countries, competition for land is greater in urban than rural areas and the resultant high land values and declining quality of life is leading to a **repopulation** of the countryside, e.g. migration out of New York and London.

Figure 16.1 shows some of the major competitors for land in a rural area. In many parts of the world, farming takes up the majority of this land and employs most of the population.

◀ Figure 16.1

Competition for rural land use and the need for management of rural resources

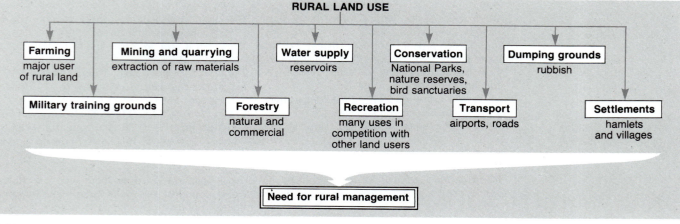

Q Using the descriptions on page 380, which of the following do you perceive or consider to be 'rural': a Lake District valley, a parish in SE England, a drought-stricken area in Ethiopia, part of the northern coniferous forest (e.g. Canadian Shield), the Ganges Valley, an oasis in the Sahara and a village in the Bolivian Andes? Explain your answers.

Farming

Environmental controls on farming

The location of different types of agriculture at all scales is a result of the interaction between physical, cultural, and economic factors. Where individual farmers in a capitalist system or the state in a centrally planned economy have a knowledge, or understanding, of these three influences then decisions may be made. How these decisions are reached involves a fourth factor: the **behavioural** element.

Physical factors affecting farming

Although there has been some movement away from the view that agriculture is controlled solely by physical conditions, it must be accepted that environmental factors do exert a major influence in determining the type of farming practised in any particular area.

In 1966 McCarty and Lindberg produced their **optima and limits** model, an adaptation of which appears in Figure 16.2. They suggested that there was an **optimum** or **ideal** location for each specific type of farming based on climate, soils, slopes and altitude. As distance increases from this optima then conditions become less ideal, i.e. too wet or dry; too steep or high; too hot or cold; or a less suitable soil. Consequently, the profitability of producing the crop or rearing animals is reduced, and the **law of diminishing returns** operates because output decreases or the cost of maintaining high yields becomes prohibitive. Eventually a point is reached where physical conditions are too extreme to permit production on an economically viable scale and later even at a subsistence level (page 397). McCarty and Lindberg applied their theory to the cotton belt of the USA (Figure 16.3) but it can equally be adapted to account for the growth of spring wheat on the Canadian Prairies (Figure 16.4). Can you suggest other regions where the theory might be successfully tested?

Temperature

This is critical for plant growth because each plant or crop type requires a minimum growing temperature and a minimum growing season. In temperate latitudes, the critical temperature is 6°C. Below this figure, members of the grass family, which include most cereals, cannot grow — an exception is rye, a hardy cereal, which may be grown in more northerly latitudes.

In Britain, wheat, barley and grass for the rearing of animals begin to grow only when average temperature rise above 6°C — this coincides with the beginning of the growing season. The **growing season** is defined as the number of days between the last severe frost of spring and the first of autumn. It is therefore synonymous with the **number of frost-free days** which are required for plant growth. Figures 16.3 and 16.4 show that cotton needs a minimum of 200, and spring wheat 90. Barley can be grown further north in Britain than wheat, and oats further north than barley because wheat requires the longest growing season of the three and oats the shortest. Frost is more likely to occur in hollows and valleys (page 179). It has beneficial effects as it breaks up the soil and kills pests in winter but it may also damage plants and destroy fruit blossom in spring.

▼ Figure 16.2

The optima and limits model (*after* McCarty and Lindberg)

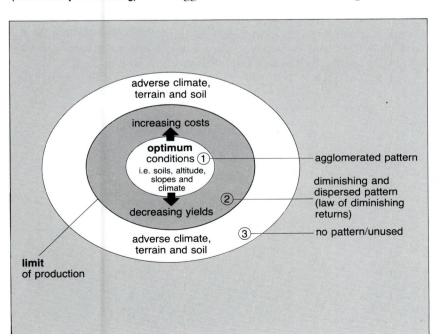

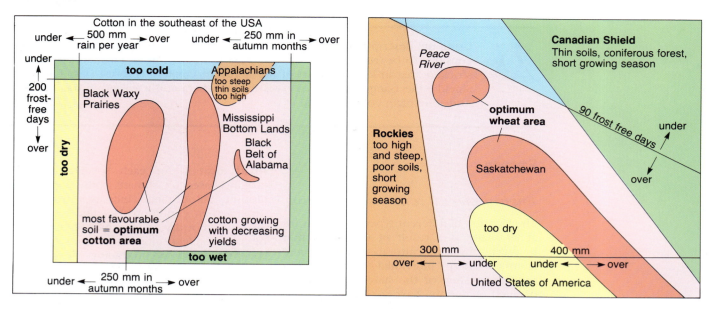

Within the tropics there is a continuous growing season, provided moisture is available. Temperatures and the length of the growing season both decrease with height above sea level. This produces a succession of natural vegetation types according to altitude, some of which have been modified for farming purposes (Figure 16.5).

Precipitation and water supply

The mean annual rainfall for an area determines whether its farming is likely to be based upon tree crops, grass or cereals, or irrigation. The relevance and effectiveness of this annual total depends on temperatures and the rate of evapotranspiration. Although few crops can grow naturally in temperate latitudes with less than 250 mm a year, the equivalent figure for the tropics is 500 mm. However, the seasonal distribution of rainfall is usually more significant for agriculture than is the annual total. Wheat is able to grow on the Canadian Prairies because there

is a summer maximum of rain meaning that water is available during the growing season. The Mediterranean lands of southern Europe have relatively high annual totals yet the growth of grasses is restricted by the summer drought. Some crops require high rainfall totals during their ripening period (e.g. maize in the American corn belt) whereas for others a dry period before and during harvesting is vital (e.g. coffee).

The type of precipitation is also important. Long, steady periods of rain allow the water to infiltrate into the soil making moisture available for plant use. Short, heavy downpours can lead to surface runoff and soil erosion and so are less effective for plants. Hail, especially in areas like the Canadian Prairies which are vulnerable to heavy convectional storms in summer, can destroy crops. Snow, in comparison, can be beneficial as it insulates the ground from extreme cold in winter and provides moisture on melting in spring. In Britain we tend to take

◄ **Figure 16.3,** *above left*
Optima and limits model applied to the former cotton belt in southeastern USA

▲ **Figure 16.4**
Optima and limits model applied to wheat growing on the Canadian Prairies

▼ **Figure 16.5**
The effect of altitude on vegetation and farming in the Andes and Alps

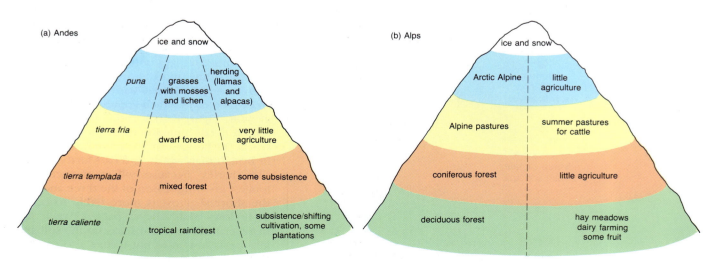

rain for granted, forgetting that in many parts of the world amounts and occurrence are unreliable. India depends upon the monsoon — if this fails then there is drought and famine. Even in the best of years the Sahel countries receive a barely adequate amount of moisture. The ecosystem is so fragile in the Sahel that should rainfall decrease even by a small amount (and some years no rain has fallen at all), then crops fail disastrously — this seems to be occurring more frequently. In Britain we would barely notice a shortfall of a few millimetres a year: in the Sahel an equivalent fluctuation from the mean will ruin harvests and cause the deaths of many animals.

CASE STUDY

Effects of precipitation and water supply on farming patterns in Kenya

The Rendille tribe live on a flat, rocky plain in northern Kenya where the only obvious vegetation is a few small trees and thorn bushes. Their traditional way of life has been to herd sheep, goats and camels, moving about constantly in search of water.

'On the government map of Kenya, the realities of the Rendille's land are summarised in a few words "Koroli Desert", it says, and just above this is the warning "Liable to Flood".

There are two rainy seasons here: the long rains in April and May and the short rains in November. But the word season suggests that the rains are much more predictable and steady than they are in reality. Add together rainfall from the long and the short rains and you arrive at only 150 mm on the Rendille's central plains in an average year. But the word average means nothing here, because normal variation from that average can bring only 35 mm of rain one year and 450 mm the next. Variation from place to place is even more erratic than variation from year to year. Rains can be heavy when they do come, and water often rushes off the baked ground in flash floods; thus the apparent contradiction of a flood-prone desert. It may suddenly rain in a valley for the first time in ten years; and it may not rain there for another decade.'

Therefore, the Rendille do not so much follow the rains as chase them, rushing to get their animals on to new grasses, which are more easily digested and converted into milk than are the drier, older shoots.

(*Source*: BBC Enterprises Ltd, *Only One Earth* by Lloyd Timberlake)
(*See also* measures of dispersion, page 208)

Wind

Strong winds increase evapotranspiration rates which means the soil dries out and becomes vulnerable to erosion. Several localised winds have harmful effects on farming: the *mistral* brings cold air to the south of France; the *khamsin* is a dry, dust-laden wind found in Egypt; hurricanes, typhoons and tornadoes can all destroy crops by their sheer strength. Other winds are beneficial to agriculture: the föhn and chinook (page 202) melt snows in the Alps and on the Prairies respectively, so increasing the length of the growing season.

Altitude

The growth of various crops is controlled by the decrease in temperature with height. In Britain few grasses, including those grown for hay, can give commercial yields at heights exceeding 300 m whereas in the Himalayas, in a warmer latitude, wheat can ripen at 3000 m. As height increases, so too does exposure to wind and the amounts of cloud, snow and rain while the length of the growing season decreases. Soils take longer to develop as there are fewer mixing agents, humus takes longer to break down and leaching is more likely to occur. Those high altitude areas which do have developed soils are prone to erosion.

Angle of slope (gradient)

Slope affects the depth of soil, its moisture content, its pH (acidity) (soil catena, page 214), and therefore the type of crop which can be grown on it. It influences erosion and is a limitation on the use of machinery. Until recently, a 5° slope was the maximum for mechanised ploughing but technological improvements have increased this to 11°. Many steep slopes in SE Asia have been terraced to overcome some of the problems of a steep gradient and to increase the area of cultivation.

Q

1 Account for the relationship between the height of the land and selected types of land use in southeast Arran as shown in Figure 16.6a.

2 Account for the relationship between the angle of slope and selected types of land use in southeast Arran as shown in Figure 16.6b.

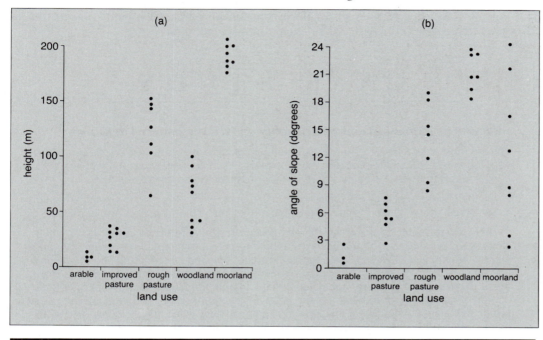

◄ Figure 16.6

Relationships between land use and

a altitude

b slope, in SE Arran

Aspect

Aspect is part of the micro-climate. **Adret** or **sonnenseite** slopes are those in the northern hemisphere which face south (Figure 9.11). These have appreciably higher temperatures than the **ubac** or **schattenseite** slopes facing north. The adret slopes receive the maximum incoming radiation and sunshine whereas the ubac may be permanently in the shade. Crops and trees both grow to a higher altitude on the adret slopes.

Soils (edaphic factors)

Farming depends upon the depth, stoniness, water retention capacity, aeration, texture, structure, pH, leaching and mineral content of the soil (Chapter 10). Three examples help to show the extent of the soil's influence on farming.

1 Clay soils tend to be heavy, acidic, poorly drained, cold and ideally should be left under permanent grass.

2 Sandy soils tend to be lighter, less acidic, perhaps too well-drained, warmer and more suited to vegetables and fruit.

3 Lime soils (chalk) are light in texture, alkaline, dry and give high cereal yields.

Although soils can be improved, e.g. by adding lime to clay, clay to sands, and applying fertiliser, there is a limit to the increase in their productivity i.e. the law of diminishing returns.

Cultural (human) factors affecting farming

Land tenure

Farmers may be owners, tenants, landless labourers or state employees on the land which they farm. The *latifundia* system is still common to most Latin American countries. The land is organised into large centrally managed estates worked by peasants who are semi-serfs. Even in the mid-1980s it was estimated that in Brazil 70 per cent of the land belonged to 3 per cent of the landowners. Land is worked by the landless labourers among the peasantry who sell their labour, when conditions permit, for sub-standard wages on the large estates or commercial plantations.

Other peasant farmers in Latin America have some land of their own held under insecure tenure arrangements. This land may be owned by the farmer but it is more likely to have been rented from a local landowner or pawned to a moneylender. This latter form of tenancy takes two forms: cash-tenancy and share-cropping. **Cash-tenancy** is when farmers have to give as much as 80 per cent of their income or a fixed pre-arranged rent to the landowner. If the farmer has a short term lease he tends to overcrop the area and cannot afford to use fertiliser or to maintain farm buildings. If the lease is long term the farmer may try to invest but this often leads to serious debt. **Share-cropping** is when the farmer has to give part of his crop direct to the landowner. As this fraction is usually a large one, the farmer works hard with little incentive and remains poor. This system operated in the cotton belt of the USA following the abolition of slavery and still persists in places. Both forms of tenancy, together with that of *latifundia*, resemble feudal systems found in earlier times in western Europe.

The plantation is a variant form of the large estate system (page 404) in that it is usually operated commercially, producing crops for the world market rather than for local use as in *latifundia*. On some plantations (e.g. oil palm in Malaysia) the labourers are landless but are given a fixed wage, on others (e.g. sugar in Fiji) they are smallholders as well as receiving a payment.

CASE STUDY

Land tenure in the Soviet Union

State ownership of land is fundamental to communist ideology. In the USSR, the belief was that large scale farms were preferable to smallholdings and that cooperative enterprise was better than that undertaken by individuals. During the 1920s, the *kolkhoz* system was created in which farmers were forced to join with their neighbours to form collective farms. Each *kolkhoz* unit was managed by an elected committee.

Since World War Two the Soviet administration has 'encouraged' the sovkhoz system. A *sovkhoz* is a state-owned farm where the workers are paid a weekly wage and where the government sets and controls production targets which the farm must meet. The individual farmer as a decision maker has not existed in the USSR for most of this century, though the situation has changed in the late 1980s. The sovkhoz are so large that the workers live in large blocks of flats (Figure 16.7).

This system has been less successful than had been hoped, possibly because of the lack of personal incentives, and it appears that output targets have not always been met. Recently, to try to increase production, farms which have exceeded their production targets have been allowed to sell their surplus produce privately — a move which presumably will continue under Gorbachev's *perestroika*.

▶ Figure 16.7

Housing on a *sovkhoz* (state-owned farm) in Byelorussia, USSR

In developed countries many farmers are **freeholders**, i.e. they own or have taken out a mortgage on the farm where they live and work. Such a system should in theory provide maximum incentives for the farmer to become efficient and to improve the land and its buildings. (Compare this with the Case Study of state farming in the Soviet Union.)

Inheritance laws and the fragmentation of holdings

In several countries, inheritance laws have meant that on the death of a farmer the land is divided equally between all his sons (except in parts of Africa where the sons share the animals and the daughters receive the land, as livestock is perceived to be more valuable). Such traditions have led to the subdivision of farms into numerous scattered and small fields. In Britain, fragmentation of land parcels may also result from the legacy of the open field system (page 338) or, more recently, from farmers buying up individual fields as they come onto the market. Fragmentation results in much time being wasted in moving from one distant field to another, and may cause problems of access.

Farm size

Inheritance laws, as described above, will tend to reduce the size of individual farms, often so that they can operate only at subsistence level or below. In most of Europe and North America the trend is for farm sizes to be increased in order to make them more efficient and profitable so that farming becomes **agribusiness**. These capital-intensive farms use much machinery, fertiliser, etc. and have a wider choice in types of production.

In SE Asia and Latin America the rapid expansion of population is having the reverse effect. Farms, already inadequate in size, are being further divided and fragmented, becoming too small for mechanisation (even if the labourers could afford machines). They are increasingly limited in the types of production possible and output in certain areas is falling. Although farms of only 1 ha can, under intensive rice production, support one family, the average plot size in many parts of Taiwan, Nepal and South Korea has fallen to less than 0.5 ha (about half the size of a football or hockey pitch). In comparison, farms of several hundred hectares are needed in marginal farming areas in other parts of the world to support just one family (e.g. upland sheep farming in Britain and cattle ranching in northern Australia).

Economic factors affecting farming

However favourable the physical environment may be, it is of limited value until human resources are added to it. **Economic man**, the term introduced by von Thünen, applies resources to maximise profits. Yet these resources are often available only in developed countries or where farming is carried out on a commercial scale.

Transport

This includes both the type of transport available and the cost of moving raw materials to the farm and produce to the market. For perishable commodities, like milk and fresh fruit, the need for speedy transport to the market demands an efficient transport network. For bulky goods, like timber or potatoes, transport costs must be lower for output to be profitable, so such items are ideally grown near to their market.

Markets

The role of markets is closely linked with transport. Perishable goods must reach the market as quickly as possible and 'economic man' limits the distance which bulky goods have to be moved. Market demand depends upon the size and affluence of the market population, its religious and cultural beliefs (e.g. fish consumed in Catholic countries, abstinence from pork by Jews), its preferred diet and changes in taste and fashion over time (e.g. increased popularity of vegetarianism).

Capital

Most developed countries, with their supporting banking systems, private investment and government subsidies, have large reserves of readily available finance, which over time have been used to build up **capital-intensive** types of farming (Figure 16.22) such as dairying, market gardening and mechanised cereal growing. Capital is often obtained at relatively low interest rates but remains subject to the law of diminishing returns. In other words, the increase in input ceases to give a corresponding increase in output, whether that output is measured in fertiliser, capital investment in machinery or hours of work expended.

Third World farmers, often lacking support from financial institutions and having limited capital resources of their own, have to resort to **labour-intensive** methods of farming (Figure 16.22). A farmer wishing to borrow money may have to pay exorbitant interest rates and may easily become caught up in a spiral of debt. The purchase of a tractor or harvester can prove a liability rather than safe investment in areas of uncertain environmental, economic and political conditions.

Technology

Technological developments such as new strains of seed, cross-breeding of animals, improved machinery and irrigation may extend the area of optima conditions and the limits of production (Green Revolution, page 403 and Figure 16.29). Lacking in capital and expertise, Third World countries are rarely able to take advantage of these advances and so the gap between them and the developed world continues to increase.

Governments

We have already seen that in centrally planned economies like the USSR the state controls what and how much each *sovkhoz* produces. In Britain, farmers have been helped by government subsidies. Organisations such as the Milk and Egg Marketing Boards have seen that farmers get a guaranteed price for their products. Today, most decisions affecting British farmers are made by the European Community (EC). Sometimes EC policy benefits British farmers, e.g. support grants to hill farmers and sometimes it reduces their income, e.g. reduction in milk quotas. Certainly countries in the EC have improved yields, as evidenced by the butter and beef 'mountains' and wine and olive oil 'lakes', and have adapted farming types to suit demand. Governments are also responsible for agricultural training schemes and for giving advice on new methods.

Von Thünen's model of rural land use

Heinrich von Thünen was a German who owned a large estate near Rostock in the Duchy of Mecklenburg for 40 years in the early nineteenth century. It was while he operated this estate on the Baltic coast that he published his ideas on farming in a book entitled *The Isolated State* (1826). He was interested in how and why agricultural land use varied with distance from a market. He illustrated this by means of a model in which he recognised that there was a pattern of land use around a market which was dependent upon competition from other types of agriculture.

Like other models, von Thünen's theory makes several basic assumptions to simplify the complex real world situation.

Von Thünen's assumptions

■ The existence of an **isolated state**, cut off from the rest of the world (it was the early nineteenth century when transport links were poorly developed). There were no external trading links.

■ The state was dominated by one large urban market (or central place). All farmers received the same price for a particular product at any one time.
■ The state occupied a broad, flat, featureless plain, or isotrophic surface (page 346), that was uniform in soil fertility and climate and over which transport was equally easy in all directions.
■ There was only one form of transport available (in 1826 this was horse and cart).
■ The cost of transport was directly proportional to distance.
■ The farmers acted as 'economic men' wishing to maximise their profits and all having equal knowledge of the needs of the market.

In his model von Thünen tried to show that with distance from the market **a**, the intensity of production decreased, and **b** the type of land use varied. Both concepts were based upon **locational rent** which von Thünen referred to as **economic rent**. Locational rent is the difference between the revenue received by a farmer for a crop grown on a particular piece of land and the total cost of producing and transporting that crop. Locational rent is therefore the profit from a unit of land, and should not be confused with **actual rent** which is that paid by a tenant to a landlord.

Since von Thünen assumed that all farmers got the same price (revenue) for their crops and that costs of production were equal for all farmers then the only variant would be the cost of transport, as this increased proportionately with distance from the market. Locational rent can be expressed by the formula:

$$LR = Y(m - c - td)$$

where:
LR = locational rent
Y = yields per unit of land (hectares)
m = market price per unit of commodity
c = production costs per unit of land (hectares)
t = transport costs per unit of commodity
d = distance from the market.

Since Y, m, c and t are constants then it is possible to work out by how much the LR for a commodity decreases as the distance from the market increases. Figure 16.8 shows that the LR (profit) will be at its maximum at 0 (the market), where there are no transport costs. LR decreases from 0 to X with diminishing returns, until at X, the **margin of cultivation**, the farmer ceases production as revenue and costs are the same, i.e. there is no profit.

▼ Figure 16.8

The relationship between locational (economic) rent and distance from the market

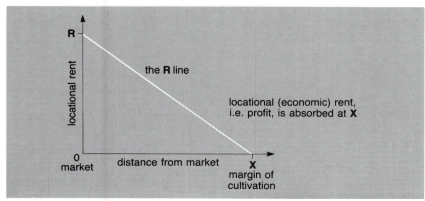

R

the R line

locational (economic) rent, i.e. profit, is absorbed at X

locational rent

0
market

distance from market

X
margin of cultivation

Q

1 Complete the table and graph in Figure 16.9 to show the *LR* for a particular crop. In this example the yield is £100 t/km², the market price is £60/t, the cost of production is £10/t and transport costs are £1/t/km.

2 At what distance from the market will the limit of cultivation be?

distance from the market (km)	Y yield tonnes/km²	m market price (£ per tonne)	c production costs (£ per tonne)	td transport costs (£ per tonne/km)		LR profit (£)
0	100	60	10	0	LR=Y(m-c-td) LR=100(60-10-0) = 5000 LR=6000-1000-0	5000
10	100	60	10	10	LR=Y(m-c-td) LR=100(60-10-10) = 4000 LR=6000-1000-1000	4000
20	100	60	10	20	LR=Y(m-c-td) LR=100(60-10-20) LR=6000-1000-2000	
30	100	60	10	30		
40						
50						

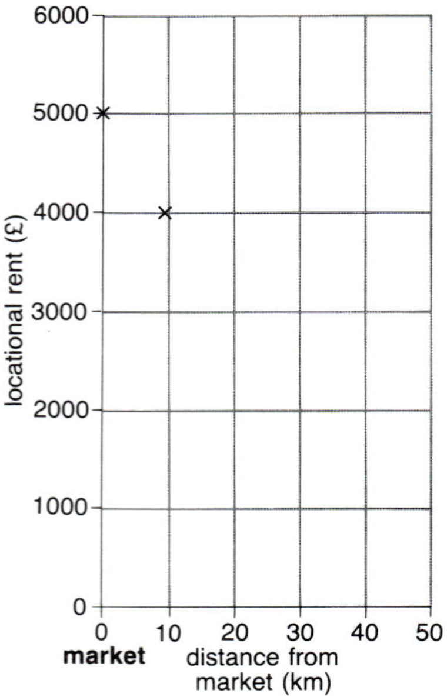

▲ Figure 16.9

Developments of von Thünen's theory

1 Intensity of production decreases with distance from the market

Two British geographers, Bradford and Kent (1977), illustrated this concept by referring to two farmers, Giles and Brown, who both wished to grow the same crop. Giles's farm was much nearer to the urban market than Brown's. Although both had similar production costs and would receive the same revenue at the market, Giles would obtain higher locational rent as his transport costs would be lower than Brown's. This is shown in Figure 16.10 together with the consequences of both farmers then wishing to intensify their production. In order to do this, costs may double from £1000 to £2000 to pay for extra labour, fertiliser and machinery. However, yields increase by only 60 per cent (from 50 t to 80 t). The higher yields mean both farmers incur increased transport

Farmer	Distance of farm from market (km)	Type of farming	Cost of production (c)	Increase in costs by intensifying (%)	Yields (Y) (tonnes)	Increase in yields by intensifying (%)	Transport costs (td) (i.e. yield × distance × cost per tonne/km)	Total costs (c + td) (costs of production + transport, £)	Total revenue (Y × m) (yield × market price per tonne)	Locational rent (total revenue − total costs)
Giles	1	extensive	1000		50		50	1050	2750	1700
		intensive	2000	100	80	60	80	2080	4400	2320
Brown	30	extensive	1000		50		1500	2500	2750	250
		intensive	2000	100	80	60	2400	4400	4400	0

Market price (m) = £55 per tonne   Transport costs (td) = £1 per tonne/km

◄ Figure 16.10

Variations in locational rent with intensity of production

▶ Figure 16.11

How locational rent varies with distance from the market for different intensities of production (*after* Bradford and Kent)

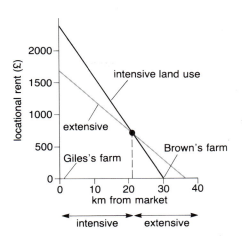

2 Type of land use varies with distance from the market

Von Thünen also tried to account for the location of several crops in relation to the market. He suggested that bulky crops, e.g. potatoes, should be grown close to the market as their extra weight would increase transport costs. He claimed, writing at a time before refrigeration had been introduced, that perishable goods should also be produced as near as possible to the market. Consequently both bulky and perishable goods will have steep *R* lines.

Figure 16.12 shows the result of two crops, potatoes and wheat, grown in competition. The two *R* lines, showing the locational rent or profit for each crop, intersect at *Y*. If a perpendicular is drawn from *Y* to *Z* then locational rent can be translated into land use. Potatoes, an intensive, bulky crop, are grown near to the market (between 0 and *Z*) as their transport costs are high. Wheat, a more extensively farmed and less bulky crop, is grown further away (between *Z* and *X*) because it incurs lower transport costs.

What happens if three crops are grown in competition?

This is the combination of von Thünen's two concepts: variation of intensity and type of land use with distance from market. Let us suppose that wool is produced in addition to potatoes and wheat:

costs but only Giles is able to increase his profit (from £1700 to £2320) because the increase in his revenue (£2750 to £4400) is greater than the increase in his costs (£1050 to £2080). Brown does not increase his profit by intensifying — indeed, any earlier profit would be lost (£250 to zero) because the increase in his revenue is less than his increase in costs. So although it is profitable for Giles to intensify, Brown is better off cultivating on an extensive scale with lower costs (Figure 16.11). When given a choice of two systems, economic man will adopt that with the greater *LR*.

Farm product	Market price per unit of commodity	Production costs per unit of land (hectares)	Transport costs per unit of commodity	Profit if grown at market
Potatoes	100	30	10	70
Wheat	65	20	3	45
Wool	45	15	1	30

▶ Figure 16.12

Locational rents for two crops grown in competition

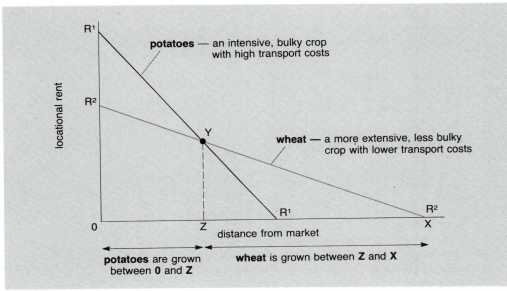

Potatoes therefore give the greatest profit grown at the market and wool the least. However, as potatoes cost £10 to transport every kilometre, after 7 km their profit will have been absorbed in these costs (£70 profit − £70 transport = £0). This has been plotted in Figure 16.13 which is a **net profit graph**. Wheat costs £3/km to transport and so can be moved 15 km before it becomes unprofitable (£45 profit − £45 transport = £0). Wool, costing only £1/km to transport, can be taken 30 km before it, too, becomes unprofitable. The three net profit curves have been drawn in Figure 16.13. This diagram shows that although potatoes can be grown profitably for up to 7 km from the market, at point A, only 3.5 km from the market, wheat farming becomes equally profitable as potatoes; beyond that point, wheat farming is more lucrative. Similarly, wheat can be grown up to 15 km from the market but beyond 7.5 km it is less profitable than and therefore replaced by wool. The point at which one type of land use is replaced by another is called the **margin of transference**.

The types of land use can now be plotted spatially. The lower part of Figure 16.13 shows three concentric circles with the market as the common central point. As on the graph, potatoes will be grown within 3.5 km of the market. This is because competition for land, and consequently land values, are greatest here so only the most intensive farming is likely to make a profit. The plan also shows that wheat is grown between 3.5 and 7.5 km from the market, while between 7.5 and 30 km, where the land is cheaper, farming is extensive and wool becomes the main product. Von Thünen's land use model is therefore based on a series of concentric circles around a central market.

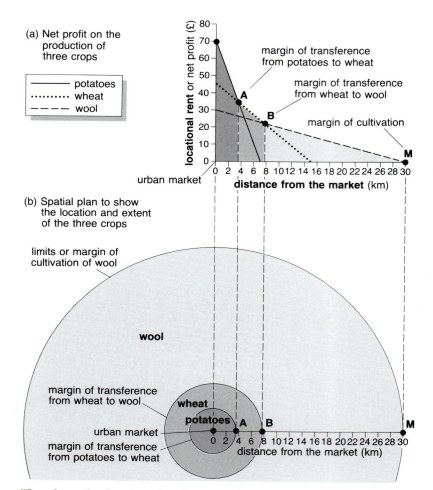

(a) Net profit on the production of three crops

— potatoes
········ wheat
– – – wool

(b) Spatial plan to show the location and extent of the three crops

The formula for locational rent, given earlier, assumed that market prices (m), production costs (c) and transport costs (t) were all constant. What would happen to the area of production of a crop if each of these in turn were to alter?

If the market price falls or the cost of production increases then there is a decrease in both the profit and the margin of cultivation of that crop (Figure 16.14a and b).

▲ Figure 16.13

a Locational rent (net profit) on the production of three crops grown in competition

b Spatial plan to show location and extent of the three crops

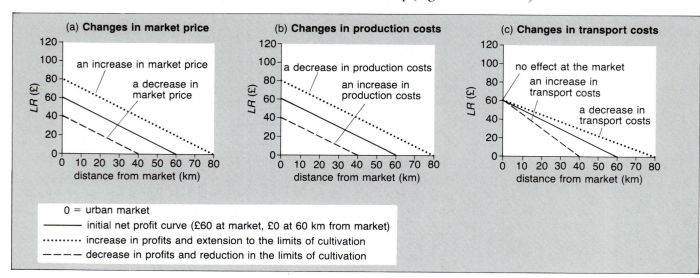

(a) **Changes in market price**
an increase in market price
a decrease in market price

(b) **Changes in production costs**
a decrease in production costs
an increase in production costs

(c) **Changes in transport costs**
no effect at the market
an increase in transport costs
a decrease in transport costs

0 = urban market
——— initial net profit curve (£60 at market, £0 at 60 km from market)
········ increase in profits and extension to the limits of cultivation
– – – decrease in profits and reduction in the limits of cultivation

Conversely, if the market price rises or the costs of production decrease then profits would rise, leading to an extension in the margin of cultivation. Changes in transport costs will not affect any farm at the market (Figure 16.14c) but an increase in transport costs reduces profits for distant farms, causing a decrease in the margin of cultivation. Conversely, a fall in transport costs makes those distant farms more profitable and enables them to extend their margin of cultivation.

Q

1 a Draw a graph to show net profit curves for market gardening, dairying and wheat using the following data:
   locational rent (profit) for market gardening is £160 at the market and £0 at 40 km,
   locational rent for dairying is £120 at the market and £0 at 60 km,
   locational rent for wheat is £80 at the market and £0 at 80 km.
  b Give the two marginal distances for market gardening and dairying.

2 Label on your graph
  a the margin of transference between market gardening to dairying
  b the margin of transference between dairying to wheat
  c the margin of cultivation for wheat.

3 Draw a spatial plan, using three concentric circles, to show the location and extent of the three types of land use surrounding the central market.

4 a Explain how the graph and plan illustrate the locational rent for each of market gardening, dairying and wheat.
  b Explain why land use changes at the two margins of transference.

Refer to the net profit graph in Figure 16.15.

5 Give a concise description of the spatial pattern of land use shown by the net profit graph.

6 What would be the cost per km of transporting one hectare's production of crop *B* to the market?

7 What might be the effects of a drop in the transport costs of crop *C*?

8 Describe the effects on the spatial pattern of land use which would result if the market price for all three crops were to double at the same time.

◄ Figure 16.15

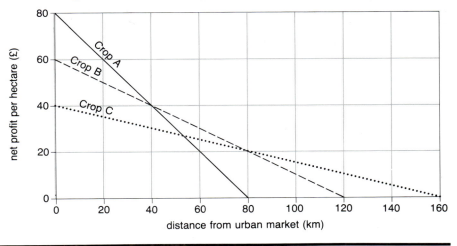

Modifications to von Thünen's land use model

◄ Figure 16.14

Effect on locational rent of changes in

a market value

b production costs

c transport costs

Von Thünen combined his conclusions on how the intensity of production decreased and the type of land use varied with distance from the market to create his model (Figure 16.16a). He suggested six types of land use which he located by concentric circles as described below.

1 Market gardening (horticulture) and dairying were practised nearest to the city due to the perishability of the produce. Cattle were kept indoors for most of the year and provided manure for the fields.

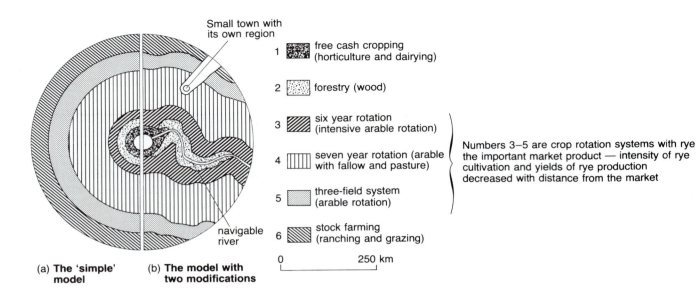

Small town with its own region

1 free cash cropping (horticulture and dairying)

2 forestry (wood)

3 six year rotation (intensive arable rotation)

4 seven year rotation (arable with fallow and pasture)

Numbers 3–5 are crop rotation systems with rye the important market product — intensity of rye cultivation and yields of rye production decreased with distance from the market

5 three-field system (arable rotation)

6 stock farming (ranching and grazing)

navigable river

0   250 km

(a) **The 'simple' model**

(b) **The model with two modifications**

▲ Figure 16.16

The von Thünen land use model

2 Wood was a bulky product much in demand as a source of fuel and building material within the town (there was no electricity when von Thünen was writing) and it was expensive to transport.

3 An area with a six-year crop rotation based on the intensive cultivation of cereals (rye, potatoes, clover, rye, barley and vetch) with no fallow period.

4 Cereal farming was less intensive as the seven-year rotation system relied increasingly on animal grazing (pasture, rye, pasture, barley, pasture, oats and fallow).

5 Extensive farming based on a three-field crop rotation (rye, pasture and fallow). Products were less bulky and perishable to transport and could bear the high transport costs.

6 Ranching with some rye for on-farm consumption. This zone extended to the margins of cultivation beyond which was wasteland.

Later, von Thünen added two modifications in an attempt to make the model more realistic (Figure 16.16b). This immediately distorted the land use pattern and made it more complex. The inclusion of a **navigable river** allowed an alternative, cheaper and faster form of transport than his original horse and cart. The result was a linear, rather than a circular, pattern and an extension of the margin of cultivation. The addition of a **secondary urban market** involved the creation of a small trading area which would compete, in a minor way, with the main city.

Later still, von Thünen relaxed other assumptions. He accepted that climate and soils affected production costs and yields (though he never moved from his concept of the isotropic plain) and that farmers do not always make rational decisions so it was necessary to introduce individual behavioural elements.

How relevant is von Thünen's theory to the modern world?

It was commented in the conclusion to Chapter 14 that although theories are difficult to observe in the real world they *are* useful because reality can be measured and compared against them. This section looks at examples at a local, national and international scale which show similarities with von Thünen's model, then considers the criticisms made of his model and the justifications for this criticism.

Local level

Many present day villages in Mediterranean lands of Europe (Figure 16.17) and in Third World countries (Figure 16.18) exhibit several principles of von Thünen's model. This usually occurs where transport links are poor, local affluence is limited and the village provides the only market. Although many villages in southern France and southern Italy occupy hilltop sites — very different from an isotrophic surface — they are surrounded by small plots in which vegetables are grown. Beyond these are first the vineyards, which need much attention, and second the citrus fruit orchards. As the distance from the village increases then the amount of land actually farmed and the yields from it both decrease.

Two critical factors appear to be the distance which farmers are prepared to walk to the fields in the absence of transport and the amount of time or intensity of attention needed to cultivate each crop.

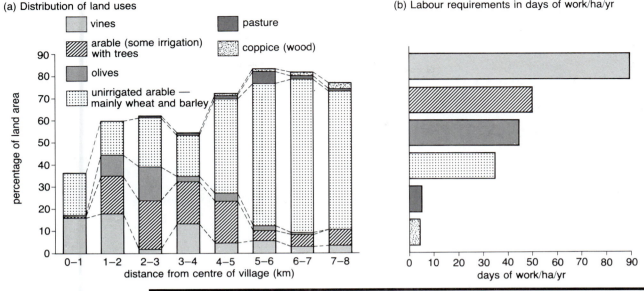

(a) Distribution of land uses

(b) Labour requirements in days of work/ha/yr

▲ Figure 16.17

Land use around a hilltop village in southern Europe

a Distribution of land use around the village

b Labour requirements in number of days work per hectare per year

Figure 16.17 shows some of the farming characteristics in an area surrounding a hilltop village in southern Europe.

1 Describe and attempt to explain the general pattern of land use.

2 To what extent and for what reasons does this general pattern of land use **a** conform to and **b** differ from the principles of the von Thünen model (i.e. how closely are changes in the intensity of production and type of land use related to distance from the market?).

CASE STUDY

Land use zonation in the semi-arid and savanna lands of Africa

Figure 16.18 is an idealised model, produced by H. Ruthenberg, showing the spatial organisation of land use common in the savannas of tropical Africa. He claimed that the intensity of cropping decreased in proportion to the distance from the village or hut. Land adjacent to the village is permanently cultivated, with no fallow period, as it is enriched by household waste and manure from the animals (cattle, sheep, goats and poultry) which are allowed to wander freely in the village. The crops grown in these 'gardens' are for one of three purposes:

1 they form part of the staple diet and so are needed frequently, e.g. fruit and indigo,

2 they required considerable attention during their growth, e.g. maize and vegetables,

3 they need manure-enriched soils, e.g. cotton and tobacco.

Beyond the gardens is a zone of intensive rotational farming. Here the fields may be used to grow the staple food or a cash crop one year, followed by pasture in the next. As distance from the village increases then more land is left fallow and for a longer period of time.

In the second outermost zone (labelled 'bush fallowing' in Figure 16.18) where crops may be grown only once in every decade, shifting cultivation (page 401) has been carried out. In the last few years as a result of increasing population pressure and the effects of drought, shifting cultivation has been replaced by a more permanent system which has fewer fallow years but also exhausts the soil more rapidly. The area of tree savanna is beyond the limits of cultivation and of wood collection for the villagers.

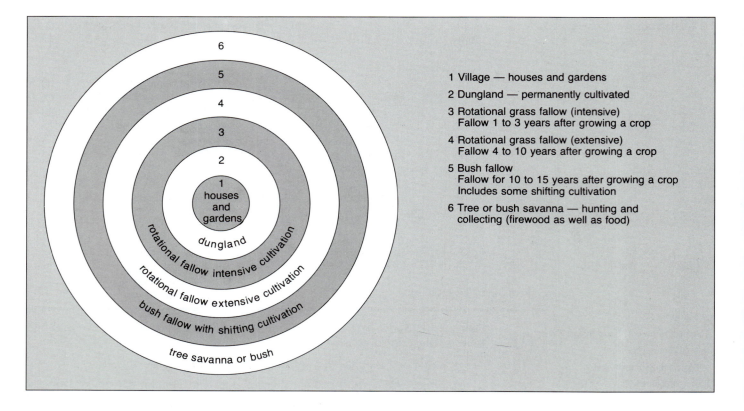

1 Village — houses and gardens

2 Dungland — permanently cultivated

3 Rotational grass fallow (intensive)
Fallow 1 to 3 years after growing a crop

4 Rotational grass fallow (extensive)
Fallow 4 to 10 years after growing a crop

5 Bush fallow
Fallow for 10 to 15 years after growing a crop
Includes some shifting cultivation

6 Tree or bush savanna — hunting and
collecting (firewood as well as food)

Q Use the land use (agriculture) map of the British Isles in an atlas to answer the following questions.

1 Draw a sketch map to show the location of the main types of farming in Britain.

2 Are there any similarities between your map and von Thünen's land use model in the following situations?
  a If London is taken to be the only market.
  b If several large urban areas are added to modify the map.
  c Suggest reasons for any differences you observe.

National level

Uruguay is one of only a few countries which broadly resemble von Thünen's isotrophic surface. Figure 16.19 maps the generalised pattern of land use within that country. Notice that the map includes the same two modifications that von Thünen made to his original model: the introduction of a navigable river as an alternative source of transport (in this case the Rio Uruguay) and a secondary market (Fray Bentos, noted for its corned beef).

You may agree that Uruguay resembles the original model but remember that perhaps it has been chosen specifically in an attempt to find an example which *does* fit that model. It has already been pointed out that as geographers we should be trying to compare reality with models rather than forcing reality to fit them (page 293).

▼ Figure 16.19

Land use in Uruguay — von Thünen's model applied at a national scale

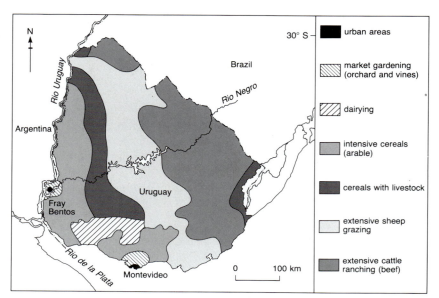

Legend:
- urban areas
- market gardening (orchard and vines)
- dairying
- intensive cereals (arable)
- cereals with livestock
- extensive sheep grazing
- extensive cattle ranching (beef)

◄ Figure 16.18

Land use around a village in the African savannas — von Thünen's model applied at a regional scale

International level

From early in the twentieth century it had been noted that zones of similar agricultural land use could be found around most large urban areas in Western Europe and North America. In 1952, van Valkenburg and Held produced a map based upon the intensity of agriculture in Europe. They took eight crops: barley, hay, maize, oats, potatoes, rye, sugar beet and wheat, none of which are typically Mediterranean, and for each crop worked out the average yield per acre for each country. The average yield for Europe as a whole was given as an index of 100. A simplified version of their map is given in Figure 16.20.

It shows a core, or optima zone (page 381), consisting of SE England, the Benelux countries, Denmark and North Germany where yields had an index of over 150. This core included several major urban markets and industrial centres which today coincide with areas of greater affluence. Beyond this core, and similar to von Thünen's predicted pattern, were a series of zones of decreasing intensity. The areas with the lowest yields lay on the periphery of Europe where, as in the optima and limits model (Figure 16.2) there is adverse climate (e.g. too cold in northern Scandinavia, summer drought in the Mediterranean), inhospitable terrain (e.g. Alps and Pyrenees) and infertile, eroded soil (e.g. Apennines).

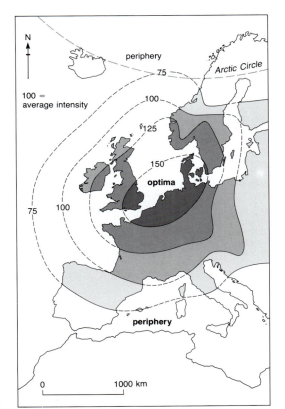

▶ Figure 16.20

Intensity of agriculture in Europe — von Thünen's model applied at an international scale

Why is it difficult to apply von Thünen's ideas to the modern world?

No model is perfect and so each is open to criticism. In the case of von Thünen these criticisms may be grouped under four headings: oversimplification, out-datedness and failure to recognise the behavioural patterns of farmers and the role of governments.

Critics have pointed out that few areas consist of flat featureless plains and if such a landscape did occur it is likely to have several markets rather than one. Few areas are homogeneous in climate and soils and therefore certain locations are more favourable than others. Similarly, the 'isolated state' is rarely found in the modern world (Albania may be the nearest to this situation) and there is much competition for markets both within and between countries. However, von Thünen did recognise that he was trying to simplify the real world situation and that when he did introduce two variables, his model became much more complex and potentially unworkable (Figure 16.16b).

Critics have also attacked the model as being out-dated and of limited value in modern farming economics — it was written over 160 years ago. Certainly since 1826 there have been significant advances in technology, changing uses of resources, pressures created by population growth, and the emergence of differing economic policies. The invention of the internal combustion engine and motorised vehicles, trains and aeroplanes have revolutionised transport, often increasing accessibility in one particular direction as well as making the movement of goods quicker and relatively cheaper. Milk tankers and other refrigerated lorries mean that perishable goods can be produced or grown further from the market and stored for longer periods of time (London uses fresh milk from Devon and the EC stockpiles butter). Electricity and gas have replaced wood as a fuel in developed countries and so trees need not be grown so near to the market, while supplies of timber in Third World countries are rapidly being consumed and not actively replaced. Improved farming techniques using fertilisers and irrigation have improved yields and extended the margins of cultivation. Elsewhere, farmland has been taken over by urban growth or used by competitors who obtain higher economic rent (Figure 16.1). In the developed world, changes in taste, increases in affluence, concern over health and the demand for 'fast foods' have all contributed to changes in the nature and patterns of agricultural output.

Governments may alter land use by granting subsidies or imposing quotas, and are empowered to limit a free market by imposing tariffs and trade barriers. The EC, by setting quotas for milk production in the late 1980s, has forced farmers to turn to other forms of agriculture. In the same way it has begun to pay farmers to take land out of production in the hope of reducing the European Community's food surpluses. Socialist governments, as in the USSR, may directly control the types and amount of production rather than manipulating market mechanisms.

Von Thünen has also been criticised for his assumption that farmers are 'rational economic men'. Farmers do not have full knowledge, they may not always make rational or consistent decisions, they may prefer to enjoy increased leisure time rather than seeking to maximise profits and may be reluctant to adopt new methods. Farmers, as human beings, may have different levels of ability, ambition, capital and experience and none can predict changes in the weather, government policies or demand for their product.

Conclusions

Von Thünen's land use model still has some modern relevance, providing its limitations are understood and accepted. However, it would appear to be more applicable to Third World countries, especially those in Africa, than to capitalist and socialist economies. His concept of locational rent, which is useful in studying urban as well as rural land use (page 355) is still applicable today. Concentric circles of land use may not exist around every urban centre, yet many areas do have patterns which show a similarity to, and can be partially explained by, principles of the von Thünen model.

The farming system

Farming is another example of a system, and one which you may have studied already (Framework 1, page 44). The systems diagram (Figure 16.21) shows how physical, cultural (human), economic and behavioural factors form the inputs. In areas where farming is less developed, physical factors are usually more important but as human-economic inputs increase then these physical controls become less significant. This system model can be applied to all types of farming, regardless of scale or location. It is the variations in inputs which are responsible for the different types and patterns of farming.

Types of agricultural economies

The simplest classification shows four contrasts between types of farming system.

▼ Figure 16.21
The farming system

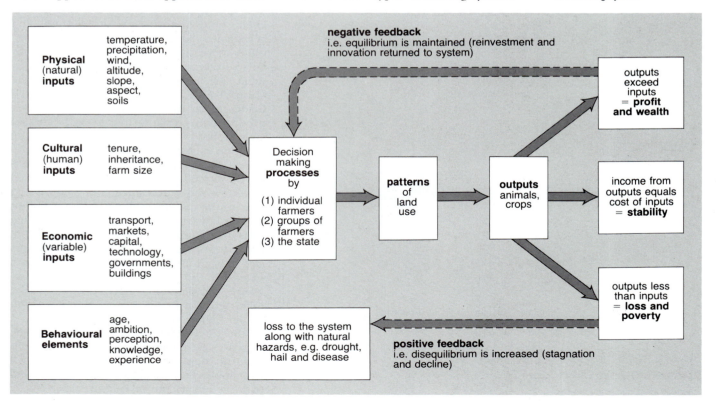

1 Arable, pastoral and mixed

Arable is the growing of crops, usually on flatter land where soils are of a higher quality. The development of new strains of cereals led to the first permanent settlements in the Tigris—Euphrates, Nile and Indus valleys (Figure 14.1). Much later, in the mid-nineteenth century, it was the building of the railways across the Prairies, Pampas and parts of Australia which led to the biggest increase in the area of the world 'under the plough' (page 408). Today there are few areas left where arable farming can potentially be extended. This fact, coupled with rapidly increasing global population, has led to concern over the world's ability to feed its inhabitants in the future — a fear first voiced by Malthus (page 319). Indeed in those parts of Africa affected by drought and soil erosion, the amount of arable land is actually decreasing.

Pastoral farming is the raising of animals (page 399). This is usually on land which is less favourable to arable farming (i.e. is colder, wetter, steeper, and higher). However, if the grazed area has too many animals on it, or the quality of the soil and grass is not maintained, then erosion and (especially in Africa) desertification may result.

Mixed farming is practised in more developed countries (page 410). Here crops are grown to feed both humans and animals as well as to reduce the risks involved by relying upon a single crop or animal (monoculture).

2 Subsistence and commercial

Subsistence farming results from farmers being able to provide just enough food for the farmer's own family or the local community — there is no surplus. The main priority of subsistence farmers is to avoid risk and uncertainty so that they can survive and so, whenever possible, they try to grow and/or rear a wide range of crops and/or animals (page 401). Such farmers are unable to improve their output mainly because they lack capital, land and technology and not because they lack effort or ability. They are the most vulnerable to famine of all farming types.

Commercial farming is done on a large, profit-making scale. Commercial farmers, or the companies for whom they work, seek to maximise yields per hectare. This is often achieved, especially within the tropics, by growing a single crop or rearing one type of animal. **Cash-cropping** operates successfully where transport is well-developed, domestic markets are large and expanding, and there are opportunities for international trade (page 410).

3 Shifting and sedentary

The earliest farmers were forced to move about as they hunted and collected their food. **Shifting** cultivation is still practised where there are low population densities and the demand for food is limited, where there are poor soils which become exhausted after three or four years of cultivation (page 402) or where there is a seasonal movement of animals in a search of pasture (page 400).

When farming becomes permanently established in one place it is referred to as **sedentary**. Once farmers begin to grow crops they have to remain in the area to look after and to harvest them. Most types of farming in developed countries as well as commercial farming in the Third World, are sedentary.

4 Extensive and intensive

These terms have already been used in describing von Thünen's model (Figure 16.11). **Extensive** farming is carried out on a large scale whereas **intensive farming** involves a high degree of concentration. Farming is extensive or intensive depending upon the relationship between three factors of production: labour, capital and land (Figure 16.22). Extensive farming occurs in the following situations.

▶ Figure 16.22
Extensive and intensive farming
(*after* K. Briggs)

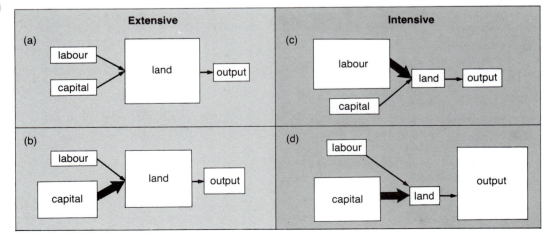

- Amounts of labour and capital are small in relation to the area being farmed, e.g. the Amazon Basin (page 402). Here the yields per hectare and the output per farmer are both low (Figure 16.22a).
- The amount of labour is still limited but the input of capital may be high, e.g. the Canadian Prairies (page 409). In this case the yields per hectare are often low but the output per farmer is high (Figure 16.22b).

Intensive farming occurs in the following situations.

- The amount of labour is high even if the input of capital is low in relation to the area farmed, e.g. Ganges valley (page 401). Here the yields per hectare may be high although the output per farmer is often low (Figure 16.22c).
- The amount of capital is high but this time the input of labour is low, e.g. the Netherlands (page 410). In this case both the yields per hectare and the output per farmer are high (Figure 16.22d).

World distribution of farming types

There is no widely accepted consensus as to how the major types of world farming should be classified or recognised. There is disagreement over the basis used in attempting a classification (e.g. intensity, land use, tropical or temperate, level of human input, the degree of commercialisation); the actual number of farming types; the nomenclature of the various types; the exact distribution and location of the major types.

You should be aware that:
1 Boundaries drawn on a map are usually very arbitary.
2 One type of farming merges gradually with a neighbouring type and there are no rigid boundaries.
3 Several types of farming may often be found within each broad area, e.g. in West Africa where sedentary cultivators live alongside nomadic herdsmen.
4 A specialised crop may be grown locally, e.g. a rubber plantation in an area otherwise used by subsistence farmers,
5 Types of farming alter over a period of time with changing economies, rainfall or soil characteristics, behavioural patterns and politics.

Figure 16.23 suggests one classification and shows the generalised location and distribution of farming types based upon the four variables described in the previous section. On a continental scale, this map demonstrates a close relationship of farming types with the physical environment and the global pattern of biomes (page 260) and major climate, soils and vegetation types: it disguises the important human–economic factors operative at a more local level.

▼ Figure 16.23
Location of the world's major farming types

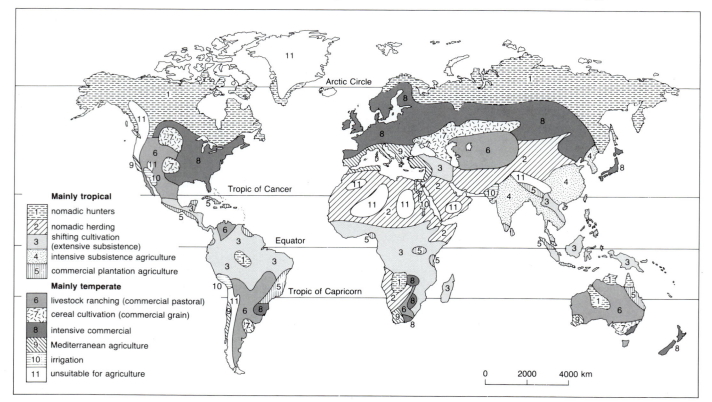

Mainly tropical
1	nomadic hunters
2	nomadic herding
3	shifting cultivation (extensive subsistence)
4	intensive subsistence agriculture
5	commercial plantation agriculture

Mainly temperate
6	livestock ranching (commercial pastoral)
7	cereal cultivation (commercial grain)
8	intensive commercial
9	Mediterranean agriculture
10	irrigation
11	unsuitable for agriculture

0   2000   4000 km

The following section describes the main characteristics of each of these categories of farming together with the conditions favouring their development. A case study is given for each (which should be supplemented by wider reading) together with an account of recent changes or problems within that rural economy.

1 Hunters and gatherers

Some classifications ignore this group on the grounds that they consider it to be a relict way of life, because the original lifestyle has largely been destroyed by contact with the outside world. Others do not consider that it constitutes a type of 'true' farming because no crops or domesticated animals are involved. It is included here as before the advent of sedentary farming all early societies had to rely upon hunting birds and animals, catching fish and collecting berries and fruit in order to survive ... and why else do we farm today? There are now very few hunter–gatherer societies remaining — the Bushmen of the Kalahari, the Pygmies of central Africa, several Amazon Indian tribes in the tropical rainforests, and the Australian Aborigines. All have a variety of diet resulting from their intimate knowledge of the environment, but each needs an extensive area from which to obtain their basic needs.

2 Nomadic herding

In areas where the climate is too extreme to support permanent settled agriculture farmers have to move about in search of grass for their animals and so are known as **nomadic pastoralists**. These pastoralists live in inhospitable environments where vegetation is sparse and the climate is arid or cold. The movement of most present day nomads is determined by the seasonal nature of rainfall and the need to find new sources of grass for their animals, e.g. the Bedouin and Tuareg in the Sahara, the Fulani in West Africa, the Masai in Kenya and the Kirghiz in central Asia. The Lapps of northern Finland have to move when their pastures become snow-covered in winter while the Fulani may migrate to avoid the tse-tse fly.

There are two forms of nomadism. **Total nomadism** means the nomad has no permanent home, while **semi-nomads** may live seasonally in a village. There is no ownership of land and the nomads may travel extensive distances, even across national frontiers, in search of fresh pasture. There may be no clear migratory pattern but migration routes increase in size under adverse conditions, e.g. the drought in the Sahel providing even less water and grass. The animals are the source of life. Depending upon the area, animals may provide milk, meat and blood as food for the tribe; wool and skins for family shelter and clothing; dung for fuel; mounts and pack animals for transport; and products which can be exchanged in return for crops and other commodities. Just as sedentary farmers will not sell their land unless in dire economic difficulty, similarly the pastoralists will not part with their animals, retaining them to generate the herd as quickly as possible when conditions improve.

CASE STUDY

The Australian Aborigines: hunters and gatherers

When Captain Cook landed in Australia in 1770 there were about 300 000 Aborigines living on the island. Of Melanesian descent, there is evidence that Aborigines had lived there for over 50 000 years. Their rock paintings are older than those at Lascaux in France and predate the Egyptian pyramids by 10 000 years. The majority lived close to the northern coast or inland near to rivers (many of which were seasonal) or to Ayers Rock which was sacred to them (Figure 6.4). They had strong religious beliefs and family ties.

The Aborigines' diet, which some suggest contained over 350 different items of food, included emus and their eggs; kangaroos and wallabies; grubs and ants; acacia seeds, roots and grasses. Those living nearer rivers caught fish and crocodiles, while others on the coast ate turtles, crabs and mussels. Their main hunting weapons were boomerangs, long spears, shields, stone axes, fishing nets and fish hooks. As the food supply was limited the population remained stable in size. After a tribe had remained some time in an area, the local food supply would begin to decrease so the tribe would move to a different location.

By 1930, only 70 000 Aborigines remained and their lifestyle was severely threatened. Recently they have been given land rights to a third of the Northern Territory (including Ayers Rock) and a fifth of South Australia. Their demands for more land and improved civil rights were vocally expressed during the bicentennial celebrations of 1988 but many non-Aborigines oppose the granting of such rights. Today relatively few survive in their natural environment as economic and population pressures (numbers have increased to 160 000) have forced many to leave the reserves and migrate to urban centres. Here, Aborigines tend to experience discrimination and poverty, an infant mortality rate twice the national average and a life expectancy 20 years less than the national average. They have poor housing and education, and with few rights, many fall victim to social problems. Indeed, Aborigines face all the problems of an immigrant ethnic minority when in fact they were the indigenous inhabitants.

The Rendille of northern Kenya: nomadic herders

Rainfall is too low and unreliable in northern Kenya to support settled agriculture (page 383). Over the years the Rendille have learnt how to survive in an extreme environment. The only things they need are their animals (camels, goats and a few cattle): all the animals need is water and grass. The tribe are constantly on the lookout for rain which usually comes in the form of heavy, localised downpours. Once the rain has been observed or reported, the tribe pack their limited possessions on to camels (a job organised by the women) and they head off, perhaps on a journey of several days, to an area of new grass growth. In the past this movement prevented overgrazing as grazed areas were given time to recover. Camels and to a lesser extent goats can survive lengthy periods without water by storing it within their bodies or by absorbing it from edible plants — food supply is as important as water. Humans, who can go longer than animals without food but much less without water, rely upon the camels for milk and blood. Indeed the main diet of blood and milk avoids the necessity of cooking and the need to find firewood.

But their way of life is changing. Land is becoming overpopulated and resources over-stretched as the numbers of people and animals increase (Kenya as a country has the world's highest birth rate) and as water supplies and vegetation become scarcer. Consequently as the droughts of recent years continue, pastoralists are forced to move to small towns, such as Korr. Here there is a school, health centre, better housing, jobs, a food supply and, most importantly, a permanent supply of water from a deep well. This deep well waters hundreds of animals, many of which are brought considerable distances each day. However, these extra animals result in overgrazing while the increased numbers of townspeople has led to the clearance of all the nearby trees for firewood.

The result has been an increase in soil erosion creating a desert area extending 150 km around the town. Although attempts are being made to dig more wells to disperse the population, travelling shops now take provisions to the pastoralists and the tribespeople have been shown how to sell their own animals at fairer prices so many Rendille are moving to Korr to live. There the children, having been educated, remain, looking for jobs, with the result that there are fewer pastoralists left to herd the animals (Figure 16.25).

◄ Figure 16.24, *far left*
Herders at a shallow hand-dug well in Kenya

▲ Figure 16.25
The Rendille of northern Kenya — camels and goats at a water hole

3 Shifting cultivation (extensive subsistence agriculture)

Subsistence farming was the traditional type of agriculture in most tropical countries before the arrival of Europeans. The inputs to this system are extremely limited. Only relatively few labourers are needed (although they may have to work intensively), technology is limited (possibly to stone axes) and capital is not involved. Over a period of years extensive areas of land may be used as the tribes have to move on to new sites. Outputs are also very low with barely sufficient being grown for the immediate needs of the family or tribe.

The most extensive form of subsistence farming is shifting cultivation. This is still practised in the tropical rainforests, e.g. the *milpa* of Latin America and *ladang* of SE Asia, and occasionally in the wooded savannas, e.g. the *chitimene* of Central Africa. It tends to be limited to less accessible tropical areas such as the Amazon Basin (Case Study), Central America, Zaire, Zambia and parts of Indonesia.

4 Intensive subsistence farming

This involves the maximum use of the land with neither fallow or any wasted space. Yields, especially in SE Asia, are high enough to support a high population density, e.g. up to 2000 per km^2 in parts of Java and Bangladesh. The highest yielding crop is rice. This is grown chiefly on river flood plains (e.g. Ganges, Figure 16.27) and in river deltas (e.g. Mekong and Irrawaddy) where peak river flows following the monsoon rain are trapped behind *bunds*, or walls (Case Study). Rice is also grown on terraces on steep hillsides as found in Java and the Philippines (Figure 16.28).

As rice requires a growing season of only 100 days, then the constant high temperatures of SE Asia enable two and sometimes even three croppings a year. However, although 'winters' are warm enough for the crop to grow, water supply may be a problem. During the rainy season from July to October, the *kharif* crops of rice, millet and maize are grown (page 200). Rice is planted as soon as the monsoon rains have flooded the padi fields and harvested in October when the rains have stopped and the land has dried out — equatorial areas are not ideal for rice cultivation because of the absence of a dry season. During the dry season from November to April, the *rabi* crops of wheat, barley and peas are grown and harvested. Where water is available for longer periods, a second rice crop may be grown. Rice needs very fertile soils. These may have developed from the annual deposition of silt as rivers flow over their flood plains, or from weathered volcanic rock. Padi or wet rice is grown in river silts, while hill or dry rice is found on higher land. Although the latter is easier to grow it gives lower yields and can support fewer people.

Rice forms up to 90 per cent of the total diet in some parts of SE Asia and has a high nutritional value. Yields per hectare are high which is essential in an area of small farms, many of which are only 1 ha. These high yields can support the large population needed to tend the crop. Rice growing is labour intensive with much manual effort needed to construct the *bunds* or terraces; to build irrigation channels; to prepare the fields; and to plant, weed and harvest the crop. It is estimated that it takes 2000 hours of labour per year to farm each 1 ha plot. The flooded padi fields may be stocked with fish which add protein to the human diet and fertiliser to the soil. Rice can be stored easily in the hot climate.

There are many problems associated with rice growing. The small farm sizes rarely allow any surplus to be produced for sale; land is often fragmented as a result of inheritance practices; and the farmers are usually tenants and have to pay a proportion of their crops to their landlord (share-cropping). There is little machinery and the animals are often overworked with their manure frequently used as a fuel rather than being returned to fertilise the land. Many areas suffer from overpopulation, a lack of capital for investment, and climatic extremes of drought and flooding. The poor transport network hinders the marketing of any surplus crops after a good season and the reception of food relief during times of shortage.

CASE STUDY

Shifting cultivation in the Amazon Basin

With the help of stone axes and machetes the Amerindians clear a small area of forest (Figure 16.26). Sometimes the largest trees are left standing to protect young crops from the sun and heavy rain, as are those which provide food, e.g. banana and kola. The buttress roots of the giant trees are left. After being allowed to dry, the felled trees and undergrowth are burnt which has given rise to the alternative name of **slash and burn** cultivation. This has the advantage of providing ash as a fertiliser and clearing the weeds but it does destroy useful organic material and bacteria. The main crop, manioc, is planted along with yams (which need a richer soil), pumpkins, beans, tobacco and coca. The productivity of the rainforests depends upon the rapid and unbroken recycling of nutrients.

◀ Figure 16.26
Shifting cultivation in the
Amazon Basin

Once the forest has been cleared the nutrient cycle is broken (Figure 12.10). The heavy afternoon rains can then hit the bare earth causing erosion and leaching. With the source of humus removed and in the absence of fertiliser and animal manure, the ferrallitic soils (page 265) rapidly lose their fertility. Within four or five years yields have declined and weeds have infested the area forcing the tribe to shift to another part of the forest. Although this appears to be a wasteful use of land, it has no long term adverse effect upon the environment. In places humus can build up sufficiently in 25 years to allow the land to be used again if necessary.

The traditional way of life of the Amerindian is being threatened by the destruction of the rainforests. As land is cleared for hydro-electric power schemes, highways, cattle ranches, mineral exploitation or timber, the Amerindians are pushed further into the forest or are forced to live on reservations. Of the estimated six million Amazon Amerindians living in the forests when the Europeans arrived in Latin America, only about 4 per cent remain today, as the result of illnesses caught from and death inflicted by the invaders. With the Amerindians coerced to live on reservations and new farms created along the highways for some of Brazil's landless, especially in Rondonia, sedentary farming is rapidly replacing shifting cultivation ... but what happens after four or five years when the soils on which they farm become infertile and they can no longer shift to fresh sites?

Rice growing in Thailand

The delta region of the Chao Phraya River, near Bangkok, is the 'rice bowl' of Thailand. The silt deposited by the river, the monsoon rains and the year-long growing season provide the optimum conditions for the crop. Fields are small but the land is intensively used and, with the use of water buffalo, two rice crops and a dry season crop can be grown within each year. The *bunds* (embankments) between the fields are stabilised by two tree crops. The tall coconut palm not only gives a source of food, drink and sugar, but it also acts as a cover crop for the smaller banana tree. The Thai farmers are therefore able to support their families and may even have a surplus of 'fragrant' (non-hybrid) rice for sale. Thailand is, in fact, an exporter of rice.

▲ Figure 16.27
Intensive subsistence farming
— rice cultivation on a flood
plain in China

▶ Figure 16.28, *above right*
Intensive subsistence farming
— rice cultivation on upland
terraces in Bali

▼ Figure 16.29
The Green Revolution

The Green Revolution

The Green Revolution refers generally to the application of modern, western-type farming techniques to Third World countries. Its beginnings were in Mexico when during the two decades after World War Two, new varieties or **hybrids** of wheat and maize were developed in an attempt to solve the country's domestic food problem. At that time there was no intention of trying to transform the agriculture of other Third World countries. The new strains of wheat produced dwarf plants capable of withstanding strong winds, heavy rain and disease (especially the 'rusts' which had attacked large areas). Yields of wheat and maize increased by three and two times respectively and the new seeds were taken to the Indian sub-continent. Later, new varieties of improved rice were developed in the Philippines. The most famous, the IR-8 variety, increased yields sixfold at its first harvest. Since then further improvements have shortened the growing season required, allowing an extra rice crop to be grown, and made strains tolerant of a less than optimum climate.

In 1964, many farmers in India were short of food, lacked a balanced diet and had an extremely low standard of living. The government, given limited resources, was faced with the choice (Figure 16.29) of attempting a land reform programme (redistributing land to landless farmers) or trying to improve farm technology. They opted for the latter.

Eighteen thousand tonnes of Mexican HYV (high-yielding varieties) wheat seeds were imported as well as considerable amounts of fertiliser. Tractors were introduced in the hope that they would replace water buffalo. Communications were improved and there was some land consolidation. The successes and failures of the Green Revolution in India have been summarised in the table on page 404. In general, it has improved food supplies in many parts of the country but it has also created adverse social, environmental and political conditions. Certainly the biochemical and mechanical innovations referred to in Figure 16.29 have been far more successful than the social changes. In the late

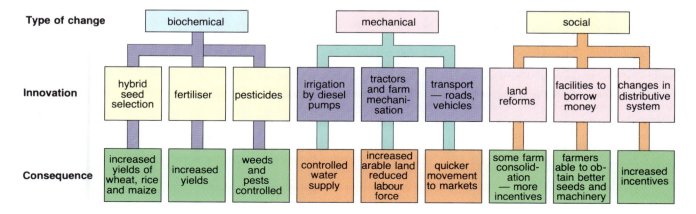

Type of change	biochemical			mechanical			social		
Innovation	hybrid seed selection	fertiliser	pesticides	irrigation by diesel pumps	tractors and farm mechanisation	transport — roads, vehicles	land reforms	facilities to borrow money	changes in distributive system
Consequence	increased yields of wheat, rice and maize	increased yields	weeds and pests controlled	controlled water supply	increased arable land reduced labour force	quicker movement to markets	some farm consolidation — more incentives	farmers able to obtain better seeds and machinery	increased incentives

1980s. there was still considerable rural overpopulation, migration to the towns continued and a low standard of living persisted for most rural dwellers. As one television commentator said: 'The poorer families can probably now afford two meals a day instead of one, while the rich can educate their children'.

An appraisal of the Green Revolution in the Indian sub-continent

Successes	Failures
Wheat and rice yields have doubled	HYV seeds need heavy application of fertiliser and pesticides which have increased costs, encouraged weed growth and polluted of water supplies
Often an extra cropping per year	
Rice, wheat and maize have widened the diet	Extra irrigation is not always possible
Dwarf plants can withstand heavy rain and wind	HYV not as drought-resistant nor as suited to waterlogged soils
Farmers able to afford tractors, seed and fertiliser now have a higher standard of living	Farmers unable to afford tractors, seeds and fertiliser have become relatively poorer
Farmers with land exceeding 1 ha have usually become more wealthy	Farmers with less than 1 ha of land have usually become poorer
The need for fertiliser has created new industries and jobs	Farmers who have to borrow money to buy seed and fertiliser likely to get into debt
Some road improvements	Still only a few tractors, partly due to cost and shortage of fuel
Area under irrigation has increased	Mechanisation has increased unemployment
Some land consolidation	Some HYV crops are less palatable to eat
Conclusions	
A production and economic success which has lessened but not eliminated the threat of food shortages	Social, environmental and political failure: bigger gap between rich and poor and between groups

5 Tropical commercial (plantation) agriculture

Plantations were developed in tropical areas, usually where rainfall was sufficient for trees to be the natural vegetation, by European and North American merchants in the eighteenth and nineteenth centuries. Large areas of forest were cleared and a single crop, either a bush or a tree, was planted in rows — hence the term monoculture (page 235). This so-called **cash crop** was grown for export and was not used or consumed locally. (Case Study, Malaysia.)

Plantations needed a high capital input to clear, drain and irrigate the land; to build estate roads, schools, hospitals and houses; and to bridge the several years before the crop could be harvested. Although plantations were often located in areas of low population density, they needed much manual labour. The owners or managers were invariably white. Black and Asian workers, obtained locally or brought in as slaves or indentured labour from other countries, were engaged as they were prepared, or forced, to work for minimal wages. They were also capable of working in the hot, humid climate. Today many plantations, producing most of the world's rubber, coffee, tea, cocoa, palm oil, bananas, sugar cane and tobacco, are owned and operated by large multinational companies.

The costs and benefits of plantation farming

Advantages	Disadvantages
Higher standards of living for the local workforce	Exploitation of local workforce, minimal wages
Capital for machines, fertiliser and transport provided initially by colonial power, now the multinational corporations	Overuse of land has in places led to soil exhaustion and erosion
	Most produce is sent overseas to the parent country
Use of fertiliser and pesticides improves output	Most profit returns to Europe and North America
Increases local employment	Dangers of relying on monoculture: fluctuating world prices and demand
Housing, schools, health services, transport provided, often electricity and a water supply	Cash crops grown instead of food crops: local population have to import foodstuffs

CASE STUDY

Rubber in Malaysia: tropical commercial agriculture

A plantation is defined in Malaysia as an estate exceeding 40 ha in size. Many extend over several thousand hectares. Rubber is indigenous to the Amazon Basin but some seeds were smuggled out of Brazil in 1877, brought to Kew Gardens in London to germinate and then sent out to what is now Malaysia. The trees thrive in a hot, wet climate, growing best on the gentle lower slopes of the mountains forming the spine of the Malay Peninsular (Figure 16.32). Rubber tends not to be grown on the coasts where the land is swampy, but near to the relatively few railway lines and the main ports. The 'cheap' labour needed to clear the forest, work in the nurseries, plant new trees and to tap the mature trees was provided by the poorer Malays or immigrants from India.

Young trees, having begun their lives in a nursery, are planted out in rows. It takes six years before they can be tapped. Rubber is made from latex, a white liquid obtained from within the trunk. Each day the tapper makes a small incision, less than 0.5 m long, into the bark of mature trees. The latex runs downwards to be collected in a cup (Figure 16.31). The ideal time to tap the tree is 5 a.m. when the internal pressure is at its greatest, the worst time is after 11 a.m. when the heat prevents the liquid from flowing, or during rain which dilutes the latex. An experienced tapper can cover 2 ha (500–600 trees) in a morning and is paid per litre of latex collected, not per tree tapped. Each tree gives on average 200 ml (7 fluid ounces) daily. Therefore, 600 trees should give 120 litres for which the tapper will receive $20 Malaysian (just under £5).

Many tappers, both male and female, try to find a second job in the afternoon to increase their income. Each plantation is likely to have its own processing factory where the latex is turned into sheets ready for transportation, often along good roads, to the nearest port. Rubber trees produce less latex after 25 years and so are felled, chopped up, and allowed to decay into a mulch which acts as a fertiliser into which new trees are planted.

▶ Figure 16.30

A rubber plantation in Malaysia

The Malaysian government has now taken over all the large estates, formerly run by such multinationals as Dunlop and Guthries, having seen them as a relict of colonialism. However, these publicly owned companies now only account for 30 per cent of the land devoted to rubber: the remainder is in the form of smallholdings. In the early 1970s the Federal Land Development Authority (FELDA) was set up. Under FELDA, the forest is cleared and land divided into 5 ha plots which are allocated to farmers. For the first four years the government does all the work, supervising planting and caring for the young trees. The farmer is then put in charge but is still provided with free fertilizer and pesticide as the trees are too young to provide any income. Once the crop is ready it is bought and marketed by the government. Most farmers live in *kampongs* (small villages) and grow subsistence crops around their homes. Some keep Brahman cattle which are allowed to graze between the rubber trees, keeping down the undergrowth. As their income increases farmers can buy their smallholding. Some smaller units have amalgamated as co-operatives to help share costs.

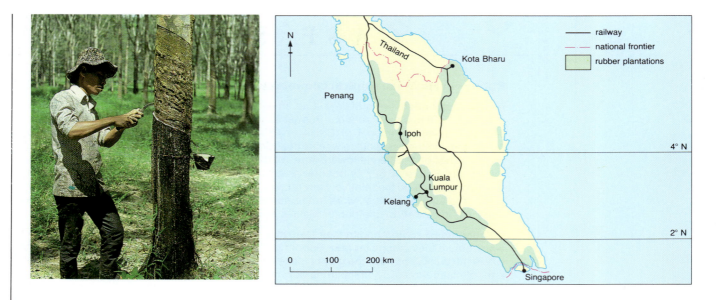

Since World War Two, the world demand for rubber had steadily declined due to competition from synthetic rubber. Ironically, this trend has been reversed since 1988 because of the need to take precautions against AIDS!

◀ **Figure 16.31,** *above left*
Tapping a rubber tree, Malaysia

Q Figure 16.33 is a model for tropical agriculture.

1 Explain what the model shows about tropical agriculture.
2 Describe, with reference to specific examples, the methods used in each of the three types of farming shown.

▲ **Figure 16.32**
The location of rubber plantations in Malaysia

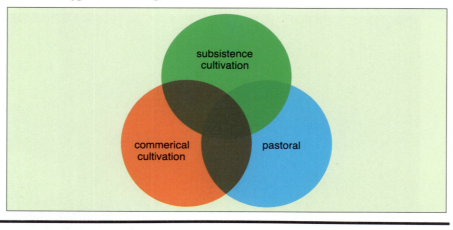

◀ **Figure 16.33**
Interrelationships of tropical agriculture

6   *Extensive commercial pastoralism (livestock ranching)*

Livestock ranching earns the lowest net profit per hectare of any commercial type of farming (Figure 16.13). It is practised in more remote areas where other forms of land use are limited. It needs extensive areas of cheaper land with sufficient grass to support large numbers of animals. Although it is found mainly in areas with a low population density it aims to give the greatest output from minimum inputs. This means that there is a relatively small capital investment in comparison to the size of the farm or ranch but output is larger per farmworker.

This type of farming includes commercial sheep farming, e.g. central Australia, Canterbury Plains, New Zealand, Patagonia, upland Britain; and commercial cattle ranching mainly for beef, e.g. Argentinian Pampas, American Midwest, northern Australia and, recently, Amazonia.

It therefore corresponds to the outer land use zone of von Thünen's model (Figure 16.16) and does not include commercial dairying which, being more intensive, is found nearer to the urban market. (Case Study, page 407)

CASE STUDY

Beef cattle on the South American Pampas

The Pampas covers Uruguay and the northern part of Argentina. The area receives 500–1200 mm of rainfall a year; enough to support a grassland vegetation. During the warmer summer months the water supply has to be supplemented from underground sources while in the cooler, drier winter much of the grass dies down. Temperatures are never too hot to dry up the grass in summer nor low enough to prevent its growth in winter. The relief is flat, and this is one of the few parts of the world which may be seen to resemble von Thünen's isotrophic surface. The soils are often deep and rich having been deposited by rivers crossing the plain (Figure 16.34). The grasses help to maintain fertility by providing humus when they die back.

Several economic improvements have been added to the natural physical advantages. Alfalfa, a leguminous, moisture-retaining crop, is grown to feed the cattle when the natural grasses die down. Barbed wire, for field boundaries, was essential where rainfall was insufficient for the growth of hedges. Pedigree bulls were brought from Europe to improve the local breeds and later British Hereford cattle were crossed with Asian Brahman bulls to give a beef cow capable of living in warm and drier conditions. At first cattle were reared for their hides because the distance from world markets meant that meat could not be transported without perishing, but the construction of a railway network linking extremities of the Pampas to the chief ports and the introduction of refrigerated wagons and ships meant beef could be exported to the industrialised countries. Later, in response to new and changing demands, corned beef, Oxo and Bovril were processed.

Many ranches, or *estancias*, exceed 100 km² and keep over 20 000 head of cattle. Most are owned by businessmen or large companies based in the larger cities and are run by a manager with the help of cowboys or *gauchos*.

▼ Figure 16.34

Commercial pastoralism (livestock ranching) on the Pampas of northern Argentina and Uruguay

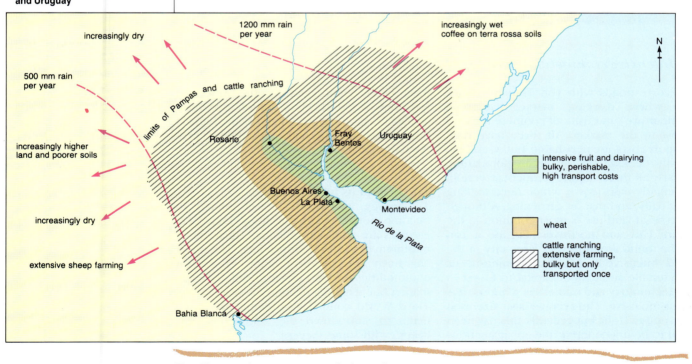

7 Extensive commercial grain

As is evident on the map of the Pampas (Figure 16.34) and in the von Thünen model (Figure 16.16), cereals utilise the land use zone closer to the urban market than commercial ranching. Grain is grown commercially on the American Prairies (Case Study, page 409) and the Russian Steppes (page 415) as well as in Australia, Argentina and parts of northwest Europe (Figure 16.23). In all those areas the productivity per hectare is low but per farmworker it is high.

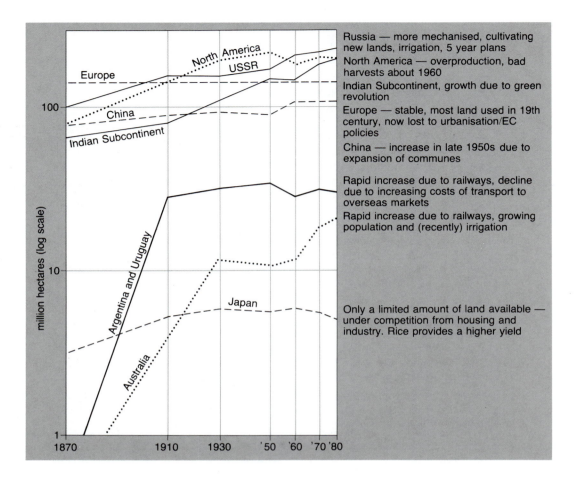

Russia — more mechanised, cultivating new lands, irrigation, 5 year plans

North America — overproduction, bad harvests about 1960

Indian Subcontinent, growth due to green revolution

Europe — stable, most land used in 19th century, now lost to urbanisation/EC policies

China — increase in late 1950s due to expansion of communes

Rapid increase due to railways, decline due to increasing costs of transport to overseas markets

Rapid increase due to railways, growing population and (recently) irrigation

Only a limited amount of land available — under competition from housing and industry. Rice provides a higher yield

▲ Figure 16.35
Changes in the world's arable areas, 1870–1980

8 Intensive commercial agriculture

This corresponds with von Thünen's inner zone where dairying, market gardening (horticulture) and fruit all compete for land closest to the market. All three have high transport costs, are perishable, bulky, and are in daily demand by the urban population. Similarly all three require frequent attention, particularly dairy cows, which need milking twice daily, and market gardening. Although this type of land use is most common in the eastern USA and northwest Europe it can also be found around every large city in the world. Intensive commercial farming needs considerable amounts of capital to invest in high technology and numerous workers: it is labour intensive. The average farm size used to be under 10 ha but recently this has been found to be uneconomic and amalgamations have been encouraged by the American government and the EC in order to maximise profits. This type of farming gives the highest output per hectare and the highest productivity per farmworker.

The EC and its food surpluses
As farming in the EC has continued to become more efficient, output has increased. Farmers are paid subsidies or a guaranteed minimum price, for their produce. The result has been the overproduction of certain commodities for the European and domestic market but at a price beyond the reach of the Third World countries. Attempts are being made to slim down the 'mountains' and 'lakes', with some success.

EC food surpluses:	Jan 1986	Jan 1989
	(figures in thousand of tonnes unless otherwise stated)	
Butter	1 400	500
Skimmed milk powder	800	0
Beef	500	1 000
Cereals	15 000	9 000
Wine/alcohol	4 000 hectolitres	3 500 hectolitres

CASE STUDY

The Canadian Prairies: extensive commercial agriculture

The Prairies have already been referred to in the optima and limits model (Figure 16.4). Although this area has many favourable physical characteristics it also has disadvantages. Wheat, the major crop, ripens well during the long, sunny summer days while the winter frosts help to break up the soil. However, the growing season is short and in the north falls below the minimum requirement of 90 days. Precipitation is low, about 500 mm, but though most of this falls during the growing season there is a danger of hail ruining the crop, and droughts occur periodically. The winter snows may come as blizzards but otherwise they insulate the ground from severe cold and provide moisture on melting in spring. The Chinook wind (page 203) melts the snow in spring and helps to extend the growing season, but tornadoes in summer can damage the crop. The relief is gently undulating, which aids machinery and transport. The grassland vegetation has decayed over the centuries to give a black (chernozem) or very dark brown (prairie) soil (pages 275–6). However, if this natural vegetation is totally removed the soil becomes vulnerable to erosion by wind and convectional rainstorms.

▶ Figure 16.36
Extensive commercial cereal farming on the Canadian Prairies

When European settlers arrived they drove out the local indians, who had survived by hunting bison, and introduced cattle. The world price for cereals increased in the 1860s and demand from the industrialised countries in western Europe rose. The trans-American railways were completed in response to the increased demands (and profits to be made) and vast areas of land were ploughed up and given over to wheat. The flatness of the land enabled straight, fast lines of communication to be built: these were essential as most of the crop had to be exported. Cultivation of cereals also led to the land being divided (page 337) into sections measuring one square mile (1.6 km^2). In the wetter east, each farm was given a quarter or a half section while in the drier west they received at least one full section.

The input of capital has always been high in the Prairies as farming is highly mechanised (Figure 16.36). This has reduced the need for labour although a migrant force, with combine harvesters, now travels northwards in late summer as the cereals ripen. Seed varieties have been improved, e.g. disease-resistance, drought resistance and faster growth have been bred. Fertilisers and pesticides are used to increase yields and the harvested wheat is stored in huge elevators awaiting transport via the adjacent railway.

In the last two decades wheat has become less of a specialist crop and the area upon which it is grown has decreased. Many farms have diversified to produce sugar beet, flax, dairy produce and beef.

CASE STUDY

Intensive farming in the western Netherlands

Most of the western Netherlands, stretching from Rotterdam to beyond Amsterdam, lies 2–6 m below sea level. Reclaimed from the sea several centuries ago, this land is referred to as the **old polders**. Today they form a flat area drained by canals which run above the general level of the land. Excess water from the fields is pumped (originally by windmills) by diesel pumps into the canals. Although ideal for agriculture, the polders are rapidly becoming urbanised. With 429 persons per km^2 in 1986 (compared to only 360 in 1975), the Netherlands have the highest population density in Europe. Consequently, with farm land at a premium, the cost of reclamation so high and the proximity of a large domestic urban market, intensive demands are made on the use of the land (Figure 16.37).

There are three major types of farming on the old polders.

■ **Dairying** is most intensive north of Amsterdam and is favoured by the mild winters, which allow grass to grow for most of the year, the evenly distributed rainfall, which provides lush grass, the flat land and the proximity of the *Randstad* conurbation. Most of the cattle are Friesians. Some of the milk is used fresh but most is turned into cheeses (the well-known Gouda and Edam) and butter. Alkmaar has the major cheese market. Most farms have installed computer systems to control animal feeding.

■ The land between The Hague and Rotterdam is a mass of glasshouses (Figure 16.38) where **horticulture** is practised on individual holdings averaging only 1 ha. The cost of production is exceptionally high. Oil-fired central heating maintains high temperatures and sprinklers provide water in the controlled glasshouse environment. Heating, moisture and ventilation are all controlled by computerised systems. Machinery is needed for weeding and to behead dead plants. The soil requires fertiliser and manure despite annual renewal. Some plants are covered in black plastic to retard growth in order to stagger the period of ripening to meet market demand. Several crops a year can be grown in the glasshouses, e.g. cut flowers in spring, tomatoes and cucumbers in summer and lettuce in autumn and winter.

■ The sandier soils between Leiden and Haarlem are used to grow **bulbs**. Tulips, hyacinths and daffodils, protected from the prevailing winds by the coastal sand dunes, are grown on farms averaging 8 ha. The flowers form a tourist attraction as well as being sent to local markets and exported all over Europe from nearby Schipol Airport.

◄ **Figure 16.37,** *below left*
Agricultural land use patterns in the Netherlands
▼ **Figure 16.38**
Intensive farming on reclaimed polder land in the Western Netherlands

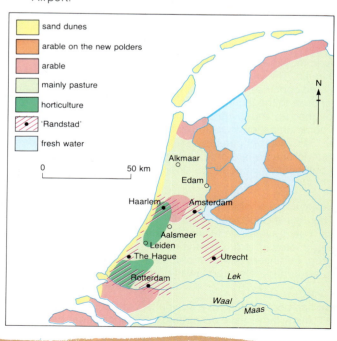

sand dunes
arable on the new polders
arable
mainly pasture
horticulture
'Randstad'
fresh water

0   50 km

N

Alkmaar
Edam
Haarlem   Amsterdam
Aalsmeer
Leiden
The Hague   Utrecht
Rotterdam
Lek
Waal
Maas

The EC has spent most money on dairy products. In 1986 it cost almost £1 million a day to store butter and when some of it was sold to the USSR at 7p/lb there were political repercussions. This action, together with some limited sales at reduced prices to the poor and the elderly within the Community, used up large amounts of the EC budget. Although the decision in 1986 to reduce milk output by 9.5 per cent by 1989 may have appeared sensible, it should be remembered that this has forced many dairy farmers to sell up and to slaughter some five million cattle. In 1988, attention was turned to cereals. It was agreed that £35/ha should be paid to arable farmers if they 'set-aside' their land instead of growing crops on it. The beef mountain and wine lake were tackled in 1989.

9 Mediterranean agriculture

A distinctive type of farming has developed in areas surrounding the Mediterranean Sea. Winters are mild and wet allowing the growth of cereals and the production of early spring vegetables or *primeurs*. Summers are hot, enabling fruit to ripen, but are too dry for cereals and grass to grow. Rainfall amounts decrease and the length of the dry season increases from west to east and from north to south, so irrigation becomes more important. River valleys and their deltas, e.g. the Po, Rhône and Guadalquivir, provide rich alluvium but many parts of the Mediterranean are mountainous with steep slopes and thin rendzina soils (page 230). Due to earlier deforestation, many of these slopes have suffered from soil erosion. Frosts are rare at lower levels though the cold Mistral and Bora winds may damage crops (Figure 12.27).

Farming tends to be labour intensive but with limited capital. There are still many absentee landlords and output per hectare and per farmworker is usually low. Land use frequently shows (Figure 16.17) that crops which need most attention are grown nearest to the farmhouse or village, and that land use is more closely linked to the physical environment than controlled by human inputs. Many gardens in the villages and fields surrounding them are devoted to citrus fruits (Figure 16.39). Oranges, lemons and grapefruit have thick waxy skins to protect their seeds and to reduce moisture loss. These fruits are also grown commercially where water supply is more reliable, e.g. oranges in Spain around Seville and on the *huertas* (irrigated farms) near to Valencia, lemons in Sicily and grapefruit in Israel.

Vines, another highly labour intensive crop, and olives, the 'yardstick' of the Mediterranean climate, are both adapted to the climate. They can tolerate thin, poor, dry soils and hot, dry summers by having long roots and protective bark. Wheat may be grown in the wetter winter period in fields further from the village as it needs less attention, while sheep and goats may be reared on the poor quality grass of the hillsides furthest from the village. Grass becomes too dry in summer to support cattle and so milk and beef are scarce in the local diet.

Except for central Chile, other areas experiencing a Mediterranean climate have a more commercialised type of farming based upon irrigation and mechanisation. Central California supports agribusiness based on a large affluent domestic market which is, in terms of scale, organisation and productivity, the ultimate in the capitalist system. Southern Australia produces dried fruit to overcome the problem of distance from world markets.

10 Irrigation

Irrigation is the provision of a constant supply of water from a river, lake or underground source to enable an area of land to be cultivated. Irrigation is needed in areas with the following climatic characteristics.

1 Little rainfall or where evapotranspiration exceeds precipitation, i.e. semi-arid or arid lands, e.g. the Peruvian desert.
2 A seasonal shortage of water due to drought, e.g. southern California with its Mediterranean climate.
3 Unreliable amounts of rainfall, e.g. the Sahel countries.
4 A high population density, e.g. the Nile Valley in Egypt.
5 Intensive farming, either subsistence or commercial, even if rainfall amounts are relatively high, e.g. rice growing areas of SE Asia.

In the developed countries, large dams may be built, pipelines or canals are dug to carry water many kilometres and a dense network of channels constructed to take computer-controlled amounts of water to the fields in highly capital intensive agricultural systems. Unfortunately, it is often the Third World countries, lacking in capital and technology, who suffer most severely from water deficiencies. Unless they can obtain funds from overseas then most of their schemes have to be constructed and operated by hand and are extremely labour intensive.

▼ Figure 16.39
Mediterranean terraced agriculture near Amalfi, Italy

CASE STUDY

The Nile Valley in Egypt

From the times of the Pharoahs to very recently, water for irrigation was obtained from the River Nile by two methods. First, each autumn the annual floodwater was allowed to cover the land, remain trapped behind small *bunds*, and deposit its silt. Second, during the rest of the year when river levels were low, water could be lifted 1 or 2 m by a *shaduf*, *saquia* (sakia) or Archimedes screw. However, the Egyptians had long wished to control the Nile so that its level would remain relatively constant throughout the year. Barrages of increasing size were built throughout the early twentieth century. But it has been the rapid increase in population, which doubled from 25 to 50 million between 1960 and 1987, and the accompanying demand for food that led to the building of the Aswan High Dam (opened 1971) and several new schemes to irrigate the desert near Cairo in the late 1980s.

The main purpose of the High Dam is to hold back the annual floodwaters generated by the summer rains in the Ethiopian Highlands (Figure 16.40). Some water is released throughout the year, allowing an extra crop to be grown, while any surplus is saved as an insurance against a failure of the rains. The régime of the river below Aswan is now more constant and allows trade and cruise ships to travel on it at all times. Two and sometimes three crops a year can now be grown. Yields have increased and extra income is gained from cash crops of cotton, maize, sugar cane, potatoes and citrus fruits. The dam incorporates a hydro-electric power station which provides Egypt with almost a third of its energy needs for domestic and industrial purposes. The lake is also important for fishing and tourism.

Egypt has modernised its methods of irrigation. Electricity is now used to power pumps which can raise water to higher levels allowing a strip of land 12 km wide on both sides of the Nile to be irrigated (Figure 16.42). **Drip irrigation** is the laying of plastic pipes, in which small holes have been made, over the ground. Water is allowed to drip onto the plants in a much less wasteful manner than in other methods as less evaporates. Between the Nile and the Suez Canal, **boom irrigation** has been introduced (Figure 16.41) creating fields several hectares in diameter.

However, the Aswan High Dam has created problems. Environmental consequences are that it has stopped the annual deposition of silt on the fields, meaning fertiliser now has to be added. The Nile delta has started to retreat having lost its supply of sediment. Increased numbers of irrigation channels have led to an increase in the bilharzia snail.

◄ Figure 16.40, *below left*
Landsat photo of the Lower Nile Valley — if this picture was in colour, it could be seen Nat shown black, the irrigated crops are magenta red, Cairo is pale blue and the extension to the southwest is the pyramids at Giza. Numerous dry valleys can be seen in the desert (Chapter 7)

▼ Figure 16.41
Boom irrigation in the Libyan Sahara — a giant sprinkler rotates around a central pivot

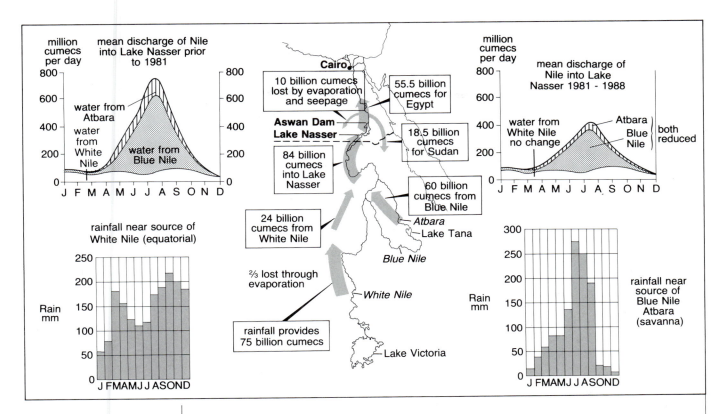

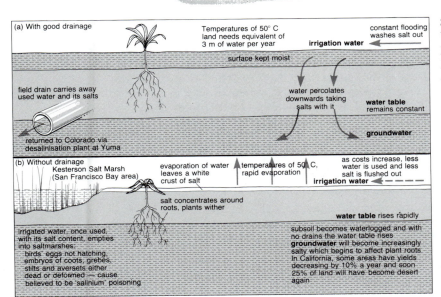

▲ **Figure 16.42**

Irrigation in the Nile valley —

causes of the Nile flood, and

reasons for a water shortage

despite the building of the

Aswan High Dam

Economically and socially it has encouraged farmers to grow cash crops instead of providing a better diet for themselves but costs have increased because they now have to buy fertiliser. Clay is no longer available for making bricks in the traditional manner. One unforeseen problem has been the drought affecting Ethiopia. Since 1981 this has meant a dramatic decrease in flood water brought down by the Blue Nile and Atbara (Figure 16.40). By 1988 Lake Nasser was only 40 per cent full, ships were having difficulty travelling up the shallow Nile, and there was talk of having to close the turbines which generate the hydro-electricity. Large areas of land reclaimed after 1971 have become saline and have been returned to the desert. The UN Food and Agriculture Organisation (FAO) claimed that between 1974 and 1984 there was actually a *decrease* of 12 per cent in land under irrigation and that 30–40 per cent of the remaining irrigated land was affected by salinisation.

▼ **Figure 16.43**

Salinisation in California

Salinisation

Unless water released by irrigation onto the land is drained away, the soil becomes water-logged. As the water table rises it brings dissolved salts into the topsoil (Figure 16.43). These affect the roots of crops which are intolerant of salt and consequently give reduced yields. Over a period of time, the salinisation kills the crops. Where water evaporates, a crust of salt is left on the surface and the area may revert to desert. To date, only rough estimates have been made as to the total area affected by salinisation, but figures suggest it may be as high as 40 per cent of the irrigated land in Pakistan, 30–40 per cent in Egypt and 25 per cent in California.

Q

1 For one (or several) of the ten types of farming listed above:
   a Describe its characteristics,
   b Describe the conditions favouring its development.

2 With reference to specific examples, discuss the extent to which the characteristics of farming systems may be influenced by each of the following:
   a *either* the amount and seasonal distribution of precipitation *or* drought,
   b relief features and soil types,
   c availability of capital,
   d mean temperatures and seasonal variations in temperature,
   e transport and marketing costs,
   f population density,
   g slope gradient,
   h land tenure.

Agriculture in socialist economies

Farming in the USSR

Despite its being the largest country in the world by area, only a small percentage of the USSR is suitable for agriculture (1985 figures):

Land area	Arable	Pastoral	Non-farming
22.27 km²	10%	17%	73% (forest, tundra, desert and semi-desert)

The climate, relief and soils are suitable for agriculture (Figure 16.44) only in the deciduous forest belt and on the Steppes. However, in these areas the type of land tenure and the organisation of farming practice is very different from that of countries in the non-communist world. In the Soviet Union the individual farmer, or the company-run estate, is replaced by the *kolkhoz* (collective farm) system or the *sovkhoz* (state farm) system.

The *kolkhoz* system

During the late 1920s, when the USSR was facing serious food shortages, especially in the rapidly growing industrialised towns, Stalin attempted to enforce **collectivisation**. This was a process by which small farms had to join with their neighbours to form a larger unit. The land was then leased from the government and managed by an elected chairperson and committee representing up to 400 families. The farm produce was collected together and sold to the state at a fixed price. Income from this was then equally divided amongst all of the families. Once each family had fulfilled their tasks they could grow and sell their own produce such as vegetables, milk and eggs on minute 'private' plots.

The *kolkhoz* were mainly confined to the Ukraine — the western part of the Steppes — which contained the country's best farmland (Figure 16.44). Despite considerable resistance from the Russian peasants, the number of individually owned farms fell from 25 million in 1928 to 0.2 million in 1940. The peasants, resentful at losing their land, found little incentive in a 'shared income economy' and often failed to meet the production targets set by the socialist state. The *kolkhoz* were overpopulated in relation to their inputs and resources and so after 1940 they were increasingly replaced by *sovkhoz*.

The *sovkhoz* system

A *sovkhoz* is a state-owned farm where the farmers are paid a weekly wage in the same way as labourers in a factory. The Soviet government favours this system as it can control and set production targets, determine which type of crop is grown or animal reared on each unit, and can control the sale and marketing of produce. The state provides the necessary capital resources, e.g. machinery and fertiliser. Each *sovkhoz* is considerably larger than a *kolkhoz*.

	% of USSR farmland	Number in 1940	Number in 1975	Size of community (families)	Average size (hectares)
Kolkhoz	38	130 000	27 000	400	500
Sovkhoz	61	4 200	24 000	4 000–10 000	1 500
Privately owned	1	2 000	0?	1	1

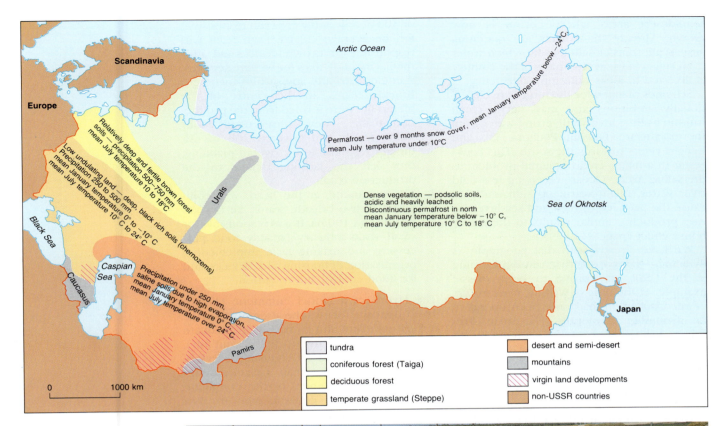

▲ Figure 16.44
Climatic controls on farming in the USSR

► Figure 16.45
A state farm grain centre at Kazakh in the Kustanai region of the USSR

The *sovkhoz* workers, who live in blocks of flats (Figure 16.7), are divided into 'brigades', each of which specialises in a particular job, e.g. mechanics, tractor drivers, accountants. Each 'farm' is self-contained with its own shops, schools, libraries and places of recreation (Figure 16.45).

Incentives were offered to farmers who exceeded production targets, but many *sovkhoz* found it difficult to attain these levels. This was mainly because the majority were developed on the 'virgin lands' (Figure 16.44) where the precipitation of less than 500 mm a year does not guarantee reliable crop yields. The original aim in developing the virgin lands in states such as Kazakhstan was to plough up the natural grassland to grow wheat and other cereals. Later, to help cereal production, irrigation schemes were begun. These have since been extended into

415

semi-desert areas where cotton is now grown. Already the Soviets are operating large scale transfer schemes by which water from rivers in the wetter parts of the country is diverted to areas suffering a deficiency.

Schemes planned for the future are exceptionally ambitious and may never reach fruition as they involve diverting water from the northward flowing Pechora, Ob and Yenisei rivers towards the south. This could result in a major environmental problem as the saline Arctic Ocean, receiving less cold river water, might warm up sufficiently to cause the pack ice to melt and sea levels to rise, producing similar results to the greenhouse effect (page 211).

We have already seen that geography is dynamic — it is constantly changing. Farming in the USSR may prove to be a classic example. At the time of writing, *perestroika*, or reconstruction, had only been operative for a number of months. The Soviet President, Mikhail Gorbachev, is attempting to revitalise the Soviet economy by 'injecting a dose of capitalism'. This could mean less state controls over Soviet agriculture ... but you will be in a better position to describe any changes and to assess their significance at the time you actually read this section.

Farming in China

Before the establishment of the People's Republic in 1949, farming in China was typical of SE Asia, i.e. it was mostly intensive subsistence (page 401). Farms were extremely small and fragmented with the many tenants having to pay up to half of their limited produce to rich and often absentee landlords. Cultivation was manual or using oxen. Despite long hours of intensive work, the output per worker was very low. The need for food meant that most farmland was arable, with livestock restricted to those kept for working purposes or which could live on farm waste, e.g. chickens. The type of crops grown changed from north to south as rainfall totals and the length of the growing season both increase (Figure 16.46).

People's communes, 1958

Immediately after taking power in 1949 the communists confiscated land from the large landowners and divided it amongst the peasants. These individual plots proved too small to support individual farmers but after several interim experiments, the government created the 'people's communes'. The communes were meant to become self-sufficient units. They were organised into a three-tier hierarchy.

1 Production teams, at the foot of the hierarchy, consisted of about 50 families or 300 people and were each responsible for, on average, 20 ha. Each team was responsible for its own finances and, after the payment of taxes for welfare services, it could distribute any surplus income.

2 Brigade teams, in the centre of the hierarchy, were formed from ten production teams. They consisted of about 3000 people and covered 200 ha. Although responsible for overall planning, they left the farming details to the production team.

3 The commune, at the head, was composed of five brigade teams, i.e. 15 000 people and 100 ha. It was responsible for small scale industry (mainly food processing and making farm implements), organisation of housing and services, and for flood control and irrigation.

The layout of a model commune (Figure 16.47 opposite) shows the attempts to make it a self-sufficient unit by having an adequate food supply (crops, livestock, fruit and fish), water control, industry, housing, services (hospital, schools) and an agricultural research station. Members of the commune elected a people's council, who in turn elected a committee to ensure that production targets were met and that machinery, fertiliser and new strains of seed were used correctly (the Green Revolution, page 403). By pooling their resources farmers were able to increase yields per hectare.

Production targets were set by the Central Planning Committee (the government) in a series of Five Year Plans.

Production Responsibility System, 1978

This system has largely replaced the commune and has transformed China's peasants into tenant farmers. Either singly or

▼ Figure 16.46
Farming in China — relationships between precipitation and types of farming

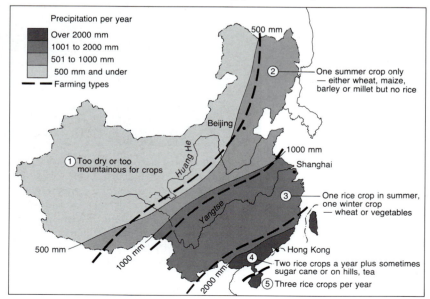

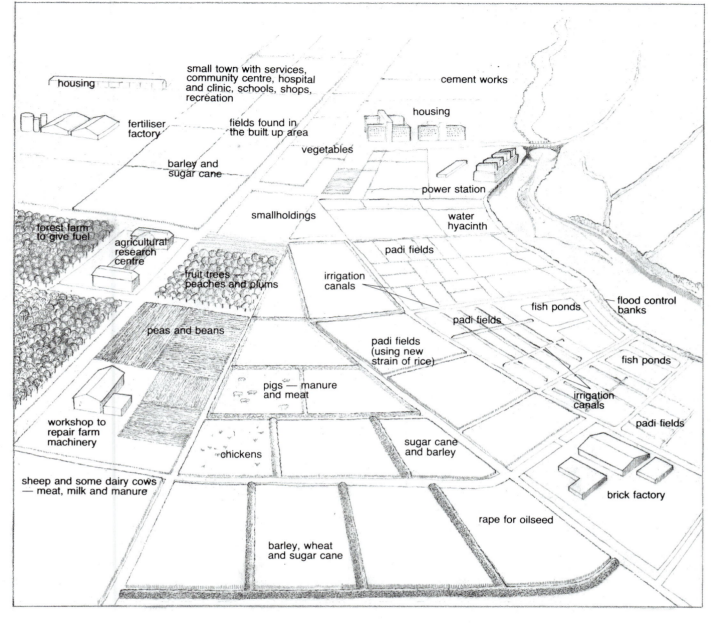

▲ Figure 16.47

A Chinese commune showing land use variations in attempts at self-sufficiency

Labels within figure: housing; small town with services, community centre, hospital and clinic, schools, shops, recreation; cement works; fertiliser factory; fields found in the built up area; housing; barley and sugar cane; vegetables; power station; smallholdings; water hyacinth; forest farm to give fuel; agricultural research centre; padi fields; fruit trees — peaches and plums; irrigation canals; fish ponds; flood control banks; peas and beans; padi fields; padi fields (using new strain of rice); fish ponds; workshop to repair farm machinery; pigs — manure and meat; irrigation canals; padi fields; chickens; sugar cane and barley; sheep and some dairy cows — meat, milk and manure; brick factory; rape for oilseed; barley, wheat and sugar cane

in small groups, farmers have to meet a fixed production target and then they can sell any surplus on the open market. To meet its quota, each working group receives the free use of land, tools and seed. Rural markets have thrived and the profits have been used to buy better seed and machinery and to create village industry. The production team is now the highest tier with smaller units, the workgroup or the family, now making the decisions previously taken at commune level. Most people have been prepared to work hard and their standard of living has improved: in some cases incomes have quadrupled. Southern China, with its advantageous climate and proximity to the open Hong Kong market, has benefitted the most from agrarian reforms.

Farming types and economic development

Throughout the section on 'types of farming' several fundamental assumptions have been made. These included the generalisations that: 'the poorest countries are those who, because they have the lowest inputs of capital and technology, have the lowest outputs'; and 'the wealthier countries are those who can afford the highest inputs giving them the maximum yield, or profit, per person'. Is it really possible to make a simple correlation between wealth (the standard of living) and the type of agriculture?

417

The table in Figure 16.48 shows 15 countries selected as representing the main types of farming — and therefore *not* chosen randomly — which have been taken as examples in the previous section. Using the five variables chosen it is possible to postulate four hypotheses.

1 'The less developed a country (i.e. the lower its GDP per capita), the greater the percentage of its population involved in agriculture.'

2 'The less developed a country, the greater the percentage of its GDP is made up from agriculture.'

3 'The less developed a country the less fertiliser it will use.'

4 'The less developed a country the less mechanised will be its farming' (e.g. fewer tractors per head of population).

As with all data, there are considerations which you should remember when drawing conclusions from Figure 16.48.

- The countries were selected with some bias in order to cover all the main types of farming.
- GDP is not the only indicator of wealth (page 325).
- GDP figures are not necessarily accurate and may be derived from different criteria (page 325).
- There may be several different types of farming in each country.

◄ Figure 16.48

Types of farming, GDP and agricultural data for fifteen selected countries

Country	Major farming type	A	B	C	D	E
1 Ethiopia	Nomadic herding	135	77	45	1	10882
2 Bangladesh	Intensive subsistence	158	82	48	26	20581
3 China	Centrally planned	218	56	45	47	1247
4 India	Intensive subsistence	233	60	29	37	1480
5 Kenya	Nomadic herding/subsistence	279	76	27	14	3006
6 Egypt	Irrigation	711	49	19	361	1117
7 Uruguay	Extensive commercial ranching	1331	11	12	3	188
8 Malaysia	Commercial plantation	1566	45	20	111	1900
9 USSR	Centrally planned	2588	14	20	38	102
10 Greece	Mediterranean	2966	34	16	69	142
11 Spain	Mediterranean	3853	14	6	47	65
12 Argentina	Extensive commercial ranching	4124	12	13	1	120
13 UK	Intensive commercial	6514	2	2	140	96
14 Netherlands	Intensive commercial	7716	4	4	340	77
15 Canada	Extensive commercial grain	13034	4	3	32	38

A Gross domestic product (GDP) per capita in US $, in rank order
B % of population engaged in agriculture
C % of GDP derived from agriculture
D Kg of fertiliser used per hectare of population
E Number of people in country for each tractor

Q

1 Draw four scattergraphs to see if there appears to be any correlation between the GDP of the 15 countries and the four other variables.

2 The countries have already been ranked in order of their GDP. Rank the other four variables and for each use the Spearman rank correlation test (page 328) to calculate the correlation coefficient (Rn) between them and the GDP.

3 Using your results to questions 1 and 2:
   a Which variable shows the closest correlation with GDP per capita?
   b How well do you consider that your results support each of the four hypotheses given above?
   c Describe and try to account for any outstanding anomalies, e.g. why do Argentina and Canada use so little fertiliser? Why does Egypt use so much fertiliser? Why does Malaysia have so few tractors?
   d With how much confidence can you say that 'Farming in Third World countries is traditional and non-mechanised with a small output per worker despite the large numbers it employs. As countries become more developed the numbers in agriculture decline, the methods become more scientific and mechanised and there is a high output per worker and per unit area'?

4 Figure 16.49 is an alternative method of showing links between various types of farming and levels of economic development. It was a group effort by a number of sixth-formers. Criticise their model and try to suggest other ways by which farming systems and levels of development may be compared.

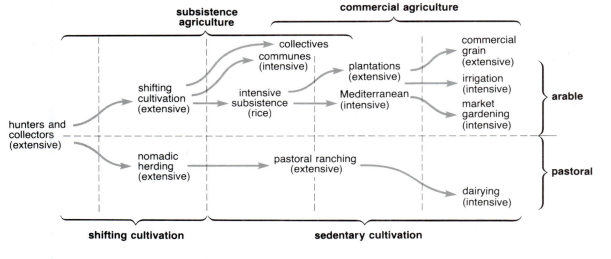

▲ Figure 16.49

Farming types and levels of
development

Problems of how to increase food supplies in the Third World

There is enough food produced each year for everybody in the world to have up to 3300 kilocalories per day (2500 kilocalories are adequate on average). However, the distribution of food supplies is uneven: output is actually declining in sub-Saharan Africa because of drought, accentuated by the greenhouse effect; and in other areas for a combination of reasons: population growth; a lack of capital; small, fragmented farms; landless farmers.

So how can food supplies be increased?

The obvious answer might appear to be by increasing the cultivated area. But there are very few potential areas left in the Third World other than in marginal areas where attempts to farm could easily lead to soil erosion and desertification. The solution is not therefore in extending the cultivated area but in using those areas already farmed more efficiently. Water supply is a major problem, yet prestigious schemes have often in the past benefitted only a few farmers and have created further problems, e.g. salinisation (page 413). What is needed, perhaps, are more wells so that people do not migrate to the few existing ones, drip irrigation as this is less wasteful than channels, and **stone lines** (Figure 16.50) or **check dams**. Stones are laid, following the contours, even on gentle slopes in Burkina Faso, while small dams built of loess (page 238) are constructed across gulleys in North China (Figure 10.31). In both cases, surface runoff is trapped instead of removing soil. This gives the water time to infiltrate into the soil as well as

▶ Figure 16.50

Stone lines in Burkina Faso

causing the silt to be held by the barriers. These simple methods have increased crop yields by over 50 per cent, even though the dams take up 5 per cent of the farmland. On steeper slopes, more terracing, as used in parts of SE Asia, could increase yields and help prevent erosion.

The introduction of new hybrid seeds has significantly increased output. Wheat and rice yields have doubled in India and the Philippines as new strains are faster growing (allowing an extra crop a year), and are drought- and disease-resistant. In some areas new crops may be a salvation in the future — the Somalian Yehab bush with nutritious peanut-sized seeds is ideal for the semi-deserts, the hairy potato repels pests and may be bred into other crops, while the winged bean is a totally edible legume (fixes nitrogen in the soil) which was unknown in the 1970s.

Many of these hybrid seeds need heavy applications of chemical fertiliser and pesticides which are expensive for Third World countries but which nevertheless improve yields considerably (Figure 16.51). Organic fertilisers from animals are much cheaper to use and the animals also provide meat and milk, but in many parts of Africa dung is needed as fuel instead of being returned to the fields. If the land is left unploughed after harvesting the crop residue can improve the soil by adding humus and increasing its moisture retention capacity (page 220).

Because of their reliance on agriculture, Third World countries often fall into the trap of growing cash crops for export instead of subsistence crops for themselves. Attempts are being made in several countries to break up the large plantations into smallholdings (Malaysia, page 405) where farmers may grow some crops for their own consumption as well as a main crop for export. Mixed farming and a variety of crops are preferable to monoculture as the soil is less likely to be exhausted and the system is less vulnerable to disease and to fluctuating world prices and demand. Intercropping can also increase yields, especially where a tree crop protects smaller plants from excessive heat and rain.

Although as viewed from developed countries the lack of mechanisation appears to hinder agricultural development in the Third World, in reality equipment such as tractors and deep ploughs can create more damage than good. What is needed is appropriate or **intermediate technology** (page 455) suited to the climate, soils and wealth of the developing country. Assistance is needed to store food against attack by rats, insects and mildew as well as in improving transport links to potential markets.

▲ Figure 16.51
Spraying young rice plants in China

Perhaps the greatest need is for land reform to overcome the inefficiencies in the use of land and labour. This may include the expropriation of large estates and distributing the land to individual farmers or communal groups; the consolidation of small, fragmented farms; increasing security of tenure; and attempting new land colonisation projects. At the same time, too many governments have regarded agriculture as subordinate to urban development and industrialisation, and have not seen fit to reject overseas aid which 'may ultimately have promoted prestigious schemes rather than helped to feed the inhabitants of that country'. In 1985, a number of African countries agreed to allocate 25 per cent of their investment to agriculture.

Farming and the environment

There are, today, numerous pressure groups alleging that the traditional British countryside is being spoilt, yet this countryside has never been 'traditional' — it has always been changing. The primeval forests, regarded as Britain's climatic climax vegetation (page 240), were largely cleared, initially for sheep farming and later for the cultivation of cereals. Much later, land was 'enclosed' by planting hedges and building dry stone walls (page 338). It is this eighteenth-century landscape which has become looked upon as the traditional or natural environment. Between 1945 and 1985, estimates suggest that 40 per cent of the remaining broad-leaved woodlands, 20 per cent of the hedgerows and 25 per cent of the semi-natural environment (wetlands, heaths and moors) 'disappeared'. While most accusing fingers point to the farmers' responsibility for these changes, it should be remembered that their land, too, is under threat from rival land users.

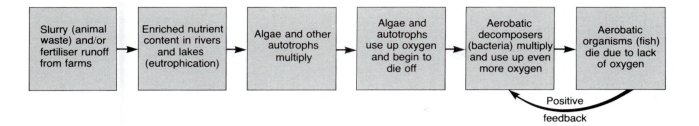

| Slurry (animal waste) and/or fertiliser runoff from farms | → | Enriched nutrient content in rivers and lakes (eutrophication) | → | Algae and other autotrophs multiply | → | Algae and autotrophs use up oxygen and begin to die off | → | Aerobatic decomposers (bacteria) multiply and use up even more oxygen | → | Aerobatic organisms (fish) die due to lack of oxygen |

Positive feedback

▲ **Figure 16.52**

Flowchart to show how eutrophication can upset the ecosystem

Threats to the farming environment

Possibly the greatest threat to farmers, not least in Britain, is urbanisation. The outward expansion of towns, the suburbanisation of villages, the creation of small businesses and the improvement in communications routes (mainly roads but sometimes airports) all take over land, often of high agricultural value, which was previously farmed. The green belts, established around the larger British cities to protect land currently used for farming and recreation, are continually under threat from developers.

At the same time, farmland is under pressure from other rural land users (Figure 16.1) including the Forestry Commission, water authorities and tourist organisations such as the Ramblers Association. The latest threat comes from industry and power stations even if they are remote from population. Some industries dump their toxic wastes onto farmland, acid rain (page 183) increases the acidity of soils and reduces crop yields, and the effects of the nuclear accident at Chernobyl in the USSR was still seriously affecting sheep farmers in North Wales and the Lake District three years and more after the event.

Farming as a threat to the environment

Fertiliser, pesticides and slurry all contribute to pollution of the environmental system. Fertiliser, in the form of mineral compounds containing elements essential for plant growth, is widely used to produce a healthy crop and increase yields. If too much nitrogenous fertiliser or animal waste (manure) is added to the soil then some remains unabsorbed by the plants and may be leached to contaminate underground water supplies and rivers. Where chemical fertilisers accumulate in lakes and rivers the water becomes enriched with nutrients (**eutrophication**) and the ecosystem is upset (Figure 16.52). In parts of France, West Germany, the Netherlands and the UK, levels of nitrates in groundwater are above the EC safety limits.

In Britain, the Water Authorities claim that slurry is now the major pollutant of rivers. Indeed, after several decades in which the quality of river water had improved, the last few years have seen levels of pollution again increasing, especially in farming areas. In 1989 water authorities asked permission to *increase* the discharge of sewage into rivers.

Pesticides and **herbicides** are chemicals applied to crops to control pests, disease and weeds. Estimates suggest that without pesticides, cereal yields would be reduced by 25 per cent after one year and 45 per cent after three. However, the Friends of the Earth claim that pesticides are injurious to health and although there have been no human fatalities reported in Britain in the last ten years, there are many incidents in the Third World resulting from a lack of instruction, fewer safety regulations and faulty equipment. Pesticides are blamed for the rapid decrease in Britain's bee population and a 60–80 per cent reduction in 800 species of fauna in the Paris Basin. Pesticides are also likely to contaminate groundwater. The most emotive outcries against farmers have been at the clearances of hedges and, latterly, ponds and wetlands. As stated earlier, over 25 per cent of British hedges were removed between 1945 and 1985 (in Norfolk the figure was over 40 per cent). The table shows some of the arguments for and against the removal of hedgerows and the drainage of ponds. Figures 16.53 and 16.54 illustrate landscapes with and without hedgerows.

▼ **Figure 16.53**

An agricultural landscape *with trees and hedges, Devon*

The case for and against hedgerows and ponds in a farming environment

For	Against
■ **Hedgerows** Form part of the attractive, traditional British landscape	Are not traditional and were initially planted by farmers
Form a habitat for wildlife: birds, insects and plants (Large Blue butterfly is extinct, 10 other species endangered)	Harbour pests and weeds
	Costly and time-consuming to maintain
Act as windbreaks (and snowbreaks)	Take up space which could be used for crops
Roots bind soil together reducing erosion by water and wind	Limit size of field machinery (combine harvesters need an 8 m turning circle)
■ **Ponds** Form a habitat for wildlife: birds, fish and plants	Takes up land that could be used more profitably
Add to the attractiveness of the natural environment	Stagnant water may harbour disease
i.e. Concern is **environmental** (aesthetic advantages)	i.e. Concern is **economic** (financial advantages)

Farming may increase soil erosion, i.e. the rate at which soil is removed by the erosive powers of running water and wind. The rate is determined by climate, topography, soil type and vegetation cover (page 234), but it is accelerated by poor farming practices (over-cropping and overgrazing) and deforestation. The conditions favouring wind erosion are given on page 149 and the processes by which material is transported are shown on Figure 7.5). In Britain this form of erosion tends to be limited to parts of East Anglia and the Fens where the natural vegetation cover, including hedges, has been removed and soils are light or peaty. Water erosion is most likely to occur not only after heavy rainfall, but on soils with less than 35 per cent clay content, in large and steeply-sloping fields and where ploughing has exposed the soil.

In parts of Europe, subsidies from the EC have encouraged wheat and maize to be grown, replacing grass and other more protective cover plants. Intensification of farming in areas of 'highly erodible soils' in the USA has led to actual decreases in the total yield, and an estimated loss of one-third of the country's topsoil. Deforestation by farmers, not only in the tropical rainforests but also in mountainous and semi-arid areas (e.g. Nepal, the Sahel) is accelerating the process of desertification (page 425). We saw earlier in this chapter (page 411) how irrigation, not a common practice in Britain, can lead to the salinisation and waterlogging of soils (page 413).

Farming: attempts to improve the environment

Between 1985 and 1990, more than 400 British farmers have reverted to **organic farming**. This means they avoid the use of artificial fertilisers and pesticides and use instead animal manure, compost and natural additives. This causes a drop in yields and

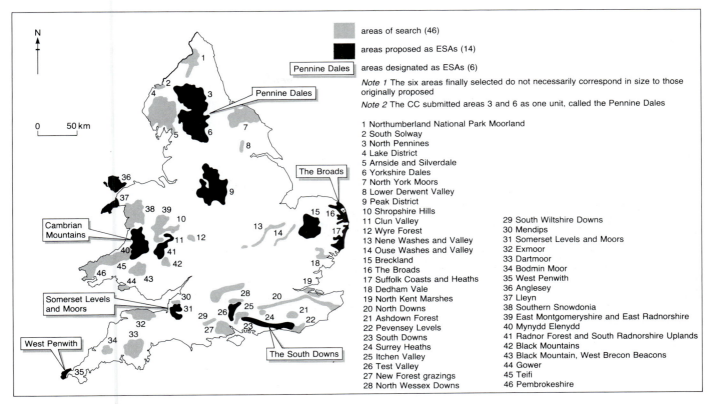

▲ Figure 16.55

'Environmentally sensitive areas' (ESAs) in England and Wales

areas of search (46)
areas proposed as ESAs (14)

Pennine Dales | areas designated as ESAs (6)

Note 1 The six areas finally selected do not necessarily correspond in size to those originally proposed
Note 2 The CC submitted areas 3 and 6 as one unit, called the Pennine Dales

1 Northumberland National Park Moorland
2 South Solway
3 North Pennines
4 Lake District
5 Arnside and Silverdale
6 Yorkshire Dales
7 North York Moors
8 Lower Derwent Valley
9 Peak District
10 Shropshire Hills
11 Clun Valley
12 Wyre Forest
13 Nene Washes and Valley
14 Ouse Washes and Valley
15 Breckland
16 The Broads
17 Suffolk Coasts and Heaths
18 Dedham Vale
19 North Kent Marshes
20 North Downs
21 Ashdown Forest
22 Pevensey Levels
23 South Downs
24 Surrey Heaths
25 Itchen Valley
26 Test Valley
27 New Forest grazings
28 North Wessex Downs
29 South Wiltshire Downs
30 Mendips
31 Somerset Levels and Moors
32 Exmoor
33 Dartmoor
34 Bodmin Moor
35 West Penwith
36 Anglesey
37 Lleyn
38 Southern Snowdonia
39 East Montgomeryshire and East Radnorshire
40 Mynydd Elenydd
41 Radnor Forest and South Radnorshire Uplands
42 Black Mountains
43 Black Mountain, West Brecon Beacons
44 Gower
45 Teifi
46 Pembrokeshire

profit for the first year or two but soon after, as the moisture retention capacity, drainage and aeration properties of the soil improve and potential erosion is reduced, output rises. In addition to maintaining a better soil, the cost of inputs into the farm is reduced as there are no fertilisers, weed killers or pesticides.

One method of encouraging farmers to operate with consideration for the environment is to offer financial incentives.

In 1985, a three-year experimental scheme was jointly set up in Britain by the Ministry of Agriculture and the Countryside Commission on the Halvergate Marshes on the Norfolk Broads. The two parties represent the conflicting interests of farmers and conservationists. At the same time, the Countryside Commission and the Nature Conservancy Council (NCC) looked at 48 'search' areas where farmed landscapes were considered to be under threat from changing farming practices (Figure 16.55). Assuming the government would have limited resources, a shortlist of 14 was drawn up. The following year the government announced that six of these farming areas in England and Wales were to be designated 'Environmentally Sensitive Areas' (ESA). Each was considered important for its landscape, historic and habitat value. Farmers in those areas are now eligible for two levels of payment: a lower level, paid on the condition that they maintain the present landscape, and a higher level if they make environmental improvements such as replanting hedges and restoring ponds. It is possible that further ESAs, including two in Scotland and one in Northern Ireland, may be created in the future.

CASE STUDY

The organic farm

Colin Hutchin farms near Taunton in Somerset. He keeps his land fertile by adding lime, slag and animal manure. When parts of the farm are left fallow, to allow it to rest and be replenished, large areas are given over to clover. Colin has a rotation pattern of two straw crops followed by a root crop and then either grass and clover (clover returns nitrogen to the soil) or a leguminous crop (peas or beans). He believes that this rotation not only maintains a natural soil fertility, it also reduces pests and disease by changing the crop before they can take hold.

Forestry

Commercial forestry in the developed and the Third World

Britain

Neolithic farmers began the clearances of Britain's primeval forests about 3000 years ago. Aided by the development of axes, some clearances may have been on a scale not dissimilar to parts of the tropical rainforests of today. In 1919, with less than 4 per cent of the country covered in trees, the Forestry Commission was set up to begin a controlled replanting scheme. Since then the policy has been to look towards an economic profit over the long term while trying to protect the environment. The area under trees is increasing, as in most other countries in Western Europe and North America, and was 8.9 per cent in 1986.

Young trees are grown in nurseries until at a height of 50 cm, they are planted in rows in land which has been cleared and drained. The trees are regularly sprayed with fertiliser and pesticide. The lower and dead branches are removed while the poorer quality trees are felled. This carefully controlled process of thinning gives more light and room for the healthier trees to grow, though over-thinning can increase the risk of wind damage.

The forester fells trees with modern light-weight chain saws and uses machinery which can uproot trunks, remove branches and bark, and cut them into standard lengths ready for transport out of the forest. The land is then cleared ready for another 'crop' of young trees to be planted. The softwoods of Northern Britain take 40–60 years to mature so afforestation is an investment for the future. A strong environmental lobby has ensured that modern plantations are carefully landscaped (Figure 16.57).

Commercial forestry is easier to operate in Britain than in Third World countries as Britain's forests are nearer the world markets, softwoods are in greater demand than tropical hardwoods and even where growing naturally there are few species within a large area which makes selection and felling easier.

Third World countries

Commercial forestry is a relatively new venture in the tropics. It is usually controlled by multinationals based overseas who look for an immediate economic profit and have little thought for the long term future or the environment. The forested areas in Latin America, tropical Africa and SE Asia are rapidly decreasing. Figures giving the rate of forest clearances are diverse and unreliable but the UN suggests that for every 29 ha deforested in Africa, only 1 ha is replanted (the ratio is 11:1 in tropical America). The forest is totally cleared by chain saw bulldozer and fire: there is no selection of trees to be felled (Figure 16.56). The large buttress roots are left protruding from the ground and a secondary succession of poorer quality trees develops as little restocking is undertaken (page 265). Where afforestation does take place, there is little money for fertiliser or pesticide. Demand for wood (as fuel) and land (for farming) is increasing as the population grows. Hardwoods take over 100 years to mature, so the urgent need for trees is by the present population with little thought for future generations. The hope for the future may lie in **agroforestry**, where trees and food crops are grown alongside each other. Forest soils, normally rated unsuitable for crops, can be improved by growing leguminous tree species.

Commercial forestry is less easy to operate in Third World countries as they are distant from world markets, demand for hardwood is less than softwoods and although there are several hundred species in a small area only a few are of economic value.

The consequences of deforestation

The threat of total destruction of the rainforests has become such a global concern that it is now well documented. Some consequences of deforestation are evident from the specific examples described in the case studies which follow.

▼ Figure 16.56

Destruction of rainforest for short term economic gain

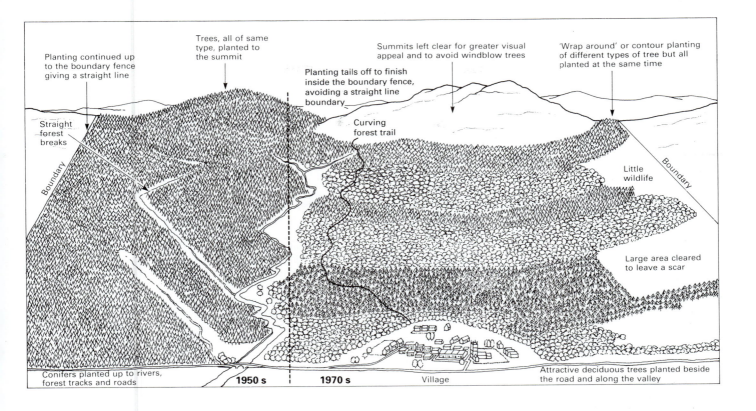

Labels on figure:
Planting continued up to the boundary fence giving a straight line
Trees, all of same type, planted to the summit
Planting tails off to finish inside the boundary fence, avoiding a straight line boundary
Summits left clear for greater visual appeal and to avoid windblow trees
'Wrap around' or contour planting of different types of tree but all planted at the same time
Straight forest breaks
Boundary
Curving forest trail
Little wildlife
Boundary
Large area cleared to leave a scar
Conifers planted up to rivers, forest tracks and roads
1950 s
1970 s
Village
Attractive deciduous trees planted beside the road and along the valley

▲ Figure 16.57
Commercial forestry and the environment

C A S E   S T U D Y

Ethiopia

Early in the twentieth century, 40 per cent of Ethiopia was forested. Today the figure is under 2 per cent. In 1901, a traveller described part of Ethiopia as being 'most fertile and in the heights of commercial prosperity with the whole of the valleys and lower slopes of the mountains one vast grain field. The neighbouring mountains are still well-wooded. The numerous springs and small rivers give ample water for domestic and irrigation purposes, and the water meadows produce an inexhaustible supply of good grass the whole year.'

In 1985 the same area was described as 'a vast barren plain with eddies of spiralling dust that was once topsoil. The mountains were bare of vegetation and the river courses dry.'

As the trees and bushes were cleared, less rainfall was intercepted and surface runoff increased resulting in less water for the soil, animals and plants.

Nepal and Bangladesh

In 1945 nearly 60 per cent of the Himalayan kingdom of Nepal was forested. However, population increases led to vast deforestation of the steep hillsides so that by 1985 only one-third of the country was wooded. Each year, as the torrential monsoon rain falls on the mountains, the water, no longer intercepted by the trees or slowed down by their roots, makes its way over the land surface. Giant landslides are an annual hazard with villages, villagers, animals, roads and farmland being swept away. The rivers, full of silt, flood with increasing frequency and severity.

Bangladesh is comprised mainly of the deltas of several Himalayan rivers including the Ganges and Brahmaputra. As an increasing amount of silt is transported from the deforested highlands, the distributaries of these rivers become blocked as their beds are raised — adding to the risk of flooding from the sea (page 17).

CASE STUDY

Panama

Ships passing through the Panama Canal have to be lifted and lowered 30 m through a series of locks as well as traversing the artificial Lake Gatun. The watersheds of rivers draining into the lake are being deforested which causes eroded red soils to fill in and discolour the lake. Without the trees, evapotranspiration has been drastically reduced and subsequently, with less moisture in the atmosphere, rainfall totals are declining. The reduced supply of water may mean that the locks, and therefore the canal, will eventually have to close and the country will lose its main source of income.

Amazonia

The clearance of the rainforests means a loss of habitat to many indian tribes, birds, insects, reptiles and animals. Over half of our drugs, including one from a species of periwinkle which is used to treat leukaemia in children, come from this region. Perhaps we are clearing away the cure for AIDS and other as yet incurable diseases. (Despite being the world's richest repository of medical plants, only 2% have so far been studied for potential health properties.) Without tree cover, the fragile soils are rapidly leached of their minerals making them useless for crops and vulnerable to erosion. Half of the world's oxygen is supplied from trees in the Amazon. The process of burning these trees reduces the amount of oxygen released into the air and releases carbon which produces carbon dioxide on contact with the atmosphere. This gas traps heat (the greenhouse effect) which is altering the world's climate. It has also been suggested that the decrease in rainfall following the reduction in evapotranspiration from the loss of trees could have serious climatic repercussions — could the Amazon Basin become another Ethiopia?

Malaysia: a model for the future?

Malaysia has several thousand species of tree, mainly hardwoods. Timber is the country's fifth largest export accounting for 35 per cent of the world's hardwoods. However, the government has imposed strict controls, and the Forestry Department insists that trees reach a specific height, age and girth before they can be felled. Logging companies are given contracts only on agreement that they will replant the same number of trees which they remove. The newly planted hardwoods are ready for harvesting within 15–20 years due to the rapid growing conditions of the tropical rainforests. Further experiments are being made with acacias which are even faster growing. Consequently, two-thirds of Malaysia is still forested and most of the remaining third is under tree crops such as rubber, oil palm and coca, so stocks are being successfully maintained.

Extraction of mineral resources

Studies of rural land use often emphasise agriculture and fail to take account of other uses to which non-urban land is put, such as transport recreation and mining, which is another important competitor for rural land.

Industry depends upon 80 major minerals of which 18 are in relatively short supply: e.g. lead, sulphur, tin, tungsten and zinc. No essential **resource** is expected to run out although **reserves** may decline.

The size of a mineral reserve may vary according to several factors.

- Mineral content, or **grade**, of a rock, e.g. high grade iron ores may have a 66 per cent iron content, low grade ores less than 33 per cent.
- Local geological conditions; output may cease if a vein is too thin to exploit or a seam is faulted.

- Accessibility to markets and transport costs.
- Costs of extraction and market price.
- Changes in world demand and competition from rival substitute products.
- Whether market demand exceeds the problems of the physical environment (e.g. Bolivian tin and Alaskan oil).
- Local levels of technology, or its introduction by a colonial power (past) or a multinational corporation (present).

The most convenient methods of mining include **open cast**, where surface soil is removed to expose the mineral, and **quarrying**. Where the mineral is exposed on valley sides then horizontal **adit** mines are used. **Shaft** mines are those sunk vertically to deeper seams or veins.

Resources *are the amount of a mineral in the earth's crust. The quantity is determined by geology.*

Reserves *are the amount of a mineral which can be economically recovered.*

CASE STUDY

Mining, the community and the environment
Blaenau Ffestiniog, North Wales,

The Oakley slate quarries were first worked in 1818. By the 1840s the most easily obtained slate had been won and mining began. The introduction of steam power and the building of the Ffestiniog railway led to the export of 52 million slates from Porthmadog in 1873. At the quarry's peak productivity, 2000 men and boys were employed on seven different levels. These were steeply inclined into the hillsides and were worked to a depth of 500 m. Apart from farming, the slate mines were the sole providers of employment, and Blaenau Ffestiniog's population peaked at 12 000. Working in candlelight in damp and dusty conditions for up to 12 hours a day, and with rock falls common (pressure release, page 31) the life expectancy of miners was short. By the turn of the century the manufacture of roof tiles heralded the beginning of the industry's decline and in 1971 the mine at Blaenau closed. A decade later, renamed Gloddfa Ganol, the underground galleries were re-opened to tourists, many of whom arrived via the narrow-gauge Ffestiniog railway.

▼ Figure 16.58
Spoil heaps above Blaenau Ffestiniog, Gwynedd, North Wales

▶ Figure 16.59, *far right*
Galleries cut in the Dinorwic slate quarries, Gwynedd, North Wales — the mill in the foreground indicates the scale

As the mines closed, many people were either made unemployed or were forced to move to seek work — the present population of Blaenau is under 500. Even so, as a tourist attraction, the slate mines are still the largest single employer. As in many British mining towns, the older houses are rows of cottages coated grey with the dust of earlier economic activity. Dust still blows into the cottages but the inhabitants no longer die from silicosis. Above the town rise the large and unsightly spoil heaps (Figure 16.58) though they appear more stable than the coal tips which affected Aberfan (Figure 2.13). On average, 10 tonnes of waste were created for every tonne of usable slate. Discarded machinery and unused buildings add to the scene of dereliction though some are being restored to add as tourist attractions. Fortunately there are few signs of the subsidence which is common to other mining areas.

CASE STUDY

Tin mining in Malaysia

Malaysia is the world's major producer of tin and the mineral is the country's third most important export. Early tin mining was typical of the colonial trade. British ex-patriots brought in the capital, machinery and technology; supervised the mining; and exported the tin for refining. Malaysia itself received few advantages. Most tin was obtained by open cast methods together with the use of hyraulic jets.

Although today mining is operated by the Malaysian government there are large tracts of land left flooded and there are disused buildings and overhead 'railways' where the mineral has been exhausted or is too uneconomic to work.

Tourism

Tourism is becoming a major source of employment and income in many rural areas of Britain and other EC countries. In these countries the rapid growth in the tourist industry in the last two or three decades has been linked to the fact that the majority of people work shorter hours, have more disposable wealth, receive longer, paid holidays, and have better mobility. The opportunities to relax in rural and coastal areas and to travel overseas have been created through package holidays, cheaper air fares, improved roads and advertising. Other reasons have been the aspiration to a break for those in employment working under increased pressures, the increased desire (perpetuated through the media) to travel to more distant places and to sample different types of activity, the fashion to take more than one holiday a year and a growth in mini-breaks.

These changes have put tremendous pressures upon many areas of countryside and careful planning and management is now essential if visitors to rural areas are to maximise their leisure time spent there without coming into conflict with other land users (Figure 16.1) or ruining the area and amenities which they want to enjoy. Existing resources of rural areas are increasingly having to be managed for specific leisure activities.

It is possible to differentiate between land use for tourism according to three levels of activity.

1 **High intensity** areas where recreation is the major consideration, e.g. the Lea Valley and various theme parks.
2 **Average intensity** areas where a balance must be kept between tourism and other land users, and between recreation and

conservation, e.g. the New Forest and the Peak District National Park.
3 **Low intensity** areas, usually of high scenic value, where conservation of land and wildlife is given top priority, e.g. upland parts of Snowdonia and the Slimbridge Wildfowl Trust.

Is it possible to plan and manage the countryside so as to provide access to places of natural beauty and wildlife and to encourage energetic and possibly noisy pursuits while at the same time protecting that environment and without imposing undue restrictions? Robin Arvill wrote in 1967:

'To do this requires sound knowledge of those features contributing to the quality of an area — the special landscapes, sites for intensive recreation, countryside treasures, nature reserves and sanctuaries, buildings of artistic, architectural or historic interest, and caves, archaeological, geological and other attractions. In sum, a 'heritage inventory' is essential to the effective planning and development of an area. And where the heritage is deficient in desirable features, then the plan should provide for the creation — as far as possible — of these values. This could embrace open spaces, public parks and gardens, woods, water areas, individual buildings of distinction and town areas of character. In general, in the development of towns and resorts, incompatible uses should be separated, traffic segregated from pedestrians, buildings zoned to create centres of quality for community activities, and much more insight shown in reconciling development with the natural constraints of land, air, water and wildlife'.

► Figure 16.60

Some possible causes of conflict in a National Park

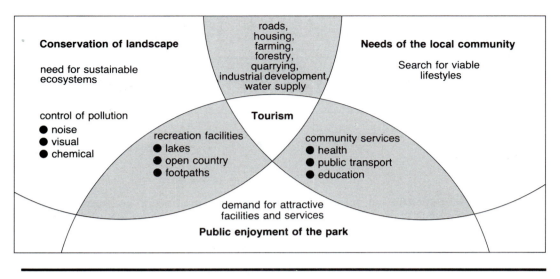

Several of the questions that follow should give you scope for individual research and the expression of your own values (Framework 9). You may wish to consult Figure 16.60.

1 What types of problems may tourism create for:
  a farmers
  b forestry workers
  c water authorities
  d ornithologists?

2 Describe the possible impact of tourism on each of the following environments:
  a a mountainous National Park
  b a lake or reservoir
  c a coastal area
  d a wetland
  e a small village.

3 What are the economic, social and environmental advantages and disadvantages of increased tourism in:
  a a British National Park
  b an Alpine ski resort
  c a Third World country?

4 Choose either a small area of countryside near to where you live or, if you live in a large urban area, a well known National Park. Carefully describe how you would plan the area to both maximise and protect its resources.

References

Agricultural Land Use and Landscape Change, Geography 16–19 Project, Longman, 1984.

Environmentally Sensitive Areas in the UK, Geofile No. 85, Mary Glasgow Publications, 1987.

Human Geography, Ken Briggs, Hodder & Stoughton, 1982.

Human Geography, M.G. Bradford and W.A. Kent, Oxford University Press, 1977.

Human Geography, Patrick McBride, Blackie, 1980.

I.D.G. Publications, The Hague, Netherlands, 1986.

Impacts of Mining Developments, Geography 16–19 Project, Longman, 1986.

Man and Environment, Robin Arvill, Penguin Books (Pelican), 1967.

New Geographical Digest, George Philip, 1986.

Only One Earth, Lloyd Timberlake, Earthscan/ BBC (Book and Television programme), 1987.

Pattern and Process in Human Geography, Vincent Tidswell, UTP, 1976.

Soil Erosion in the UK, Geofile No. 92, Mary Glasgow Publications, 1987.

The Fight for the Amazon, Sting and Jean-Pierre Dutilleux, Barrie & Jenkins, 1989.

The Third World, M. Barke and G. O'Hare, Oliver & Boyd, 1984.

Third World Studies: Pastoralism in the Sahel, Open University, 1983.

Third World Studies: The Green Revolution in India, Open University, 1985.

World Resources 1987, World Resources Institute, Basic Books, 1988.

CHAPTER 17

Industry

'We need methods and equipment which are cheap enough so that they are accessible to virtually everyone; suitable for small scale production; and compatible with man's need for creativity. Out of these three characteristics is born non-violence and a relationship of man to nature which guarantees permanence. If one of these three is neglected, things are bound to go wrong.'

Small is Beautiful, E.F. Schumacher, 1974

What is meant by industry? In its widest sense the word industry is used to cover all forms of economic activity: **primary** (farming, fishing, mining and forestry), **secondary** (manufacturing and construction), **tertiary** (back-up services such as administration, retailing and transport) and **quaternary** (hi-tech and information services). In this section the use of the term will be confined to manufacturing industry, i.e. the processing of raw materials (e.g. iron ore, timber) and of semi-processed materials (e.g. steel, pulp). As with farming, it is useful to draw a flow diagram to show linkages in an industrial system (Figure 17.1).

The location of industry

The processes which contribute to determine the location and distribution of industry are more complex than those affecting agriculture and making generalisations is less easy. Some reasons for this complexity include the following.

- Some locations were sited before the Industrial Revolution and many more during it. Initial factors favouring a location may no longer apply today. For example, the original raw materials may have been exhausted (e.g. iron ore and coal in West Cumbria), or replaced by new innovations (e.g. cotton replaced by synthetic fibres) and sources of energy (e.g. electricity has replaced water power).
- New locational factors which were not applicable last century include cheaper and more efficient transport systems, the movement of energy (in the form of electricity), new techniques and automation.
- Some industries have developed from older industries and are linked to these

▼ Figure 17.1
Types of industry

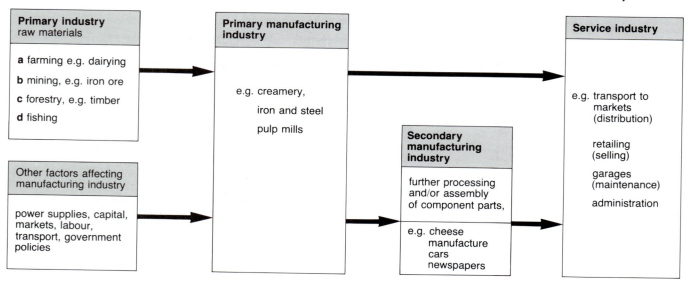

430

former patterns of production though the modern product may be barely related to that previously made on the site.

- Before the twentieth century, industry was usually financed and organised by individual entrepreneurs but decisions concerning modern industry are often made by the state or **multinational companies** (MNCS or **transnationals**). Increasingly these decisions are being made far away from the location of a particular factory.
- Many factories now produce a single component and therefore are a part of the much larger organisation which they supply.
- The sites of many early factories were chosen by individual preference or chance, i.e. the founder of a firm just happened to live at, or to like, a particular location. Other industries have found their present location as a result of 'trial and error'.

Factors affecting the location of manufacturing industry

Raw materials

Nineteenth-century industry was often located close to raw materials (ironworks near iron ore) or sources of power (coalfields) mainly due to the immobility of the raw materials which were heavy to move as transport was expensive and inefficient. Today's industries are rarely tied to the location of raw materials: they are described as **footloose**. There is a greater efficiency in the use of raw materials, power is more mobile, transport is more efficient and relatively cheaper, and some firms may simply assemble component parts made elsewhere.

Industries which still need to be located near to raw materials are those using materials which are heavy, bulky or perishable, are low in value in relation to their weight or which lose weight or bulk during the manufacturing process. **Alfred Weber**, whose theories on industrial location are referred to later, introduced the term **material index**

or *MI*. He defined the material index:

$$MI = \frac{\text{total weight of raw materials}}{\text{total weight of the finished product}}$$

There are three possible outcomes.

1 If the *MI* is greater than 1, then there must be a weight loss in manufacture. In this case the raw material is said to be **gross** and the industry should be located near to that raw material, e.g. for iron and steel:

$$MI = \frac{6 \text{ tonne raw material}}{1 \text{ tonne finished steel}} = 6.0$$

2 If the *MI* is less than 1, then there must be a gain in weight during manufacture. This time the industry should be located near to the market, e.g. brewing:

$$MI = \frac{1 \text{ tonne raw material}}{5 \text{ tonne beer}} = 0.2$$

3 Where the *MI* is exactly 1 then the raw material must be **pure** as it does not lose or gain weight during manufacture. This type of industry could therefore be located at the raw material, the market or any intermediate point.

Industries which lose weight during manufacture include food processing (e.g. butter has only one-fifth the weight of milk, refined sugar is only one-eighth of the cane), smelting of ores (e.g. copper ore is less than 1 per cent copper) and forestry (e.g. paper has much less mass than trees). Industries which gain weight in manufacture include those adding water (e.g. brewing and cement) and assembling component parts (e.g. cars and electrical goods). In this case the end product is more bulky and expensive to move than its many smaller constituent parts.

Power supplies

Early industry tended to be located near to sources of power which in those days could not be moved. However, as newer forms of power were introduced and the means of transporting made easier and cheaper, this locational factor has become less important.

Power supplies and changing locations of iron and steel works

Time	Source of power	Examples of location
Early iron industry	charcoal	wooded areas (The Weald, Forest of Dean)
Later iron industry	waterwheels	fast-flowing rivers (Sheffield)
Early steel industry	coal	coalfields (South Wales, NE England)
Present day	electricity	coastal (Port Talbot, Newport)

When water was a prime source of power during the Middle Ages, mills had to be built alongside rivers. When steam power took over as the Industrial Revolution began in Britain, factories had to be built on or near to coalfields, as coal was bulky and expensive to move. As canals and railways were constructed to move coal then new industries were located along transport routes. By the mid-twentieth century, oil (relatively cheap before the 1974 Middle East War) was being increasingly used as it could be easily transported by tanker or pipeline. This freed industry from the coalfields and gave it a freer choice of location (except for such oil-based industries as petrochemicals). Today oil, coal, natural gas, nuclear and sometimes hydro-electric power can all be used to produce electricity to feed the National Grid. Electricity, in addition to its cleanliness and flexibility has the advantage that it can be transferred economically over considerable distances, to the long-established industrial areas, where activity is maintained by **geographical inertia** or to new areas of growth. (In 1900, electricity could be economically transmitted only 50 km; today the distance is over 1500 km.) Under certain conditions, industry is carried out at hydro-electric plants, e.g. aluminium smelting at Akosombo in Ghana and Kitimat in Canada.

Transport

Transport costs were once a major consideration when locating an industry. Weber based his industrial location theory on the premise that transport costs were directly related to distance (von Thünen's assumptions, page 387). Since then, new forms of transport have been introduced, including lorries (for door-to-door delivery), railways (preferable for bulky goods) and air (where speed is essential). Meanwhile, transport networks have improved with the building of motorways and methods of handling goods have become more efficient through containerisation. For the average British firm, transport costs are now only 2–3 per cent of their total expenditure. Consequently, raw materials can be transported further and finished goods sold in more distant markets without any considerable increase in costs. Firms previously tied to coastal sites relying on overseas materials or markets, or earlier nodal points, now have a freer choice of location — although many still favour **growth point axes** such as the M4 corridor (Figure 17.17).

There are two types of transport costs from an economic point of view. **Terminal costs** refer to the time and equipment needed to handle and store goods, e.g. cranes and warehouses, and costs of providing the transport system, e.g. building railway tracks, docks and motorways. **Long haul costs** refer to the cost of actually moving the goods, e.g. lorries, petrol and salaries. The advantages and disadvantages of different forms of transport are summarised in Figure 18.1.

Markets

Today the pull of a large market is more important than the location of raw materials and power supplies.

Industries will locate near to markets if:
1 The product becomes more bulky with manufacture or there are many linkage industries involved, e.g. the assembling of motor vehicles.
2 The product becomes more perishable after processing (bread is more perishable than flour), if it is sensitive to changing fashion (clothes) or if it has a short lifespan (newspapers).
3 The market is very large (e.g. the north-eastern states of the USA, or SE England).
4 The market is wealthy.
5 Prestige is important, e.g. publishing.

Labour supply

In the nineteenth century, a huge force of unskilled workers operated in large-scale heavy industries doing manual jobs in steelworks, shipyards and textile mills. Today there are fewer numbers of semi-skilled, together with some highly skilled, workers operating in small-scale 'light' industries. Their jobs increasingly involve machines, computers and robots. The cost of labour, especially in Britain, is high and can account for up to 60 per cent of total production costs — hence the introduction of mechanisation to reduce human inputs, and the exploitation of female labour and 'cheap' Third World workers.

Traditionally, labour has been largely immobile. Although there was a drift to the towns during the Industrial Revolution, since the First World War British people have expected respective governments to bring jobs to them rather than themselves having to move for jobs. Certain industries need specialist training (e.g. cutlery, furniture-making and electronics) and so are difficult to develop in a new area with an untrained labour force — if you were to open a cutlery business, for example, you would probably choose Sheffield in preference to Exeter or Aberdeen.

However, as industries become more mechanised and need a similar workforce they can locate more freely. A problem of the early 1990s is that while Britain still has 1¾ million unemployed it is facing an increasing shortage of highly skilled people.

Similarly the roles of women are changing as more seek career jobs, and the part played by trade unions, now with reduced membership numbers, has altered considerably.

The labour market has been polarised, therefore, to a demand for unskilled workers at low cost (supplied by female, part-time, Third World labour) and highly skilled workers for research and development, etc (supplied by graduates). Both sources are in limited supply in Britain as people tend to be semi-skilled and therefore too 'expensive' or insufficiently qualified. The need to obtain labour at one of these extremes of ability can be a major influence on industrial location.

Capital
Capital may be in three forms.
1 **Working capital** (money) which is acquired from a firm's profits, shareholders or financial institutions such as banks. Money is mobile and can be used within and exchanged between countries. Location is rarely constrained by working capital unless money is to be borrowed from the government who might direct industry to certain areas (Government policies below). Capital is more readily available in the SE of England where most of the financial institutions are based.

2 **Physical** or **fixed capital** refers to buildings and equipment. This form of capital is not mobile, i.e. it was invested for a specific use. Although some nineteenth-century textile mills have now been converted to other uses, most mills, shipyards and steelworks have become derelict following closure.

3 **Social capital** is linked to the workforce's out-of-work needs rather than to the factory itself. Houses, hospitals, schools, shops, and recreational amenities are social capital which may attract a firm, particularly its management, to an area.

Government policies
These should aim at evening out differences in employment, income levels and investment within a country. In 1934, **regional assistance** was granted in Britain to Clydeside, West Cumberland, NE England and South Wales (compare these areas with the Development Areas of the late 1980s on Figure 17.2). This was followed by the setting up of **trading/industrial estates** in areas of high unemployment such as Team Valley (Gateshead), Trafford Park (Manchester) and Treforest (north of Cardiff). Since then successive governments have tried various schemes to try to restrict firms from locating in certain areas (mainly the South East) and giving inducements if they locate in others.

In the early 1980s the government set up Enterprise Zones where unemployment and the state of the environment posed serious problems (Figure 17.2). Firms locating in EZs are 'exempt from rates, receive 100 per cent capital allowances on industrial and commercial property, are subject to simplified planning procedures (so long as health, safety and pollution control standards are guaranteed), and are exempt from industrial training levies'. In the 1980s further policies were introduced to try to help badly hit docklands (London and Liverpool) and other inner city areas (page 369). Governments have also made direct attempts to attract foreign firms, e.g. Nissan and Toyota.

In centrally planned economies (eastern Europe) the state is the sole decision-maker in the location of industry. So the state determines where industrial plants will be located and which units will close down.

▼ Figure 17.2
Assisted areas and Enterprise Zones in the UK, 1989 (*Source: Department of the Environment*)

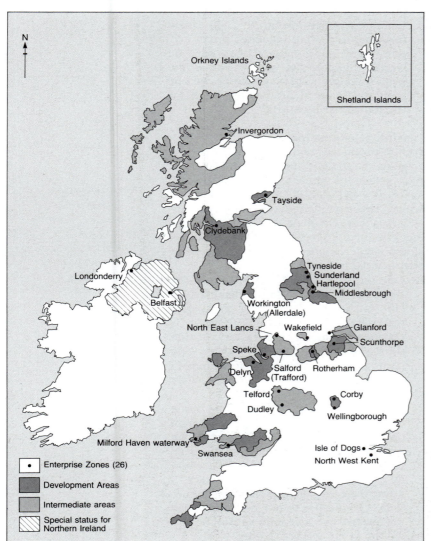

Land

In the nineteenth century, extensive areas of flat land were needed for the large factory units. Today, although modern industry is usually smaller in terms of land area occupied, it prefers cheaper land, less congested and cramped sites with improved accessibility, as found on green field sites on the edges of cities. British government policy is at present trying to attract industry to derelict and underused sites, especially in the inner cities (page 369) so as to utilise the existing infrastructure and reduce pressure on green field sites.

Environment

The 1980s saw an increasing demand by both managers and workforce to live and work in a more pleasant environment. This has led to firms seeking locations in smaller towns within easy reach of open countryside and away from the commuting problems and high house prices of SE England. Yet there is still a common perception that 'the north' is all Coronation Streets and 'the south' quiet rural villages. Consequently potential employers have to work hard to 'sell' any social or environmental advantages which their area possesses.

Chance factors

As described on page 431 this includes industries begun by individuals in their home towns (Morris cars in Oxford, Rowntree's chocolate in York) or at sites which appealed to them (Unilever at Port Sunlight).

Theories of industrial location

There are two main approaches to locational theory in a 'free market' or capitalist system.
- The industrialist seeks the lowest cost location (LCL) — Weber.
- The industrialist seeks the area which will give the highest profit — Smith.

Weber's model of industrial location

Alfred Weber was a German 'spatial economist' who, in 1909, advanced a model to try to explain and predict the location of industry. Like von Thünen before him and Christaller later, Weber tried to find a sense of order from apparent chaos and made assumptions to simplify the real world in order to produce his model.

These assumptions were that:
- There was an isolated state with flat relief, a uniform transport system in all directions, a uniform culture, climate, political and economic system.

- Most of the raw materials were not evenly distributed across the plain (this differs from von Thünen). Those which were evenly distributed (e.g. water, clay) he called **ubiquitous** materials. As these did not have to be transported, firms using them could locate as near to the market as was possible. Those which were not evenly disributed he called **localised** materials. He divided these into two types; gross and pure (page 431).
- The size and location of markets were fixed.
- Transport costs were a function of the mass of the raw material and the distance it had to be moved. This was expressed in tonnes per kilometre (t/km).
- Labour was found in several fixed locations on the plain. At each point it was paid the same rates, had equivalent skills, was immobile and in large supply. Similarly, entrepreneurs had equal knowledge, related to their industry, and motivation.
- Perfect competition existed over the plain, i.e. markets and raw materials were unlimited which meant that no single manufacturer could influence prices (i.e. there was no monopoly). As revenue would therefore be similar across the plain then the best site would be the one with the minimal production costs, i.e. the **least cost location**.

Possible least cost locations

Weber produced two types of locational diagram. A straight line was sufficient to show examples where only one of the raw materials was localised (it could be pure or gross). However, when two localised raw materials were involved then he introduced the locational triangle. Figure 17.3 shows nine possible variations based on the type of raw material involved.

1 One gross localised raw material. As there is weight loss during manufacture (the material index for a gross raw material is more than 1) then it is cheaper to locate the factory at the source of the raw material — there is no point in paying transport costs if some of the material will be left as waste after production (Figure 17.4a).

2 **a** One ubiquitous raw material or **b** one pure localised raw material gaining weight on manufacture (*MI* less than 1). If the raw material is found all over the plain (ubiquitous) then transport is unnecessary as it is already found at the market. If a pure material gains mass on manufacture then it is cheaper to move it rather than the finished product and so again the LCL will be at the market (Figure 17.4b).

Agglomeration is when several firms choose the same area as their location in order to minimise costs. This can be achieved by linkages between firms (where several join together to buy in bulk or to train a specialist workforce), within firms (individual car component units) and between firms and supporting services (banks, utility services — gas, water and electricity).

Deglomeration is when firms disperse from a site, possibly due to increased land prices, labour costs or a declining local market.

3 One pure localised raw material. If this neither gains nor loses weight during manufacture ($MI = 1$) then the LCL can either be at the market, the raw material or at any intermediate point (Figure 17.4c).

4 Two ubiquitous raw materials. As these are found everywhere then the LCL is the market as the raw materials, whether gross or pure, do not have to be transported.

5 Two raw materials: one ubiquitous and one pure and localised. The LCL is at the market because the ubiquitous material is already there and so only the pure localised material has to be transported (Figure 17.5a). It will be cheaper to move one raw material than the more cumbersome final product.

6 Two raw materials: one ubiquitous and the other gross and localised. Here, one materal is available at every location. As the gross material loses weight then, theoretically, it could be produced at any intermediate point (example 1 above). However, if the mass of the product is greater than the raw material then the LCL is at the market; if it is less than the raw material the LCL is at the raw material; and if it is the same then the LCL could be at the mid-point (Figure 17.5b).

7 Two raw materials: both localised and pure. In the unlikely event of the two raw materials lying to the same side of and in line with the market, the LCL will be at the market. If the materials do not conform with this arrangement but form a triangle with the market (Figure 17.6), then the LCL is at an intermediate point near to the market. This is because the weight and therefore the transport costs of the raw material are the same as, or less than, those of the product.

8 Two localised raw materials: one pure and one gross. In this case the industry will locate at an intermediate point (Figure 17.7a). The LCL will move increasingly nearer to the source of the gross material the greater the loss of weight during production.

9 Two raw materials: both localised and gross. If both raw materials have an equal loss of weight then the LCL will be equidistant between these two sources but closer to them than to the market (Figure 17.7b). However, if one raw material loses more mass than the other then the industry is more likley to be located closer to it (Figure 17.7c).

▶ Figure 17.3

Least cost locations dependent upon types of raw material

Type(s) of raw material (RM) MI = material index	LCL at raw material	LCL at market	LCL at any intermediate point
1 one gross localised RM>MI1	✔		
2 one raw material gaining weight or one obiquitous RM MI<1		✔	
3 one pure localised RM MI = 1			✔
4 two ubiquitous RM (pure or gross)		✔	
5 two RM (one ubiquitous and one pure)		✔	
6 two RM (one ubiquitous and one gross)	(according to amount of weight loss could be any site)		✔
7 two RM (both pure)		✔	
8 two RM (one pure, one gross)	✔ (if big weight loss)		✔ (if a small weight loss)
9 two RM (both gross)	✔ (at RM with greatest weight loss)		✔ (equal weight loss)

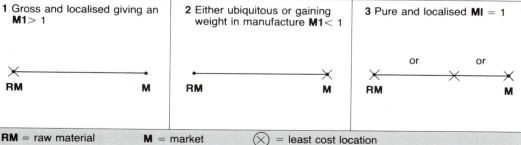

1 Gross and localised giving an M1> 1	2 Either ubiquitous or gaining weight in manufacture M1< 1	3 Pure and localised MI = 1
✕———————• RM · · · · · · · · · M	•———————✕ RM · · · · · · · · · M	or · · · · or ✕———✕———✕ RM · · · · · · · · · M

RM = raw material · · · · **M** = market · · · · ⊗ = least cost location

▶ Figure 17.4

Least cost locations with one raw material

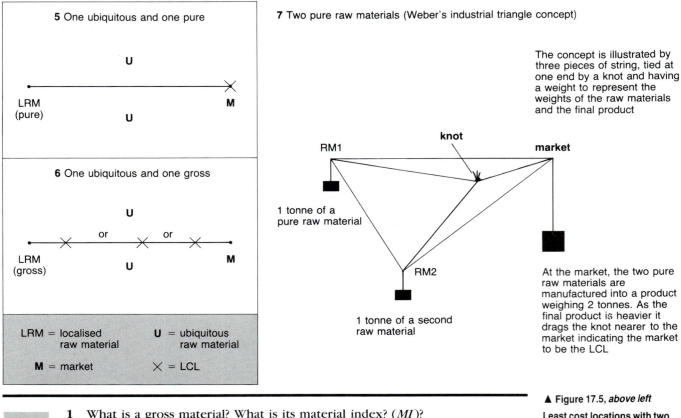

5 One ubiquitous and one pure

6 One ubiquitous and one gross

LRM = localised raw material   U = ubiquitous raw material

M = market   ✕ = LCL

7 Two pure raw materials (Weber's industrial triangle concept)

The concept is illustrated by three pieces of string, tied at one end by a knot and having a weight to represent the weights of the raw materials and the final product

1 tonne of a pure raw material

1 tonne of a second raw material

At the market, the two pure raw materials are manufactured into a product weighing 2 tonnes. As the final product is heavier it drags the knot nearer to the market indicating the market to be the LCL

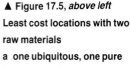

▲ Figure 17.5, *above left*

Least cost locations with two raw materials

a one ubiquitous, one pure

b one ubiquitous, one gross

▼ Figure 17.6

Least cost locations with two localised pure raw materials

Q

1 What is a gross material? What is its material index? (*MI*)?

2 What is a pure material? What is its material index?

3 Which of the following materials are gross and which are pure: iron ore; sand and gravel; oil; television components; sugar beet?

Weber claimed that four factors affected production costs: the cost of raw materials, the cost of transporting them and the finished product, labour costs and **agglomeration/deglomeration economies** (page 434).

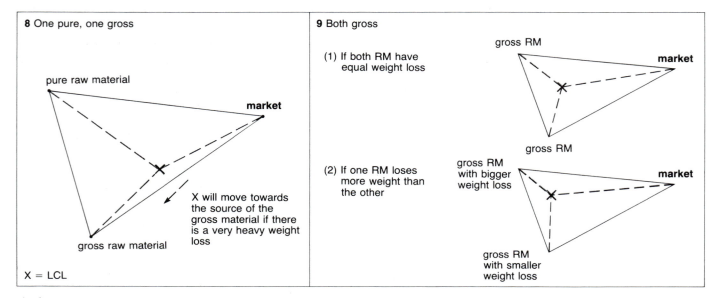

8 One pure, one gross

X will move towards the source of the gross material if there is a very heavy weight loss

X = LCL

9 Both gross

(1) If both RM have equal weight loss

(2) If one RM loses more weight than the other

Spatial distribution of transport costs

*An **isotim** is a line joining all places with equal transport costs for moving either the raw material (Figure 17.8a) or the product (Figure 17.8b).*

*An **isodapane** is a line joining all places with equal total **transport costs**, i.e. the sum of the costs of transporting the raw material and the product (Figure 17.8c).*

Transport costs lie at the heart of his model so Weber had to devise a technique which could both measure and map the differences in these costs over space in order to find the LCL. His solution was to produce a map with two types of lines or circles similar to contours which he called **isotims** and **isodapanes**.

Figure 17.8a shows the costs of transporting 1 t of a raw material (*R*) as concentric circles. In this example it will cost 5 t/km (tonne/kilometres) to transport the material to the market. Figure 17.8b shows, also by concentric circles, the cost of transporting 1 t of the finished products (*P*). The total cost of

moving the product from the market to the source of the raw material is again 5 t/km. By superimposing these two maps it is possible to show the total transport costs (Figure 17.8c).

If a factory were to be built at *X* then its transport costs would be 7 t/km (i.e. 2 t/km for moving the raw material plus 5 t/km for moving the product). A factory built at *Y* would have lower transport costs of 6 t/km (4 t/km for the raw material plus 2 t/km for the product). However, the LCL in this instance may be at the source of the raw material, the market or any intermediate point in a straight line between the two because all these points lie on the 5 t/km isodapane.

▶ Figure 17.8

Isotims and isodapanes

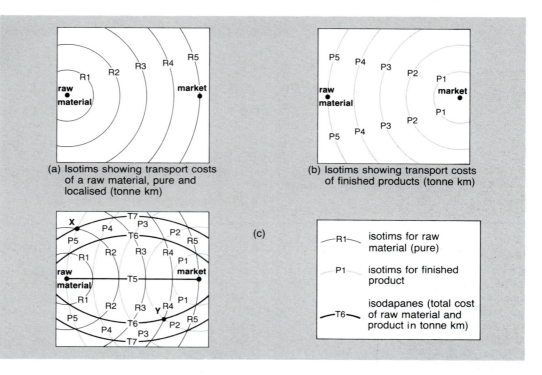

(a) Isotims showing transport costs of a raw material, pure and localised (tonne km)

(b) Isotims showing transport costs of finished products (tonne km)

(c)

—R1—	isotims for raw material (pure)
P1	isotims for finished product
—T6—	isodapanes (total cost of raw material and product in tonne km)

◀ Figure 17.7

Least cost locations with two localised raw materials

a one pure, one gross

b both gross

Q

1 The statements below describe the raw materials used by four different industrial plants, *A*, *B*, *C*, *D*.
   Plant *A* uses two gross, ubiquitous raw materials.
   Plant *B* uses two pure, localised raw materials.
   Plant *C* uses one gross, localised raw material.
   Plant *D* uses one pure, localised raw material and one gross, ubiquitous raw material.
   a State the material index for each of the four plants (i.e. is it >1, <1, or = 1?).
   b Describe the location of each of the four plants (i.e. has it a market location, a raw material location or an intermediate location?).

2 Show with the aid of appropriate diagrams, how the LCL in Weber's locational triangle may be:
   a near to the market,
   b near to that of a raw material,
   c at an intermediate point.

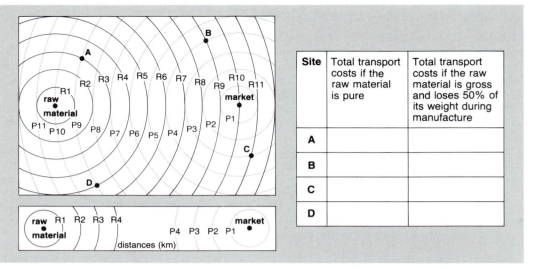

◀ Figure 17.9

Site	Total transport costs if the raw material is pure	Total transport costs if the raw material is gross and loses 50% of its weight during manufacture
A		
B		
C		
D		

3   Study Figure 17.9 which shows an area with one single raw material and one market.

Assume two different situations:

1   the raw material is pure,

2   the raw material is gross, losing 50 per cent of its mass (weight) during manufacture.

a   Complete the table to show the total transport costs (in tonnes/kilometre) for an industry located at each of the four sites A to D.
b   For each situation, 1 and 2, describe the least cost location (LCL). Give reasons for your answer.
c   What will be the lowest value isodapane for situations 1 and 2?

The effects of labour costs and agglomeration economies

It has been stated that Weber considered that four factors affected production costs: we have seen the effects of the costs of raw materials and transport — let us now look at labour costs and agglomeration economies.

■ **Labour costs** Weber considered the question of whether any savings made by moving to an area of cheaper or more efficient labour would offset the increase in transport costs incurred by moving away from the LCL. He plotted isodapanes showing the increase in transport costs resulting from such a move. He then introduced the **critical isodapane** as being the point at which savings made by reduced labour costs equalled the losses brought about by extra transport costs. If the cheaper labour lay within the area of the critical isodapane then it would be profitable to move away from the LCL in order to use this labour.

■ **Agglomeration economies** Figure 17.10 shows the critical isodapanes for three firms. It would become profitable for all the firms to locate within the area

formed by the intersection of the three critical isodapanes (the shaded area in Figure 17.10). It may be slightly more profitable for firms A and B, but less profitable for firm C, to locate within the dark grey area. However, it would not be additionally profitable for any firm to move if none of the isodapanes overlapped.

▼ Figure 17.10

Critical isodapanes and agglomeration economies

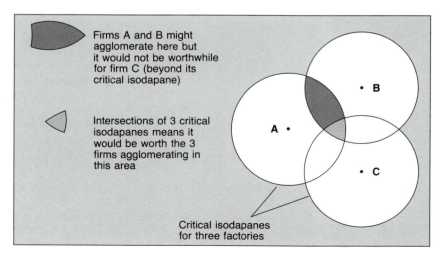

Firms A and B might agglomerate here but it would not be worthwhile for firm C (beyond its critical isodapane)

Intersections of 3 critical isodapanes means it would be worth the 3 firms agglomerating in this area

Critical isodapanes for three factories

Criticisms of Weber's model

The point has already been made with previous examples, (e.g. Framework 8) that no model is perfect and all have their critics. Criticisms of Weber's industrial location model include:

- It no longer relates to modern conditions, i.e. Weber underestimated the role of government intervention (e.g. grants and aid to Enterprise Zones), improvements in and reduced costs of transport, technological advances in processing raw materials, development of new types of industry other than those directly involved in the processing of raw materials, increased mobility of labour and the increased complexity of industrial organisation (multinationals instead of single product firms).
- Each country evolves its own industrial patterns and may be in different stages of economic development (Figure 17.29).
- There are basic misconceptions in his original assumptions, e.g. there are changes over time and space in demand and price, there are variations in transport systems, perfect competition is unreal as markets vary in size and change over a period of time, and decisions made by industrialists (who do not all have the same knowledge) may not always be rational (von Thünen's 'economic man', page 387).
- His material index was a crude measure and applicable only to primary processing or to industries with a very high or low index.

Bradford and Kent claim that Weber's work, apart from its overemphasis on transport costs and leaving some gaps in our understanding of the system as a whole, has not yet been superseded. Rather, 'others have added important principles which, taken together with those of Weber, help to explain a much more complex industrial world'.

D. Smith's area of maximum profit

An alternative approach to that of Weber was put forward by **David Smith** in 1971. He suggested that as profits could be made anywhere when total revenue exceeded total costs, then although there would be a point of **maximum profit**, there would be a wider area where production was still profitable (Figure 17.11). To introduce his **space–cost curves**, Smith also used isodapanes. By taking a cross-section he was able to identify, spatially, margins of profitability. He reasoned that firms rarely located at the ideal, or LCL site, but somewhere between the two profit margins. In other words, firms choose a **suboptimal** location because they have imperfect knowledge about production and market demand, they have imperfect decision-makers who do not always act rationally, and they may be tempted or encouraged to locate in areas of high unemployment.

Industrial location: changing patterns

Four different types of industry have been selected as examples to demonstrate changes in the location of industry through time.

1. A primary industry — the production of pulp and paper in Sweden.
2. An industry initially tied to raw materials — iron and steel in the UK.
3. A market-orientated industry — car manufacture in Japan.
4. Market-orientated hi-tech industries — the M4 corridor in England.

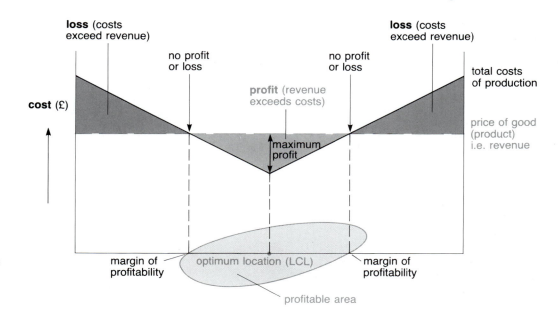

▶ Figure 17.11

Space-cost curve to show the area of maximum profit (*after* D. Smith)

1 Primary industry: wood pulp and paper in Sweden

There are three stages in this industry: the felling of trees, the processing of wood pulp, (primary processing), and the manufacture of paper, (secondary processing). In Sweden, many pulp and paper mills are located at river mouths on the Gulf of Bothnia (Figure 17.12). Timber is a gross raw material which loses much of its mass during processing. It is bulky to transport and requires much water to turn it into pulp. Towns such as Sundsvall and Kramfors are ideally situated with the natural coniferous forests providing the timber, the fast-flowing rivers of the Ljungan, Indals and Angerman initially having provided cheap water transport for the logs and now the necessary and cheap hydro-electricity, and the Gulf of Bothnia providing an easy export route. Paper has a higher value than pulp and it is convenient and cheaper to have integrated mills.

Weber's agglomeration economies seem to operate with the clustering of so many mills. Smith's concept of an area of profitability rather than an individual least cost location also seems applicable, because the limits of the forest may alter, and there are several processing centres.

2 Industry tied to raw materials: iron and steel in the UK

Although the early iron and later steel industry was tied to raw materials, modern integrated iron and steelworks have adopted new locations as the sources of ore and energy have changed (page 430). The following chronology describes the economic history of the industry up to the present.

■ **Before AD 1600** Iron-making was originally sited where there were outcrops of iron ore on the earth's surface and abundant wood for use as charcoal (the Weald, Forest of Dean, Figure 17.13a).

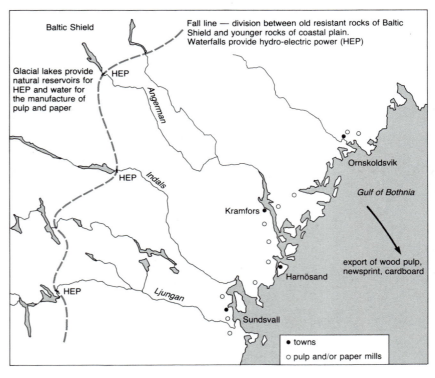

▲ Figure 17.12

Location of wood pulp and paper factories in central Sweden

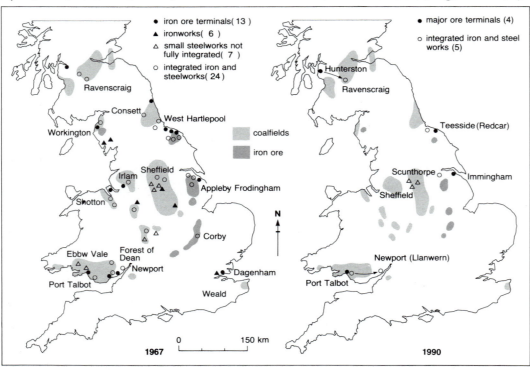

◄ Figure 17.13

Location of iron and steelworks in Scotland, Wales and England

a 1967

b 1990

Period of time		A Location of early 19th century iron foundries in South Wales, e.g. Ebbw Vale	B Disadvantages of these early locations by 1960, e.g. Ebbw Vale	C Location of integrated steelworks of the 1980s at Port Talbot and Llanwern (Newport)
Physical				
Raw materials	Coal	mined locally in valleys	older mines closing	limited reserves in West Wales, some now imported, little now needed
	Iron ore	found within the coal measures	had to be imported — long way from coast	imported from North Africa and North America
	Limestone	found locally	found locally	found locally
	Water	for power and effluent — local rivers	insufficient for cooling	for cooling — coastal site
Energy/fuel		charcoal for early smelting, later rivers to drive machinery and then coal	electricity from national grid	electricity from national grid using coal, oil, natural gas and nuclear power
Natural routes		materials mainly on hand, export routes via the valleys	poor, restricted by narrow valleys	coastal sites
Site and land		small valley floor locations	cramped sites, little flat land	large areas of flat, low potential farmland
Human and economic				
Labour		large quantities of unskilled labour	still large numbers of unskilled workers	still relatively large numbers but with a higher level of skill, fewer due to high-tech/mechanisation
Capital		local entrepreneurs	no investment	government, EC incentives
Markets		local,	difficult to reach Midlands and ports,	tin plate industry (Llanelli) and the car industry
Transport		little needed, some canals, low transport costs	poor, old fashioned, isolated	M4, purpose-built ports
Geographical inertia		not applicable	not strong enough	tradition of high quality goods
Economies of scale		not applicable	worked against the inland sites	two large steelworks more economical than numerous small iron foundries
Government policy		not applicable	Ebbw Vale kept open by government help	having the capital governments can determine locations and closures and provide heavy investment
Technology		small scale — mainly manual	out of date	high technology — computers, lasers, etc.

▲ Figure 17.14
Growth, decline and changing
location of iron and steelworks
in South Wales

▼ Figure 17.15
Vehicle manufacture in Japan

▶ Figure 17.16, *below right*
A car factory built on reclaimed
land in Japan — the Mizushima
Works of the Kawasaki Steel
Corporation

Locations were at the source of these two raw materials as they had a high material index, were bulky and expensive to transport, had a limited market and could not be moved far owing to the poor transport system. Such sites were isolated and the supply of ore was limited.

■ **Before AD 1700** Local ores in the Sheffield area were turned into iron by using fast-flowing rivers to turn waterwheels. These provided a cheaper source of energy than charcoal.

■ **After AD 1700** In 1709, Abraham Darby discovered that coke could be used to efficiently smelt iron ore. At this time it took 8 t of coal and 4 t of ore to produce 1 t of iron so the new furnaces were located on coalfields. The first areas to develop were in South Wales and the West Midlands where bands of iron ore (blackband ores) were found between seams of coal. The advantages possessed by South Wales at that time are given in

Figure 17.14a. Later the industry extended to other coalfields in North Wales, northern England and central Scotland. When these local ores became exhausted, the industry continued in the same locations because of:

□ geographical inertia,
□ a pool of local skilled labour,
□ a local market using iron as a raw material,
□ improved techniques reducing the amount of coal needed to 2 t per tonne of final product,
□ improved and cheaper transport systems (rail and canal) which brought distant mined iron ore,
□ the beginnings of agglomeration economies.

In time, these areas lost many initial advantages (Figure 17.14b) and many iron- and later steelworks were increasingly forced to close.

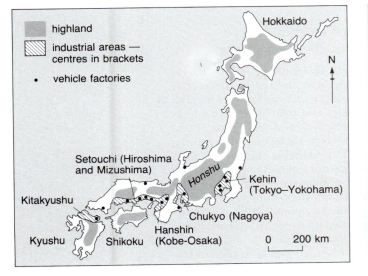

■ **After 1850** Until this time, the low ore and high phosphorous content of deposits found in the Jurassic limestone, extending from the Cleveland Hills to Oxfordshire, had not been touched. After 1879 the **Gilchrist-Thomas process** allowed this ore to be smelted economically. By now iron ore had a higher material index than coal and so was more expensive to move. As a result, new steelworks were opened in the next few decades on Teesside, near to the Cleveland Hills deposits, and at Scunthorpe and Corby, on the ore fields. Scrap iron was increasingly used and skilled workers were brought in from declining areas. However, the major markets remained on the coalfields.

■ **After 1950** With iron ore still the major raw material (less than 1 t of coal was now needed to produce 1 t of steel) but with deposits in the UK largely exhausted, Britain became increasingly reliant upon imported ores. This meant that new **integrated** steelworks were located on coastal sites, e.g. Port Talbot, Newport and Redcar (Figure 17.13b). By 1980 the only two remaining inland sites, at Ravenscraig and Scunthorpe, had been linked to new, nearby ore terminals. The new advantages of the South Wales steelworks are given in Figure 17.14c.

Until the 1950s the iron and steel industry satisfied much of Weber's theory. Since that date, unforeseen by him 40 years earlier, three new elements have become increasingly important in the location of new steelworks: government intervention, improved technology and reduced costs of transport. The location of modern integrated plants and decisions regarding which are to be kept open and which are to close have been made by the government since the industry has been nationalised. Improved technology has meant less reliance upon raw materials and labour. Reduced transport costs have allowed raw materials to be imported. The introduction of the oxygen furnace in the 1950s has meant that relatively little energy (coal) is now needed while the reduction in the number of workers needed has cut production costs but increased unemployment — an economic gain but a social loss.

3 Market-orientated industry: car manufacture in Japan

Japan produced 7.1 million cars in 1985, making it second only to the USA (7.6 million) in the league of world output and well ahead of West Germany (third with 3.8 million). This has been achieved despite the lack of basic raw materials.

Most of the iron ore and coking coal needed to manufacture steel has to be imported and hence the logic of the steelworks being located around deep, sheltered natural harbours. Only 15 per cent of Japan is flat enough for economic development (for homes, industry and agriculture) which means that most of the country's population also live around these harbours (Figure 17.15). The five major conurbations provide the workforce and large local markets for such steel-based products as cars. Within these conurbations, especially Keihin, Chukyo, and Setouchi, are the many firms engaged in making component parts. This agglomeration of firms limits transport costs and conforms with Weber's concept that industries gaining weight through processing (in this case the assembling of parts) are best located at the market.

The local labour force is large and contains both skilled and semi-skilled workers. It is well-educated, industrious and workers are very loyal to their firms. As numerous small firms have amalgamated into large scale companies, the extra space required for their factories has had to come from land reclaimed from the sea (Figure 17.16). Despite the extra cost these make excellent sites from which to export finished cars to all parts of the world.

Japanese industry has a high level of automation and production — it produces three times the number of cars per worker than Western Europe. Labour relations are good and the cars produced are both reliable and universally acceptable in design. Consequently they have gained a strong foothold in world markets. To further expand in their overseas markets the Japanese have built assembly plants in several countries (e.g. Nissan at Washington in Tyne and Wear) and have amalgamated with foreign manufacturers (Honda–Rover in the UK).

4 Market-orientated hi-tech industries: the M4 corridor

The term **high technology** refers to industries developed in the last twenty years, whose processing techniques often involve micro-electronics and which demand a high input of information and expertise. They have also been given the terms **quaternary** (page 430) and **sunrise** industries and, because they are not linked to raw materials, are said to be **footloose** or to have a free choice of location. They form what Weber would have called agglomerated economies and locate according to Smith's terminology in areas of maximum profit.

The major concentration of hi-tech industries in Britain is the 'Sunrise Strip' which

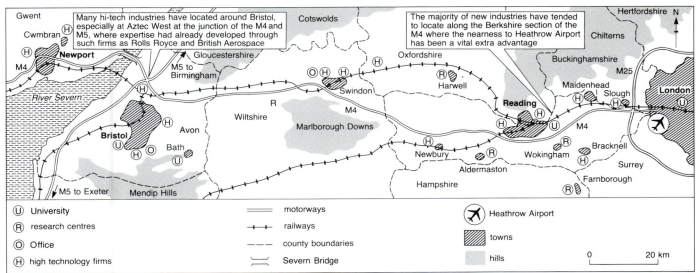

▲ Figure 17.17

The M4 'Sunrise strip'

▼ Figure 17.18

A science park — the 'City of Science and Industry', Paris. Science park developers tend to emphasise the creation of a pleasant working environment

follows the route of the M4 from London, westwards to Newbury ('Video Valley'), Bristol (Aztec West) and into South Wales (Figure 17.17). Because of the presence of the motorway, mainline railway and, in the east, Heathrow airport, transport is convenient. Costs are relatively insignificant anyway, as the raw materials (e.g. silicon chips) are lightweight and final products (e.g. computers) are high value/small bulk).

Most firms which have located here claim that the major factor affecting their decision was the availability of three types of labour: highly skilled, female (for delicate work and part-time flexibility) and research staff. (Often the highly skilled research staff can dictate areas where they want to live and work, e.g. areas of high environmental and cultural quality because their abilities are in short supply.) The previous location and existence of government and other research centres (universities in Bath, Bristol, Reading, London, etc.) had created a pool of skilled labour. There were many females, often career-minded, who had recently moved out of London into new and overspill towns. Nearby were universities with research facilities and staff — indeed, the government has encouraged the development of science parks, adjacent to university campuses, where hi-tech firms can locate (Figure 17.18). The success of early firms led to newer ones moving in and agglomeration economies. Sunrise Strip is also surrounded by attractive countryside (Figure 17.17).

The location quotient

This is a quantitative method to show the degree of concentration of a specific industry within a particular area.

The **location quotient** (*LQ*) is expressed by the formula:

$$LQ = \dfrac{\left(\dfrac{\text{Number of people employed in industry } A \text{ in area } X}{\text{Number of people employed in all industries in area } X}\right)}{\left(\dfrac{\text{Number of people employed in industry } A \text{ nationally}}{\text{Number of people employed in all industries nationally}}\right)}$$

The national average always has an *LQ* of 1.00. The higher the index then the greater the degree of concentration within that area. Figure 17.19 gives employment figures for two industries in Britain in 1981. What was the degree of concentration of shipbuilding in the north of England and SE England for that year?

Industry	Region										Great Britain
	South East	East Anglia	South West	West Midlands	East Midlands	Yorkshire and Humberside	North West	North	Wales	Scotland	
Shipbuilding	39.5	3.5	19.5	0	1.5	7.3	10.0	48.3	1.6	43.0	174.2
Textiles	19.6	2.6	8.5	16.6	85.4	65.0	64.2	11.0	8.5	41.7	323.1
All manufacturing	1683.5	186.0	395.7	800.7	533.4	578.9	799.8	339.4	238.2	502.0	6057.6

◀ Figure 17.19
The location quotient — employment figures for shipbuilding and textiles by region in the UK, 1981

north of England:

$$LQ = \frac{\left(\dfrac{48.3}{339.4}\right)}{\left(\dfrac{174.2}{6057.6}\right)} = \frac{0.141}{0.029} = \mathbf{4.90}$$

This means that the north has a concentration of shipbuilding almost five times greater than the national average while that of the southeast is marginally below.

southeast:

$$LQ = \frac{\left(\dfrac{39.5}{1683.5}\right)}{\left(\dfrac{174.2}{6057.6}\right)} = \frac{0.023}{0.029} = \mathbf{0.79}$$

Q **1** What are the location quotients for the textile industry in:
   a the northwest of England
   b Wales?

Industrial linkages and the multiplier

When Weber introduced the term 'agglomeration economies' he acknowledged that many firms made financial savings by locating close to, and linking with, other industries. It may be the success of one firm attracting a range of associated or similar type industries (e.g. cutlery in Sheffield) or several small firms combining to produce component parts for a larger unit (e.g. car manufacture in Coventry). **Industrial linkages** may be divided into backward and forward linkages:

Backward linkages		**Forward linkages**
to firms providing raw materials or component parts	←FACTORY→	to firms further processing the product or using it as a component part

A more detailed classification of industrial linkages is given in Figure 17.20 opposite.

The more industrially advanced a region or country the greater is the number of its linkages. Developing countries have few linkages partly because of their limited number of industries and partly because few industries go beyond the first stage in processing — the simple chain in Figure 17.20a. Industrial linkages may result in:

- energy savings,
- reduced transport costs,
- waste products from one industry forming a raw material for another,
- energy given off by one process being used elsewhere,
- economies of scale where several firms buy in bulk or share distribution costs,
- improved communications, services and financial investment,
- higher levels of skill and further research,
- a stronger political bargaining position for government aid.

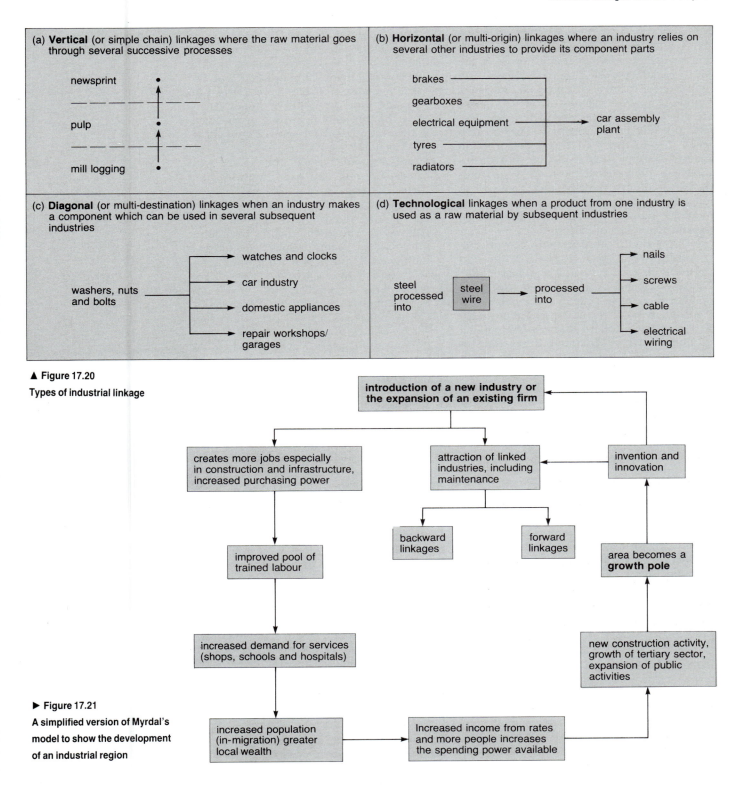

(a) **Vertical** (or simple chain) linkages where the raw material goes through several successive processes

newsprint

pulp

mill logging

(b) **Horizontal** (or multi-origin) linkages where an industry relies on several other industries to provide its component parts

brakes
gearboxes
electrical equipment → car assembly plant
tyres
radiators

(c) **Diagonal** (or multi-destination) linkages when an industry makes a component which can be used in several subsequent industries

washers, nuts and bolts →
- watches and clocks
- car industry
- domestic appliances
- repair workshops/garages

(d) **Technological** linkages when a product from one industry is used as a raw material by subsequent industries

steel processed into → steel wire → processed into →
- nails
- screws
- cable
- electrical wiring

▲ Figure 17.20

Types of industrial linkage

introduction of a new industry or the expansion of an existing firm

creates more jobs especially in construction and infrastructure, increased purchasing power

attraction of linked industries, including maintenance

invention and innovation

backward linkages

forward linkages

improved pool of trained labour

area becomes a **growth pole**

increased demand for services (shops, schools and hospitals)

new construction activity, growth of tertiary sector, expansion of public activities

▶ Figure 17.21

A simplified version of Myrdal's model to show the development of an industrial region

increased population (in-migration) greater local wealth

Increased income from rates and more people increases the spending power available

The multiplier effect and Myrdal's model of cumulative causation

If a large firm, or a specialised type of industry, is successful in an area it may generate a **multiplier effect**. Its success will attract other forms of economic development creating jobs, services and wealth — a case of 'success breeds success'. This circular and cumulative process was used by **Gunnar Myrdal**, a Swedish economist writing in the mid 1950s, to explain why inequalities were likely to develop between regions and countries. A simplified version of his model is drawn in Figure 17.21.

Myrdal suggested that a new or expanding industry in an area would create more jobs and so increase the spending power of the local population. If, for example, a firm employed a further 200 workers and each worker came from a family of four then there would be 800 people demanding housing, schools, shops and hospitals. This created more jobs in the service and construction industries as well as attracting more firms linked to the original industry. As **growth poles**, or points, develop there will be an influx of migrants, entrepreneurs and capital. Myrdal's multiplier model may be used to explain a number of patterns.

1 The growth of nineteenth-century industrial areas such as South Wales and the Ruhr (*see* Industrial regions).

2 Development of growth poles in Third World countries, e.g. Rio de Janeiro–São Paulo in Brazil, the Damodar Valley, India.

3 Creation of modern government regional policies which encourage the siting of new, large, key industries (e.g. Nissan in NE England) in a peripheral, less developed or a high unemployment area in the hope of stimulating economic growth (Figure 17.21). This policy is more likely to succeed if the industry is labour intensive.

Industrial regions

Much of Britain's early industrial success stemmed from the presence of basic raw materials for the early iron industry and later the integrated iron and steel industry, mass production of materials using the processed iron and steel, and the development of overseas markets. The coalfields became the core industrial regions of the last century — especially those in South Wales, northern England and central Scotland.

South Wales

Pre-1920: industrial growth

The growth of industry in South Wales was based upon readily obtainable supplies of raw materials. Coking coal and blackband iron ore were frequently found together, exposed as horizontal seams outcropping on steep valley sides. Their proximity to each other meant that the area around Merthyr Tydfil was ideally suited as Weber's least cost location for two gross raw materials (Figure 17.22). Added to this was the presence of limestone, only a few kilometres to the north, and the expertise of the local population in iron making because water wheels, driven by fast-flowing rivers, had earlier been used to power the blast furnace bellows. By the time the more accessible coal had been used up, mining techniques had improved sufficiently to allow shafts to be sunk into the valley floors. When local supplies of iron ore became exhausted there were ports into which substitute ore could be imported.

'. . . Thus began the spread of the well-known industrial landscape of the Valleys. Pits crammed themselves into the narrow valley bottoms, vying for space with canals, housing and, later, railways and roads. Housing began to trail up the valley sides, line upon line of terraces pressed against the steep slopes (Figure 17.23, page 448). The opening-up of the underground coal seams resulted in massive immigration, much of it from rural areas. Working conditions, living conditions and wages were deplorable while health and safety standards underground were poor. Housing was overcrowded as the provision of homes, financed by the local entrepreneur ironmasters, lagged far behind the supply of jobs.'

The rapid increase in coalmining and iron working partly resulted from the growth of large overseas markets as both products were mainly exported. Transport to the Welsh ports first involved simply allowing trucks to run downhill under gravity. Later canals and then railways were used to move the bulky materials. While Barry, Cardiff and Newport developed as exporting ports, Swansea and Neath grew from smelting the imported ores of copper, nickel and zinc.

The interwar years: depression and industrial decline

Just as the existence of raw materials and overseas markets had led to the growth of local industry, so did their loss hasten its decline. Iron ore had long since been exhausted and now it increasingly became the turn of coal. The steelworks which had replaced the iron foundries had been built upon the same sites. However, these were distant from the ports and too cramped for expansion. Overseas markets were lost as rival industrial regions with lower costs were developed overseas.

The post-war years: industrial diversification in a peripheral area

Steel-making and non-ferrous metal smelting have been maintained despite a significant fall in output and workers, due to geographical inertia. Two of Britain's five remaining integrated steelworks are at Port Talbot and Llanwern (Newport). Tin plate, using local steel, is produced at Trostre near Llanelli (the Felindre works near Swansea closed in 1989) while the Mond nickel works near Swansea are the world's largest (Figure 17.24, page 448).

However, the major factor to have affected industry in the region in the last 50 years has been government intervention (or lack of it, depending upon your political views). The Special Areas Act of 1934 saw the first government assistance which set up industrial estates at Treforest, Merthyr Tydfil and Rhondda (Figure 17.25, page 449). Much of the former coalfield remains a Development Area in 1989 (Figure 17.2) while two local areas of exceptionally high unemployment, Swansea and Milford Haven Waterway, are two of Britain's 27 Enterprise Zones. As such, together with the development areas, they qualify for grants, loans and other incentives.

The Ford Motor Company has taken advantage of these incentives to build two plants in the region and in 1988 announced a large extension to their Bridgend works. Many Japanese and American firms, especially hi-tech, have moved into the region (the Rhondda Valley is now known locally as 'Honda Valley'!). These industries are subject to decisions made outside Wales and employ only a small percentage of those previously engaged in mining and the steel industry. Cwmbran was designated one of Britain's first new towns in 1949.

It was government policy which built an integrated steelworks at Ebbw Vale in 1938, and which closed it in 1979. The future for Port Talbot (highly modernised and hopefully secure) and Newport (under threat during the recession of the early 1980s) is also in government hands. A policy to decentralise some government departments has seen vehicle licensing moved to Swansea and the Royal Mint to Llantrisant (west of Cardiff). Improvements in communications include the M4, the Heads of Valleys Road, the InterCity rail link and Cardiff airport — some of which were financed by EC funds. Most new industrial development has taken place on the edge of the coalfield where communications are better and access to markets is easier.

The region has a large pool of skilled and semi-skilled labour although people need retraining for the new style industry. The Pembrokeshire Coast and Brecon Beacons National Parks and the Gower Peninsular provide many environmentally attractive areas favourable for tourism. Money has also been spent on landscaping old industrial areas such as the lower Swansea valley which were previously scarred by metal smelting industries and former slag and colliery waste tips, and in refurbishing other areas for recreation and tourism, e.g. the Welsh Industrial and Maritime Museum in Cardiff's docklands.

The Ruhr

Although today the Ruhr forms Europe's largest coal and steel producing area, its origin as an industrial region lies long before the 1850s when coal was mined here for the first time in large quantities and steel was processed economically.

The growth of industry
The Ruhr area had a number of early advantages.
- Deposits of iron ore and large reserves of several types of coal.
- Rivers which provided water originally for power and washing and later for cooling and effluent.
- Although the Rhine was the only navigable river, canals were easily constructed within the region and to the North Sea (Figure 17.26, page 449).
- An agglomeration of industries with many linkages.
- The multiplier effect leading to in-migration and a pool of skilled labour in manufacturing.
- After the unification of Germany in the late nineteenth century, industry grew to meet the increasing political and military demands.
- Its position on the Hellweg, the major east–west trade route in northern Europe. Market towns of Duisburg, Essen, Bochum and Dortmund had grown on fertile loess soils (page 107) between the uplands to the south and glacial deposits to the north.

Textiles
The area had favourable conditions for the textile industry.
- Sheep on surrounding uplands provided wool for the early industry in Wuppertal (south of the Ruhr valley).
- Flax grown on loess soils.
- Soft water from rivers.
- River Ruhr provided early water power.
- Coal from the exposed coalfield later provided steam power.
- Large domestic market developed.
- Cotton was imported via the Rhine and synthetic fibres were produced locally.

Iron and steel
Several factors led to the growth of the early iron industry.
- Local iron ore from nearby Sauerland Plateau.
- Charcoal from local forests.
- Water power from rivers.
- Limestone quarries were relatively close.
- Coking coal from adit mines in Ruhr valley.

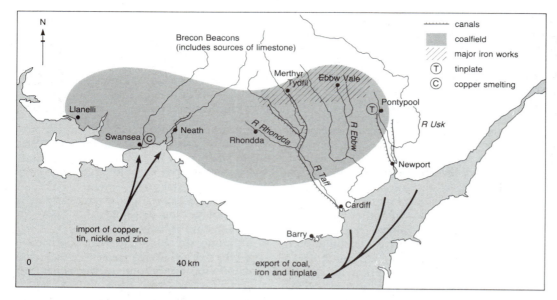

◄ Figure 17.22
South Wales industrial area in the eighteenth and nineteenth centuries

◄ Figure 17.23
Industry and terraced housing along the valley side — Llanilleth (Ebbw Vale), South Wales (compare Figure 15.18K)

▼ Figure 17.24
South Wales industry in the late 1980s (compare Figure 17.22 above)

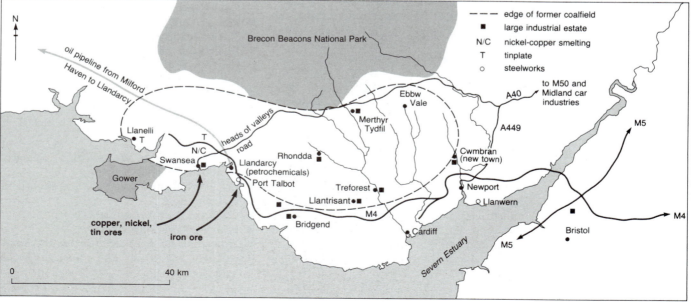

▶ Figure 17.25

A modern industrial estate in South Wales — compare this with the characteristic nineteenth-century industrial development in Figure 17.23

The later steel industry had the following advantages.

- Coking coal from shaft mines in Ruhr and Emscher valleys.
- Iron ore imported from Sweden via River Rhine and Dortmund-Ems canal.
- Rivers and canals for moving bulky goods and reducing transport costs.

Agglomerated manufacturing industries developed with considerable specialisation between areas, e.g. swords, scythes and later cutlery at Solingen; machine tools at Remscheid and heavy electrical equipment at Mulheim. Essen, still a major centre, was the headquarters of the Krupp armament works.

Chemicals

- These were originally based on coal and oil by-products.
- There was early demand by the textile industry for detergents, bleaches and dyes.
- Local farming needed fertiliser.
- The Rhine was invaluable for moving bulky products.

Reconstruction and rationalisation

- Decline and stabilisation of coal production (Figures 17.27 and 17.28). Output fell from 125 million tonnes to 70 million and the numbers of miners decreased by a third between 1957 and 1981. Since then both figures have remained constant.
- Rationalisation in the steel industry. Production has fallen rapidly due to the world recession of the early 1980s, competition from other EC countries (especially those with coastal locations), Third World countries, and the use of rival products. There are only five large integrated steelworks left in the Ruhr (Figure 17.26).
- Petro-chemicals have increased since the opening of the oil pipeline from the North Sea port of Wilhelmshaven. Most factories are near to the Rhine.
- Textiles have adapted to synthetic, oil-based fibres.
- The need for low labour costs and unskilled labour has been overcome by using cheap 'guest workers', especially those from Turkey (page 309) and east European countries.

▼ Figure 17.26

The Ruhr, the ECs 'capital' for coal and steel production

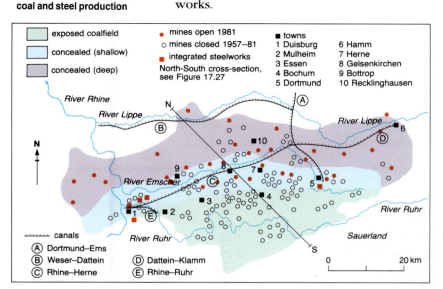

Advantages
Centrally located within the EC
Large proven reserves which should last up to 300 years
Large variety (over 50 types) of coal being suitable for several purposes
New mines have a high level of productivity due to mechanisation
EC policy includes a greater reliance on coal by AD 2000
West German law requires 40 million tonnes of coal a year to be produced for electricity and 30 million tonnes for steel, giving a large, guaranteed market
Water transport by the Rhine and numerous canals, e.g. Dortmund–Ems for iron ore from Sweden

Disadvantages
The most easily obtainable coal is exhausted
Increasing depth of shafts with new coalfaces 1000 m below the surface
Increasing depth means increasing costs of production
During the 1960s and 1970s many mines closed due to competition from the then cheaper imports of oil and natural gas
Competition from cheaper coal from the USA and eastern Europe
Mine closures have led to large scale social problems where over 300 000 miners were made redundant between 1960 and 1980: many towns, e.g. Bochum had relied on mining as their major source of employment — there were 65 000 miners at its peak, only 200 in 1981
Problems of a scarred landscape with atmospheric and river pollution
Newer colleries are in semi-rural and rural areas causing conflicts with conservationists and other rural land users

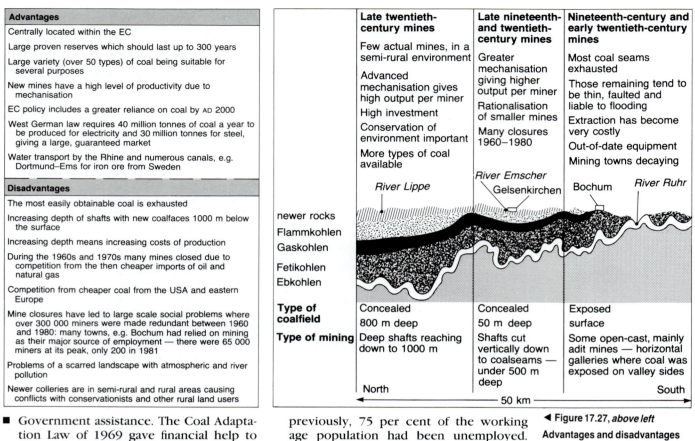

	Late twentieth-century mines	Late nineteenth- and twentieth-century mines	Nineteenth-century and early twentieth-century mines
	Few actual mines, in a semi-rural environment	Greater mechanisation giving higher output per miner	Most coal seams exhausted
	Advanced mechanisation gives high output per miner	Rationalisation of smaller mines	Those remaining tend to be thin, faulted and liable to flooding
	High investment	Many closures 1960–1980	Extraction has become very costly
	Conservation of environment important		Out-of-date equipment
	More types of coal available		Mining towns decaying
Type of coalfield	Concealed 800 m deep	Concealed 50 m deep	Exposed surface
Type of mining	Deep shafts reaching down to 1000 m	Shafts cut vertically down to coalseams — under 500 m deep	Some open-cast, mainly adit mines — horizontal galleries where coal was exposed on valley sides
	North		South

◄ Figure 17.27, *above left*
Advantages and disadvantages of present day coalmining in the Ruhr area

▲ Figure 17.28
Changes to the coal industry in the Ruhr

■ Government assistance. The Coal Adaptation Law of 1969 gave financial help to those parts of the coalfield suffering high unemployment and needing to diversify. As in Britain's Development Areas, low interest loans and employment grants have attracted new industries, e.g. two Opel car factories at Bochum where, previously, 75 per cent of the working age population had been unemployed. Attempts have been made to improve the environmental image of the Ruhr by creating green belts, parks, leisure zones (which now account for 60 per cent of the region's land use) and clearing and landscaping derelict industrial land.

Q

1 a Describe the location of coalmines in the Ruhr (Figures 17.27 and 17.28) at the following periods:
   the late nineteenth century
   the early twentieth century
   the late twentieth century.
   b Account for these changes.

2 Why are there relatively few mines and mining settlements in the present day coal producing area?

3 Why is the Ruhr a good example of the multiplier effect?

4 Attempt to show how the model of cumulative causation (Figure 17.21) could be applied to the Ruhr. Use a diagram or flow chart for your answer.

Stages in economic growth

The Rostow model

In 1960, following a study of fifteen countries, mainly in Europe, W.W. Rostow put forward his model of economic growth. He suggested that all countries had the potential to break the cycle of poverty and develop through five linear stages (Figure 17.29).
Stage 1 Traditional society A subsistence economy based mainly on farming with very limited technology or capital to process raw materials or develop industries and services.

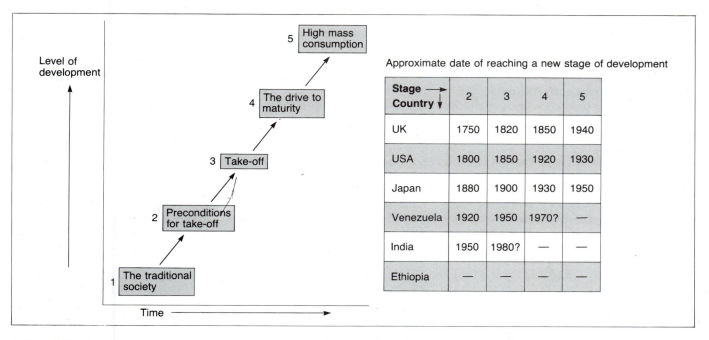

Approximate date of reaching a new stage of development

Stage → Country ↓	2	3	4	5
UK	1750	1820	1850	1940
USA	1800	1850	1920	1930
Japan	1880	1900	1930	1950
Venezuela	1920	1950	1970?	—
India	1950	1980?	—	—
Ethiopia	—	—	—	—

▲ Figure 17.29

Rostow's 'linear stages' model of economic growth (Rostow, 1960) and his 'learning curve' — Rostow suggested that the time taken for a country to develop diminishes as it learns from already developed nations and can tap their accumulated expertise

Stage 2 Preconditions for take-off A country often needs an injection of external help to move into this stage. Extractive industries develop. Agriculture is more commercialised and becomes mechanised. There are some technological improvements and a growth of infrastructure. The development of a transport system encourages trade. A single industry (often textiles) begins to dominate. Investment is about 5 per cent of GDP.

Stage 3 Take-off Manufacturing industries grow rapidly. Airports, roads and railways are built. Political and social adjustments are necessary to adapt to the new way of life. Growth is usually limited to one or two parts of the country (**growth poles**) and to one or two industries (**magnets**). Numbers in agriculture decline. Investment increases to 10–15 per cent of GDP or capital is borrowed from wealthier nations.

Stage 4 The drive to maturity By now growth should be self-sustaining. Economic growth spreads to all parts of the country and leads to an increase in the number and types of industry (the multiplier effect). More complex transport systems develop and manufacturing expands as technology improves. Some early industries may decline. There is rapid urbanisation.

Stage 5 The age of high mass-consumption Rapid expansion of tertiary industries and welfare facilities. Employment in service industries grows and in manufacturing declines. Industry shifts to the production of durable consumer goods.

Criticisms of Rostow's model

Rostow suggested that capital was needed to advance a country from its traditional society. Despite relatively large injections of aid many of the less developing countries of Africa and Asia still remain at the traditional stage and those who have moved towards the take-off stage have done so by incurring huge national debts, e.g. Brazil and Mexico. Another major criticism is the short time predicted between when growth begins and when it becomes self-sustaining. There seems decreasing evidence to substantiate Rostow's **learning curve**, i.e. the time taken for a country to develop diminishes through history as countries learn from already developed nations (Singapore, Hong Kong, South Korea and Taiwan are notable exceptions which support Rostow's claim). Economists point out that growth is more complex than the model indicates while historical evidence suggests that the sequence is not universal. Also, as pointed out by Barke and O'Hare, the model is Eurocentric.

Barke and O'Hare's model for West Africa

M. Barke and G. O'Hare claimed in 1984 that although developed industrial countries may have moved through Rostow's five stages, it seems increasingly unlikely that Third World countries will follow the same pattern. This may be because capital alone is insufficient to promote take-off. Perhaps what is needed is a fundamental structural change in society which encourages people to save and invest and to develop an entrepreneurial, business class. Perhaps the process allowing transition from traditional agriculture to advanced industry is a relict one, being applicable only to the early industrialised countries which had unlimited use of the world's resources and markets. Barke and O'Hare have suggested a four-stage model for industrial growth in Third World countries such as those in West Africa.

Stage 1 Traditional craft industries These were in existence before European colonisation. Northern Nigeria (Kano) had cloth weaving, iron working, wood carving and leather goods.

Stage 2 Colonialism and the processing of primary products Raw materials were initially exported in an unprocessed form (e.g. cocoa and palm oil) while the chief imports (textiles and machinery) came from the colonial power and, being cheaper, destroyed many local craft industries. Later, some processing took place, usually in ports or the primate city (page 342), if it reduced the weight of the product (vegetable oils), if it was too bulky to import (cement) or if there was a large local market (textiles). To help obtain raw materials from their colonies, the European powers built ports (e.g. Accra and Lagos), but railways were only constructed if there were sufficient local resources to make them profitable (Figure 18.20). Education, along with the development of industrial and management skills, was neglected.

Stage 3 Import substitution Following independence, West African countries had to replace the import of textiles, furniture, hardware and simple machinery with their own manufactured goods. Production was in small units using limited amounts of capital and technology.

Stage 4 Manufacture of capital and durable consumer goods As standards of living rose in several countries (notably in Nigeria with its oil revenue) there was an increased demand for heavier industry and 'western' style durable consumer goods. These industries, often because of the investment and skills needed, were developed by multinational companies wishing to take advantage of cheap labour, tax concessions and entry to a large local market. The American Valco company, for example, constructed a dam on the Volta River, a hydro-electric power station at Akosombo and an aluminium smelter at Tema in return for duty and tax exemptions on the import of bauxite and the export of aluminium, together with the purchase of cheap electricity, granted by the Ghanaian government. Multi-nationals in Nigeria include Shell, Honda (in Lagos) and Mitsubishi (in Ibadan). Projects developed by multinationals are usually prestigious, of little value to the country and may be withdrawn (e.g. Volkswagen stopped operating in Nigeria) should world sales drop.

Employment structures

Although per capita income is generally regarded to be the best single indicator showing the level of development of a country (page 325), one of several alternative criteria relates to the numbers involved in primary, secondary and service industries. An interpretation of Rostow's model might show the following **employment structures**:

Proportion of workforce employed in			
	1 Primary	2 Secondary	3 Tertiary (services)
Stage 1	vast majority	very few	very few
Stage 2	vast majority	few	very few
Stage 3	declining	rapid growth	few
Stage 4	few	stable	growing rapidly
Stage 5	very few	declining	vast majority

Figure 17.30 shows the employment structures for 15 selected countries. Nepal, Uganda and Tanzania would presumably fit into stage 1. They have a high percentage (over 80 per cent) of their workforce in primary industry, most of whom are farmers. As agriculture is mainly at a subsistence level, and as few people are employed in industry (less than 6 per cent) there is little capital to provide services. Industry may not have developed for several reasons: a lack of capital; a limited education system producing a less highly-skilled worker; lack of technical knowledge; the export of unprocessed raw materials; a lack of mechanisation; a lack of energy supplies; a limited local market unable to afford manufactured goods. The small numbers in service industries are related to limited developments in education, health, commerce, transport, recreation and tourism.

India, Nigeria and Egypt are harder to categorise because parts of each country are more developed and industrialised (growth poles of Stage 3) than others. Farming may have become more mechanised (India's Green Revolution, Egypt's improved irrigation system) while industry has developed through the harnessing of important energy resources — coal in India's Damodar valley,

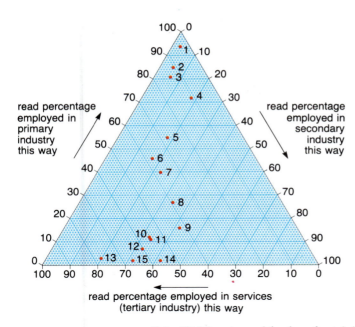

Number	Country	GDP per capita (1984) US $
1	Nepal	141
2	Uganda	238
3	Tanzania	299
4	India	233
5	Nigeria	730
6	Egypt	711
7	Brazil	2232
8	Portugal	2344
9	Italy	5549
10	New Zealand	6736
11	Japan	9928
12	Netherlands	7716
13	Kuwait	12709
14	UK	6514
15	USA	13968
Rank order based on percentage employed in primary industry		

▲ Figure 17.30

Employment structures for fifteen selected countries

oil in SE Nigeria and hydro-electricity from Egypt's Aswan Dam. With increased wealth (albeit from a low base), service industries are growing in response to demands of local wealthier business people and foreign tourists. Brazil, regarded to be a more developed Third World country, shows a more evenly balanced employment structure.

The more developed countries have a low percentage, often under 10 per cent, engaged in primary activities — this is possibly due to the intensive nature of farming (Netherlands, page 410) or a reliance upon imports. The decline in the secondary sector may be caused by a loss of overseas markets or by heavy capital investment (robots and computers) in labour intensive domestic industries. The service sector accounts for over half of the working population — a reflection of a higher standard of living and a demand for comfort and material possessions.

Q The countries listed in Figure 17.30 have been ranked according to the percentage of the working population employed in primary industry (*see* Frameworks for thinking 11).

1 Give the numbers employed in primary, secondary and tertiary activities for each of the 15 countries.

2 Calculate the Spearman rank correlation coefficient (page 328) between the GDP of the 15 countries and the numbers employed
   a in the primary sector
   b in the service sector.

3 Using the graph provided in Figure 13.45 comment upon the statistical significance of the two correlation coefficients which you calculated in question 2.

Employment in Third World countries

Employment structures only include those people who have a permanent **formal** job. In the mid 1980s it was estimated that there were over 400 million unemployed persons in Third World countries, compared with 15 million in developed countries. Many others were underemployed. The prospects for increased industrial growth in Third World countries seems remote following the world recession of the early 1980s, import embargoes by several developed countries, a lack of capital and a rapidly growing young population. The UN suggests that only about 40 per cent of the working population in Third World countries have formal jobs (e.g. in the police, civil service, services and manufacturing). The remaining 60 per cent, and a growing percentage, seek work in the **informal** sector. The differences between the formal and informal sectors are described in Figure 17.31 overleaf.

Formal	Informal
Description	
Employee of a large firm	Self employed
Often a multinational	Small scale/family enterprise
Much capital involved	Little capital involved
Capital intensive with relatively few workers, mechanised	Labour intensive with the use of very few tools
Expensive raw materials	Using cheap or recycled waste materials
A guaranteed standard in the final product	Often a low standard in quality of goods
Regular hours (often long) and wages (often low)	Irregular hours and uncertain wages
Fixed prices	Prices rarely fixed and so negotiable (bartering)
Jobs done in factories	Jobs often done in the home (cottage industry) or on the streets
Government and multinational help	No government assistance
Legal	Often outside the law (illegal)
Usually males	Often children and females
Type of job	
Manufacturing — both local and multinational industries	Distributive, e.g. street pedlars and small stalls
Government-created jobs such as the police, army and civil service	Services, e.g. shoecleaners, selling clothes and fruit
	Small scale industry, e.g. food processing, dress repairs, furniture repairs
Advantages	
Uses some skilled and many unskilled workers	Employs many thousands of unskilled workers
Provides permanent jobs and regular wages	Jobs may provide some training and skills which might lead to better jobs in the future
Produces goods for the more wealthy (cars, food) within their own country so that profits may remain within the country	Any profit will be used within the city
Waste materials provide raw materials for the informal sector	Uses local and waste materials — the products will be for local use by the lower paid people

◄ Figure 17.31
Differences between the 'formal' and 'informal' sectors of the economy in a Third World country

◄ Figure 17.32
Re-using vehicle tyres in western Java—recycling waste materials (tins, bottles, plastic containers) to save energy and raw materials may be the only means of scraping a living

CASE STUDY

Don Mariano — an electrician in Guatemala

Don Mariano, a refugee from Nicaragua, lives with his family in a shanty settlement in San Jose, Guatemala. His hut, which was built with boards and roofed with corrugated iron, also contains his 'workshop'. It is a separate room with a worktable and an electric outlet. In this workshop he repairs all kinds of electrical appliances, heating plates, electric heaters, radios and television sets. His equipment is totally inadequate — two screwdrivers and one pair of pliers. For a Voltmeter he uses an old lamp — he can tell the voltage from its brightness. He does not own a blowtorch. Don Mariano's primary problem is not a lack of skill but a lack of equipment. If he had the money he could buy a blowtorch and a voltmeter. If he had those he could do more repairs and earn more money. If . . .

Intermediate technology

Dr E.F. Schumacher presented the concept of intermediate technology as an alternative course for development for poor people in the 1960s. He founded the Intermediate Technology Development Group (ITDG) in 1966 and published his ideas in his book, *Small is Beautiful* (1973). Schumacher himself wrote:

'If you want to go places, start from where you are.
If you are poor, start with something cheap.
If you are uneducated, start with something relatively simple.
If you live in a poor environment, and poverty makes markets small, start with something small.
If you are unemployed, start using labour power, because any productive use of it is better than letting it lie idle.
In other words, we must learn to recognise boundaries of poverty.
A project that does not fit, educationally and organisationally, into the environment, will be an economic failure and a cause for disruption.'

In 1988 the ITDG stated that:
'Essentially, this alternative course for development is based on a local, small-scale rather than national, large-scale approach. It is based on millions of low-cost workplaces where people live — in the rural areas — using technologies which can be made and controlled by the people who use them and which enable those people to be more productive and earn money.'

These ideas challenged the conventional views of the time on aid. Schumacher said 'The best aid to give is intellectual aid, a gift of useful knowledge . . . The gift of material goods makes people dependent, but the gift of knowledge makes them free — provided it is the right kind of knowledge, of course.'
To illustrate this he quoted an old proverb:
'Give a man a fish and you feed him for a day; teach him how to fish and he can feed himself for life.'
The first part of this might be seen as the traditional view of aid where 'giving' leads to dependency. The second part, 'teaching', is a move in the direction of self-sufficiency and self-respect.
Schumacher added a further dimension to the proverb by saying 'teach him to make his own fishing tackle and you have helped him to become not only self-supporting but also self-reliant and independent.'

1 Why do ITDG consider that aid is not the solution for poor people?

2 Explain what is meant by 'giving leads to dependency, teaching to self-sufficiency and self-respect'.

3 Consider several specific national, large scale development schemes in Third World countries (e.g. Aswan Dam in Egypt page 412, the Akosombo Dam and Tema aluminium plant in Ghana, the Union Carbide factory at Bhopal in India, etc.). Who benefits most from such schemes? Has poverty been diminished in these areas?

4 Study Figure 17.33 overleaf. Why have these schemes, according to ITDG, proved to have been successful?

◀ **Figure 17.33**

c, *centre left* Carpenters in Malawi are taught how to make their own tools so that they may return to their villages to make furniture and agricultural equipment. Thus the carpenter earns a living and the village gets a craftsman

d A dyeing winch in Bangladesh — the machine was designed by an IT engineer working in the country, was made locally and allows coloured dyes to be spread evenly over the cloth

◀ Figure 17.33 (ITDG)
Intermediate technology at work
a, *top left* Pottery stoves made
by local potters in Sri Lanka —
each is mud-covered to prevent
it cracking and to improve
durability and fuel efficiency.
This stove uses only half the
wood needed when cooking on
an open fire. This saves the
women time and money
(collecting and buying wood)
and reduces deforestation
b, *right* Making roofing tiles in
Kenya using local materials — a
battery-operated vibrating table
is needed to strengthen the tiles
by forcing out air bubbles. Tiles
are used in self-help building
projects

▼ Figure 17.34
Industrial development in China

Industry in a centrally planned economy: China

When the People's Republic of China was
declared in 1949 the country's economy was
still feudal with virtually no agricultural or
industrial development. There was a very
low standard of living, most of the land and
wealth was in the hands of a small minority,
and life expectancy was merely 32 years.

1949–1976: the commune system

State control was overseen by Mao Zedong.
The government dictated where a particular
factory had to be located, how many workers
would be employed there, who would
have jobs and what products would be made.
The state controlled who was responsible for
each stage in the manufacturing process,
how much was to be produced, how much
the wages would be, where the product
would be sold and at what price. There were
no individual entrepreneurs or decision
makers.

The first Five Year Plan was introduced in
1953. It concentrated on developing heavy
industry and the production of consumer
goods was actively discouraged. China's huge
coal reserves became the basis for an iron
and steel industry from which ships, textile
machinery, tractors, locomotives and railway
track were produced. These associated industries were labour intensive. Progress was
slow and in 1958 Mao launched the 'Great
Leap Forward' in an attempt to mobilise the
people and resources of China towards rapid
economic growth. In an attempt to meet
production targets workers were organised,
like their counterparts in agriculture, into
self-sufficient communes (page 416). Many
smaller scale 'cottage industries' were set up
but the leap proved a disaster.

In 1966 Mao embarked upon the 'Cultural
Revolution' which aimed to create a class-
less society. Attempts were made to elimate
differences between farmers and industrial
workers (the latter tended to receive more
wages for an easier job), between white
collar and manual workers, and between the
city, where most industry was located, and
the countryside. Those with education and
skills (engineers, technicians and teachers)
were sacked and sent to the countryside to
work in the fields. The next ten years are
now referred to as the 'decade of destruction'. The Cultural Revolution lasted until
Mao's death in 1976.

1979 to the present: the responsibility scheme

In 1979 the government began to replace
the regimentation of farm production through
communes with the **responsibility system**
which turned China's peasants into tenant
farmers (page 416). This system was later
applied to industry with the result that total
production and its quality improved. The
increased momentum of change under Deng
Xiaoping has led to the growth of a mixed
economy (some Marxists would describe it
as capitalism). Encouragement has been given
to individual entrepreneurs to set up their

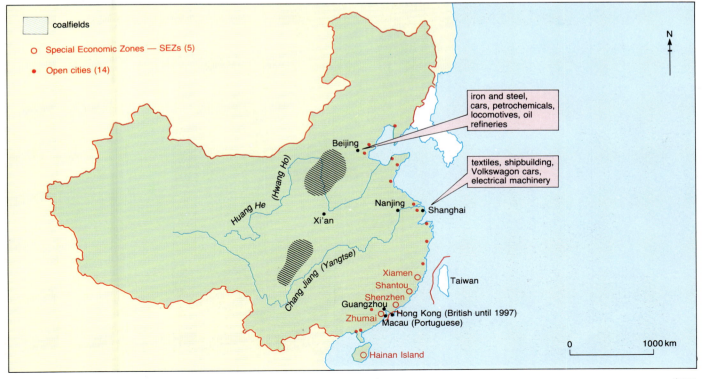

- coalfields
- ○ Special Economic Zones — SEZs (5)
- • Open cities (14)

iron and steel,
cars, petrochemicals,
locomotives, oil
refineries

textiles, shipbuilding,
Volkswagon cars,
electrical machinery

Beijing

Huang He (Hwang Ho)

Xi'an

Nanjing

Shanghai

Chang Jiang (Yangtse)

Xiamen
Shantou
Shenzhen
Guangzhou
Zhumai
Hong Kong (British until 1997)
Macau (Portuguese)

Taiwan

Hainan Island

0   1000 km

N

own firms, especially in consumer goods. State-owned companies, still accounting for 70 per cent of the total firms, are allowed to sell any products surplus to their production targets. Bonuses are paid for meeting production targets, and profits made from selling extra stock are shared between workers. Competition between firms is encouraged. An 'open door' policy has allowed overseas firms to invest and settle in the country, e.g. Pepsi-Cola at Shenzhen and Volkswagen in Shanghai (Figure 17.34).

Similarly, China has begun to invest in overseas countries with a 50 per cent share in an American oil company, a 12½ per cent share in Cathay Pacific Airways, and involvement in the Anglo–French Channel tunnel. Chinese are encouraged to work overseas to acquire skills and capital.

Since 1980, five **special economic zones** (SEZs) and 14 **open cities** have been created (Figure 17.34). These, rather like free trade areas and ports in Third World countries (page 465), offer tax concessions and low labour and land costs to overseas firms — especially those involved in hi-tech industries. It is not coincidental that the five SEZs are in the southeast of the country near to Hong Kong which handles over 40 per cent of China's trade.

Although the open cities are more widespread, they too are all in coastal areas. China accepts that these coastal areas will become more prosperous (indeed they have already) than the remainder of the country, acting as growth poles, but the hope is that in time economic development will diffuse inland. At present, and in apparent contradiction of socialist beliefs, the Chinese Communist Party is placing economic growth before the building of an equal society. One major problem is that the country is having difficulty selling its goods overseas in the face of competition from Japan, South Korea, Taiwan, Hong Kong and Singapore — and so most inland areas remain very poor.

Development of the interior is limited by a still inadequate transport system and the vast distances involved. Most intercity journeys are made by steam train, by rivers where convenient, and increasingly (but still on a limited scale) by aeroplane. Few people own cars but 100 million own bicycles (Figure 17.35). In 1988, China produced more steel than any European country and was the world's leading manufacturer of cotton textiles. Even so, the standard of living remains low and, despite two problems described in the case study, industry is still largely state controlled.

◀ Figure 17.35
Bicycles are major form of transport in Shanghai, China

CASE STUDY

China's dilemma

'China is asking how to dismiss 30 million surplus workers, accustomed to cradle-to-grave security, and persuade them that they and the country will be better off if they work somewhere else. Chinese economists say that reallocation of these workers is crucial to the success of ambitious reforms to change overmanned and inefficient state firms into streamlined and well-managed enterprises. The official press has admitted that 30 million out of the urban workforce of 130 million have nothing to do at work. Workers cannot be dismissed for fear of unrest and the social consequences.' [Perhaps the events in Beijing's Tiananmen Square in June 1989 have affected this fear of potential unrest.]

'In Chinese cities a factory provides not only a job but an apartment, medical insurance, nurseries and a pension. When a man leaves a factory he loses all these. Meanwhile the factory manger is constantly being told by one side to lower costs and increase productivity while, by the other that he faces go-slows from workers if they do not get similar pay to other firms.'

'Up to one-fifth of China's urban workers have taken second jobs to keep pace with soaring inflation. Millions of factory workers, unable to maintain their living standards, have taken up semi-legal extra work in shops, restaurants and as street vendors'.

(Source: *South China Morning News*, July, 1988.)

▶ Figure 17.36
The industrial area of Anjan, NE China

References

High Technology Industry (UK), Geofile No. 126, Mary Glasgow Publications, April 1989.

Human Geography, Ken Briggs, Hodder & Stoughton, 1982.

Human Geography, M.G. Bradford and W.A. Kent, Oxford University Press, 1977.

Human Geography, Patrick McBride, Blackie, 1980.

Industrial Location, M. Day, F. Meyer and S. Day, Thomas Nelson, 1985.

Intermediate Technology, IT Publications (ITDG), 1987.

London Docklands, Geofile No. 106, Mary Glasgow Publications, 1988.

North−South: a Programme for Survival, (Report of the Brandt Commission), Pan, 1980.

Pattern and Process in Human Geography, Vincent Tidswell, UTP, 1976.

Small is Beautiful, E.F. Schumacher, Abacus, 1974.

The Overcrowded Island, Industry (Japan), BBC Television, 1974.

The Third World, M. Barke and G. O'Hare, Oliver & Boyd, 1984.

The World, David Waugh, Thomas Nelson, 1987.

CHAPTER 18

Transport and trade

'Progress be dammed. All this will do will be to allow the lower classes to move around unnecessarily ...'

The Duke of Wellington (on seeing the first train)

'I will build a motor car for the great multitude so low in price that no man will be unable to own one ... and enjoy with his family the blessing of hours of pleasure in God's great open spaces.'

attributed to Henry Ford c. 1900

Transport

The following references to the role of transport have already been made at various stages of this text.

1 It may be viewed as an indicator of wealth and economic development, as measured by the number of cars or tractors per 1000 people. While the more developed countries have only 25 per cent of the world's population they have 88 per cent of its rail traffic and 72 per cent of its cars and lorries.

2 Transport is essential in linking people, resources and activities and to help in the exchange of goods (trade) and ideas (information).

3 Transport was considered to be a major factor in industrial location (Weber) and in determining agricultural land use (von Thünen). The relative decrease in transport costs to the average firm or farm has made this a less significant locational and land use factor since the 1950s (page 432).

4 In early economic/geographical theory costs were thought to be proportional to distance (von Thünen's central market and Christaller's central place) — although this was when applied to an isotrophic surface with transport being equally easy and cheap in all directions. Later costs were regarded as a function of a raw material's weight as well as the distance it had to be moved (Weber).

5 The development of an area or country may be significantly influenced by transport (Rostow).

Characteristics of modern transport systems

Early tribespeople had to travel on foot or by animal. The first civilisations, after 6000 BC, used river transport (Egypt, Mesopotamia and China). By 4000 BC the wheel, arguably one of the most important inventions ever, was being used in ceremonial processions in Mesopotamia. The next major technological advance did not come until the use of steam power in the late eighteenth century and with it the invention of the steam railway engine and steam ship. 1885 saw the first journey by a petrol-driven car and 1903 the first powered flight. A comparison of the characteristics of the major forms of present day transport — canal, ocean-shipping, rail, road, air and pipeline — is given in Figure 18.1. Each type has advantages and disadvantages over rival forms of transport. Figure 18.2 shows the relative costs and Figure 18.3 the amount of freight and passengers moved by the various types of transport in Britain.

Transport costs

Figure 18.1 refers to terminal and haulage costs (page 432). **Terminal costs** are fixed regardless of the length of time of journey and are highest for ocean transport and lowest for road transport. **Haulage costs**, which increase with distance but decrease with the amount of cargo handled, are lowest for water transport and highest for road and air (Figure 18.2).

▶ Figure 18.1

Comparable characteristics of transport systems

		Canals and rivers	Ocean transport and deep sea ports	Rail	Road	Air	Pipelines
P H Y S I C A L	**Weather**	Canals can freeze in winter Drought/heavy rains make rivers unnavigable	Storms, fog Icebergs in North Atlantic	Very cold (frozen points) Heavy snow (blocks line) Heavy rain can cause landslides	Fog, ice both can cause accidents, pile ups Cross winds for big lorries, snow blocks routes, Sun can dazzle	Fog, icing and snow — less since planes have had automatic pilots Airports better if sheltered from wind away from hills and areas of low cloud	Not greatly affected
	Relief	Width of channels, needs, flat land or gentle gradients Soft rock/soil for digging, problems with deltas Rivers must be slow, without rapids and of a constant discharge	Harbours need to be deep, wide and sheltered Tidal problems	Cannot negotiate steep gradients so have to avoid hills Estuaries can be obstacles Flooding in valleys	Avoids/takes detours around highland Valleys may flood May go around estuaries	Large areas of flat land Firm foundations Ideally cheap farmland or land needing reclamation Seas and mountains not a barrier Flatland for terminal buildings and warehousing	Difficult to lay, then relief is not a problem
E C O N O M I C	**Speed/time**	Slowest form of transport Long detours and possible delays at locks	Slow form of transport, yet most economical	Fast over medium length distances	Fast over short distances and on motorways (autobahns) Urban delays	Fastest over long distances, not over short ones due to delays getting to and passing through airports	Very fast as continuous flow
	Running or haulage costs (wages and fuel) these increase with distance	Often family barge Limited fuel use means the cheapest form of transport over lengthy journeys	Expense of oil (fuel) increases with distance	Relatively cheap over medium length journeys Fuel and wages quite high	Cheapest over shorter distances Haulage costs rapidly increase with distance	Very expensive yet speed makes it competitive over very long distances	Cheapest as no labour is involved (provided diameter is large)
	Terminal costs (loading and unloading costs and dues) no change with distance	Canals expensive to build and to maintain, unless natural waterways used	Ports expensive Harbour dues/taxes Expensive to build specialised ships Less since containers Cheapest with distance	Building and maintenance of track/ stations/signalling/rolling stock is very expensive Not used at off peak times	Expensive building and maintenance costs, especially motorways Car/tax instead of dues, but roads built by community taxation therefore lower overheads	Very expensive to build and maintain airports High airport dues Planes expensive to purchase and maintain	Very expensive to build
	Number of routes	Relatively few Inflexible	Relatively few ports, inflexible due to increased specialisation of ships, also linked to hinterlands	Decreasing — limited to main cities and power stations Not very flexible 'InterCity'	Many and at different grades Great flexibility, most in urban and industrial areas	Often only a few — internal — international airports Not very flexible because of safety	Limited to key routes, inflexible
	Goods and/or passengers carried Congestion	Little congestion Heavy, bulky non-perishable, low value goods Present day tourists	Little congestion Heavy, bulky, non-perishable low value goods Cruise passengers	InterCity passengers Heavy and bulky goods, chemicals, coal, also rapid (mail) Little track congestion other than commuter trains Can carry several hundred passengers Dependent and safe	Many passengers Perishable, smaller loads by lorry Relatively few people carried by one bus or car Congestion in urban areas and at holiday times	Mainly passengers Usually little other than at peak holiday times Freight is light (mail), perishable (fruit) or high value (watches)	Bulk liquid (oil, gas, slurry, liquid cement) no congestion
	Convenience and comfort	Neither very convenient, unless for leisure/relaxation, nor very comfortable	Not very convenient Cruise liners very comfortable	Not very, other than InterCity Needs transport to and from station Comfortable for passengers	Door to door — most convenient and flexible Safety is questionable Strain for drivers but independent and private	Country to country Relaxing over longish journeys Jet lag after very long journeys	Raw material or port to industry
E N V I R O N M E N T A L	**Environmental problems**	Some oil discharged but relatively few problems	Tankers discharging oil — relatively few problems	Noise and visual pollution limited to narrow belts - electric trains are almost pollution free	Major noise and air pollution Affect on ozone layer, acid rain, therefore greenhouse effect Uses up land, especially farmland, structural damage caused by vibrations	High noise level Some air pollution Uses up much land for airports	Few once 'buried' Eyesore on surface

461

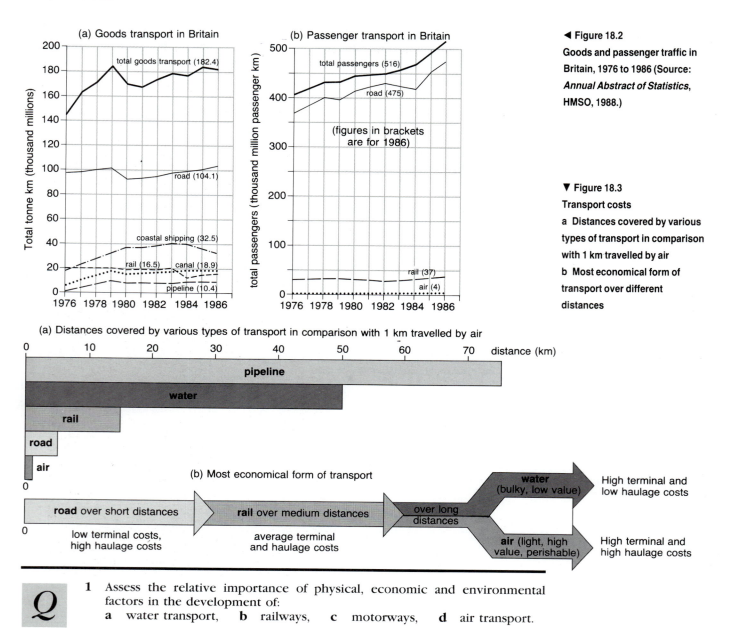

Figure 18.2
Goods and passenger traffic in Britain, 1976 to 1986 (Source: *Annual Abstract of Statistics*, HMSO, 1988.)

Figure 18.3
Transport costs
a Distances covered by various types of transport in comparison with 1 km travelled by air
b Most economical form of transport over different distances

Q 1 Assess the relative importance of physical, economic and environmental factors in the development of:
a water transport,   b railways,   c motorways,   d air transport.

Types of transport

Inland waterways

The major canal building period in western Europe was in the late eighteenth and early nineteenth centuries. The virtual monopoly by the canals was shortlived because of competition, first from rail and later from road. However, since the 1970s there has been a rejuvenation of the inland waterways. This began with the increase in road and rail charges caused by the rise in oil prices following the 1974 Middle East war and an appreciation that, despite its slowness and inflexibility of routes, water transport re- mains the cheapest form for bulky, non- perishable goods. Huge 4400 t barges, propelled by push-barges have been intro- duced by the EC which can carry the equivalent of 110 railtrucks or 220 × 20 t lorries. To accommodate these 'super-barges' the EC governments have financed the widening and deepening of canals and the linking of the Mediterranean, North and Black Seas. Smaller canals, including those in Venice and Amsterdam, are benefiting from increased tourism.

CASE STUDY

Water transport in southern Thailand

The Chao Phraya River provides a major waterway as it passes through Bangkok on its way to its delta in the Gulf of Thailand. Large flat-bottomed boats carry cargo from ocean ships which cannot negotiate the shallow water. These wooden boats, resembling a modern Noah's Ark, provide permanent homes to large Thai families. Much smaller vessels, also acting as family homes, pull up to a dozen barges each (Figure 18.4). Between these commercial ships ply passenger ferries and numerous 'long boats' (similar to long rowing boats but powered by a car's engine), which carry the daily goods of the local people. Parts of the older city are a maze of smaller *klongs* (canals) along the banks of which the homes of Bangkok's inhabitants are built on stilts.

Some 80 km west of Bangkok is one of Thailand's floating markets. Here women from all over the delta meet each day to exchange their produce (Figure 18.5). They paddle their small boats along the network of *klongs* which provide the only form of transport between the villages. Their sales are, today, being increased by selling to tourists.

▼ Figure 18.4
Barges on the Chao Praya River, Bangkok, Thailand

▶ Figure 18.5
The floating market, Dannoensaduak, Thailand

Ocean shipping

Many ports in western Europe and eastern North America developed either by trading with each other across the Atlantic, or by importing raw materials from colonies and other developing countries, then exporting manufactured goods in return.

Recently, several major changes have occurred in British ports which have led to the growth of those in the southeast of the country (with the exception of London) and a corresponding decline elsewhere. The increase in ship size, especially purpose-built **bulk carriers** for oil and iron ore, has meant that wider and deeper estuaries are needed. The recession in world trade has led to a general decrease in the number of ships and ports needed.

A ship berthed at a quayside is not earning money. Two innovations have managed to shorten appreciably the turn round time (the time taken to unload and load cargo). The first was the development of **roll-on/ roll-off** (Ro-Ro) methods, where lorries loaded with freight are driven on board so reducing the need for cranes. The second was the introduction of **containerisation** (unitised traffic) in which goods are packed into containers of a specified size at the factory, taken by train or lorry to the container port, and easily and quickly loaded by specialist equipment. Some of the larger container ships (containers are usually stacked on deck) can hold upwards of 4000 units. Unfortunately these technological advances have decimated the dock labour force and reduced the number of required ships (Figure 18.6).

◀ Figure 18.6
Containers at Felixstowe
Harbour

Felixstowe was only the ninth largest port in Britain in 1960 handling 1 per cent of the country's seaborne trade. By 1986 it was second with a 15 per cent share. Reasons for this growth include: the presence of deep water at all stages of the tide necessary for handling large vessels; its coastal site so that no time is lost sailing up an estuary; its position on the major trade routes with the EC and Scandinavia; the availability of Ro-Ro facilities and four container terminals; its being sited on marshy, poor quality land which was cheap and available for purchase with land space for future expansion; it has good road links with its hinterland, and good industrial relations.

CASE STUDY

◀ Figure 18.7
Hong Kong harbour — notice
the number of ships entering
and leaving the harbour

Free ports: Singapore and Hong Kong

On founding the modern port of Singapore in 1819, Sir Stamford Raffles decreed that it was open to all maritime nations. By 1988, over 700 shipping lines had taken advantage of that decree. At any given time over 400 ships may be in port with one arriving or weighing anchor every ten minutes (about 30 000 vessels per year). One tonne of cargo is handled in less than a second, and during every second of the year. Warehouses are automated and computerised. Since 1981 the port has outstripped London, New York and Yokahama to become the world's second largest port after Rotterdam. Vessels vary from modern supertankers and container ships to the more traditional bumboats and junks. To save docking time, harbour pilots now fly out to incoming vessels by helicopter.

Like Singapore, Hong Kong (Figure 18.7) is a **free port**, open to all countries and with several **free trade zones**. These are where goods can be made or assembled without payment of import or export duties, and where profits can be sent back to the parent company without being taxed. Many hi-tech firms assemble their products here and sell them at very competitive prices (a Japanese student visiting Singapore was surprised to find his own country's brand names selling more cheaply in Singapore than in Japan). Hong Kong not only has strong British connections but is the outlet for 40 per cent of trade from the People's Republic of China.

Rail transport

Railways became the dominant form of transport in Britain during the Industrial Revolution and their popularity rapidly spread. Although they are still the most convenient method of moving bulky goods and large numbers of passengers (Figure 18.1) the inflexibility of their routes and terminals have left them at a considerable disadvantage to road transport as cities have grown and industry has changed its location. Railways were constructed when cities were much smaller and industry was sited near to the mainline station in what is now the inner city. Modern industry in the developed world now tends to seek an edge-of-city or small town location which does not have a rail terminal. Rail transport has regained some trade recently in North America and western Europe partly because of congestion and other road problems and partly because of successful attempts to modernise the systems. France (the *TGV*), West Germany, Italy and Spain have all built new high-speed routes.

CASE STUDY

British Rail

Three major problems which faced British Rail in the late 1980s were the need to improve its dowdy image, to obtain the necessary finances for safety and modernisation, and to meet a hitherto unpredicted increase in use by the public. The east coast main line between London and Edinburgh will be electrified by 1991. The entire route will be controlled from just seven modern signalling centres. Curves are being straightened, bridges and stations built or rebuilt, new coaches designed and a new marketing strategy devised. Much depends upon a new high speed electric locomotive, 'Class 91', which should be easier to maintain than the existing, more complicated, diesel engine.

The introduction of freightliners has enabled containers of internationally agreed standard size to be carried by rail to specialist ports. This saves time and labour costs, reduces the risk of theft and damage. Construction has started on the Channel (Fixed Link) Tunnel project (page 468). This scheme, funded (in Britain) entirely by the private sector, will allow passengers to travel from London to Paris in 2 hours 30 minutes. The British Rail network, which, if plans go ahead, will have been privatised by then, will link with the continental high speed routes. As yet no decision has been made as to whether the trains to run between London and Paris will be the French *TGV* or the British 'Class 91'.

Road transport

The major advantage of road transport is its flexibility which allows it to operate from door-to-door and especially over short distances, at the most competitive price (Figure 18.3). Yet while the building of motorways and urban roads is essential if an area wishes to improve its accessibility and stimulate economic activity, the results of such construction often create environmental problems (Figure 18.1 and page 469). Older urban areas, built in a pre-automobile era, experience the modern problems of congestion:

□ unplanned, narrow streets (unless redeveloped),
□ car parking difficulties,
□ delays caused by buses and unloading delivery lorries,
□ air pollution from exhausts (also contributing to acid rain and the greenhouse effect page 211),
□ noise and visual pollution,
□ roadworks (often repairing old utility services),
□ old buildings being shaken by passing traffic.

Even where redevelopment has occurred there is still heavy pollution, land is taken up by road widening, land values next to urban motorways are reduced and the risk of accidents seems as great (Figure 18.8).

Newer urban centres, such as Milton Keynes and Brasilia (Figure 18.9), have been built primarily with the motorist in mind.

◄ Figure 18.8, *below left* São Paulo — the problems of urban traffic congestion

▲ Figure 18.9 Brasilia, a city built for the motorist — the twin towers of the congress building are over 2 km away from the point where this photograph was taken

C A S E   S T U D Y

Brasilia

Brasilia's first brick was laid in 1957. Thirty years later its population, affluent by Brazilian standards, had risen to 1.3 million. Although the design resembles an aeroplane in shape, its use is intended for the motorist! The two wings, each 6.5 km long, are used for housing (superblocks) with a three-lane expressway extending along the axis. The 'fuselage', 9 km long, is divided into sections for commerce, culture, hotels, local and national government, each connected by additional three-lane roads.

When the city planner's wife was killed in a car crash he vowed to make Brasilia as accident-free as possible. His road plan avoided the use of traffic lights and road junctions with the result that there are considerable distances between places (too far to walk) and a greater danger to any potential pedestrian. As traffic increased, so did the number of accidents, and lights were introduced in the mid-1980s (Figure 18.9).

Air transport

Air transport has the highest terminal charges, high haulage costs (fuel) and tends to arouse more opposition from environmental groups than other forms of transportation (page 470). Its advantages (Figure 18.1) include speed over long distances for passengers, both tourists and business people, and for freight if it is of high value, light in weight and perishable. Air transport is also important in countries of considerable size (Brazil) and with difficult terrain (Australia) or to bring relief following a major disaster such as earthquake or flood.

Government policies

British governments have a major say in transport just, as we have seen previously, as in agriculture and industry. The then Secretary of State for Transport, Mr Paul Channon, said in October 1988:

> 'We have to look forward now to the twenty-first century and to the needs of our country then. Let me tell you, my vision is of a Britain with a first-rate transport system — which combines public sector investment with the initiative and imagination of the private sector ... To keep our industry competitive with our neighbours, we must provide the people of Britain with a first class transport network for the future.'

Note: To quote government transport policies, which seem to change daily, and to use official data, which like all secondary information is already out of date before it is printed, immediately dates any geography book. However, the most up-to-date ideas and information have been included at the time of writing (autumn 1989) and it is left to you, the reader, to:

1 Update your information by finding data issued after autumn 1989.
2 Decide for yourself (Framework 9) whether transport systems in Britain should combine the public and private sector, and if not, which sector would you prefer?
3 Assess the relative success or failure of 1989 policies as we pass through the 1990s.

Roads

At present only 45 per cent of families have one car and 18 per cent two cars in Britain. This leaves a tremendous capacity for future increases. The 17 million cars (22 million vehicles in total) on the roads is likely to increase by 30 per cent to 23 million by the turn of the century — the car is a symbol of economic success for most British people. In the decade after 1976 the length of Britain's motorways increased by just under 100 km per year and that of principal roads by 150 km per year. This has barely kept up with the continued rise in car ownership.

The precedent of private funding to improve access to business centres has been set at Gateshead's MetroCentre (page 364) and Sheffield's Meadowhall development. The private sector is now involved in the Dartford River Crossing (M25, page 469) and is being considered for a proposed extension, with tolls, of the M11 via the Humber Bridge to the northeast of England. The government has to decide whether 'a road free of congestion would make tolls acceptable to business'. If a private consortium were to finance a road by itself then it would need to levy a charge to recoup its capital outlay. Although some drivers would use alternative routes (spreading out the congestion?), in countries where tolls are used (e.g. Italy) drivers seem prepared to pay them.

Rail

The accidents at Clapham, Purley and Glasgow, within the space of a few weeks in 1988/9, raised the question of safety on what had previously been considered the safe railway network. Certainly, the tragedies have highlighted the problems of overcrowding and poor service on commuter trains. The proposed London–Channel Tunnel route awaits final approval (page 468). The government admitted, following loud protests from pressure groups opposing different rail links, that each of four proposals had disadvantages.

London Transport's underground has seen a dramatic increase in its patronage (up 70 per cent between 1982–1986) causing severe overcrowding. The Central Line is to be re-equipped and the Fennell Report, following the King's Cross fire of November 1987, is being implemented as quickly as is possible. It is thought (at the time of writing) that plans for a new underground line may soon be announced.

Air

The increase in air routes and availability of lower fares are encouraging more people to travel. Scheduled international traffic handled by British carriers has increased by over 50 per cent in the last ten years, and domestic traffic by 80 per cent. This has led to congestion at airports, especially at peak holiday weekends and competition for airspace. The Civil Aviation Authority (CAA) is undertaking studies of airport use and capacity and traffic distribution between London's airports (page 469) while its investment should see the modernisation of air traffic control and an increase of at least 30 per cent in airspace capacity over SE England by 1995.

Transport and the environment

As an increasing number of people continue to become more mobile so the need for improved traffic systems also rises. While suggested schemes may make travel quicker, easier and safer, they invariably have some impact on the environment. The unanswerable question is 'how can we accommodate this ever growing demand for travel, particularly leisure travel, without causing serious and lasting damage to the environment?' Three modern issues in SE England involve rail, road and air.

1 The Channel Tunnel rail link

Now that building of the tunnel is underway, the main economic/environmental conten-

tion has become the location of the rail link between the tunnel and London. Initially British Rail offered four proposals — the most favoured was to cost an estimated £1.2 million. However, these options cut through some of the most attractive scenery and residential areas of Kent and aroused considerable opposition. The route eventually accepted — mainly as it was considered by the government to have a less adverse impact on the environment (and, of course, on their local political standing) — is to cost £1.7 million. Although opposition groups felt that the government should foot the bill for this extra £0.5 million, the government announced that the extra cost would eventually have to be met by the rail link users themselves (Figure 18.10).

Q Figure 18.10 shows the likely Channel Tunnel Rail Link route. Using the information provided, or researched by yourselves, answer the following questions.

▼ Figure 18.10
The proposed Channel Tunnel rail link and its potential environmental impact (April 1989)

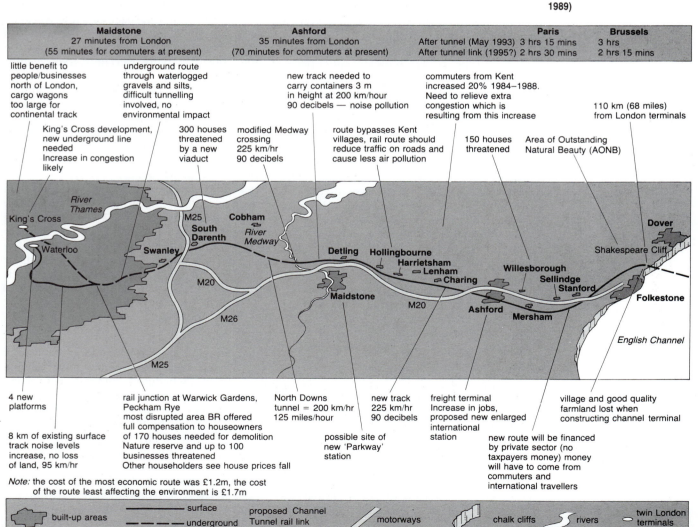

1 To what extent do you think that the rail link is of economic advantage:
   a to Britain as a whole,
   b to SE England?

2 How successfully does the proposed route minimise environmental problems?

3 Which do you consider to be the more important: an efficient rail link to benefit business people and tourists or the preservation of the environment? Explain your answer.

---

2 The M25 orbital road

There were many reasons for the building of this motorway, e.g. to reduce the volume of traffic in London, to relieve the overcrowded North and South Circular (outer) ring roads, to link motorways converging on the capital and to link the four London airports, especially Gatwick and Heathrow. The M25 was built with three lanes in anticipation of its being used by 80 000 vehicles a day. To reduce objections from environmentalists, landowners and residents, the route meandered, was in parts landscaped and, between the M11 and A121, had bridges and underpasses to allow the movement of deer.

The M25 is now used by 160 000 vehicles daily, confirming the theory that new roads attract more traffic which only increases congestion — the 'record' jam on the M25 is a 35 km three-lane standstill. Already a fourth lane is being constructed on the Heathrow–Gatwick section on land set aside for this purpose but not initially used as planners felt such a proposal would have been defeated on environmental grounds. The extra vehicle-released fumes are adding to levels of acid rain and accelerating the greenhouse effect (page 211): such concerns probably led to the Chancellor reducing tax on lead-free petrol in the 1989 Budget. One major bottleneck is the Dartford Tunnel under the river Thames. A road bridge is at present being built and financed by the private sector. This money will be reclaimed by the imposition of tolls, after which the bridge will revert to public ownership.

The M25 is not alone in being threatened by ever larger lorries. The EC uses 40 t compared to Britain's legal maximum of 38 t for lorries. The EC wishes Britain to use the larger size, claiming that fewer would be needed because of their greater capacity and it would avoid the added cost of transferring goods on entry to Britain. However, except for motorways, most of our road bridges are too small and narrow to accommodate the larger juggernauts.

3 London's airports

Delays for passengers at airports, especially at Gatwick, and the safety factor involved both in the increasing use of airspace over London and the limited number of available runways is of major concern. As shown in the table, these problems are likely to get worse before they can be improved.

Heathrow and Gatwick will be at full capacity before the year 2000, by which time a new runway will have had to be built 'somewhere'. The airlines say a decision is needed soon and they anticipate lengthy public enquiries and fights to get planning permission accepted.

	Passengers (millions) 1976	1988	Percentage increase 1976–1988	Estimate for AD 2000
Heathrow	23.2	37.5	160	
Gatwick	5.7	20.8	365	
Luton	1.8	2.7	150	
Stansted	0.4	1.0	250	
London Airports (total)	31.1	62.0	200	123.0

What are some of the alternatives?

1 To use off-peak times at Gatwick and Heathrow — but this would mean night flights which are not favoured by travellers or local residents.

2 To build a fifth terminal at Heathrow — but this area is already congested both on the ground and in the air.

3 To open a second runway at Gatwick — the North Terminal has been opened on land previously set aside for this.

4 To try to raise the present 5 million passenger limit imposed at Luton.

5 To use the extra potential available at Stansted especially as the new terminal building will increase the airports capacity by ten times — aviation authorities

consider the air would be safer and London quieter if the extra runway was built to the south of the capital.

6 Build a new airport near Shoeburyness on land reclaimed from the Maplin Sands in the Thames estuary — this was rejected in the 1970s on both environmental and economic grounds.

7 Use larger aircraft.

The problem was highlighted by events in March 1989.

20 March An all-party committee at Westminster recommended that a second runway should be built at Gatwick.

21 March Local residents had already set up an action group and, along with environmentalists and local politicians, claimed that land set aside for a new runway had partly been built upon (the North Terminal). To build it would mean the destruction of the village of Charlwood.

22 March Airline pilots accused environmentalists of 'misleading the public' as the better site for a runway, having fewer residents and being environmentally 'softer', lies to the south of the present airport. They added that 'Gatwick is now far more dangerous than almost any other airport because of its single runway. A second runway is essential before an accident leaves bodies strewn across the Sussex countryside.'

The question of the extra runway is a classic example of the NIMBY syndrome — it is generally accepted as essential, but 'Not In My Back Yard'. This is a major transport issue for the 1990s: the safety and convenience of air travellers in conflict with the quality of life of local residents and the protection of the environment.

Transport routes and networks

A single route location

The shortest distance between two points is a straight line, yet in practice, direct routes have rarely been constructed, even by the Romans. P. Haggett (1969, 1977) has suggested that deviations from a straight line route were for either positive or negative reasons.

Positive deviations are when routes are diverted to gain more traffic. Figure 18.11a shows that although (assuming no physical obstacles) the straight line route is cheapest to build, having the shortest distance, it is the most expensive to the user. In comparison, the **least cost** route for the user (Figure 18.11b) is likely to be the longest and most expensive to build. The selected route is likely to be a compromise between the two extremes (Figure 18.11c), provided it gains acceptance from a third party — the environmental lobby.

Negative deviations result from routes being diverted to avoid physical obstacles, e.g. a mountain range (Figure 18.12) or a river estuary.

Builder and user costs
The least cost network to the **builder** is likely to be the one with the shortest overall length (Figure 18.13a). These are more likely to be found in areas with a sparse population and low traffic density (e.g. Australia), in highland areas where routes are funnelled along valley floors (e.g. Lake District) or where, as in the case of motorways, the construction cost per metre is extremely high.

The least cost network to the **user** is the one where the traveller can move as quickly and directly as possible between any point in the network (Figure 18.13b). They are likely to occur in conurbations where high population and traffic densities demand numerous alternative routes.

A **network** is defined as a group of places joined together by a number of routes to form a **system**.

▼ Figure 18.11
Single route location — a positive deviation

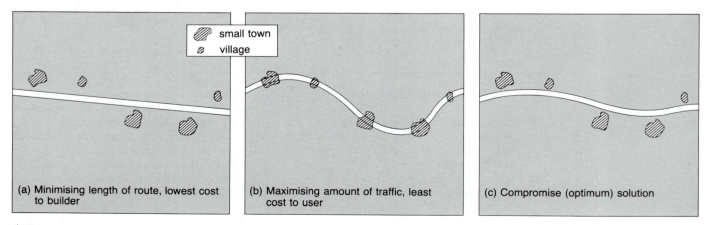

(a) Minimising length of route, lowest cost to builder

(b) Maximising amount of traffic, least cost to user

(c) Compromise (optimum) solution

small town
village

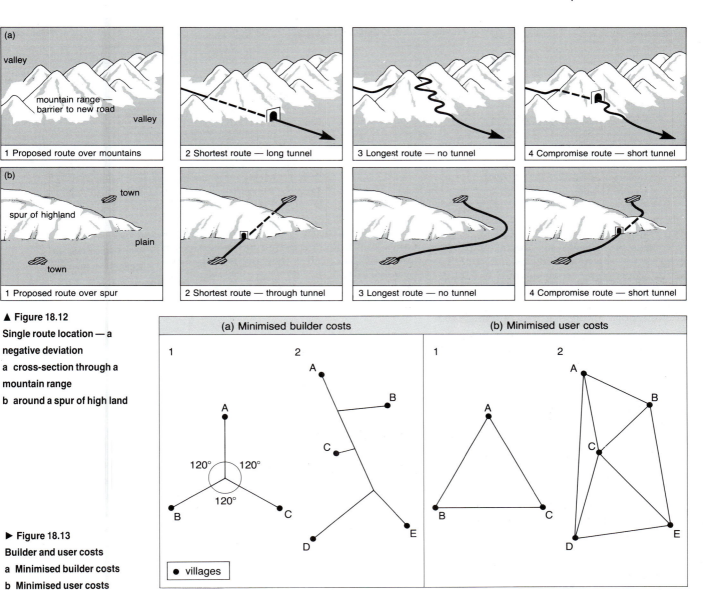

▲ **Figure 18.12**

**Single route location — a
negative deviation**

**a cross-section through a
mountain range**

b around a spur of high land

▶ **Figure 18.13**

Builder and user costs

a Minimised builder costs

b Minimised user costs

Transport networks

Network patterns vary greatly between and within countries so it is necessary to adopt a universally acceptable method of description to allow comparison between different systems. Network analysis is achieved by showing the network as a series of straight lines. Drawing a **topological map** preserves the pattern of routes and junctions but distorts directions and distances (e.g. the London Underground and the Tyne and Wear Metro maps). The resultant map (Figure 18.14) consists of vertices and edges. **Vertices** or **nodes** are points in the network. **Edges** or **links** form the direct lines of transport between these points.

Vertices (Figure 18.14) may be a point of origin or destination (Swansea), a significant place *en route* (Cardiff) or a junction of two or more routes (Bristol). This division into vertices and edges helps to describe the **accessibility** of a place and the **efficiency** of the network. Figure 18.14 also gives the accessibility matrix for points on the topological map. The aim is to follow a route between two places, passing through as few vertices as possible, since they can cause congestion and delay. The **Shimbel index** (which is 14 for Bath) is the total number of edges needed to connect any one place with all the other vertices. The place(s) with the lowest Shimbel index (Newport and Bristol with 11) has the highest **centrally** and is the most accessible. If the Shimbel index for all the vertices are plotted it is possible to draw an isopleth map to show the relative accessibility of places (e.g. isopleths for 12, 15, 18 and 21 could be added to Figure 18.14, page 472).

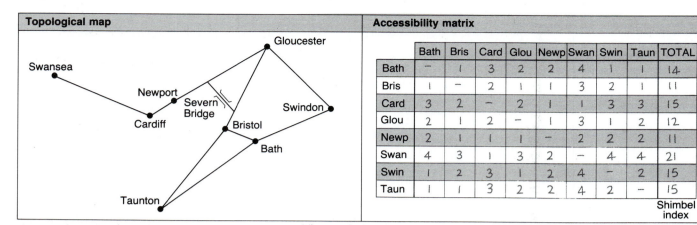

Topological map

Accessibility matrix

	Bath	Bris	Card	Glou	Newp	Swan	Swin	Taun	TOTAL
Bath	–	1	3	2	2	4	1	1	14
Bris	1	–	2	1	1	3	2	1	11
Card	3	2	–	2	1	1	3	3	15
Glou	2	1	2	–	1	3	1	2	12
Newp	2	1	1	1	–	2	2	2	11
Swan	4	3	1	3	2	–	4	4	21
Swin	1	2	3	1	2	4	–	2	15
Taun	1	1	3	2	2	4	2	–	15

Shimbel index

Types of network

Networks are shown by means of **planar graphs**, i.e. where all vertices and edges are in the same plane so that while two edges may meet at a vertex they cannot cross over each other. Figure 18.15 shows six different methods of linking, or not linking, four vertices.

a In a **null graph**, none of the vertices is linked — they may be divided by mountain barriers or political frontiers.

b A **connected graph** is when every vertex is linked to the network but only indirectly to other vertices, e.g. a cross-city bus route or the Central Line in the London Underground.

c A **circuit**, or **closed path**, follows a 'circular' route linking up all the vertices, e.g. a milkround beginning and ending at a dairy or the Circle Line in the London Underground. It is the shortest possible distance linking all points.

d In a **tree graph**, all the vertices are linked but there are no circuits. Each edge, leading to a terminal point, acts as a branch line, e.g. motorways radiating from London.

e A **complete graph** is where each vertex is directly linked to every other vertex. It is more likely to be found in a developed country and equates with Christaller's $k = 3$ model, page 347.

f A **subgraph** is when part of the network is detached from the rest. This incomplete system may be associated with the pre-take-off stage of the Rostow model (Figure 17.29).

The connectivity of networks

Connectivity is a means of **measuring** the **efficiency** of a network (the more vertices which are connected with each other, the more efficient the network) and **comparing**, quantitatively, different networks. Before looking at four methods of measuring connectivity, two basic principles should be noted.

■ The **minimum** number of edges, e (minimum), needed to link all the vertices in a network is one less than the total number of vertices (v).

Therefore: $e_{(min)} = v - 1$

e.g. in the tree and connected graphs with 4 vertices, $e = 3$.

■ The **maximum** number of edges, e (maximum), which can exist in a network without a duplication of routes (linkages) is found by subtracting 2 from the number of vertices and then multiplying by 3.

Therefore: $e_{(max)} = 3(v - 2)$

e.g. in a complete graph with 4 vertices, $e = 3(4 - 2) = 6$.

▲ Figure 18.14

A topological map and accessibility matrix

▼ Figure 18.15

Types of network

a null graph

b connected graph

c circuit   d tree graph

e complete graph

f subgraph

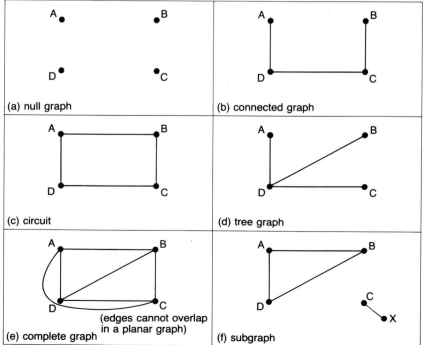

(a) null graph

(b) connected graph

(c) circuit

(d) tree graph

(e) complete graph
(edges cannot overlap in a planar graph)

(f) subgraph

Measuring connectivity

■ **The beta index** (β) This is the easiest method of measuring connectivity as it simply involves dividing the number of edges (e) in the network by the number of vertices (v).

Therefore: $\beta = \dfrac{e}{v}$

Three situations can result:

1 In simple networks, the beta index is less than 1.0,

    e.g. Figure 18.15a $0 \div 4 = 0$ (null)
    Figure 18.15b and d
    $3 \div 4 = 0.75$ (tree and connected)

2 In a network with one circuit the beta index is 1.0,

    e.g. Figure 18.15c $4 \div 4 = 1.0$
                   (circuit)

3 In a complete network the beta index is greater than 1.0,

    e.g. Figure 18.15e $6 \div 4 = 1.5$
                   (complete)

As can be seen, the higher the value of the beta index the greater the degree of connectivity and the more efficient the system.

■ **The cyclomatic number** (c) This refers to the number of circuits in a given network. The more edges there are in a network the greater the number of circuits is likely to be. The formula is derived from the fact that a tree graph plus one edge gives a circuit. Therefore it can be assumed that 'the number of circuits = the number of edges (e) minus the number of edges needed to form a tree ($v - 1$)'. This is the first of the two basic principles given above.

Therefore: $e - (v - 1)$

which is the same as $c = e - v + 1$

In a tree graph where there are no circuits (Figure 18.15d):

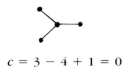

$c = 3 - 4 + 1 = 0$

In a complete graph where there are three circuits (Figure 18.15e):

$c = 6 - 4 + 1 = 3$

A disadvantage of this method is that networks with very different forms may have the same cyclomatic number.

■ **The alpha index** (α) This is very useful when comparing networks. It is obtained by expressing the number of actual circuits in a network (the cyclomatic number) as a percentage of the maximum possible number of circuits in that network.

This can be expressed by the formula:

$$\alpha = \frac{(e - v + 1)}{2v - 5} \times 100$$

This can give a value between 0–100 per cent. A low alpha index means there is little or no connectivity whereas one of 100 per cent indicates that the network is completely connected.

In this network:

$$\alpha = \frac{2}{8 - 5} \times 100$$

Therefore: $\alpha = 66\%$

■ **The gamma index** (γ) This index, considered less useful than the alpha index, compares the actual number of edges in a network with the maximum possible number of edges in that network.

This is expressed as:

$$e_{(max)} = \frac{e}{3(v - 2)} \times 100$$

Using the same two diagrams as for the cyclomatic number, the maximum links in the circuit will be:

$$e_{(max)} = 3(4 - 2) = 6$$

However, in the tree graph where there are only three edges, then:

$$e_{(max)} = \frac{3}{3(4 - 2)} = 100 = 50\%$$

The gamma index also ranges from 0–100 per cent. It is used when comparing the efficiency of networks of different sizes.

The detour index (DI)

When using topological maps, the edges which link the vertices are drawn as straight lines. Yet it has been shown that in reality communications rarely follow the most direct route. The efficiency of a route can be measured by determining how far it deviates from the straight line route.

This is calculated by using the formula:

$$DI = \frac{\text{Shortest possible actual route distance}}{\text{Direct (straight line) distance}}$$
(also known as the 'desire line')

The lowest possible index is 1.00 (some texts suggest this is given as a percentage). A **detour index** of 2.00 means the actual route is twice as long as the direct route.

Q **1** Figure 18.16 shows four transport networks with five vertices. Describe briefly the main advantages and disadvantages to the user and builder of each of these networks.

▼ Figure 18.16

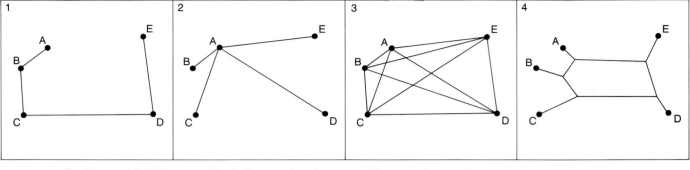

2 Figure 18.17 is a topological map showing part of a main line rail network.
  a Using the beta index calculate the connectivity of the network in 1963 and 1982.
  b Explain what the difference in the indices between the two dates means.
  c Which vertices were the most accessible in 1963 and 1982?
  d For the rail network in 1982 work out the alpha index and the cyclomatic number.

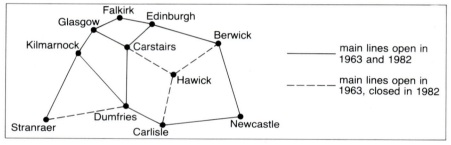

◀ Figure 18.17 (*after* J. Howells)

main lines open in 1963 and 1982

main lines open in 1963, closed in 1982

3 Figures 18.18 and 18.19 opposite refer to Hong Kong's Mass Transit Railway (MTR).
  a Using the MTR stations given in 18.18, name the vertices labelled *A* to *F* on Figure 18.19.
  b Using three different indices describe the level of connectivity of the MTR in Hong Kong.
  c With so much commuter traffic between Hong Kong Island and the Kowloon Peninsula it may be sensible to construct a second tunnel. If this were to be built between vertices *E* and *F*, how would this affect the connectivity and accessibility of the network as a whole?

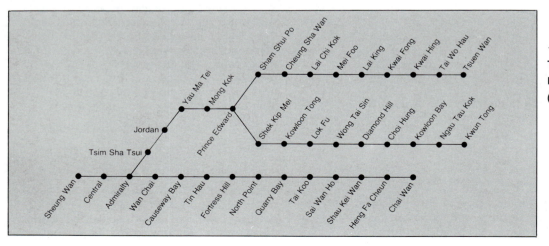

◀ Figure 18.18

Topological map of Hong Kong's mass transit railway (MTR)

▶ Figure 18.19

Location map of Hong Kong's MTR

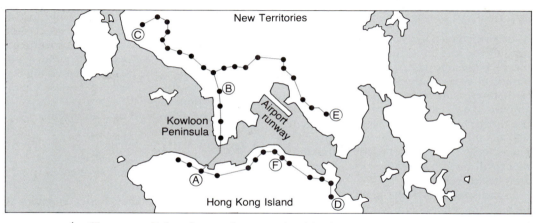

4 'Transport links often make considerable detours from the straight line route between two places.'

With reference to specific examples, discuss the statement in relation to each of the following: water transport; motorways; air routes; railways. (You may wish to consider physical, climatic, economic, human, political and conservation factors.)

A model showing the evolution of a transport network in a Third World country

In 1963 Taaffe, Morrill and Gould put forward a sequential model to show how a transport network might evolve in a Third World country. Their model, based on studies in Nigeria and Ghana, has been summarised in Figure 18.20. The early stages reflected the needs of a controlling colonial power to:

1 politically and militarily link the centres of administration, which were on the coast, with inland centres,
2 exploit mineral resources,
3 develop plantations and increase agricultural production,
4 carry freight in preference to passengers.

The model has since been applied to other Third World countries, especially in South America, Africa and Malaysia. In many cases, only stage two has been reached, mainly because following independence these countries had insufficient money to improve and modernise their networks beyond the main urban areas, except, perhaps to a prestigious airport. To reach stages 3 and 4 (Brazil and Malaysia), it appears that a country needs the development of a manufacturing industry and the emergence of an affluent sector in the population.

▼ Figure 18.20

Model showing the evolution of a transport network in a Third World country (after Taaffe, Morrill and Gould)

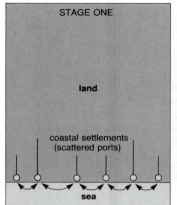

STAGE ONE

land

coastal settlements (scattered ports)

sea

Pre-colonial days, reliance upon local sea-borne trade with tracks leading inland. Port development limited by shallow seas, heavy surf, few natural inlets and deltas at river mouths

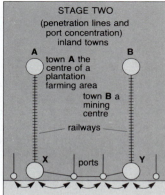

STAGE TWO
(penetration lines and port concentration) inland towns

town A the centre of a plantation farming area

town B a mining centre

railways

X ports Y

Early colonial development included routes inland, usually rail, to newly developed towns associated with the collection of a cash crop or the exploitation of a mineral — both of which were exported to the colonial power

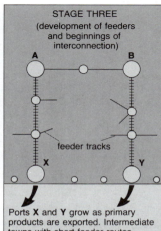

STAGE THREE
(development of feeders and beginnings of interconnection)

feeder tracks

X Y

Ports X and Y grow as primary products are exported. Intermediate towns with short feeder routes grow up along the major inland routes (very few imports)

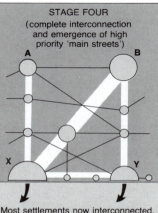

STAGE FOUR
(complete interconnection and emergence of high priority 'main streets')

X Y

Most settlements now interconnected. Increase in passenger traffic as opposed to earlier goods movements. Major centres connected by all-weather roads, air and rail. Major link is between the capital B and major port, X. Some feeder tracks may be abandoned

Interaction (gravity) models

As described on page 348, these models try to predict the degree of interaction, or movement, between pairs of places. They are based on two premises. First, as the size of both towns increases then so too will the amount of movement between them. Second, the further apart the two towns are, the less will be the movement between them (distance decay). The model predicts that movement will be greatest where two large cities are close together.

In transport terms the gravity model may be used to estimate two features of a system.

■ The **boundary** between two towns beyond which people find it easier (less distance, less time or less cost) to travel to one centre rather than the other (Reilly's breaking point, Figure 14.22).

■ The **volume** of traffic (passengers, freight or, as initially suggested by Christaller, telephone calls) between various towns.

Geographers are interested not only in the direction of flow of people and goods but also in predicting the volume of movement. The latter can be estimated by multiplying the population of the towns and dividing the product by the square of the distance between them.

This can be expressed by the formula:

$$\text{movement } A \text{ to } B = \frac{pA \times pB}{d^2}$$

where: pA is the population of town A
pB is the population of town B
d is the distance between towns A and B.

A worked example is given in Figure 18.21. It shows that on the basis of the gravity model, more buses or trains (public transport) and better roads (private transport) are needed, for example, to cope with the volume of movement between Montrose and Arbroath than between Montrose and Forfar.

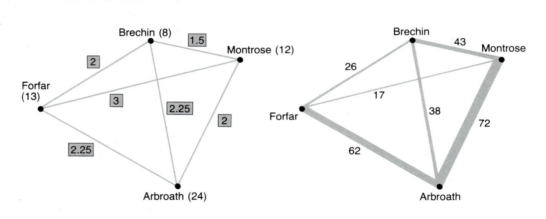

population (thousands)
i.e. Brechin (8) = 8000

distance (tens of km)
i.e. Brechin to Forfar [2] = 20 km

Brechin to Forfar = $\dfrac{8 \times 13}{2^2}$ = 26

Brechin (8) 1.5 Montrose (12)
2
Forfar (13) 3   2.25   2
2.25
Arbroath (24)

Brechin 43 Montrose
26
17
Forfar 38 72
62
Arbroath

◄ Figure 18.21
Interaction (gravity) model
applied to an area of eastern
Scotland

International trade

The development of international trade

Trading results from the uneven distribution of raw materials over the earth's surface. Today trade plays a major role in the economy of most countries as none has large enough supplies of the full range of minerals, fuels and foods to make it self-sufficient.

While early peoples bartered in order to exchange goods, the first civilisations found that making transactions with money proved a method by which wealth could be accumulated. By 1500 BC the Canaanites had learnt to become 'middlemen' in the trade

between Egypt, Asia Minor and Mycenae (Greece) while their descendants, the Phoenicians, traded five centuries later around and beyond the Mediterranean Sea. Until this time, trading was mainly the simple exchange of raw materials:

Area A grew wheat $\underset{\longleftarrow}{\longrightarrow}$ Area B reared animals

Subsequently certain communities proved to be more inventive than others and began to process some of these raw materials:

Country A grew $\rightarrow$ Country B
cotton and mined $\leftarrow$ manufactured textiles
iron ore   and smelted iron

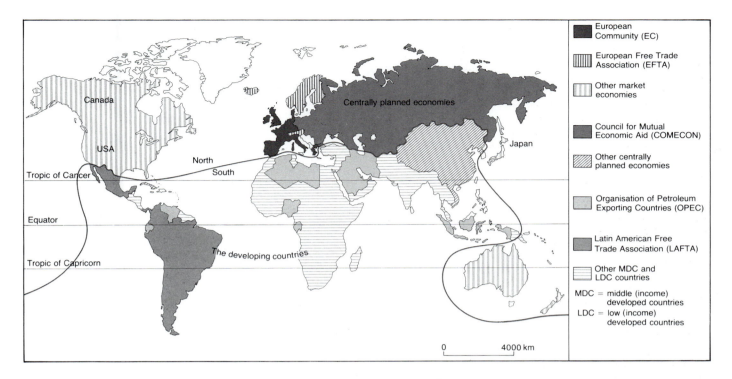

▲ Figure 18.22

Major global trading groups

Such exchanges led to the emergence of European colonial powers who used the raw materials from their colonies to establish their own domestic industries. This saw the beginnings of modern international trade and the widening gap between those countries producing the raw materials (the pre-oil days) and those who made a much larger profit by manufacturing goods. Trade developed further as new and improved forms of transport were introduced, such as transcontinental railways and refrigerated ships.

Since the 1960s there has been, especially among the industrialised nations of Europe and North America, a tendency to specialise in particular aspects of manufacturing. This creates greater benefits (because of the **law of comparative advantage**) than trying to compete with other countries who have equal, or better, opportunities. This can be seen in the modern car industry:

Country *A*   Country *B*   Country *C*
brakes,   → tyres,   → clutch, oil
distributors ← carburettors ← pumps and
and glass   and   pistons
                       headlights

The law of comparative advantage operated in nineteenth-century Britain. Britain could have produced more of its own food but, having a greater comparative advantage over other countries in manufacturing, opted to concentrate on industry where greater profits could be made.

In recent decades there has also been a grouping together of trading partners to form **common markets**, e.g. the EC, EFTA, OPEC, LAFTA CARICOM and COMECON — the definitions and their locations are given in Figure 18.22. These trade associations enable goods to pass more easily and cheaply within an enlarged 'domestic' or local market.

Directions of international trade

The following points may be noted from Figure 18.23 overleaf.

- Nearly half of the world's trade is between the **advanced market economies**.
- The advanced market economies are involved in 85 per cent of total world trade.
- There is relatively little trade between the centrally planned economies and the rest of the world.
- The advanced market economies receive more trade from the **OPEC** countries than is returned to them.
- The advanced market economies have relatively little trade with the Third World but export more to them than they import.
- There is very little trade between Third World countries.

Changing trends in world trade during the 1980s

1 Volumes of trade slumped, especially in the earlier period, because of worldwide recession. Exports from the Third World to the advanced market economies were the most affected (Figure 18.24, page 478).

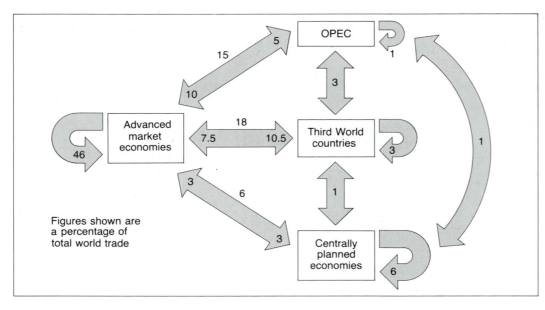

◀ Figure 18.23
Simplified pattern of world trade in the mid-1980s

Figures shown are a percentage of total world trade

2 The older industrialised nations of Europe and North America faced increasingly strong competition from the newly industrialising countries (NICs) in SE Asia — Japan, South Korea, Singapore, Hong Kong and Taiwan.

3 Although there appeared to be an increase in trading between the Third World countries, eight of these accounted for over 60 per cent of the 'south–south' trade in 1986 (e.g. Brazil and Mexico).

4 OPEC increased its share by 150 per cent (Figure 18.24).

5 The trade deficit for most Third World countries increased as they continued to export unprocessed raw materials and to import manufactured goods — in 1970,

Honduras needed to export 3 t of bananas to buy one tractor: by 1985, it had to produce 15 t.

6 Primary products, exported from the Third World, became increasingly more vulnerable to changes in market prices and world demand.

7 Several Third World countries continued to rely upon one or two major raw materials as their source of income, e.g. Zambia 90 per cent from copper, Nigeria 89 per cent from oil, Uganda 86 per cent from coffee and Mauritius 90 per cent from sugar cane.

8 World trade continued to become increasingly dominated by large multinational companies.

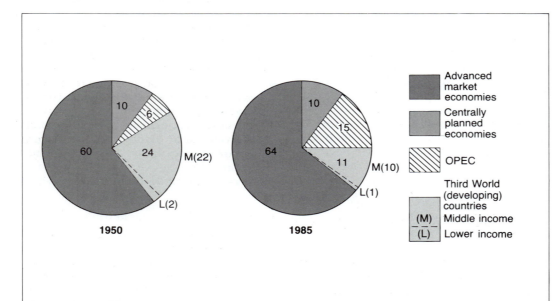

1950

1985

Advanced market economies

Centrally planned economies

OPEC

Third World (developing) countries
(M) Middle income
(L) Lower income

◀ Figure 18.24
Exports in percentage terms (UN figures) of major trading groups

OPEC is the Organisation of Petroleum-Exporting Countries. Its members are: Algeria, Ecuador, Gabon, Indonesia, Iran, Iraq, Kuwait, Libya, Nigeria, Qatar, Saudi Arabia, the United Arab Emirates and Venezuela.

Comparison of trading links between developed, Third World and OPEC countries

Figure 18.25 gives an indication of the nature and value of exports of three contrasting economic systems. The developed countries of the EC, North America and Japan often have manufactured goods, machinery and transport equipment accounting for over half their exports. These countries have accumulated the capital and technology to buy and to process the necessary raw materials. In contrast, while most Third World countries have some manufacturing, it is often primary processing or is operated by multinationals taking advantage of cheap labour rates.

The world market in fuels (of which oil is 75 per cent) is dominated by the OPEC countries. Most is exported to the energy-short advanced market economies in Europe (excluding the UK and Norway) and to Japan — North America is relatively self-sufficient. Third World countries have not the wealth to afford these fuels and this further retards their economic development. The pattern of mineral exports is less obvious: both developed (Australia and Canada) and developing (Jamaica, Zambia) being major exporters. Once again it is the advanced market economies who are the importers.

Agricultural products usually account for over 75 per cent of a Third World country's exports, although an increasing number of African states are having to import cereals as their food production decreases (page 322). Developed countries have to import tropical goods but, on balance, export almost as much agricultural produce as they import. The USA, Canada and Australia, with their extensive agricultural systems (page 406), are net exporters accounting for over 75 per cent of the world's wheat, while others, like the Netherlands and Denmark, use their farmland intensively (page 410) to obtain high yields. Most OPEC countries have to rely on imports because of their desert location which does not favour farming.

Despite supporting three quarters of the world's population, in the mid 1980s Third World countries earned only 13 per cent of its export income. This disparity in income is illustrated in the table below and in Figure 18.21.

Value of exports per capita (1985) in US$

Advanced market economies		OPEC countries		Third World countries	
Belgium	5 039	Libya	4 189	Nepal	6
Netherlands	4 567	United Arab Emirates	11 752	Somalia	8
Canada	3 455	Venezuela	915	Burkina Faso	9
West Germany	2 758	Kuwait	6 437	Bangladesh	10
*Singapore	9 567			India	11
*Hong Kong	5 283			Ethiopia	12

(*Any visit to these countries will show them to be highly developed in most areas)

Future trade and Third World countries

Third World countries are making demands for a fairer trading system. One request is for higher and fixed prices for primary products as this might prevent a further widening of their trade gap. A second priority is to be given better access to markets within the developed countries. At present, if there is a recession or if a Third World country begins to increase its trade 'too much' with an industrialised country, then the latter may impose quotas to limit the number of goods imported or add tariffs so that the price of the imported goods increase and become less competitive in the domestic market (e.g. EC). While the developed country thus protects its jobs it causes increased depression within the Third World country. This may result in a drop in the income of the Third World country, meaning it has less money to spend on goods manufactured in the industrial country. Both parties may therefore suffer from a trade recession.

Other demands include changes in the international monetary system to eliminate fluctuations in currency exchange rates; encouraging industrialised countries to share their technology; to stop developed countries 'dumping' their unwanted and sometimes untested products cheaply; a reduction in interest rates and an increase in aid free of economic, political and bureaucratic strings. At the same time, more efforts might be made to develop trading links with other Third World countries.

Britain's trade

Why have Britain's trading partners changed since 1951?

In 1951 most of Britain's trade was with colonies and countries within the British Empire. As the colonies gained independence, many have sought markets elsewhere

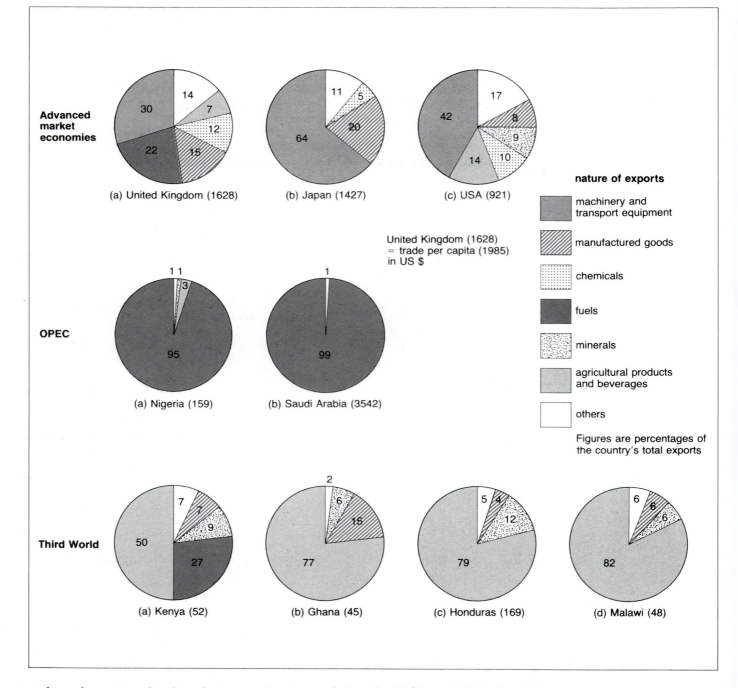

nature of exports

- machinery and transport equipment
- manufactured goods
- chemicals
- fuels
- minerals
- agricultural products and beverages
- others

Figures are percentages of the country's total exports

United Kingdom (1628) = trade per capita (1985) in US $

Advanced market economies

(a) United Kingdom (1628)
14, 7, 12, 15, 22, 30

(b) Japan (1427)
11, 5, 20, 64

(c) USA (921)
17, 8, 9, 10, 14, 42

OPEC

(a) Nigeria (159)
1, 1, 3, 95

(b) Saudi Arabia (3542)
1, 99

Third World

(a) Kenya (52)
7, 7, 9, 50, 27

(b) Ghana (45)
2, 6, 15, 77

(c) Honduras (169)
5, 4, 12, 79

(d) Malawi (48)
6, 6, 6, 82

▲ Figure 18.25

Nature of exports for selected countries (1985)

or have begun to develop their own industries. Not only has this meant a reduction in their imports from Britain, but by producing similar products at a cheaper price, they have succeeded in taking other traditional British markets. Other former trading partners have been lost due to distance decay and the loss of Commonwealth preference (Australia, New Zealand), politics (South Africa) and Britain joining the European Community in 1973. Although the volume of trade with the USA has declined it still remains a major trading partner.

After an appreciable increase in importance during the 1960s and 1970s the OPEC countries slipped back as Britain developed its North Sea oilfields. Trade is minimal with the Third World and the centrally planned economies. Other reasons suggested for the loss of British overseas markets include: the changing value of the pound; poor production quality; late deliveries; strikes and poor labour relations; poor management; bad sales and marketing; high prices; the world recession and a decline in domestic production. An important corollary of Britain's declining role in world trade has been the associated drop in the merchant fleet.

Figure 18.26 shows the direction and changes in the direction of trade in 1951 and 1986.

For each of the countries listed in the two import columns name *one* commodity which it exports to Britain.

▶ Figure 18.26

Changes in the source and direction of Britain's trade, 1951 and 1986 (Source: *Annual Abstract of Statistics*, HMSO, 1988)

	Direction of Britain's exports		Source of Britain's imports	
Rank	1951	1986	1951	1986
1	Australia	USA	USA	West Germany
2	South Africa	West Germany	Canada	USA
3	Canada	France	Australia	France
4	USA	Netherlands	Netherlands	Netherlands
5	India	Belgium/Lux	New Zealand	Japan
6	New Zealand	Irish Republic	India	Italy
7	Sweden	Italy	Malaya	Belgium/Lux
8	Malaya	Sweden	Sweden	Norway
9	Netherlands	Spain	France	Irish Rep
10	Denmark	Canada	Denmark	Switzerland
11	Belgium/Lux	Switzerland	Belgium/Lux	Sweden
12	France	Saudi Arabia	Argentina	Spain
13	West Germany	Australia	Italy	Denmark
14	Pakistan	Denmark	W Germany	Hong Kong
15	Italy	Japan	Norway	Canada

1951

- British Empire
- Europe
- USA
- Third World countries

1986

- EC 48%
- EFTA 10%
- Commonwealth 5%
- USA 16%
- OPEC 8%
- other developed countries 6%

Britain's Balance of Payments

The term **Balance of Trade** refers to the difference between the income received from visible exports and the cost incurred in paying for visible imports. For Britain this means the profit on the sale of, for example, manufactured goods, machinery and transport equipment, less the cost of paying for raw materials. In recent years only 1980, 1981 and 1982 have shown a Balance of Trade surplus — this was the peak period of North Sea Oil production.

The term **Balance of Payments** includes the Balance of Trade together with any **invisible** earnings, i.e. income from banking and insurance, tourism, professional advice and air/sea transport. This has normally cancelled out any **visible** trade deficit and, until 1986, left a surplus. Since then, the Balance of Payments has increasingly fallen 'in the red' (Figure 18.27).

References

Annual Abstract of Statistics 1988, HMSO, 1989.
Europe, David Waugh, Thomas Nelson, 1985.
Human Geography, Ken Briggs, Hodder & Stoughton, 1982.
Human Geography, M.G. Bradford and W.A. Kent, Oxford University Press, 1977.
Human Geography, Patrick McBride, Blackie, 1980.
Innovation and Change, BBC Radiovision, 1988.
New Geographical Digest, George Philip, 1987.
Pattern and Process in Human Geography, Vincent Tidswell, UTP, 1976.
Singapore 1988, Singapore Information Office, 1988.
The Third World, M. Barke and G. O'Hare, Oliver & Boyd, 1984.
The World, David Waugh, Thomas Nelson, 1987.
Transport and the Environment, Dept. of the Environment, HMSO, 1988.

▼ Figure 18.27

Britain's Balance of Payments, 1976 to 1986

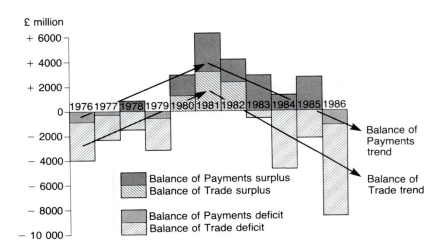

£ million

- Balance of Payments surplus
- Balance of Trade surplus
- Balance of Payments deficit
- Balance of Trade deficit
- Balance of Payments trend
- Balance of Trade trend

Index